ITALIAN PHYSICAL SOCIETY

PROCEEDINGS

OF THE

INTERNATIONAL SCHOOL OF PHYSICS « ENRICO FERMI »

COURSE LXIV

edited by N. BLOEMBERGEN

Director of the Course

VARENNA ON LAKE COMO

VILLA MONASTERO

30th JUNE - 12th JULY 1975

Nonlinear Spectroscopy

1977

NORTH-HOLLAND PUBLISHING COMPANY, AMSTERDAM - NEW YORK - OXFORD

SOCIETA' ITALIANA DI FISICA

RENDICONTI

DELLA

SCUOLA INTERNAZIONALE DI FISICA «ENRICO FERMI»

LXIV CORSO

a cura di N. BLOEMBERGEN

Direttore del Corso

VARENNA SUL LAGO DI COMO

VILLA MONASTERO

30 GIUGNO - 12 LUGLIO 1975

Spettroscopia non lineare

1977

SOCIETÀ ITALIANA DI FISICA

BOLOGNA - ITALY

PUBLISHER: SOCIETÀ ITALIANA DI FISICA - BOLOGNA

SOLE DISTRIBUTORS: NORTH-HOLLAND PUBLISHING COMPANY -
AMSTERDAM - NEW YORK - OXFORD

ISBN NORTH-HOLLAND 0 7204 0568 8

INDICE

S. A. Akhmanov – Statistical effects in resonant nonlinear optics.

J. Ducuing – Optical nonlinearities in conjugated one-dimensional systems.

R. Loudon – Nonlinear optics with polaritons.

F. De Martini – Nonlinear spectroscopy of bulk and surface polaritons.

H. Haken – Nonlinear interaction between excitons and coherent light.

F. Bassani, J. J. Forney and A. Quattropani – Structure of biexcitons and two-photon processes.

Preface to Nonlinear Spectroscopy.

N. BLOEMBERGEN

Harvard University - Cambridge, Mass.

The field of nonlinear spectroscopy is concerned with a systematic investigation of the frequency dependence of the nonlinear response of matter to electromagnetic radiation. The appearance of tunable dye lasers has made available coherent high-intensity radiation throughout the near-ultraviolet, visible and near–infra-red regions of the spectrum in many laboratories. In addition, parametric down converters and numerous molecular lasers have made available powerful tunable radiation farther in the infra-red. Thus, the field of nonlinear spectroscopy has grown rapidly during the past five years, and it was appropriate and timely to devote an Enrico Fermi course to this subject matter during the first half of July, 1976. The Villa Monastero in Varenna on lake Como, aided by favorable weather, provided a delightful setting for instructive discussion on this new branch of spectroscopy by faculty and students. The lecturers were drawn from leading active contributors from many countries, as a perusal of the contents of this volume will show. They took pains to present the basic ideas and principles of nonlinear optical response from a pedagogical point of view. The relatively small size of the student body was also conducive to intensive discussions and scientific contacts among all participants.

In addition to a general tutorial introduction to nonlinear spectroscopy in atoms, molecules and condensed matter, this volume contains many recent results of advanced research. The nonlinear response of elementary excitations in crystals has been given special emphasis. The subject matter is arranged in the following sequence of topics:

Introduction to Nonlinear Spectroscopy.

Nonlinear High-Resolution Spectroscopy of Atoms and Molecules.

Transient Nonlinear Response.

Doppler-Free Two-Photon Spectroscopy and Nonlinear Properties of Autoionizing States.

Nonlinear Spectroscopy of Molecular Liquids, Liquid Crystals and Crystalline Solids.

Higher-Order Nonlinearities and Raman-Active Coherent Spectroscopy.

Nonlinear Response of Conjugated Chains.

Nonlinear Response of Polaritons.

Nonlinear Response of Excitons and Biexcitons.

Picosecond Dynamical Response of Vibrations.

The scientific task of the Director of this course was facilitated by the efforts of the Scientific Secretary, Prof. F. DE MARTINI, and the administrative aspects were taken care of with high professional competence by Ms. G. WOLZAK. The assistance of my wife at several social events, and especially her delivery of the after-dinner speech at the farewell banquet, was also highly appreciated. The editing of this volume was aided by the splendid co-operation of the authors, the Italian Physical Society and the publishers. To all these people, as well as to the faithful audience of active participants, I wish to express my gratitude.

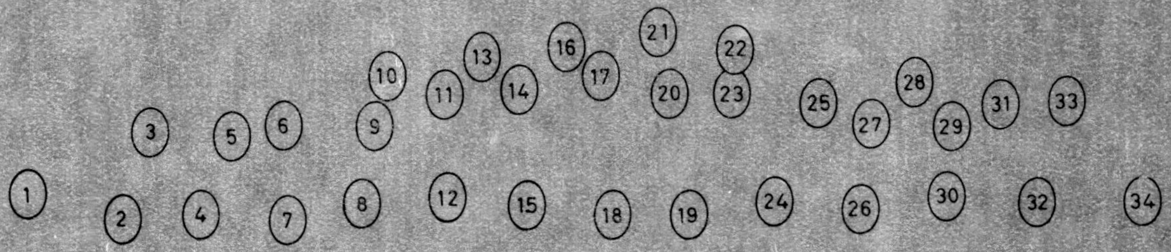

1. J. E. Bjorkholm
2. A. Tanga
3. F. Capasso
4. G. Giuliani
5. R. Boscaino
6. F. Simoni
7. M. Brai
8. Y. R. Shen
9. N. Paraire
10. H. Autwete
11. A. Dulčić
12. J. A. Armstrong
13. E. Ostertag
14. M. A. Aegerter
15. S. Akhmanov
16. L. Pardi
17. M. Baldo
18. G. A. Wolzak
19. N. Bloembergen
20. M. G. Migliorini
21. P. R. Salvi
22. R. Righini
23. M. L. Cangeloni
24. R. Loudon
25. R. Querzoli
26. G. Vetri
27. N. Pearlman
28. E. G. Hanson
29. I. Iwasawa
30. F. De Martini
31. M. Colocci
32. R. G. Brewer
33. A. Ricca
34. T. W. Hänsch

SOCIETÀ ITALIANA DI FISICA

SCUOLA INTERNAZIONALE DI FISICA « E. FERMI »

LXIV CORSO - VARENNA SUL LAGO DI COMO - VILLA MONASTERO - 30 Giugno - 12 Luglio 1975

Nonlinear Spectroscopy.

N. BLOEMBERGEN

Division of Engineering and Applied Physics, Harvard University
Cambridge, Mass. 02138

1. – Historical introduction.

The present Enrico Fermi course is concerned with a subfield of physics that has grown rapidly during the past few years. It evolved naturally during the renaissance of optics which has been spurred by the development of many kinds of lasers. Precursors of this course on « nonlinear spectroscopy » are course XXXI, Quantum Electronics and Coherent Light, held in Varenna in 1963, and course XLII, Quantum Optics, held in 1967. In this introduction the subject matter of nonlinear spectroscopy will be defined by tracing the historical development of quantum optics and electronics.

The observation of many discrete spectral lines, characteristic for each atom or molecule, by spectroscopists of the 19th century provided, of course, a strong support for the development of the Bohr model of electronic orbitals. The photoelectric effect also played an important role in establishing the quantum-mechanical concept of wave-particle duality. By 1930, the basic principles of the quantum mechanics of the electronic structure of matter were well established. The quantization of electromagnetic interactions between charged particles and of electromagnetic fields had been accomplished. About thirty years later quantum mechanics started to repay its indebtedness to optics. Although optical resonators were known since the work of PEROT and FABRY in 1905, and EINSTEIN had clearly formulated the phenomenon of stimulated emission in 1917, a long historical detour occurred before these two ideas were usefully and successfully merged into that marvelously fruitful combination of the laser. First the phenomenon of stimulated emission of radiation was used in microwave devices called masers. This word acronym standing for microwave amplification by stimulated emission of radiation was coined by TOWNES, who shared the Nobel prize with BASOV and PROKHOROV for this invention. The pumping mechanisms necessary to develop inverted populations were first established in the lower-frequency region of the electromagnetic spectrum. Originally devices working on the same principles in the optical region were called « optical masers ». Word usage, however, cannot be dictated. The word « laser » has taken hold in many different languages, although the Chinese have their own different expression.

The devices of classical electronics, such as vacuum tubes, transistors, etc., were usually based on the utilization of free electrons, either as electrons in a beam in vacuum or in a gas, or as carriers of electricity in the conduction band of a crystal. The term quantum electronics originally emphasized new devices based on the existence of discrete energy levels. Bound electrons were used in masers, in atomic clocks, in atomic magnetometers and other radio- and microwave devices developed during the fifties. The quantum nature of the electromagnetic field was not essential. The phenomena could be described readily in terms of classical coherent fields with well-defined amplitude and phase, except for the relatively small but important effects of spontaneous emission noise.

In conventional light sources the spontaneous emission noise had been dominant, until lasers were developed. The dominance of stimulated emission in lasers led to a much better definition of the phase of optical fields and a renewed interest in the concept of coherence. In quantum optics the word « quantum » refers especially to the quantum nature of the electromagnetic field. The existence and utilization of discrete bound electronic levels was not a novelty in the optical range. In the XLII Enrico Fermi Course in 1967, a large effort was devoted to describing the coherent-light fields, with the maximum definition of both amplitude and phase permitted by the uncertainty principle, in quantum-mechanical language. Attention was focused on the statistics of photon counting and the noise characteristics of laser fields and the quantum-mechanical definition of degree of coherence. Although these fluctuation phenomena are of fundamental importance, it is nevertheless true that for most other purposes the laser fields can be described very accurately in terms of classical waves with well-defined amplitudes and phases [1].

The well-defined wave vector of laser beams has three important characteristics:

1) high degree of directionality,

2) high degree of monochromaticity,

3) high degree of brightness, *i.e.* very high power flux density per unit solid angle per unit frequency interval.

These characteristics made many nonlinear optical phenomena accessible to experimental investigation. It should be noted that both Courses XXXI and XLII devoted a substantial fraction of the lectures to the field of nonlinear optics.

This field may be broadly defined as optical phenomena in which the material response to applied electromagnetic fields is not a linear function of the applied amplitude [2]. The electric polarization or the induced electric current density may, for example, be a quadratic, cubic or exponential function of the electric-field amplitudes of one or more applied electromagnetic waves. Alternatively, one may say that the electrical dielectric susceptibility and conductivity are themselves functions of the field strengths. With this broad definition, the field of nonlinear optics has a longer history than that of the lasers.

The linear electro-optic, or Pockels effect and the quadratic electro-optic, or Kerr effect were known in the 19th century and are described by a polarization proportional to the product of the light field amplitude and the d.c. electric-field amplitude for the Pockels effect, or the square of the d.c. amplitude for the Kerr effect. The d.c. field may be replaced by an a.c. field to obtain modulation of the light beam. If the frequency of the a.c. field is increased to the infra-red or visible region of the spectrum one has a polarization proportional to the product of two light field amplitudes. It is clear that a polarization which is a quadratic function of the field amplitudes will exhibit the characteristics of any quadratic device. Nonlinear devices in the radio- and microwave region of the electromagnetic spectrum have, of course, been known for a long time and have widespread usage as diode detectors, modulators, demodulators, heterodyne mixers, harmonic and parametric generators, etc. In a similar manner a quadratic response in the optical region will lead to

a) second-harmonic generation of light,

b) rectification of light,

c) sum frequency generation,

d) difference frequency generation.

An early example of an optical quadratic response is the photoelectric effect. The photocurrent is, of course, proportional to the intensity of the incident light $i_{\text{ph}} = \sigma^{\text{NL}} EE^*$. The proportionality constant may be considered as a nonlinear conductivity. Although Einstein's theory of the photoelectric effect played an important role in the early development of quantum theory, it is now generally recognized that the effect can be fully explained by the semi-classical theory [3] in which the field quantities are considered classically as *c*-numbers. Time-dependent perturbation by a harmonic perturbation with angular frequency ω leads to a transition rate from the initial bound state of the electron to a final-state free wave $\exp[i\boldsymbol{K}\cdot\boldsymbol{r}]$. This transition rate according to Fermi's golden rule is proportional to $|E|^2$ and to $\delta(W_K - W_i - \hbar\omega)$. The final states make a contribution to the photocurrent $e\hbar K/m$; $W_K - W_i$ is the energy difference between the final and initial states.

Although it is perhaps unusual to consider the photoelectric detection of light as a nonlinear optical effect, it becomes eminently practical to do so, when one considers the demodulation of light. The Fourier component in the current at $\omega_1 - \omega_2$ is calculated when two incident Fourier components at the adjacent light frequencies ω_1 and ω_2 are incident. FORRESTER *et al.* [4] considered this problem before the advent of lasers and performed a difficult experiment with conventional light sources which did not have a high degree of coherence. The effect is now used routinely for heterodyne detection of scattered laser light fields.

Holography may be considered as a related nonlinear optical phenomenon. In this case spatial beats between two light waves with the same frequency, but different directions, are detected. The complex index of refraction in the hologram is proportional to the product of the amplitudes of the reference beam and a spatial Fourier component from the illuminated object. The experimental pursuit and exploitation of holography again had to wait for the advent of lasers.

Other nonlinear processes which had been recognized and discussed many decades before the advent of the laser are the Raman effect and two-photon absorption process [2]. In the former one incident photon at ω_1 is absorbed, while another photon at frequency ω_s is emitted, so that $\hbar(\omega_1 - \omega_s)$ corresponds to an excitational energy of the material system. Frequently, this is a vibrational or rotational excitation in the infra-red, but it may just as well be an electronic excitation. The process may be formally described by a term in the Hamiltonian proportional to $a_s^\dagger a_1 c_i c_f^\dagger$, where $a_s^\dagger$ is a creation operator which puts one photon in the Raman mode, a_1 is an annihilation operator, which takes a photon out of incident beam, c_i takes the electron out of the initial state $|i\rangle$ and $c_f^\dagger$ puts an electron in the final state $|f\rangle$. For spontaneous Raman scattering the quantization of the electromagnetic modes is essential. The spontaneous Raman scattering was discovered in the late twenties. The scattered intensity is proportional to the intensity of the incident beam at ω_1, and according to our classical definition it is not a nonlinear effect. The stimulated Raman effect is nonlinear in the semi-classical sense, because the gain constant is in that case proportional to the product of intensities at both frequencies ω_1 and ω_s. It may appear rather artificial, and illogical from the point of view of field quantization, to classify spontaneous Raman scattering as a linear process and stimulated Raman scattering as a nonlinear process. Nevertheless, there is an advantage in doing so from the experimental point of view. For the detection of the nonlinear stimulated process high light intensities are required. This explains why it was discovered more than thirty years after spontaneous Raman scattering, because high intensities obtainable only in powerful laser beams were necessary for its detection.

The usefulness of our semi-classical definition of nonlinear processes becomes even more obvious, if two-photon absorption processes are considered. They are described by a Hamiltonian $a_2 a_1 c_i c_f^*$; two photons are destroyed which may or may not be taken out of the same mode, while the electron makes a transition from the state $|i\rangle$ to the state $|f\rangle$. Energy conservation now requires $W_f - W_i = \hbar(\omega_1 + \omega_2)$. The quantum theory of the two-photon absorption process is, of course, very similar to that for the Raman process. It was developed by GOEPPERT-MAYER [5] in her Ph.D. thesis at the University of Göttingen. The probability for the process is proportional to the product of the intensity of the two fields, or the square of the intensity, if only one mode is applied. The process is always nonlinear, and this explains why its experimental demonstration occurred so much later than that of spontaneous Raman scattering. Not only are high incident light intensities required, but the frequencies must be tuned to the difference between existing energy levels. The frequency of spontaneously emitted Raman light automatically adjusts its Stokes shift to correspond to the required energy difference. It is understandable why two-photon absorption spectroscopy could only flourish until tunable coherent-light sources became generally available in the form of dye lasers and optical parametric oscillators.

The systematic investigation of the frequency dependence of nonlinear optical phenomena has thus become accessible to experimental investigation. It is the purpose of the present Enrico Fermi Course to review the recent advances in nonlinear spectroscopy. The frequency dependence of nonlinear phenomena can be studied in atoms, molecules, plasmas and fluids, and in the solid state. A large variety of excitations can be studied and are discussed in this volume.

All materials exhibit certain nonlinear responses because the valence electrons cannot strictly be represented by the classical Lorentz model of harmonic oscillator. The electrons are bound in the Coulomb field of the ion core, and their energy levels are, of course, not equidistant, as they would be for a harmonic oscillator. The effects of the nonlinear response will become evident when the applied field amplitudes are a reasonable fraction of the Coulomb fields responsible for the binding of the electrons. In that case the valence orbitals will be noticeably distorted, and nonlinear optical phenomena result.

It is clear that the theory could have been developed along purely classical lines in the first decade of this century. If LORENTZ in his electron theory had allowed for some anharmonicity in his model of harmonically bound electrons, he would not only have explained the linear optical properties of matter, but would also have predicted the nonlinear ones. Again, a few decades later, quantum mechanics around 1930 was capable of predicting the nonlinear electromagnetic properties of matter inherent in the electron structure. At that time there was no stimulus to devote much attention to these problems because they were not accessible to observation and experimentation. Stimulated emission of light from lasers provided the trigger to bring the field of nonlinear optics to fruition. An important new branch is the subfield of nonlinear spectroscopy concerned with the detailed frequency dependence of nonlinear optical properties of matter.

The field of linear spectroscopy is more than a century old and still flourishing.

It may therefore be predicted with some confidence that nonlinear spectroscopy, which can augment linear spectroscopy in significant and nontrivial ways, will have a long and useful life.

2. – General theoretical framework.

In the semi-classical treatment which is adequate to describe the majority of phenomena in nonlinear optics, the electromagnetic field is considered as a classical quantity. The evolution of the material system is described quantum mechanically by the Liouville equation for the density matrix of the material system [2]

$$\dot{\varrho} = -i\hbar^{-1}[\mathscr{H}, \varrho]. \tag{1}$$

The Hamiltonian consists of three parts:

$$\mathscr{H} = \mathscr{H}_0 + \mathscr{H}_{\text{damping}} + \mathscr{H}_{\text{int}}. \tag{2}$$

Here $\mathscr{H}_0$ is the Hamiltonian for the atom or molecule, leading to a set of energy eigenstates $|n\rangle$ with energy eigenvalues W_n. The effect of time-independent applied electric and magnetic fields may be incorporated into $\mathscr{H}_0$. The material system is supposed to be weakly coupled to a thermal reservoir of harmonic oscillators. Collisions in a gas or liquid, or interactions with lattice vibrations in a solid constitute examples of such interactions, which lead to phenomenological damping terms for the density matrix elements. They also ensure the establishment of a steady-state response of the system under the influence of a stationary perturbation.

The interaction of a one-electron system with electromagnetic fields derivable from a vector potential $\boldsymbol{A}$ is given by

$$\mathscr{H}_{\text{int}} = -\frac{e}{2c}(\boldsymbol{p}\cdot\boldsymbol{A} + \boldsymbol{A}\cdot\boldsymbol{p}) + \frac{e^2}{2mc^2}\boldsymbol{A}\cdot\boldsymbol{A}. \tag{3}$$

An equivalent representation is the multipole expansion

$$\mathscr{H}_{\text{int}} = -e\boldsymbol{r}\cdot\boldsymbol{E} - \tfrac{1}{2}e\boldsymbol{Q}_m\nabla\boldsymbol{E} + \boldsymbol{m}\cdot\boldsymbol{H} + \dots. \tag{4}$$

In many cases the electric-dipole approximation represented by the first term on the right-hand side of eq. (4) is adequate.

The expectation value of the electric polarization is obtained from

$$\boldsymbol{P} = N\,\mathrm{Tr}\,(e\boldsymbol{r}\varrho), \tag{5}$$

where N is the number of one-electron systems per unit volume. The set of equations must be made self-consistent by requiring that $\boldsymbol{P}$ and $\boldsymbol{E}$ satisfy Maxwell's equations. In the electric-dipole approximation this condition is embodied by the wave equation

$$\nabla \times \nabla \times \boldsymbol{E} + \frac{1}{c^2}\frac{\partial^2 \boldsymbol{E}}{\partial t^2} = -\frac{1}{c^2}\frac{\partial^2 \boldsymbol{P}}{\partial t^2}. \tag{6}$$

The set of equations (1)-(6) is, of course, too general to provide much physical insight, but they clearly show how the material responds to the fields, some of which may have been generated by sources external to the material system, and how the induced polarization in turn generates fields. To proceed further, attention is usually restricted to a very small number of modes of the electromagnetic field. One may, for instance, select two plane waves:

$$\boldsymbol{E} = \tfrac{1}{2}\hat{e}_1 E_1 \exp[i\boldsymbol{k}_1\cdot\boldsymbol{r} - i\omega_1 t] + \tfrac{1}{2}\hat{e}_2 E_2 \exp[i\boldsymbol{k}_2\cdot\boldsymbol{r} - i\omega_2 t] + \text{c.c.} \tag{7}$$

Here one can recognize the complex field amplitude, the unit polarization vector, the wave and the frequency of each of the waves. The real physical amplitudes are $|E_1|$ and $|E_2|$.

At the same time the Hamiltonian for the material system is often truncated and attention is restricted to only two or three energy levels. This is a reasonable approximation, if the population in the other levels is negligible and they are far off resonance, so that their energy intervals do not correspond to the frequencies of the electromagnetic modes under consideration.

The simplest nontrivial case is a two-level system and one electromagnetic mode. If the total population of the material is constant, the trace of the density matrix is unity. Equation (1) can then be written as one real differential equation for the population difference $\varrho_{22} - \varrho_{11}$, and one complex equation for the off-diagonal element $\varrho_{12} = \varrho_{21}^*$. These three equations are in a one-to-one correspondence with the classical Bloch equations for a precessing magnetization. Equation (6) gives one equation for the complex amplitude of the mode under consideration, which may be split into a real equation for the real amplitude and one for the real phase. Even this simplest case thus leads to five real coupled differential equations. Solutions which describe, for example, single-mode laser operation and self-induced transparency have been extensively discussed in the literature [1].

Explicit solutions for a three-level system with two electromagnetic modes at or near resonance are also available. A recent example, discussed by BREWER and HAHN [6], involves the coherent Raman beats.

In many situations one is interested in off-resonance as well as resonant behavior. There is a systematic perturbation procedure which keeps account of all the energy levels in the material system. It develops the expectation value of the polarization in an ascending power series of the amplitudes of a small number of electromagnetic modes. The steady-state response is then described by the linear and some nonlinear susceptibilities [2]. This permits a survey of a large variety of steady-state nonlinear phenomena.

3. – Nonlinear susceptibilities.

3'1. $\chi^{(2)}$. – The lowest-order nonlinearity is described by a polarization which is a quadratic function of the field amplitudes

$$P_i(\omega_3, \boldsymbol{r}) = \tag{8}$$
$$= \tfrac{1}{2}\chi^{(2)}_{ijk}(-\omega_3, \omega_1, \omega_2)\, E_j(\omega_1)\, E_k(\omega_2) \exp[i(\boldsymbol{k}_1 + \boldsymbol{k}_2)\cdot\boldsymbol{r} - i(\omega_1 + \omega_2)t] + \text{c.c.}$$

The subscripts i, j, k stand for Cartesian co-ordinates. A summation over the indices j, k on the right-hand side is implied. The polarization at the sum frequency $\omega_3 = \omega_1 + \omega_2$ will serve as a source term in the wave equation (6). Radiation will be emitted, proportional to $|\chi^{(2)}|^2$ at the sum frequency. The third-rank tensor elements $\chi^{(2)}$ can only be nonvanishing in materials that lack inversion symmetry. The perturbation expansion for a system of N independent molecules in the ground state $|g\rangle$ per unit volume would yield

$$\chi^{(2)}_{xxx}(-\omega_3, \omega_1, \omega_2) = \frac{1}{2} N e^3 \hbar^{-2} \sum_{n,n'} \frac{x_{gn} x_{nn'} x_{n'g}}{(\omega_{ng} - \omega_1 - i\Gamma_{ng})(\omega_{n'g} - \omega_3 - i\Gamma_{n'g})} + \text{perm}(1, 2, 3)\,. \tag{9}$$

There are similar expressions for the other components of the third-rank tensor. The permutation of the indices 1, 2 and 3 corresponds to the different order in which the photons at ω_1 and ω_2 are absorbed and the photon at ω_3 is emitted. Resonances occur when any one of the three frequencies involved corresponds to a separation between energy levels, $\hbar\omega_{ng} = W_n - W_g$, etc. Near such resonances the nonlinear susceptibility becomes complex. The width Γ_{ng} of the transitions is determined by $\mathscr{H}_{\text{damping}}$. The expression (9) is a nonlinear generalization of the Kramers-Heisenberg dispersion formula. Since the operator x has odd parity, it is clear that the product $x_{gn}\, x_{nn'}\, x_{n'g}$ would vanish, unless at least one of the states $|g\rangle$, $|n\rangle$ or $|n'\rangle$ has mixed parity. The system must lack a center of inversion. If one takes $\omega_1 = \omega_2$ in eqs. (8) and (9), a description of second-harmonic generation is obtained.

Since the real fields in the electromagnetic modes contain both positive- and negative-frequency components, we may choose some of the frequencies to be negative. By replacing ω_2 by $-\omega_2$, one obtains difference frequency generation at $\omega_1 - \omega_2$. If one chooses $\omega_2 = -\omega_1$, a nonlinear polarization at zero frequency is obtained. If an intense light beam traverses a piezoelectric crystal, a d.c. voltage is developed. This may be called rectification of light.

If one takes $\omega_2 = 0$ and $\omega_3 = \omega_1$, a description of the linear electro-optic (Pockels) effect is obtained. If ω_2 is a radio- or microwave frequency, modulation of the light frequency is described.

The quadratic nonlinearity provides a coupling between a set of three electromagnetic modes. This forms the basis of the description of optical parametric oscillators.

In optically dense media there are important local field corrections, which can change the magnitude of $\chi^{(2)}$, but do not alter the phenomenological description of the phenomena.

3'2. $\chi^{(3)}$. – In media with inversion symmetry the lowest-order nonvanishing nonlinearity is a polarization cubic in the electric-field amplitudes:

$$P_i(\omega_4, \boldsymbol{r}) = \tag{10}$$
$$= \tfrac{1}{2}\chi^{(3)}_{ijkl}(-\omega_4, \omega_1, \omega_2, \omega_3)E_j(\omega_1)E_k(\omega_2)E_l(\omega_3)\exp[i(\boldsymbol{k}_1+\boldsymbol{k}_2+\boldsymbol{k}_3)r - i\omega_4 t] + \text{c.c.}$$

This nonlinearity exists even in media with the highest symmetry such as monoatomic noble gases. Again a summation over Cartesian indices j, k, l is understood. The polarization at the sum frequency $\omega_4 = \omega_1 + \omega_2 + \omega_3$ will again serve as a source term in eq. (6). Higher-order perturbation theory now yields the result for the $xxxx$-element of the fourth-rank susceptibility tensor

$$\chi^{(3)}_{xxxx}(-\omega_4, \omega_1, \omega_2, \omega_3) = \tfrac{1}{4}Ne^4\hbar^{-3}\sum_{n,n',n''} x_{gn}x_{nn'}x_{n'n''}x_{n''g}\cdot \tag{11}$$
$$\cdot\big((\omega_{ng}-\omega_1-i\Gamma_{ng})(\omega_{n'g}-\omega_1-\omega_2-i\Gamma_{n'g})(\omega_{n''g}-\omega_4-i\Gamma_{n''g})\big)^{-1}+$$
$$+\text{perm}\,(1, 2, 3, 4)\,.$$

It should be noted that resonances in the denominators occur not only when the frequencies ω_1, ω_2, ω_3 or ω_4 correspond to an energy separation between a pair of levels, but also when a combination of frequencies such as $\omega_1 + \omega_2$ in eq. (11) corresponds to such an energy difference. The other permutations of the four electromagnetic waves which are coupled by this cubic nonlinearity contain, of course, additional resonant combinations. There are again important local field corrections in dense optical media [7], and in media which lack inversion symmetry there are additional contributions to the effective $\chi^{(3)}$ proportional to $\{\chi^{(2)}\}^2$. These contain retardation effects [8], due to propagation of waves at $\omega_1 + \omega_2$, $\omega_1 + \omega_3$ or $\omega_2 + \omega_3$.

The complete frequency dependence or dispersion of the cubic nonlinearity is described in a three-dimensional frequency space. Nonlinear spectroscopy is concerned with the detailed study of this frequency dependence. A large variety of nonlinear phenomena is described by eqs. (10) and (11). It depends on the choice of frequencies, some of which may be taken equal to each other or to their negative values. Attention may be focused on the real or imaginary part of $\chi^{(3)}$, and particular resonant terms may be singled out. An enumeration of some important cases follows. While the quoted examples of these effects will pertain mostly to vapors of alkali atoms, it should be kept in mind that these effects occur in any material.

a) Third-harmonic generation.

By taking $\omega_1 = \omega_2 = \omega_3$ and $\omega_4 = 3\omega_1$, eqs. (10) and (11) describe a polarization at the third-harmonic frequency. When substituted into the wave equation (6), this will lead to the generation of radiation at $3\omega_1$ with an intensity proportional to $|\chi^{(3)}|^2$. In fig. 1*c*) the situation is diagrammatically depicted

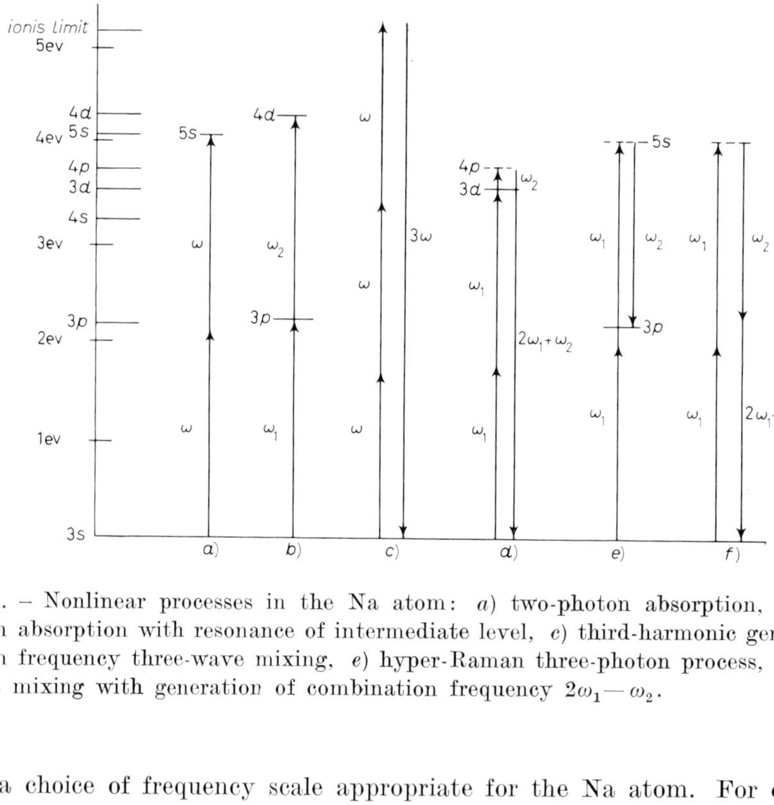

Fig. 1. – Nonlinear processes in the Na atom: *a*) two-photon absorption, *b*) two-photon absorption with resonance of intermediate level, *c*) third-harmonic generation, *d*) sum frequency three-wave mixing, *e*) hyper-Raman three-photon process, *f*) parametric mixing with generation of combination frequency $2\omega_1 - \omega_2$.

with a choice of frequency scale appropriate for the Na atom. For efficient generation momentum-matching conditions are very important. Due to the presence of resonant lines with high oscillator strength in the visible range, it is possible to match the index of refraction in the ultraviolet region to that in the infra-red. Furthermore, it is possible to enhance the value of $|\chi^{(3)}|^2$ by having $2\omega_1$ correspond to an energy separation between the $3s$ ground state and an excited s- or d-state. This resonant behavior is explicitly displayed in eq. (11), and has practical importance [9].

b) Two-photon absorption.

When two levels $|f\rangle$ and $|g\rangle$ of the same parity have an energy separation $\omega_{fg} = \omega_1 + \omega_2$, one may keep only the first resonant term with $n' = f$ on the right-hand side of eq. (11), which in this case is pure imaginary. If one takes in addition $\omega_3 = -\omega_1$ so that $\omega_4 = \omega_2$, the double summation over n and n'' may be written as

$$\chi^{(3)''}(-\omega_2, \omega_1, \omega_2, -\omega_1) = \frac{1}{4} N e^4 \hbar^{-3} \Gamma_{fg}^{-1} \left| \sum_n \left(\frac{x_{gn} x_{nf}}{\omega_{ng} - \omega_1} + \frac{x_{gn} x_{nf}}{\omega_{ng} - \omega_2} \right) \right|^2 . \tag{12}$$

The damping Γ_{ng} has been assumed small compared to $|\omega_{ng} - \omega_1|$. This corresponds to a nonlinear polarization

$$P^{\mathrm{NL}}(+\omega_2) = i\chi^{(3)''} |E(\omega_1)|^2 E(\omega_2) \exp[i\boldsymbol{k}_2 \cdot \boldsymbol{r} - i\omega_2 t] . \tag{13}$$

It is 90° out of phase with the field at ω_2, and proportional to the intensity at ω_1. It describes the two-photon absorption which is proportional to the imaginary part of the nonlinear susceptibility $\chi^{(3)}$ for this choice of frequencies and proportional to the product of the intensities in the two beams at ω_1 and ω_2 respectively. One recognizes that the transition rate for two-photon absorption processes is proportional to the square of the second-order matrix element, which appears on the right-hand side of eq. (12). The same result was already obtained by GOEPPER-MAYER in 1931, utilizing Fermi's golden rule with second-order perturbation theory [5]. The special case of two-photon absorption from a single beam may be obtained by taking $\omega_1 = \omega_2$. Recently two-photon absorption spectroscopy has undergone a rapid development due to the availability of tunable dye lasers. The two-photon absorption between the $3s$- and $5s$-level of atomic sodium [10, 11] is schematized in fig. 1*a*). Figure 1*b*) indicates the resonant enhancement that may be obtained according to eq. (12) if the frequency ω_1 approaches a one-photon resonance [12], for example between the $3s$- and $3p$ level.

The two-photon process gradually merges with a cascade of two one-photon absorption processes. The distinction is rather subtle and is experimentally evident in different line widths and transient relaxation behavior. It depends on the role played by the diagonal density matrix element ϱ_{nn}, representing the population in the intermediate state.

c) Stimulated Raman effect.

Very similar formulae result when $\omega_1 - \omega_2 = \omega_{fg}$. In this case the difference between the two light frequencies corresponds to the energy splitting of the levels

$|g\rangle$ and $|f\rangle$. The replacement of ω_2 by $-\omega_2$ in eqs. (10) and (12) leads to a change in sign of the imaginary part of the susceptibility:

$$\chi^{(3)''}(-\omega_2, \omega_1, \omega_2, -\omega_1) = -\chi^{(3)''}(\omega_2, \omega_1, -\omega_2, -\omega_1)\,.$$

Eq. (13) then predicts a negative absorption or stimulated emission, at the frequency ω_2, which is proportional to the intensity at ω_1. The stimulated Raman effect has been discussed in detail in several reviews. The behavior of the resonant Raman effect, when ω_1 approaches a one-photon resonant frequency ω_{ng}, has been observed in potassium and barium vapor [13, 14]. The process then merges with that of resonant fluorescence, in which the state $|n\rangle$ is excited by one-photon absorption and subsequently a photon $\omega_2 = \omega_{nf}$ is emitted, as discussed by SHEN [15].

d) Parametric generation by three-wave light mixing.

If one takes $\omega_3 = \omega_1$ in eqs. (10) and (11) a polarization at the frequency $\omega_4 = 2\omega_1 + \omega_2$ is obtained. The intensity generated at this frequency is proportional to $|\chi^{(3)}(-2\omega_1 - \omega_2, \omega_1, \omega_2, \omega_1)|^2$. It is possible to have a near-resonant enhancement if $2\omega_1 - \omega_{n'g}$, and an additional enhancement if $\omega_2 \sim \omega_{n'n''}$. This situation is depicted in fig. 1*d*) and was utilized by HARRIS and coworkers [16] to obtain efficient up-conversion in sodium vapor from the infra-red frequency ω_2, corresponding to 10.6 μm wavelength, to the ultraviolet. For exact resonance the parametric process becomes mixed with the process that was originally described as an infra-red quantum counter. The two-photon absorption produces a real population in the state 4*d*. When an infra-red quantum is incident, it takes the atom to the state 5*p*, whence ultraviolet fluorescence to the ground state is observable.

If one replaces ω_2 by $-\omega_2$, the generation of the combination frequency $2\omega_1 - \omega_2$ is obtained. Intermediate resonances may occur if either $2\omega_1$ or $\omega_1 - \omega_2$ correspond to a splitting between the ground state and an excited state of the same parity. The process shown in fig. 1*f*), with resonant enhancement at $2\omega_1$, has been observed by MATSUOKA *et al.* [17] in Rb vapor.

It is possible for both $2\omega_1$ and $\omega_1 - \omega_2$ to be at resonance simultaneously. The resonant two-photon absorption process and the resonant Raman process interfere with each other and with the nonresonant terms in eq. (11). Such effects have been extensively studied in condensed matter [18, 19]. In CuCl, for example, $2\omega_1$ may correspond to a sharp exciton resonance, while $\omega_2 - \omega_1$ may correspond to the optical phonon-polariton resonance in the infra-red.

The behavior of the nonlinear susceptibility near resonance is truly the subject matter of nonlinear spectroscopy.

e) Saturable absorption and intensity-dependent index of refraction.

Consider the special case that $\omega_3=\omega_1$ and $\omega_2=-\omega_1$ and consequently $\omega_4=\omega_1$

$$P^{\text{NL}}(\omega_1)=\chi^{(3)}(-\omega_1,\omega_1,-\omega_1,\omega_1)E(\omega_1)E^*(\omega_1)E(\omega_1)\,. \tag{14}$$

Off resonance, when $\chi^{(3)}$ is real, this corresponds to an intensity-dependent index of refraction. As a resonance with a large oscillator strength is approached from the low-frequency side, $\omega_{ng}-\omega_1>0$, $\chi^{(3)}$ increases and may attain large positive values. Since the index of refraction is highest in the center of a beam where the intensity is highest, a beam of finite diameter may be self-focused and self-trapped. This phenomenon has been observed in rubidium and potassium vapor [20, 21].

Above the resonance, $\omega_{ng}-\omega_1<0$, $\chi'^{(3)}$ is negative and defocusing occurs. In the vicinity of resonance, $\chi^{(3)}$ will acquire a large imaginary part. Closer inspection reveals that this imaginary part is negative. It reduces the ordinary linear absorption by an amount which increases with increasing intensity. This is the well-known phenomenon of saturation. The absorption tends to equalize the populations in the states $|g\rangle$ and $|n\rangle$. The absorption saturates as in bleachable dye cells.

3'3. *Higher-order nonlinearities.* – In media lacking a center of inversion a polarization term which is a quartic function of the field amplitudes may exist; and all materials will, in principle, have polarization terms which are a fifth-power polynomial in the applied-field amplitudes. The hyper-Raman process in which two photons at ω_1 are absorbed and a photon at ω_2 is emitted, as depicted in fig. 1*e*), can be described by the imaginary part of $\chi^{(5)}$. The transition rate is proportional to the square of the intensity at ω_1 and the intensity at ω_2. Some parametric processes due to $\chi^{(4)}$ and $\chi^{(5)}$ are described by AKHMANOV [22]. They are difficult to observe because the field strengths required to observe these high-order nonlinearities usually lie close to the break-down limit of the materials [23].

If there is linear absorption, the materials will heat up rapidly to the melting and boiling point when irradiated by a high-intensity light field. In nonabsorbing, transparent materials, the atoms or molecules may be ionized by multiphoton absorption. In most materials this nonlinear effect becomes dominant for power flux densities of about $10^{12}\ \text{W/cm}^2$, corresponding to field strengths in excess of $3\cdot10^7\ \text{V/cm}$. Field ionization may occur for very low frequencies, or even d.c. fields. KELDYSH [24] has shown that there is a continuous transition from the low-frequency tunneling to the high-frequency multiphoton ionization. The same processes occur in solids, where conduction electrons are generated by tunneling or multiphoton absorption.

As soon as a certain, low concentration of carriers is available, the number of carriers may grow more rapidly by avalanche or cascade ionization. The quasi-free electrons are accelerated by the «inverse bremsstrahlung» process, or, in classical language, the Joule heating of the electron gas is caused by the absorption of radiation by colliding electrons. Some of them acquire enough energy to produce secondary electrons in ionizing collisions. The number of carriers thus obeys a differential equation of the form [23, 25]

$$\frac{\partial N_e}{\partial t} = C|E|^{2m} + \eta(|E|)\,N_e - D\nabla^2 N_e - \beta N_e\,. \tag{15}$$

The first term on the right-hand side corresponds to ionization requiring m photons. The second term represents avalanche or cascade ionization. The ionization coefficient is a function of the field strength. The last two terms represent loss of carriers due to diffusion out of the high-intensity region or due to trapping of electrons.

Condensed matter usually contains enough shallow impurity traps which are ionized at room temperature to provide an initial concentration of carriers N_0 to start the avalanche process. Priming by multiphoton ionization or tunneling is not required, except in ultra-clean gases or solids at very low temperatures. The pulse duration t_p of the high light intensity is often short enough so that electron losses during the laser pulse may be neglected. The electron density is then an exponential function of the intensity and so is the plasma contribution to the index of refraction [23]

$$\Delta n_{p1} = -\frac{2\pi e^2 N_0 \exp\left[\int_0^{t_p}\eta(|E|)\,\mathrm{d}t\right]}{m\omega(\omega + i\tau^{-1})}\,, \tag{16}$$

where τ is the collision time for the electrons. As soon as the plasma density has reached the value for which the plasma frequency equals the light frequency, very strong absorption of light occurs, ultraviolet light is emitted in a spark, and the material may vaporize and transform into a high-temperature plasma. This laser-plasma interaction is of great technological interest, but in nonlinear spectroscopy one is concerned with properties of matter before such plasma formation takes place. Multiphoton ionization and its enhancement via quasi-resonances with bound excited states may be studied, but the field strength should remain below the threshold for dielectric break-down.

It is an interesting experimental fact that the break-down threshold in many optical transparent materials, including glass, quartz, KDP, alkali halides, is nearly constant from zero frequency throughout the infra-red to near-visible wavelengths. The break-down threshold appears to obey approximately the relationship

$$E^{\text{th}}_{\text{r.m.s.}}(\omega) = E^{\text{th}}_{\text{d.c.}}(1 + \omega^2\tau^2)^{\frac{1}{2}}$$

with $\omega\tau \approx 1$ near a wavelength of 1 μm. The threshold appears to increase in going from 1 μm to the green wavelength of 0.53 μm. The threshold field is typically on the order of $(1 \div 3) \cdot 10^6$ V/cm for nanosecond pulses and $(1 \div 2) \cdot 10^7$ V/cm for picosecond pulses. It varies systematically throughout the periodic system for several alkali halides investigated [26]. This behavior is consistent with the avalanche process and rules out multiphoton ionization as a threshold-determining process in these wide-band-gap transparent optical materials.

Each material has a certain characteristic limiting threshold value beyond which the light intensity cannot grow without plasma formation. This value puts a lower limit to the diameter of self-focused spots and filaments. In certain materials other limiting mechanisms, such as depletion of the incident light intensity by stimulated Raman scattering, may set a lower threshold, but the universal avalanche break-down mechanism will always provide an upper bound.

4. – Conclusion.

The generality of this introductory discussion does not do justice to the wide variety of phenomena that can occur in individual systems. The nonlinear spectroscopy of molecules in the infra-red region will reveal different features than the excited levels of atoms near or beyond the ionization limit. The nonlinear properties of various excitations in condensed matter will show many different aspects; excitons, spin waves, plasmas, optical phonons, polaritons, F-centers will all show individual nonlinear characteristics. The nonlinear properties vary widely in ionic crystals, liquid crystals, or molecular fluids. The numerical values of nonlinear susceptibility display large differences between aliphatic chains and chains with conjugated double bonds.

Transient phenomena display new features which are not apparent in steady-state response. Cases in point include the energy transfer and relaxation after vibrational excitation in a molecular fluid with picosecond pulses, coherent quantum beats and optical free induction decay after excitation of a three-level system by two coherent-light waves.

There is a lot of physics hidden behind the general equations (1)-(6) of nonlinear electromagnetic response of matter. The present volume aims to reveal this diversity and the scope of nonlinear spectroscopy.

REFERENCES

[1] See, for example, M. SARGENT, M. O. SCULLY and W. E. LAMB jr.: *Laser Physics* (Reading, Mass., 1974).
[2] See, for example, N. BLOEMBERGEN: *Nonlinear Optics* (New York, N. Y., 1965).
[3] See, for example, M. SARGENT, M. O. SCULLY and W. E. LAMB jr.: *Laser Physics* (Reading, Mass., 1974), p. 29.

[4] A. T. Forrester, R. A. Gudmundsen and P. O. Johnson: *Phys. Rev.*, **99**, 1961 (1955).
[5] M. Goeppert-Mayer: *Ann. d. Phys.*, **9**, 273 (1931).
[6] R. G. Brewer and E. L. Hahn: *Phys. Rev. A*, **11**, 1641 (1975).
[7] D. Bedeaux and N. Bloembergen: *Physica*, **69**, 57 (1973).
[8] Chr. Flytzanis and N. Bloembergen: *Progress in Quantum Electronics*, Vol. **5**, edited by J. H. Sanders and S. Stenholm (London, 1975).
[9] R. B. Miles and S. E. Harris: *IEEE Journ. Quantum Electr.*, QE-**9**, 470 (1973).
[10] M. D. Levenson and N. Bloembergen: *Phys. Rev. Lett.*, **32**, 645 (1974).
[11] B. Cagnac, G. Grynberg and F. Biraben: *Phys. Rev. Lett.*, **32**, 643 (1974).
[12] J. E. Bjorkholm and P. F. Liao: *Phys. Rev. Lett.*, **33**, 128 (1974).
[13] P. P. Sorokin, J. J. Wynne and J. R. Lankard: *Appl. Phys. Lett.*, **22**, 342 (1973).
[14] J. L. Carlsten and P. C. Dunn: *Opt. Comm.*, **14**, 8 (1975).
[15] Y. R. Shen: *Phys. Rev. B*, **9**, 622 (1974).
[16] D. M. Bloom, J. T. Yardley, J. F. Young and S. E. Harris: *Appl. Phys. Lett.*, **24**, 427 (1974).
[17] H. Nakatsuka, J. Okada and M. Matsuoka: *Journ. Phys. Soc. Japan*, **37**, 1406 (1974).
[18] M. D. Levenson and N. Bloembergen: *Phys. Rev. B*, **10**, 4447 (1974).
[19] S. D. Kramer, F. G. Parsons and N. Bloembergen: *Phys. Rev. B*, **9**, 1853 (1974).
[20] S. A. Akhmanov, A. I. Kovrigin, S. A. Maksimov and V. E. Ogluzdin: *JETP Lett.*, **15**, 129 (1972).
[21] D. Grischkowsky and J. A. Armstrong: *Phys. Rev. A*, **6**, 1566 (1972).
[22] S. A. Akhmanov, A. N. Dubovik, S. M. Saltiel, I. V. Tomov and V. G. Tunkin: *JETP Lett.*, **20**, 117 (1974).
[23] N. Bloembergen: *IEEE Journ. Quantum Electr.*, QE-**10**, 375 (1974).
[24] L. V. Keldysh: *Sov. Phys. JETP*, **21**, 1135 (1965).
[25] P. Braunlich, A. Schmid and P. Kelly: *Appl. Phys. Lett.*, **26**, 150 (1975).
[26] W. L. Smith, J. H. Bechtel and N. Bloembergen: *Phys. Rev. B*, **12**, 706 (1975).

Nonlinear High-Resolution Spectroscopy of Atoms and Molecules (*).

T. W. HÄNSCH

Department of Physics, Stanford University - Stanford, Cal. 94305

1. – Introduction.

Atomic and molecular gases of low density are among the simplest subjects of spectroscopic studies. Since collisional interactions can often be ignored, gases permit one to investigate the interaction of electromagnetic radiation with rather elementary quantum-mechanical systems with well-defined energy levels and corresponding sharp spectral lines.

Much of our present knowledge about atomic and molecular structure has been gained by linear spectroscopy, and it is useful to remember some of its basic features, before we start to discuss nonlinear spectroscopy, an exciting new area of research which has been opened by the advent of tunable lasers.

In linear-absorption spectroscopy one observes the attenuation of a (weak) light wave in a medium. The intensity I as a function of absorption path length z is given by the familiar exponential law

$$I = I_0 \exp[-\alpha z] , \tag{1}$$

and the experimental goal is the mapping of the (intensity independent) absorption coefficient α as a function of the light frequency. If Doppler broadening can be neglected (as in the transverse observation of a collimated atomic beam), the absorption coefficient α for light of angular frequency ω in the neighborhood of a spectral line, *i.e.* of a resonance transition between two quantum states a and b with energies U_a and U_b, is given by the Lorentzian function [1]

$$\alpha = 4\pi \frac{\omega}{c} \operatorname{Im} \chi , \tag{2}$$

(*) Work supported in part by the National Science Foundation under Grant MPS76-14786A01.

where

$$\chi = (N_a - N_b)\,|\mu_{ab}|^2 \frac{1}{\hbar}\,\frac{1}{\omega_{ab} - \omega + i\gamma_{ab}}\,. \tag{3}$$

Here, N_a and N_b are the number densities of atoms in the lower and upper state, μ_{ab} is the electric-dipole matrix element of the transition, $\omega_{ab} = (U_b - U_a)/\hbar$ is the atomic resonance frequency, and γ_{ab} is the natural line width. In the absence of collisional line broadening, this width is determined by the radiative decay rates γ_a and γ_b of the two levels according to

$$\gamma_{ab} = \tfrac{1}{2}(\gamma_a + \gamma_b)\,. \tag{4}$$

Light polarization and level degeneracy have been neglected for the sake of simplicity.

The mechanism of the absorption can be easily visualized in a semi-classical model, treating the light waves classically. Let us consider a single monochromatic wave, traveling along the z-axis, with an electric field

$$E(z, t) = \operatorname{Re} E \exp\left[-i[\omega t - kz]\right]. \tag{5}$$

This field will induce oscillating electric dipoles in the atoms and hence create a macroscopic polarization of (complex) amplitude

$$P = \chi E\,. \tag{6}$$

The magnitude and phase of this polarization depend on the (complex) linear susceptibility χ, as given by eq. (3). The atoms, in effect, form an array of small antennas with such a distribution of phases that they in turn emit a monochromatic plane wave traveling along the z-axis. Depending on the phase of this wave relative to the original driving wave, the radiation of the atoms will either reduce the light intensity by destructive interference (absorption, $N_a > N_b$) or increase it by constructive interference (stimulated emission, $N_a < N_b$). Since there is a 90° phase shift between the oscillating dipole moment and the emitted wave, the absorption coefficient is determined by the « out of phase » or « in quadrature » component of the polarization, *i.e.* by the imaginary part of the susceptibility. The « in phase » component of the polarization, corresponding to the real part of the susceptibility, leads to a phase change of the original wave and is responsible for the refractive index n according to the well-known dispersion formula

$$n^2 - 1 = 4\pi \operatorname{Re} \chi\,. \tag{7}$$

The interferometric mapping of the refractive index of a gas is another technique of linear spectroscopy, and can be used to determine the oscillator strengths of spectral lines [1].

The absorption coefficient (2) and the refractive index (7) are independent of the light intensity, as long as the linear relationship (6) between dielectric polarization and driving-field amplitude holds. In this linear approximation the atoms behave essentially like small driven damped harmonic oscillators. At higher intensities, however, this approximation is no longer valid, and a large variety of interesting nonlinear phenomena has been observed since the advent of strong laser light sources. Many of these effects are analogous to phenomena observed earlier in nuclear-magnetic-resonance experiments [2] and in radiofrequency spectroscopy. A few have already proven to be exceedingly useful as the basis of powerful new methods of nonlinear spectroscopy, and these phenomena will be the main topic of our lectures.

It is not difficult to make some qualitative predictions about the kind of nonlinear phenomena which can be expected. Perhaps the simplest one to understand is the saturation of an absorbing transition: atoms, which absorb light, are undergoing transitions between two quantum states. The population of the lower absorbing level is partially depleted, and that of the upper state is increased. As a result, the absorption coefficient α is diminished at high light intensity, *i.e.* the absorbing medium is bleached by the light field.

Other effects are expected, because the atoms at high driving-field amplitudes will no longer behave like simple harmonic oscillators, and higher harmonic frequencies will occur in the atomic oscillations. BLOEMBERGEN has shown in his lectures how time-dependent perturbation theory can be used to calculate higher-order nonlinear susceptibilities, which give corrections to the linear relationship (6). Because of symmetry, the lowest such term for gases is of third order. It makes it possible to generate a new coherent light wave at the third harmonic frequency of the incident light, or, in the case of several incident laser frequencies, at a threefold sum or difference frequency. Nonlinear frequency mixing in gases is of considerable interest, not only because it can be used to extend the wavelength range of tunable dye laser into the vacuum ultraviolet and infra-red, but also because it permits a new kind of spectroscopy. By studying the resonant enhancement of the nonlinear susceptibility at certain light frequencies, one can spectroscopically study perhaps otherwise inaccessible quantum states, such as highly excited autoionizing states of atoms. The lectures of ARMSTRONG will discuss this exciting field in detail.

A closely related phenomenon is two-photon excitation: at high light intensities an atom can undergo transitions from a quantum state a to an excited state c of the same parity by absorption of two photons, whose energies add up to the required transition energy. This process is of considerable interest for spectroscopy, because it is possible to reach high-lying states with photons of smaller energy, and one can study transitions which are forbidden by the ordinary electric-dipole selection rules. BLOEMBERGEN has pointed out that two-photon absorption can also be described in terms of a third-order nonlinear susceptibility. The formal description is very similar to that of stimulated Raman scattering in which one photon is absorbed and a second one emitted under the influence of a stimulating light field. (Spontaneous Raman scattering, though it involves two photons and is closely related to the stimulated case, is generally not considered to be a nonlinear process.)

At still larger intensities, higher-order processes need to be considered, which give rise to higher harmonic frequencies in the atomic oscillations, or lead to three and multiphoton transitions. For very high intensites perturbation theory no longer provides an adequate description. Analytic closed-form solutions of the Schrödinger equation, which are valid up to very high intensities, unfortunately can only be given for very few simple situations. The best-known example is a hypothetical atom with only two energy levels, interacting with a monochromatic light field [3]. Its Schrödinger equation can be rewritten in such a form that it is completely equivalent to the equations describing a classical spinning magnet in an external homogeneous magnetic field (corresponding to the atom) with an additional transverse oscillating magnetic field (corresponding to the light field). When viewed from a frame rotating with the frequency of the transverse field, the magnetic gyroscope will nutate around the stationary component of the driving field. In complete analogy, the two-level atom will undergo optical nutations in a strong light field with the intensity-dependent « Rabi frequency », *i.e.* it will for instance nutate periodically between the lower and upper state, with alternating absorption and stimulated emission of radiation. Optical nutations are responsible for the a.c. Stark effect, *i.e.* the splitting and shift of a spectral line in an intense radiation field. They can also explain a number of fascinating « coherent transient effects », observed in the interaction of short, intense light pulses with atoms or molecules. Self-induced transparency and photon echoes are perhaps the best-known examples. A systematic description of these phenomena will be given in the lectures by BREWER. While these effects provide valuable insight into the dynamics of the interaction of radiation with matter, their usefulness for spectroscopy has so far been limited to the measurement of certain collisional relaxation rates. The complexity of these phenomena and the difficulty of controlling the laser intensity in time and space make accurate measurements of other atomic parameters rather difficult.

So far, the most useful nonlinear phenomena for spectroscopy of atoms and molecules are perhaps saturated absorption and two-photon excitation. Both have become the bases of new methods of very-high-resolution spectroscopy, which permit one to study the spectral lines of gases without the Doppler broadening that normally masks the important details of line shape and structure. The principles and applications of these new techniques will be the main concern of our present lectures.

The spectral lines of gases in the visible region are generally several orders of magnitude wider than their natural line width γ_{ab} because of Doppler broadening due to the random thermal motion of the gas particles. From atoms with a velocity component v along the direction of light propagation, the light frequency ω' will exhibit a linear Doppler shift according to

(8) $$\omega' = \omega(1 - v/c) = \omega - kv\,.$$

The number density of atoms within any velocity interval dv is described by the normalized Maxwellian distribution

(9) $$f(v) = \sqrt{1/\pi\bar{v}^2}\exp\left[-v^2/\bar{v}^2\right],$$

where the mean velocity is

(10) $$\bar{v} = \sqrt{2\varkappa T/M}\,.$$

Here, M is the atomic mass, $\varkappa$ the Boltzmann constant and T is the gas temperature. The absorption profile of a Doppler-broadened spectral line can be calculated by a convolution of the Lorentzian profile (2) with a Gaussian distribution (9). The result can be expressed in terms of the numerically tabulated plasma dispersion function. If the natural line width is much smaller than the Doppler width, the line profile near the center is essentially a Gaussian of width

(11) $$\Delta\omega_{\text{D}} = \omega_{ab}\sqrt{2\varkappa T/Mc^2}\,.$$

It can easily be shown that the same line profile is also observed in the spontaneous-emission spectrum of excited gas atoms.

A number of « classic » spectroscopic techniques are available to circumvent or eliminate this problem of Doppler broadening. If it is possible to induce radiative transitions between two closely spaced energy levels, their splitting can be measured with high precision by radiofrequency spectroscopy or optical-radiofrequency double-resonance spectroscopy, in which the Doppler width (11) becomes negligible by the use of low frequencies. Small level splittings can also be measured by observing optical spectral lines with the (linear) methods of level-crossing spectroscopy or quantum-beat spectroscopy. Here, one studies interference effects between different atomic radiative transition channels, which lead to characteristic changes or modulations in the angular intensity distribution and polarization of the spontaneously emitted light. All these techniques avoid the need for high-resolution optical spectroscopy. They are extremely valuable for accurate measurements of the fine structure or hyperfine structure and the Zeeman or Stark splitting of atomic and molecular levels. They are rendered useless, however, if one is interested in a measurement of the precise wavelength of an optical transition, or in resolving closely spaced spectral lines without any common level such as in measurements of isotope shifts. For such purposes it is sometimes possible to achieve a moderate reduction of the Doppler width in the optical spectrum by cooling the gas sample. Transverse observation of a well-collimated atomic or molecular beam can effectively eliminate Doppler broadening, but such beams are not always easy to prepare, and are hardly feasible for very rare or expensive sample materials. It is also difficult to study transitions from short-lived exciteds tates by the atomic-beam method.

The new nonlinear techniques of high-resolution saturation spectroscopy and Doppler-free two-photon spectroscopy can eliminate Doppler broadening of gaseous absorption lines without suffering from these restrictions. They are therefore fast becoming valuable complements to the older spectroscopic methods, and they have already led to a number of novel spectroscopic studies which would hardly have been possible otherwise.

The basic principles of these two new approaches to high-resolution spectroscopy of atomic and molecular gases and some of their most important characteristics can be understood with rather simple-minded models.

In saturation spectroscopy Doppler broadening is eliminated by observing a signal which originates only from a small group of atoms within a narrow range of axial velocities, just as in the transverse observation of an atomic beam.

Let us consider a laser beam traveling through a gas sample, with a frequency ω close to the resonance frequency ω_{ab} of a transition between two levels a and b (fig. 1a)). The laser light will resonantly interact only with those atoms which

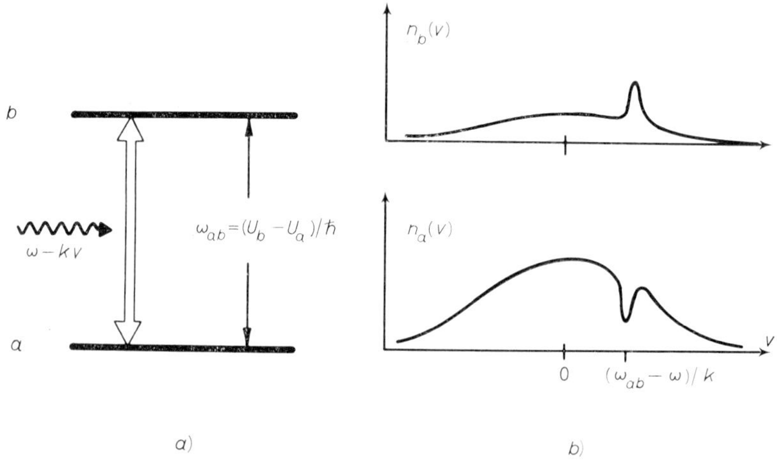

Fig. 1. – Saturation of a Doppler-broadened absorption line: a) Doppler-shifted monochromatic laser beam induces transitions between two atomic levels a and b. b) level population densities *vs.* axial atomic velocity v. The Maxwellian velocity distribution is disturbed by hole burning.

have the right axial velocity to be Doppler shifted into resonance, *i.e.* those with

$$kv \approx \omega - \omega_{ab} \,. \tag{12}$$

Radiative transitions will partly deplete the population of the lower absorbing state at this velocity, *i.e.* the laser light will burn a hole in the velocity distribution of this state, and it will produce a corresponding bump of excess population in the distribution of the upper level, as illustrated in fig. 1*b*). At low intensities the width of these narrow structures is essentially given by the natural line width γ_{ab}.

A narrow-band saturation signal can now be observed if the absorption profile of this line is scanned with a second (weak) coaxial probe laser beam. Such a measurement scans the velocity-dependent population difference of the two levels with a resolution which is again only limited by the natural line width. A narrow dip in the Doppler-broadened absorption line is expected for the probe frequency, where both light beams are interacting with the same atoms.

It is possible to operate with only one laser, *i.e.* to use the same frequency for saturating beam and probe, if the two beams travel in opposite directions, so that they experience Doppler shifts of opposite sign. The bleached and probed regions in the velocity distribution will then both move in opposite directions during a scan. A particularly sensitive saturation spectrometer, working according to this scheme, is shown in fig. 2. The output of a tunable laser is divided into a strong saturating beam and a weak probe, which are passing in near opposite directions through the absorbing-gas sample. At resonance both beams are interacting with the same atoms, those with essentially

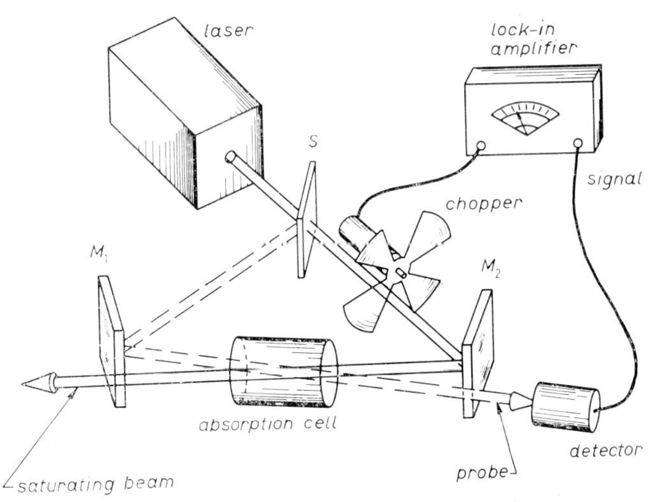

Fig. 2. – Scheme of saturation spectrometer.

zero axial velocity, and the saturating beam can bleach a path for the probe. The saturation signal can be detected with high sensitivity by periodically blocking the saturating beam with a chopper, and by recording the corresponding intensity modulation of the probe with a phase-sensitive detector.

So far we have restricted our attention to two atomic levels. Narrow-band saturation resonances are also expected in the absorption profiles of transitions from the lower level a or the upper level b to a third level c. Their observation will in general require a second laser near the resonance frequency ω_{ac} or ω_{bc}. Another possibility is high-resolution spectroscopy of the axial spontaneous fluorescence, emitted on such a coupled transition. Three-level saturation spectroscopy permits one to study the perturbations in levels a and b separately, which is of particular interest for investigations of relaxation processes.

Hole burning in Doppler-broadened lines was first discussed by BENNETT in early studies of gas laser modes [4]. It is also responsible for the well-known Lamb dip in the power *vs.* frequency characteristic of a single-mode gas laser, which was theoretically predicted by LAMB [5] and observed by JAVAN and co-workers soon after [6]. Here, the two counter-propagating traveling-wave components inside the laser cavity are both simultaneously acting as saturating and probing beams. LEE and SKOLNICK [7] were the first to place a gaseous absorber inside the cavity of a gas laser and to observe the saturation resonances as « inverted Lamb dips ». This phenomenon has subsequently been used in many laboratories to stabilize the frequency of gas lasers or to spectroscopically study molecular absorption lines in accidental coincidence with existing gas laser lines with unprecedented resolution [8-10]. The detection of saturation signals via the intensity modulation of a weak probe was first reported by HÄNSCH and TOSCHEK [11] in a study of coupled gas laser transitions. The particularly useful, convenient and sensitive method of saturation spectroscopy

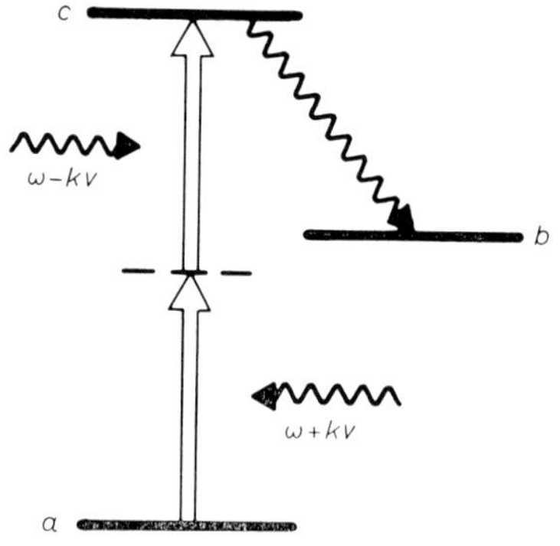

Fig. 3. – Two-photon transitions between levels a and c, induced by two counter-propagation monochromatic laser beams, whose linear Doppler shifts cancel. The spontaneous fluorescence c-b can be used to monitor the excitation.

illustrated in fig. 2 was first used by BORDÉ [12] and, independently, by SMITH and HÄNSCH [13] and by HÄNSCH *et al.* [14]. The first observation of an atomic-resonance line by this method was reported by HÄNSCH *et al.* [15], who had developed the first dye laser with a line width of only a few MHz. Saturation spectroscopy thus became a general spectroscopic technique, applicable to arbitrarily chosen visible atomic and molecular absorption lines.

An entirely different approach to Doppler-free high-resolution spectroscopy is offered by the more recently introduced two-photon method. Let us consider a gas sample, irradiated by two counter-propagating laser beams of equal frequency ω, capable of inducing two-photon transitions from a level a to an excited state c (fig. 3). From an atom moving with any velocity in any direction, the two beams have equal and opposite linear Doppler shifts, *i.e.* their sum frequency will be constant, independent of the atomic velocity. That means it is possible to observe a Doppler-free two-photon resonance line without any need for velocity selection.

The two-photon excitation can often be conveniently monitored by observing the subsequent spontaneous fluorescence from the excited state, as indicated

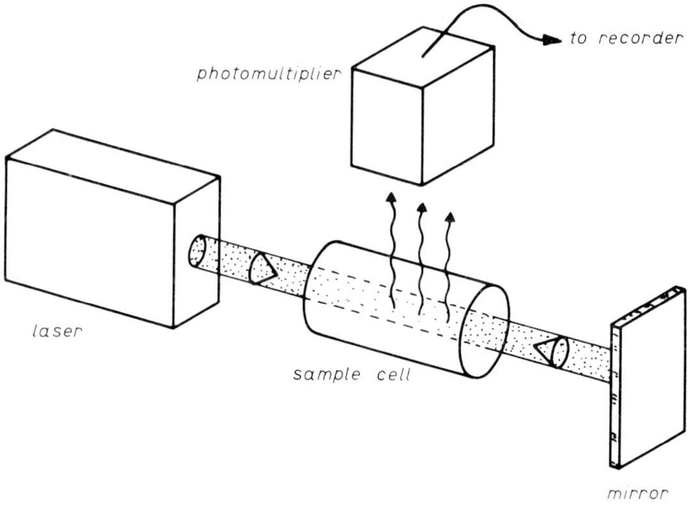

Fig. 4. – Scheme of Doppler-free two-photon spectrometer.

in fig. 4. The expected signal as a function of laser frequency is plotted in fig. 5. The narrow resonance at the center of the spectrum corresponds to excitation with two counter-propagating photons. Its width is essentially limited by the natural line width γ_{ac}. The additional Doppler-broadened pedestal corresponds to excitation by two photons traveling in the same direction. This background can be eliminated, if necessary, *e.g.* by the use of two beams of slightly different frequencies, so that each by itself is tuned off resonance. Excitation with two counter-propagating beams not only gives a dramatic improvement in spectral resolution, but also has the advantage of a strong signal enhancement, because all atoms participate in the interaction, not just a select few, as in the excitation with a single traveling wave.

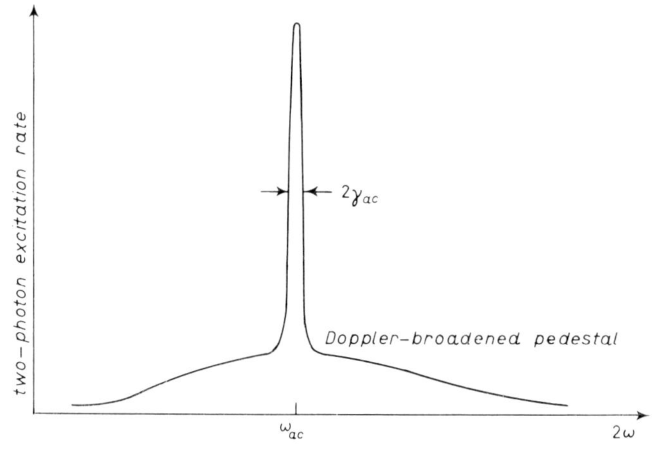

Fig. 5. – Signal expected from Doppler-free two-photon spectrometer. The narrow central resonance corresponds to absorption of one photon from each beam. The Doppler-broadened pedestal corresponds to absorption of two photons from the same beam.

An extension of this principle to higher-order multiphoton transitions has also been suggested. Suppose we want to excite atoms by absorption of N photons of respective wave vectors $\boldsymbol{k}_i$. The sum frequency will be independent of the atomic velocity vector $\boldsymbol{v}$, if the Doppler shifts $\boldsymbol{k}_i \cdot \boldsymbol{v}$ always add to zero. This can be achieved by choosing light frequencies and beam directions so that the wave vectors add to zero, *i.e.*

$$\sum_{i=1}^{N} \boldsymbol{k}_i = 0 . \tag{13}$$

The possibility of Doppler-free two-photon spectroscopy was first pointed out by CHEBOTAEV and co-workers [16] in 1970. But even though two-photon absorption had been observed before in many laboratories since the advent of strong laser sources, it was not until 1974 that the feasibility of Doppler-free two-photon spectroscopy was practically demonstrated. Successful experiments with Na vapor and tunable dye lasers were first reported independently by CAGNAC and co-workers [17], by LEVENSON and BLOEMBERGEN [18] and by HÄNSCH *et al.* [19]. In the short time span since its introduction, this powerful nonlinear spectroscopic method has already made possible numerous novel and interesting experiments.

The subsequent four sections will discuss the principles of high-resolution saturation spectroscopy and Doppler-free two-photon spectroscopy in more detail.

Section **2**, on the theory of saturation spectroscopy, is subdivided into two subsections, dealing with two-level and three-level systems respectively. Each subsection begins with a quantitative description, using a simple hole-burning model. We then turn to a more rigorous semi-classical model, describing the atoms by a density matrix, to point out a number of atomic coherence effects, such as optical nutations or coherent two-photon processes, which are ignored in the simple-minded model, but which will generally affect the observable signals. It is shown, however, that the effects of these coherence phenomena can cancel under certain circumstances in the average over the atomic velocity distribution, so that the hole-burning model can give valid results.

Experimental aspects of saturation spectroscopy are discussed in sect. **3**. The potential of the probe modulation technique is illustrated by reviewing a number of experimental studies of visible atomic and molecular absorption lines, using c.w. and pulsed tunable lasers. The advantages and limitations of alternative techniques of saturation spectroscopy (intermodulated fluorescence, saturated dispersion) are also pointed out. The possible importance of atomic coherence effects is illustrated with some older experiments in atomic three-level systems.

Section **4** is devoted to the theory of Doppler-free two-photon spectroscopy. The relationship between this technique and saturation spectroscopy for the special case of a near resonant intermediate atomic state is investigated. The third-order nonlinear susceptibility provides a general expression for the two-photon transition rate of multilevel atoms. Time-dependent perturbation theory is used to calculate the evolution of the atomic quantum state under two-photon excitation by a short light pulse.

Finally, sect. **5** will review a number of recent interesting experimental studies of atoms and molecules by Doppler-free two-photon spectroscopy.

2. – Theory of saturation spectroscopy.

A number of research papers and several review articles and books have been published on the theory of saturation spectroscopy. It is certainly impossible, within the limited scope of these lectures, to give a complete account of all this work. We shall instead attempt to present a limited number of relatively simple derivations and considerations, which illuminate some of the most important aspects of this field of nonlinear spectroscopy. The reader will be referred to the original papers for a more complete and rigorous treatment of the subject.

2·1. *Two-level atoms.* – Initially we shall restrict our attention to a single optical transition between two energy levels a and b as in fig. 1a). A good review of the extensive literature on saturation spectroscopy in two-level systems can be found in a recent paper by SHIRLEY [21]. We are considering a sample of gas atoms, irradiated by a monochromatic saturating laser beam and a second probe laser beam, traveling collinearly in the same or in opposite directions. The two laser light fields have frequencies ω_1 and ω_2 close to the atomic resonance frequency ω_{ab}, and their intensities are I_1 and I_2 $\big(I = (c/8\pi)E^2\big)$. We want to calculate the change in the optical properties of the absorbing-gas sample for the probe beam that is brought about by the presence of the saturating light field.

We start with a quantitative formulation of the simple hole-burning model which has been outlined in the introduction, and which (in hindsight surprisingly) provides the correct final results under certain limiting conditions. (Its predictions agree with quantum-mechanical third-order perturbation theory in the limit of low intensities, of a large Doppler width, *i.e.* $\Delta\omega_{\rm D} \gg \gamma_{ab}$, and if the two beams are traveling in opposite directions.)

The populations of upper and lower state in our gas sample under steady-state conditions are the result of an equilibrium between populating and depopulating processes. We can write down two rate equations for the population densities of the two levels as a function of axial atomic velocity, neglecting light polarization and a possible level degeneracy:

$$\frac{\mathrm{d}}{\mathrm{d}t}\, n_a(v) = \lambda_a(v) - n_a(v)\gamma_a - [I_1\sigma_1(v) + I_2\sigma_2(v)]\big(n_a(v) - n_b(v)\big)\,, \tag{14}$$

$$\frac{\mathrm{d}}{\mathrm{d}t}\, n_b(v) = \lambda_b(v) - n_b(v)\gamma_b + [I_1\sigma_1(v) + I_2\sigma_2(v)]\big(n_a(v) - n_b(v)\big)\,. \tag{15}$$

Here, λ_a and λ_b describe collisional excitation, and γ_a and γ_b are the spontaneous-decay rates of the levels a and b. (The lower state can thus be different from the atomic ground state.) The expressions at the right-hand side are the transition rates due to light absorption and stimulated emission, assuming an absorption cross-section per atom for the light field of $\sigma_i \cdot \hbar\omega_i$.

In the absence of the radiation fields we expect a Maxwellian velocity distribution, *i.e.*

$$n^0_\alpha(v) = N^0_\alpha f(v)\,, \qquad \alpha = a \text{ or } b\,, \tag{16a}$$

where $f(v)$ is defined in eq. (9), and hence we have

$$\lambda_\alpha(v) = N^0_\alpha \gamma_\alpha f(v)\,, \qquad \alpha = a \text{ or } b\,. \tag{16b}$$

By comparing the absorption cross-section $\sigma_i \cdot \hbar\omega_i$ with the linear absorption coefficient (2) we obtain

$$\sigma_1(v) = \frac{4\pi}{\hbar^2 c} |\mu_{ab}|^2 \frac{\gamma_{ab}}{(\omega_{ab} - \omega_1 + k_1 v)^2 + \gamma_{ab}^2}, \tag{17}$$

and a corresponding expression for σ_2. The Doppler-broadened linear absorption coefficient for the probe light in the limit of very low intensities is given by

$$\alpha_2 = \int_{-\infty}^{\infty} \big(n_a^0(v) - n_b^0(v)\big)\, \sigma_2(v)\, \hbar\omega_2\, \mathrm{d}v\,. \tag{18}$$

Let us now consider the change in population densities that is caused by switching on the saturation beam I_1. From the rate equations (14) and (15) we obtain for small intensites, *i.e.* in a linear approximation,

$$\Delta n_a(v) = -\, I_1(N_a^0 - N_b^0)\, f(v)\, \sigma_1(v)\, \gamma_a^{-1} \tag{19}$$

and

$$\Delta n_b(v) = I_1(N_a^0 - N_b^0)\, f(v)\, \sigma_1(v)\, \gamma_b^{-1}\,. \tag{20}$$

The first of these equations describes the hole in the velocity distribution of the lower state, and the second one the corresponding bump in the distribution of the upper state. We can now calculate the corresponding change in the absorption coefficient for the probe light:

$$\begin{aligned}\Delta\alpha_2 &= \int \big(\Delta n_a(v) - \Delta n_b(v)\big)\, \sigma_2(v)\, \hbar\omega_2\, \mathrm{d}v = \\ &= -\, I_1(N_a^0 - N_b^0)\, \hbar\omega_2(\gamma_a^{-1} + \gamma_b^{-1}) \int f(v)\, \sigma_1(v)\, \sigma_2(v)\, \mathrm{d}v\,.\end{aligned} \tag{21}$$

In the limit of large Doppler width the Maxwellian distribution $f(v)$ varies only slowly near resonance, and can be taken out of the integral, yielding a factor $(1/\sqrt{\pi}\bar{v}^2) \exp[-(\omega_2 - \omega_{ab})^2/\Delta\omega_{\mathrm{D}}^2]$.

The remaining integral over the product of the two absorption cross-sections corresponds to a convolution of the Lorentzian hole in the velocity distribution with the Lorentzian natural line shape of the probe transition. The result is again a Lorentzian function, whose width is given by the sum of the two individual line widths. We find, assuming $|k_1| \approx |k_2| \approx \omega_{ab}/c$,

$$\int \sigma_1(v)\sigma_2(v)\, \mathrm{d}v = \left(\frac{4\pi}{\hbar^2 c} |\mu_{ab}|^2\right)^2 \pi \frac{c}{\omega_{ab}} \frac{2\gamma_{ab}}{[(\omega_1 - \omega_{ab}) \pm (\omega_2 - \omega_{ab})]^2 + (2\gamma_{ab})^2}, \tag{22}$$

where the $+$ sign refers to counter-propagating beams and the $-$ sign to beams traveling in the same direction.

If the frequency of the saturating beam is fixed, and only the probe frequency is tuned, we obtain a Lorentzian saturation signal of twice the natural line width, centered at

$$\omega_2 = 2\omega_{ab} - \omega_1 \tag{23}$$

for opposing beams, and at

$$\omega_2 = \omega_1 \tag{24}$$

for beams traveling in the same direction. Note that the saturation resonance in this latter case does not provide any information about the exact resonance frequency, *i.e.* this configuration is not suitable to resolve closely spaced lines.

If we use counter-propagating beams of the same frequency ω, both the hole and the probed region move simultaneously in opposite directions during frequency tuning, and we obtain a Lorentzian signal of single natural line width, centered at

$$\omega = \omega_{ab} . \tag{25}$$

Obviously, the last scheme is more desirable for spectroscopic applications. The change of the probe absorption by the saturating field in this case is

$$\Delta\alpha_2 = -I_1(N_a^0 - N_b^0)\hbar\omega_2 \frac{1}{\sqrt{\pi}\,\Delta\omega_D} \exp[-(\omega_{ab} - \omega)^2/\Delta\omega_D^2] * \\ * \left(\frac{4\pi}{\hbar^2 c}|\mu_{ab}|^2\right)^2 (\gamma_a^{-1} + \gamma_b^{-1})\pi \frac{\gamma_{ab}}{(\omega - \omega_{ab})^2 + \gamma_{ab}^2} . \tag{26}$$

Division by the background absorption coefficient (18) yields

$$\frac{\Delta\alpha_2}{\alpha_2^0} = -\frac{1}{2}\frac{I_1}{I_{sat}}\frac{\gamma_{ab}^2}{(\omega - \omega_{ab})^2 + \gamma_{ab}^2} , \tag{27}$$

where the saturation parameter is given by

$$I_{sat}^{-1} = \frac{4\pi}{\hbar^2 c}|\mu_{ab}|^2 \frac{1}{\gamma_{ab}}(\gamma_a^{-1} + \gamma_b^{-1}) . \tag{28}$$

Equation (27) describes the response of any given thin layer of the absorbing gas. If the medium is optically thick, one has to take into account the intensity changes of the laser beams due to absorption by an integration over the optical path. To find the exact signal amplitude, it would also be necessary to average over the transverse-intensity variations of the saturating and probing beams.

If the two laser beams are not exactly collinear, but are at a small crossing angle θ, as illustrated in fig. 2, the probe will no longer see a purely Lorentzian hole in the atomic velocity distribution, but it will experience a residual Doppler broadening of width $\Delta\omega_D \sin\theta$.

If the lower state of our absorption line is a stable ground state, we have a zero decay rate γ_a, and eq. (28) predicts an infinitely small saturation intensity. This is not strictly true in practice, since we always have a finite interaction time between the moving atoms and the saturating beam with its limited cross-section, thus providing a nonzero effective rate γ'_a. But it is correct that very small intensities can be sufficient to accumulatively deplete the absorbing ground state by optical pumping, if the excited atoms do not return to the same level by spontaneous decay. Spontaneous transitions $b \to a$ have been neglected in our simple rate equations (19) and (20). If they are taken into account we obtain a modified saturation parameter

$$I_{sat}^{-1} = \frac{4\pi}{\hbar^2 c} |\mu_{ab}|^2 \frac{1}{\gamma_{ab}} (\gamma_a^{-1} + \gamma_b^{-1} - A_{ab}\gamma_a^{-1}\gamma_b^{-1}) , \tag{29}$$

where A_{ab} is the Einstein coefficient of spontaneous emission, *i.e.* the spontaneous-transition rate per atom.

The finite interaction time of the moving atoms with the laser beams can be the dominating cause of line broadening in the case of an extremely narrow natural line width, and optical-beam expansion may be necessary for very-high-resolution spectroscopy [20].

At extremely high resolution it becomes also necessary to take the recoil for radiative absorption and emission into account [20]. Momentum conservation requires that atoms which absorb a photon from the saturating beam arrive in the upper state with a slightly different axial velocity, *i.e.* the bump in the upper level will be displaced relative to the hole in the lower state. The corresponding recoil shift for probe transitions has different signs for absorption and stimulated emission. As a result, the saturation signal will appear as an extremely close line doublet, separated by

$$\Delta\omega_{recoil} = \hbar\omega^2/Mc^2 . \tag{30}$$

In the limit of a very large Doppler width, the rate equations (19) and (20) can be solved analytically up to arbitrary saturation intensities. The hole in the velocity distribution remains Lorentzian, but its width Δv becomes power broadened according to

$$k\,\Delta v = \gamma_{ab}\sqrt{1 + I/I_{sat}} . \tag{31}$$

This broadening is caused by the reduction of the net radiative transition rate at the center of the hole due to saturation, resulting in a relatively stronger net rate for the off-resonance atoms in the wings of the hole. The hole-burning model predicts then a power-broadened line width of the observed saturation signal

$$\Delta\omega = \gamma_{ab}(1 + \sqrt{1 + I/I_{sat}})/2 \, . \tag{32}$$

Unfortunately, this result does not agree with the results of quantum-mechanical calculations, whereas the predicted hole shape in the velocity distribution for a single strong monochromatic traveling wave remains valid.

The detection of the velocity-selective hole burning via the absorption of a probe beam requires a gas sample of sufficiently large single-pass absorption αl. Too small a relative probe intensity change due to saturation $\Delta I_2/I_2 \leqslant \Delta\alpha l \ll \alpha l$, will result in a poor signal-to-noise ratio. To study samples of very low density, the saturation signal can be observed in fluorescence rather than in absorption. The intensity of the fluorescent light, spontaneously emitted from the excited state b, is proportional to the total population density N_b of this level. If the atoms are excited by two counter-propagating laser beams of frequency ω and of intensites I_1 and I_2, and if collisional excitation can be neglected, this population is given, according to eq. (15), by

$$N_b = (\alpha_1 I_1 + \alpha_2 I_2)/\hbar\omega\gamma_c \, . \tag{33}$$

The radiative excitation rate and hence the population is reduced by saturation, if the two beams interact with the same atoms. Since each of the two beams is, in effect, acting both as a probe and as a saturating beam, the mutual saturation can be described in a linear approximation by eq. (27), and the population change ΔN_b, caused by this mutual saturation, is

$$\Delta N_b = (\Delta\alpha_1 I_1 + \Delta\alpha_2 I_2)/(\hbar\omega\gamma_c) = -\alpha^0 \frac{I_1 I_2}{I_{sat}} \frac{\gamma_{ab}^2}{(\omega - \omega_{ab})^2 + \gamma_{ab}^2} \frac{1}{\hbar\omega\gamma_c} \, . \tag{34}$$

The relative change at resonance for $I_1 = I_2 = I$ is

$$\Delta N_b/N_b = -I/(2I_{sat}) \, , \tag{35}$$

and can provide a saturation signal without unduly large background even in the limit of very small total absorption.

We now turn to a more rigorous semi-classical model, which describes the atoms quantum mechanically and hence takes atomic coherence properties into account which have been ignored in the rate equation description.

The two atomic quantum states of energies U_a and U_b are eigenstates of the Hamiltonian H^A of the unperturbed atom, *i.e.* we have

$$H^A|\alpha\rangle = U_\alpha|\alpha\rangle \, , \qquad \alpha = a \text{ or } b \, . \tag{36}$$

In general, any « pure » quantum state of our two-level atoms can be described as a linear superposition

$$c_a|a\rangle + c_b|b\rangle . \tag{37}$$

To properly describe a statistical ensemble of atoms of axial velocity v at position z and time t we use an ensemble-averaged density matrix [22]

$$\varrho(v, z, t) = \begin{pmatrix} \varrho_{aa} & \varrho_{ab} \\ \varrho_{ba} & \varrho_{bb} \end{pmatrix} = \varrho^\dagger , \tag{38}$$

whose elements are averaged over all possible initial conditions according to

$$\varrho_{ij} = \langle c_i c_j^* \rangle_{\text{av}} . \tag{39}$$

We chose the normalization so that the diagonal elements ϱ_{aa} and ϱ_{bb} are equal to the velocity-dependent population densities $n_a(v)$ and $n_b(v)$. The off-diagonal elements $\varrho_{ab} = \varrho_{ba}^*$ describe a coherent superposition of the two states, *i.e.* a coherent oscillation of the atomic charge distribution for the ensemble average. If the atomic density matrix is known, we can easily calculate the expectation value of any physical quantity A according to

$$\langle A \rangle = \mathrm{Tr}\,(A\varrho) . \tag{40}$$

In particular, we obtain the velocity-averaged macroscopic polarization of our gas medium from

$$P = \int_{-\infty}^{\infty} \mathrm{Tr}\,(\mu\varrho)\,\mathrm{d}v , \tag{41}$$

where μ is the matrix of the electric-dipole operator, *i.e.*

$$\mu = \begin{pmatrix} 0 & \mu_{ab} \\ \mu_{ba} & 0 \end{pmatrix} . \tag{42}$$

The saturating and probe laser beams are described, as before, by classical electric plane waves, traveling in the $\pm z$-direction. The combined field is

$$E(z, t) = \tfrac{1}{2} E_1 \exp[-i(\omega_1 t - k_1 z)] + \tfrac{1}{2} E_2 \exp[-i(\omega_2 t - k_2 z)] + \text{c.c.} , \tag{43}$$

where E_1 and E_2 are the complex field amplitudes of the two beams.

In the laboratory frame the atomic density matrix obeys the Schrödinger equation

$$\left(\frac{\partial}{\partial t} + v\frac{\partial}{\partial z}\right)\varrho = \frac{1}{i\hbar}[H, \varrho] + \Lambda . \tag{44}$$

The Hamiltonian

$$H = H^A + H^I + H^R \tag{45}$$

includes the interaction energy H^I with the radiation fields. In the electric-dipole approximation we set

$$H^I = \mu E(z, t) . \tag{46}$$

Relaxation processes are introduced phenomenologically by the term H^R. Using our earlier notation we set

$$\frac{1}{i\hbar}[H^R, \varrho] = -\begin{pmatrix} \varrho_{aa}\gamma_a & \varrho_{ab}\gamma_{ab} \\ \varrho_{ba}\gamma_{ab} & \varrho_{bb}\gamma_b \end{pmatrix} . \tag{47}$$

Correspondingly, the term

$$\Lambda = \begin{pmatrix} \lambda_a & 0 \\ 0 & \lambda_b \end{pmatrix} \tag{48}$$

introduces collisional excitation processes. The Doppler effect due to the atomic motion is taken into account by the spatial derivative on the left-hand side of eq. (44).

For low intensities it is possible to solve the equations of motion (44) by a perturbation expansion, setting

$$\varrho = \varrho^{(0)} + \varrho^{(1)} + \varrho^{(2)} + \dots , \tag{49}$$

where

$$\left\{\begin{aligned} & i\hbar\left(\frac{\partial}{\partial t} + v\frac{\partial}{\partial z}\right)\varrho^{(0)} - [H^A + H^R, \varrho^{(0)}] = i\hbar\Lambda , \\ & i\hbar\left(\frac{\partial}{\partial t} + v\frac{\partial}{\partial z}\right)\varrho^{(1)} - [H^A + H^R, \varrho^{(1)}] = [H^I, \varrho^{(0)}] , \\ & \dots\dots \\ & i\hbar\left(\frac{\partial}{\partial t} + v\frac{\partial}{\partial z}\right)\varrho^{(n)} - [H^A + H^R, \varrho^{(n)}] = [H^I, \varrho^{(n-1)}] , \\ & \dots\dots \end{aligned}\right. \tag{50}$$

It is somewhat tedious, though straightforward, to solve these equations explicitly up to third order [5]. Therefore, we shall only try to gain a qualitative understanding of the solution by inspecting the individual equations. If we consider the driving terms at the right-hand side of these inhomogeneous differential equations, it is not difficult to see that the matrices $\varrho^{(n)}$ under steady-state conditions will have vanishing off-diagonal elements for even n and zero diagonal elements for odd n, *i.e.* each of the matrix equations (50) has only two nontrivial components. Since $\varrho_{ab} = \varrho^*_{ba}$, it is sufficient to consider only one component for odd orders.

To lowest order we obtain simply the background population densitites $\varrho^{(0)}_{aa} = n^0_a(v)$ and $\varrho^{(0)}_{bb} = n^0_b(v)$, with the same Maxwellian velocity distribution as the excitation rates λ_a and λ_b.

The first-order correction due to the presence of the light fields is calculated from

$$(51) \qquad i\hbar\left(\frac{\partial}{\partial t} + v\frac{\partial}{\partial z} + i\omega_{ab} + \gamma_{ab}\right)\varrho^{(1)}_{ab} = (\varrho^{(0)}_{aa} - \varrho^{(0)}_{bb})\mu_{ab}E(z, t)\,.$$

The driving term on the right contains the four Fourier components of the oscillating electric field (43), and the solution $\varrho^{(1)}_{ab}$ will be the sum of four corresponding terms. Two of these are small off-resonance contributions, which are usually neglected in the « rotating-wave approximation », leaving a solution of the form

$$(52) \qquad \varrho^{(1)}_{ab} = c_1 E_1 \exp[-i(\omega_1 t - k_1 z)] + c_2 E_2 \exp[-i(\omega_2 t - k_2 z)]\,,$$

where c_1 and c_2 are velocity- and frequency-dependent amplitude factors. Equation (52) describes the linear response of the oscillating atomic dipoles in accordance with eqs. (1) and (7).

The next step gives the radiation-induced changes of the level population densities to lowest approximation. We have

$$(53) \qquad i\hbar\left(\frac{\partial}{\partial t} + v\frac{\partial}{\partial z} + \gamma_a\right)\varrho^{(2)}_{aa} = (\varrho^{(1)}_{ab} - \varrho^{(1)}_{ba})\mu_{ab}E(z, t)\,,$$

and a corresponding equation for $\varrho^{(2)}_{bb}$. This time, the source term at the right is given by the product of the oscillating light field and the oscillating atomic-dipole moments, as described by $\varrho^{(1)}_{ab}$. If we again neglect rapidly oscillating terms of small amplitude, we expect solutions of the form

$$(54) \qquad \varrho^{(2)}_{aa} = d_1|E_1|^2 + d_2|E_2|^2 + d_3(E_1 E^*_2 \exp[-i[(\omega_1 - \omega_2)t - (k_1 - k_2)z]] + \text{c.c.})$$

with velocity- and frequency-dependent amplitude factors d_i. The first two terms describe the holes burnt by the two traveling light waves, and agree with the prediction of the rate equation model. The third term describes slow population pulsations at the beat frequency of the two laser beams, if $\omega_1 \neq \omega_2$. If two counter-propagating beams of the same frequency are used, this term describes a spatial modulation of the population densities with a period of half a wavelength, corresponding to the nodes of the resulting standing wave. This additional term was neglected in our earlier rate equation model.

If the probe beam is weak compared to the saturating beam, *i.e.* if $E_2 \ll E_1$, the modulation depth of this latter perturbation is small compared to the hole burnt by the saturating beam. Nonetheless it can contribute with comparable magnitude to the observed probe signal, as is apparent in the next step of our perturbation approach:

$$i\hbar\left(\frac{\partial}{\partial t}+v\frac{\partial}{\partial z}+i\omega_{ab}+\gamma_{ab}\right)\varrho_{ab}^{(3)}=(\varrho_{aa}^{(2)}-\varrho_{bb}^{(2)})\mu_{ab}E(z,t)\,. \tag{55}$$

To find the third-order correction to the response of the atomic dipoles at the probe frequency, we need to calculate the component of $\varrho_{ab}^{(3)}$ oscillating with $\exp[-i[\omega_2 t-k_2 z]]$. One driving term for this component is obtained by multiplying the oscillating probe field with the stationary hole-burning part of $\varrho_{aa}^{(2)}-\varrho_{bb}^{(2)}$. For a weak probe it is proportional to $|E_1|^2E_2$ and provides the same probe signal as calculated in the rate equation model. But now we get an additional driving term, also proportional to $|E_1|^2E_2$, by multiplying the oscillating saturating field with the pulsating or spatially modulated part of $\varrho_{aa}^{(2)}-\varrho_{bb}^{(2)}$.

To interpret the corresponding additional probe signal, let us first consider the case of two beams traveling in the same direction. The small population pulsations generated by the beating of the saturating beam and probe will modulate the absorption for the saturating beam, and hence create side bands. One will be generated at a new frequency $2\omega_1-\omega_2$, the other will coincide in frequency with the probe beam, and modify the probe signal by interference. Detailed calculations have been carried out by CLOSE [23]. After integration over the atomic velocity distribution, one obtains a probe signal that is considerably narrower and more peaked than the Lorentzian saturation signal calculated with the simple hole-burning model [24].

In the spectroscopically more interesting situation of two counter-propagating beams we also expect an additional probe signal, because light from the saturating beam will experience « Bragg reflection » from the spatially modulated atomic densities, and the reflected wave will interfere with the probe beam. The resulting deviation from the predictions of the rate equation model can be substantial for atoms with a small spread of axial velocities, such as in a transverse atomic beam. If we integrate over a wide velocity distribution, however, the Bragg waves reflected from different velocity groups will completely cancel by mutual interference, leaving the same probe signal as predicted by the hole-burning model.

This quantitative agreement between hole-burning model and semi-classical model breaks down, however, at higher intensities of the saturating beam, for which third-order perturbation theory is no longer valid. BAKLANOV and CHEBOTAEV [25] and HAROCHE and HARTMANN [26] have analytically calculated the line shape of the saturation signal for a saturating beam of arbitrary intensity and a weak counter-propagating probe beam. This is accomplished by first solving the much simpler problem of atoms interacting with only one single strong monochromatic traveling wave, and then using the result as a starting point to calculate the response to the second probe wave by perturbation theory.

If only the saturating beam is present, eq. (44) can easily be solved exactly. Looking for a steady-state solution of the form

$$\left\{\begin{aligned} \varrho_{aa} &= n_a(v)\,, \\ \varrho_{bb} &= n_b(v)\,, \\ \varrho_{ab} &= \hat{\varrho}_{ab}(v)\exp[-i(\omega_1 t - k_1 z)]\,, \end{aligned}\right. \tag{56}$$

one obtains, by inserting these expressions into eq. (44), a system of algebraic equations, which can be solved to give

$$\hat{\varrho}_{ab}(v) = -i\,\frac{\mu_{ab}E_1}{2\hbar}\left(n_a^0(v) - n_b^0(v)\right)\frac{\gamma_{ab} - i(\omega_{ab} - \omega_1 + k_1 v)}{(\omega_{ab} - \omega_1 + k_1 v)^2 + \gamma_{ab}^2(1 + I_1/I_{\text{sat}})} \tag{57}$$

and

$$n_b(v) - n_b(v) = \left(n_a^0(v) - n_b^0(v)\right)\left(1 - \frac{\gamma_{ab}^2(I_1/I_{\text{sat}})}{(\omega_{ab} - \omega_1 + k_1 v)^2 + \gamma_{ab}^2(1 + I_1/I_{\text{sat}})}\right). \tag{58}$$

The power-broadened hole burning described in eq. (58) is the same as that predicted by the rate equation model. The simplicity of this result is somewhat deceiving, however. We obtain stationary population densities, because we are looking at an ensemble average over all possible initial conditions. If we restrict our attention to atoms, which are in a well-defined state, say in level a, at time $t = t_0$, we would obtain an entirely different answer.

As discussed elsewhere [3] in detail, the Schrödinger equation of a two-level atom, interacting with a monochromatic light field, can be written in a form describing the precession of a pseudospin vector **s** around a « torque » vector **Ω**, if relaxation is neglected:

$$\frac{\mathrm{d}}{\mathrm{d}t}\,\mathbf{s} = \mathbf{\Omega}\times\mathbf{s}\,, \tag{59}$$

where

$$\left\{\begin{aligned} s_1 &= \varrho_{ab} + \overset{*}{\varrho}_{ab}\,, \\ s_2 &= i(\varrho_{ab} - \overset{*}{\varrho}_{ab})\,, \\ s_3 &= \varrho_{aa} - \varrho_{bb} \end{aligned}\right. \tag{60}$$

and

$$\left\{\begin{aligned} \Omega_1 &= -\mu_{ab}E(t)/2\hbar\,, \\ \Omega_2 &= 0\,, \\ \Omega_3 &= \omega_{ab}\,. \end{aligned}\right. \tag{61}$$

From a frame rotating with the light frequency ω_1, the motion of the state vector $\boldsymbol{s}$ is a precession around a stationary torque vector $\boldsymbol{\Omega}$ with the Rabi nutation frequency

$$\Omega_{\text{Rabi}} = \sqrt{(\omega_{ab} - \omega_1)^2 + (\mu_{ab} E_1/2\hbar)^2}\,. \tag{62}$$

The « optical nutations », which are hidden in the results (57) and (58), become observable, however, in the absorption of a second weak probe beam, as shown in detail by HAROCHE and HARTMANN. The probe absorption profile for atoms of given axial velocity v is no longer a simple Lorentzian, but is split into two line components, separated by twice the Rabi nutation frequency, as illustrated in fig. 6. This « a.c. Stark effect » causes a rather noticeable modification of the expected probe signal, even after integration over a wide atomic velocity distribution. The saturation dip will be wider and shallower than ex-

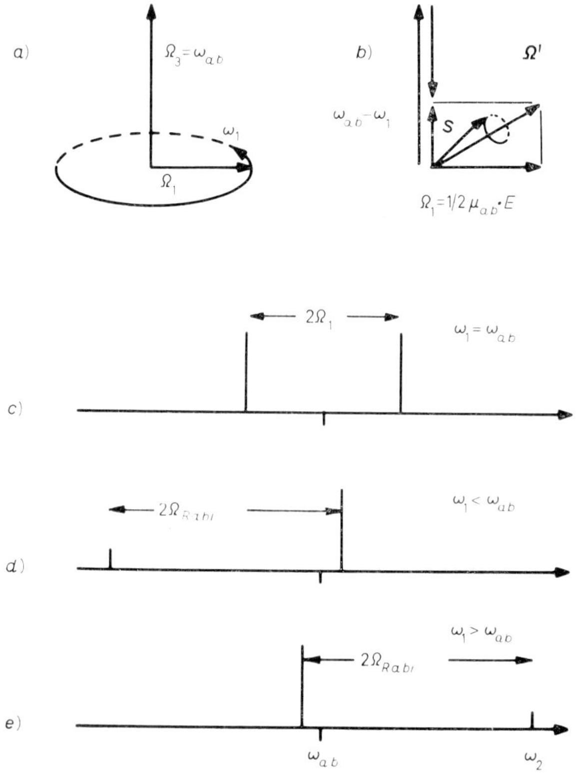

Fig. 6. – Dynamic Stark splitting explained in a two-level system [26]. *a*) Components of « torque » vector $\boldsymbol{\Omega}$ in the laboratory frame. *b*) Torque vector $\boldsymbol{\Omega}'$ in the rotating frame. The pseudospin vector $\boldsymbol{s}'$ precesses around this stationary vector with the Rabi nutation frequency $\Omega_{\text{Rabi}} = \sqrt{(\omega_{ab} - \omega_1) + \Omega_1^2}$. *c*), *d*) and *e*) Spectrum of probe resonances at frequencies $\omega_{ab} \pm \Omega_{\text{Rabi}}$ for different values of pump detuning.

pected from the population changes alone (see fig. 7). At exact resonance, for instance, the probe light at high saturating intensities will see a substantial absorption from atoms with nonzero axial velocity, whose Doppler shift is compensated by the a.c. Stark shift. As a consequence, the absorption for the probe beam can never be completely bleached, even by an arbitrarily strong saturating beam, contrary to the predictions of rate equations.

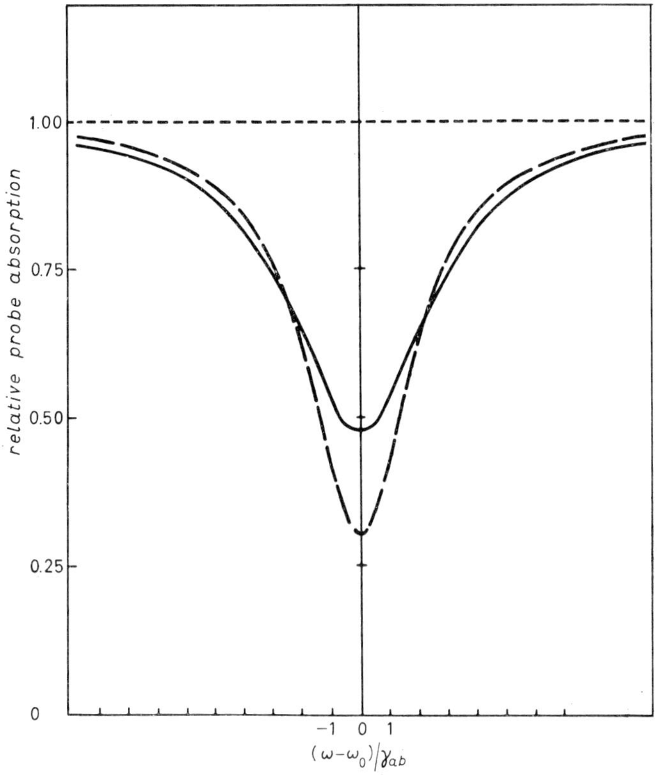

Fig. 7. – Saturated absorption line shape in the two-level system. The relative probe absorption is plotted for infinite Doppler width. The long-dashed curve gives the signal expected within the rate equation model for $I_1/I_{sat} = 10$ and $\gamma_{ab} = \gamma_a = \gamma_b$. The solid curve represents the wider and shallower signal predicted by the semi-classical model [26].

Haroche and Hartmann have also pointed out the role of additional, more subtle nonlinear phenomena, such as hyper-Raman processes, involving the absorption of two photons from the saturating beam and the emission of one photon at the probe frequency. In the case of a small Doppler width these processes can even provide some small amplification for the weak probe beam, although there is no population inversion between the two levels. For details of these rather involved calculations the reader is referred to the exceptionally clear original paper [26].

While it is possible to analytically calculate the response of a Doppler-broadened absorption line to a strong saturating light wave and a weak counter-propagating probe wave, one must in general resort to numerical computer calculations for the more complicated situation of two strong counter-propagating waves, as *e.g.* found inside the cavity of a single-mode gas laser.

Stenholm and Lamb [27] and Feldmann and Feld [28] have investigated this situation in their theory of a high-intensity gas laser. The latter authors expand the components of the ensemble-averaged atomic density matrix into spatial Fourier components, and solve the resulting set of coupled differential equations. The results are expressed in terms of infinite continued fractions, and numerically evaluated by computer. The most striking deviation from the

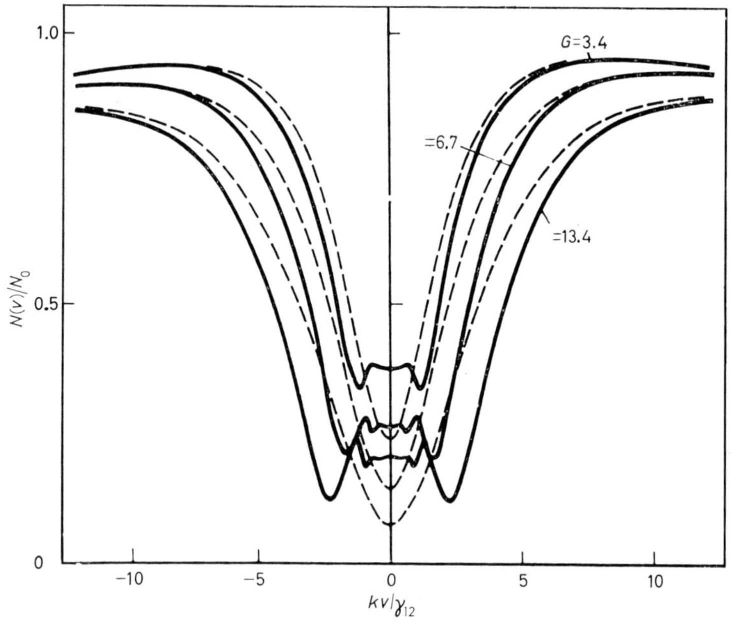

Fig. 8. – Saturation of the Doppler-broadened transition *a-b* by a strong standing wave at resonance [28]. The space-averaged population difference $n_a - n_b$ is plotted *vs.* the axial atomic velocity v.

predictions of the hole-burning model is a fine structure or modulation of the holes in the space-averaged population difference $n_a(v) - n_b(v)$, as illustrated in fig. 8. Berman [29] has shown more recently that the same fine structure can also be obtained by a « dressed atom » approach. This modulation does not show up in the absorption of the strong « probe » field E_2, however. Its detection requires the scanning of a second, coupled transition *a-c* or *b-c* with an additional probing light field, as predicted by Feldmann and Feld [30].

All the discussed deviations from the predictions of the rate equation model certainly provide interesting insights into the interaction of light with matter. But for many spectroscopic applications, the cancellation of these coherence phenomena in the limit of weak, counter-propagating beams and of a large Doppler width is more than welcome, because it permits a very simple interpretation of the saturation spectra.

2'2. *Three-level atoms.* – We next investigate the interesting situation where the probe laser beam is in resonance not with the saturated transition *a-b* itself, but with a second coupled transition *b-c* (or *a-c*) to a third atomic level *c* (fig. 9).

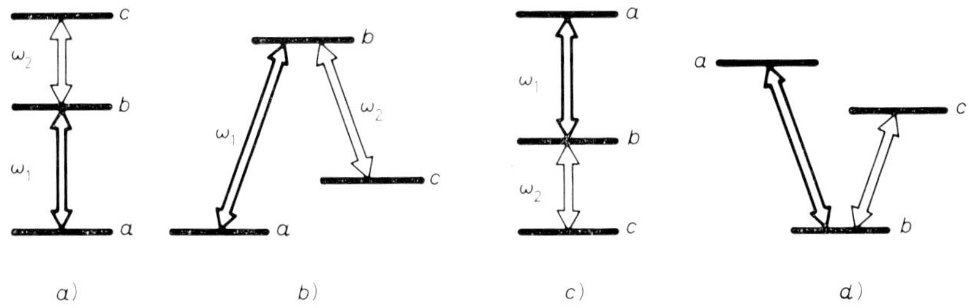

Fig. 9. – Saturation spectroscopy in atomic three-level systems. The saturating beam of frequency ω_1 is close to the resonance frequency ω_{ab} of the transition *a-b*, and the probe beam frequency ω_2 is interrogating the transition *b-c*.

Such studies will in general require a second probe laser of frequency $\omega_2 \neq \omega_1$, unless the two coupled spectral lines overlap within their Doppler width. In this latter special case, a three-level saturation signal can be observed with two counter-propagating laser beams of the same frequency, tuned half-way in between the two resonance lines. The beams can then interact via atoms of nonzero axial velocity, which are Doppler tuned into resonance with the saturating beam on their transition *a-b*, and simultaneously shifted into resonance with the probe on their transition *b-c*. The resulting signal is often referred to as « cross-over signal ».

In the study of three-level gas systems in interaction with two monochromatic traveling light waves, we will encounter new atomic coherence effects which can strongly modify the probe signal, but which can also cancel under certain limiting conditions, when we average over all atomic velocities, so that a simple rate equation model can give quantitatively correct predictions. In addition, we will be able to clarify the relationship between stepwise excitation and two-photon excitation with a near-resonant intermediate level, and thus lay the groundwork for a later discussion of Doppler-free two-photon spectroscopy. Many other aspects have been investigated in a large number of research papers. An excellent review on three-level gas systems and their interaction with radiation has recently been written by BETEROV and CHEBOTAEV [31].

To be definite, let us consider a cascade level scheme as shown in fig. 9a). Other level configurations can be treated in exactly the same way, and require only small changes of signs or subscripts in our equations.

We again look first at a simple rate equation model. Let us assume that the probe intensity is very weak, so that any hole burning is caused solely by the saturating beam. The resulting population changes $\Delta n_a(v)$ and $\Delta n_b(v)$ have already been calculated earlier in a linear approximation (eqs. (19) and (20)). But this time only the bump $\Delta n_b(v)$ in the upper, common level will contribute to the saturation signal. It will increase the absorption for the probe light, if the probe beam is tuned into resonance with the same velocity group of atoms as the saturating beam. The change of the probe absorption coefficient is simply

$$\Delta\alpha_2 = \int_{-\infty}^{\infty} \Delta n_b(v)\sigma_2'(v)\hbar\omega_2\,\mathrm{d}v = I_1(N_a^0 - N_b^0)\hbar\omega_2 \frac{1}{\gamma_b}\int_{-\infty}^{\infty} f(v)\sigma_1(v)\sigma_2'(v)\,\mathrm{d}v\,, \tag{63}$$

where $\sigma_1(v)$ is given by eq. (17) and

$$\sigma_2'(v) = \frac{4\pi}{\hbar^2 c}|\mu_{bc}|^2 \frac{\gamma_{bc}}{(\omega_{bc} - \omega_2 + k_2 v)^2 + \gamma_{bc}^2}\,. \tag{64}$$

In the limit of a large Doppler width, the slowly varying Maxwellian velocity distribution $f(v)$ can again be replaced by a constant factor near resonance, leaving a convolution integral of two Lorentzians:

$$\Delta\alpha_2 \propto \int_{-\infty}^{\infty} \sigma_1(v)\sigma_2'(v)\,\mathrm{d}v = \tag{65}$$

$$= \text{const}\,\frac{\gamma_{bc} + |k_2/k_1|\gamma_{ab}}{[(\omega_2 - \omega_{bc}) - (k_2/k_1)(\omega_1 - \omega_{ab})]^2 + (\gamma_{ab} + |k_2/k_1|\gamma_{ab})^2}\,.$$

If the frequency ω_1 of the saturating beam is fixed, and only the probe laser frequency ω_2 is varied, the saturation signal is a Lorentzian line of width $\gamma_{bc} + |k_2/k_1|\gamma_{ab}$ centered at $(k_2/k_1)(\omega_1 - \omega_{ab})$.

The factor k_2/k_1 accounts for the different Doppler shift and Doppler width of the Lorentzian bump in the atomic velocity distribution, when seen by the probe.

The limitations of this model will again become apparent in the subsequent semi-classical treatment. In analogy to our two-level calculations, we describe the atoms quantum mechanically by an ensemble-averaged density matrix

$$\varrho(v, z, t) = \begin{pmatrix} \varrho_{aa} & \varrho_{ab} & \varrho_{ac} \\ \varrho_{ba} & \varrho_{bb} & \varrho_{bc} \\ \varrho_{ca} & \varrho_{cb} & \varrho_{cc} \end{pmatrix}. \tag{66}$$

The saturating beam and probe beam are described by two classical traveling waves

$$(67)\qquad \begin{cases} E_1(z,t) = \frac{1}{2}E_1 \exp[-i(\omega_1 t - k_1 z)] + \text{c.c.}\,, \\ E_2(z,t) = \frac{1}{2}E_2 \exp[-i(\omega_2 t - k_2 z)] + \text{c.c.} \end{cases}$$

We assume that frequency ω_1 is close to ω_{ab} and that ω_2 is close to ω_{bc}. And we assume that the two atomic resonance frequencies are sufficiently different, so that we can ignore the off-resonant interaction of the saturating beam with the probe transition, and of the probe beam with the saturating transition. The Hamiltonian of the interaction energy can then be written as

$$(68)\qquad H' = \begin{pmatrix} 0 & \mu_{ab}E_1(z,t) & 0 \\ \mu_{ba}E_1(z,t) & 0 & \mu_{bc}E_2(z,t) \\ 0 & \mu_{cb}E_2(z,t) & 0 \end{pmatrix}.$$

The Schrödinger equation (44) of our three-level density matrix can again be solved by the perturbation method (50). To lowest order one obtains the background population densities of the three levels, and to first order the linear atomic-dipole oscillations at the two driving frequencies.

In the second order we obtain the population changes $\Delta n_a(v)$ and $\Delta n_b(v)$ in linear approximation. The pulsation and spatial modulation due to interference of the two beams, which were important for two-level systems, can now be ignored because of the large frequency difference. But this time we obtain an additional nonzero off-diagonal element

$$(69)\qquad \varrho_{ac}^{(2)} = dE_1E_2 \exp\left[i[(\omega_1+\omega_2)t - (k_1+k_2)z]\right],$$

which corresponds to a coherent superposition of the two atomic states a and c, and describes an oscillation of the atomic charge distribution at the sum frequency of the two light waves. For symmetry reasons, these oscillations will generally not result in an oscillating electric-dipole moment. But they will influence the atomic response at the probe frequency, as calculated in third order:

$$(70)\qquad i\hbar\left(\frac{\partial}{\partial t} + v\frac{\partial}{\partial z} + i\omega_{bc} + \gamma_{bc}\right)\varrho_{bc}^{(3)} = -\varrho_{bb}^{(2)}\mu_{bc}E_2(z,t) + \varrho_{ab}^{(2)}\mu_{ab}E_1(z,t)\,.$$

In addition to the first driving term on the right-hand side, which corresponds to the hole-burning effect, we obtain a second term, also oscillating at the probe frequency, which results from a « mixing » of the saturating light field with the atomic oscillations at the sum frequency.

For the present three-level system it is actually not necessary to resort to perturbation theory. We can solve the Schrödinger equation (44) up to arbitrary intensities of the two laser light fields, using the rotating-wave approximation, *i.e.* looking for a steady-state solution of the form

$$(71)\qquad \begin{cases} \varrho_{\alpha\alpha} = n_\alpha(v)\,, & \alpha = a,\, b,\, c\,,\\ \varrho_{ab} = \hat{\varrho}_{ab}\exp[-i(\omega_1 t - k_1 z)]\,, \\ \varrho_{bc} = \hat{\varrho}_{bc}\exp[-i(\omega_2 t - k_2 z)]\,, \\ \varrho_{ac} = \hat{\varrho}_{ac}\exp[-i[(\omega_1+\omega_2)t - (k_1+k_2)z]]\,. \end{cases}$$

By inserting these expressions into (44) and ignoring counter-rotating components of the light field, we obtain a linear system of algebraic equations, which can be readily solved, to yield [32]

$$(72)\qquad \hat{\varrho}_{bc}(v) = \frac{1}{2\hbar}\,\mu_{bc}E_2\,\frac{(n_b - n_c)\big(\varDelta_{ab}\varDelta_{ac} - (\mu_{bc}E_2/2\hbar)^2\big) - (n_a - n_b)(\mu_{ab}E_1/2\hbar)^2}{\varDelta_{ab}\varDelta_{bc}\varDelta_{ac} - \varDelta_{bc}(\mu_{bc}E_2/2\hbar)^2 - \varDelta_{ab}(\mu_{ab}E_1/2\hbar)^2}\,,$$

where

$$(73)\qquad \begin{cases} \varDelta_{ab} = \omega_{ab} - \omega_1 + k_1 v + i\gamma_{ab}\,, \\ \varDelta_{bc} = \omega_{bc} - \omega_2 + k_2 v + i\gamma_{bc}\,, \\ \varDelta_{ac} = \omega_{ac} - (\omega_1+\omega_2) + (k_1+k_2)\,v + i\gamma_{ac}\,. \end{cases}$$

The contribution of atoms within an interval $\mathrm{d}v$ of axial velocities to the nonlinear susceptibility of the probe transition is then given by

$$(74)\qquad \mathrm{d}\chi = (2\mu_{cb}\hat{\varrho}_{bc}/E_2)\,\mathrm{d}v\,.$$

For the subsequent discussion we shall be content with a weak probe beam. Expanding (72) in powers of the saturating field and keeping only the lowest-order contributions, in accordance with the results of third-order perturbation theory, we obtain [32]

$$(75)\qquad \mathrm{d}\chi = \mathrm{d}\chi^{(1)} + \mathrm{d}\chi^{(3)}|E_1|^2\,,$$

where

$$(76)\qquad \mathrm{d}\chi^{(1)} = \frac{1}{\hbar}\,|\mu_{bc}|^2\,(n_b^0 - n_c^0)\,\frac{1}{\varDelta_{bc}}\,\mathrm{d}v$$

and

$$(77)\qquad \mathrm{d}\chi^{(3)} = \frac{1}{4\hbar^3}\,|\mu_{bc}|^2|\mu_{ab}|^2\{(n_a^0 - n_b^0)(S+T) + (n_b^0 - n_c^0)D\}\,\mathrm{d}v$$

with

(78) $$S = \frac{2\gamma_{ab}}{\gamma_b} \frac{1}{\Delta_{ab}\Delta^*_{ab}\Delta_{bc}},$$

(79) $$T = -\frac{1}{\Delta_{ab}\Delta_{bc}\Delta_{ac}},$$

(80) $$D = \frac{1}{\Delta^2_{bc}\Delta_{ac}}.$$

The background susceptibility (74) and the last contribution to the third-order susceptibility (77) can be ignored if we assume that all atoms are initially in the lowest state, *i.e.* if we set $n^0_b = n^0_c = 0$. In this limit we obtain a radiation-induced transition rate $b \to c$ per atom

(81) $$\Gamma_{bc} = \frac{1}{n^0_a}\frac{1}{\hbar\omega_2} I_2 4\pi|k_1| \operatorname{Im}\chi^{(3)}|E_1|^2 = \frac{1}{16}\left|\frac{\mu_{ab}E_1}{\hbar}\right|^2 \left|\frac{\mu_{bc}E_2}{\hbar}\right|^2 \operatorname{Im}(S+T)$$

with

(82) $$\operatorname{Im} S = \frac{2}{\gamma_b} \frac{\gamma_{ab}}{(\omega_{ab}-\omega_1+k_1v)^2+\gamma^2_{ab}} \frac{\gamma_{bc}}{(\omega_{bc}-\omega_2+k_2v)^2+\gamma^2_{bc}}$$

and

(83) $$\operatorname{Im} T = \\ = \operatorname{Im}\frac{-1}{(\omega_{ab}-\omega_1+k_1v+i\gamma_{ab})(\omega_{bc}-\omega_2+k_2v+i\gamma_{bc})(\omega_{ac}-(\omega_1+\omega_2)+(k_1+k_2)v+i\gamma_{ab})}.$$

In this formulation it is clear that the term S in the third-order susceptibility describes a two-step excitation process. The corresponding transition rate $b \to c$ is proportional to the time γ_b^{-1} which the atom spends in the intermediate state b after absorption of a photon from the saturating beam. The first contribution is identical with the predictions of the rate equation model.

The additional term T gives a correction due to atomic coherence effects. To see its significance, let us first investigate the case where both laser frequencies are tuned away from their respective atomic resonances, while their sum frequency $\omega_1 + \omega_2$ is kept close to the resonance frequency ω_{ac}. In this limit we have

(84) $$\operatorname{Im} S \to \frac{2}{\gamma_b} \frac{\gamma_{ab}\gamma_{bc}}{(\omega_{ab}-\omega_1)^4}$$

and

(85) $$\operatorname{Im} T \to \frac{1}{(\omega_{ab}-\omega_1)^2} \frac{\gamma_{ac}}{(\omega_{ac}-(\omega_1+\omega_2)+(k_1+k_2)v)^2+\gamma^2_{ac}}.$$

While the stepwise excitation rate drops off with the fourth power of the detuning from the intermediate resonance, the rate due to the correction T

drops off only with the square, *i.e.* it dominates for large frequency detuning and describes a « Raman-like » coherent two-photon excitation.

For a more detailed interpretation we add the stepwise excitation term S and the two-photon excitation term T. With the assumption $\gamma_{ab} = (\gamma_a + \gamma_b)/2$, *i.e.* ignoring phase-interrupting elastic collisions, we can rewrite the result as

$$\text{(86)} \qquad \operatorname{Im}(S+T) = $$

$$= \frac{1}{\pi}\int\limits_{-\infty}^{\infty} \frac{\gamma_a/2}{x^2+\gamma_a^2/4}\,\frac{\gamma_b/2}{(x+\omega_1-\omega_{ab})^2+\gamma_b^2/4}\,\frac{\gamma_c/2}{(x+\omega_1+\omega_2-\omega_{ac})^2+\gamma_c^2/4}\,.$$

For the sake of simplicity we have assumed $v = 0$. This integral has been known for a long time in the theory of resonance fluorescence [33]. It lends itself to an intuitive interpretation: The three Lorentzian factors of its integrand describe the probability distributions of the energy in the three quantum states a, b and c. The finite widths of these levels are determined by their radiative decay rates γ_a. The first factor describes the probability for finding an atom in level a with an energy deviation $\hbar x$ from the center. After absorption of a photon $\hbar\omega_1$, it will arrive in level b with an energy offset $\hbar(x+\omega_1-\omega_{ab})$, and the corresponding probability is described by the second factor. Absorption of the second photon $\hbar\omega_2$ finally brings the atom up to level c with an energy offset $\hbar(x+\omega_1+\omega_2-\omega_{ac})$, which is taken into account in the third probability factor. The actual transition rate is obtained by integrating the combined probability over all possible initial conditions.

It is interesting to consider transitions from a stable ground state to a long-lived excited state c, via a short-lived level b. The integral (86) predicts a sharp resonance of the transition rate $a \to b \to c$ at $\omega_1 + \omega_2 = \omega_{ac}$, even though the two individual absorption lines are each much broader because of the short-lived state b. This phenomenon is actually required by the principle of energy conservation, but it is clearly ignored, if we consider only the process of stepwise excitation.

So far we have neglected elastic atomic collisions, which destroy the phases of the coherent atomic oscillations, without quenching the excited levels. If such processes are important, one expects

$$\text{(87)} \qquad \gamma_{\alpha\beta} \gg (\gamma_\alpha + \gamma_\beta)/2\,,$$

and the coherent two-photon transition term T near resonance can become negligible compared to the rate equation term S.

The combination $S + T$ can also be rewritten in another revealing form, if elastic collisions can be neglected, and if the lowest state is stable, *i.e.* if $\gamma_a = 0$ and $\gamma_{ab} = \gamma_b/2$:

$$\text{(88)} \qquad S+T = \frac{1}{\Delta_{ab}\Delta_{ab}^*\Delta_{bc}} - \frac{1}{\Delta_{ab}\Delta_{bc}\Delta_{ac}} = \frac{1}{\Delta_{ac}\Delta_{ab}\Delta_{ab}^*}\,.$$

Inserting this result into eq. (77) we obtain an expression, which is of the same form as the more general third-order nonlinear susceptibility, derived by BLOEMBERGEN. The agreement becomes apparent after we recognize that there is only one large, near-resonant term in the latter expression, when specialized to the present three-level system. But unlike this expression, our eq. (77) can correctly describe the atomic response near an intermediate resonance, even if elastic collisions are important.

Insertion of the same result (88) into eq. (81) yields a radiative transition rate

$$(89)\qquad \Gamma_{bc} = \frac{1}{16}\frac{E_1^2 E_2^2}{\hbar^4}\frac{\gamma_{ac}}{(\omega_{ac}-(\omega_1+\omega_2))^2+\gamma_{ac}^2}\left|\frac{\mu_{ab}\mu_{bc}}{\omega_{ab}-\omega_1+i\gamma_{ab}}\right|^2 ,$$

which is equivalent to a formula derived by SHEN for resonance Raman scattering in the case of a folded level array, as in fig. 9*c*) (eq. (7) of ref. [34]). A comparison with Shen's paper reveals that our stepwise excitation process has its analogue in the phenomenon of « hot luminescence », whereas the coherent two-photon excitation corresponds to « coherent resonance Raman scattering ».

Up to now we have completely ignored the third contribution D in eq. (77). This term gives an additional correction to the atomic response at the probe transition, not expected in the rate equation model. But it contributes only if $n_b^0 \neq n_c^0$, *i.e.* if there is a background absorption. (Such a situation is *e.g.* realized, if we invert our cascade level array, so that we probe the transition from the ground state c to an (empty) excited state b, while the saturating beam is in resonance with a transition from b to another excited state a.)

To better see the significance of this third term, we return to eq. (72), setting $n_b = n_a = 0$, and again assuming a weak probe beam. We then obtain

$$(90)\qquad \mathrm{d}\chi = \frac{1}{\hbar}|\mu_{bc}|^2\frac{(-)n_c\Delta_{ac}}{\Delta_{bc}\Delta_{ac}-(\mu_{ab}E_1/2\hbar)^2} = \frac{1}{\hbar}|\mu_{bc}|^2\frac{1}{2}\left\{\frac{1+\delta/\xi}{\delta-\xi+\omega_{bc}-\omega_2+i\gamma_{bc}}+\frac{1-\delta/\xi}{\delta+\xi+\omega_{bc}-\omega_2+i\gamma_{bc}}\right\},$$

where

$$(91)\qquad \delta = \tfrac{1}{2}\big(\omega_{ab}-\omega_1+i(\gamma_{ab}-\gamma_{bc})\big)\,,\qquad \xi = \sqrt{|\delta|^2+|\mu_{ab}E_1/4\hbar|^2}\,,$$

i.e. the probe line is split by a strong saturating light field into two Lorentzian components, separated by 2ξ, which is equal to the Rabi nutation frequency (56) at the saturated transition. This line splitting is nothing but the a.c. (or dynamic) Stark effect [35, 36]. A corresponding line splitting has been discussed in the previous subsection for two-level systems. The third term D in eq. (77) describes a modification of the probe line shape due to the dynamic Stark effect to lowest order.

To further clarify its significance, let us assume that the saturating beam is tuned away from the resonance frequency ω_{ab}. If only level c is populated, eqs. (75)-(80) predict a susceptibility of the probe transition

$$\text{(92)}\qquad \mathrm{d}\chi = (-)\frac{1}{\hbar}|\mu_{bc}|^2 n_c^0 \left\{\frac{1}{\Delta_{bc}} + \left|\frac{\mu_{ab}E_1}{2\hbar}\right|^2 D\right\}\mathrm{d}v \approx$$

$$\approx (-)\frac{1}{\hbar}|\mu_{bc}|^2 n_c^0 \frac{1}{\Delta_{bc} - |\mu_{ab}E_1/2\hbar|^2\left(1/(\omega_{ab}-\omega_1)\right)},$$

i.e. the correction term D produces an intensity-dependent shift

$$\text{(93)}\qquad \Delta\omega_{bc} = -\,|\mu_{ab}E_1/2\hbar|^2/(\omega_{ab}-\omega_1)$$

of the probe resonance frequency ω_{bc}. This dynamic Stark shift corresponds to the interaction energy of the atomic dipole moment, oscillating at frequency ω_1, with the off-resonance driving field E_1. It can be interpreted by assigning a corresponding energy shift to the probed level b.

It is interesting to compare our interpretation of the result (77) with that given by FELD and JAVAN for corresponding calculations [37]. These authors describe the saturation signal, *i.e.* change of the probe transition rate *b-c* by saturation of the transition *a-b*, in the form

$$\text{(94)}\qquad R = n_a^0 J_{ac} + n_b^0 J_{bc} - n_c^0(J_{cb} + J_{ca}),$$

where $J_{ac} = J_{ca}$ is the « double quantum-transition rate » per atom, while $J_{bc} = J_{cb}$ is a « modified single quantum-transition rate ». From a comparison with our eq. (77) we find the equivalence

$$\text{(95)}\qquad J_{ac} = \frac{1}{16}\left|\frac{\mu_{ab}E_1}{\hbar}\right|^2\left|\frac{\mu_{bc}E_2}{\hbar}\right|^2 \operatorname{Im}(S+T)$$

and

$$\text{(96)}\qquad J_{bc} = \frac{1}{16}\left|\frac{\mu_{ab}E_1}{\hbar}\right|^2\left|\frac{\mu_{bc}E_2}{\hbar}\right|^2 \operatorname{Im}\left(D-(S+T)\right).$$

Finally, we turn to the integration of the contributions (77) over the atomic velocity distribution, to find the observable saturation signal, *i.e.* the radiation-induced change of the probe absorption coefficient

$$\text{(97)}\qquad \Delta\alpha_2 = |E_1|^2 4\pi|k_2| \operatorname{Im}\int_{v=-\infty}^{\infty}\mathrm{d}\chi^{(3)}.$$

In the limit of large Doppler widths, the Maxwellian velocity distribution of the populations n_α can again be regarded as constant near resonance, and complex contour integration yields for the remaining integrals [32]

$$\int_{-\infty}^{\infty} S\,\mathrm{d}v = \frac{2\gamma_{ab}}{\gamma_b}\frac{2\pi}{|k_1|}\frac{1}{\Omega + i\Gamma_B}, \tag{98}$$

$$\int_{-\infty}^{\infty} T\,\mathrm{d}v = \begin{cases} -\dfrac{2\pi}{|k_1|}\min\left(1, \dfrac{\omega_2}{\omega_1}\right)\dfrac{i}{(\Omega + i\Gamma_B)(\Omega + i\Gamma_N)}, & \text{if } \dfrac{k_1}{k_2} < 0\,, \\ 0\,, & \text{otherwise,} \end{cases} \tag{99}$$

$$\int_{-\infty}^{\infty} D\,\mathrm{d}v = \begin{cases} -\dfrac{2\pi}{|k_1|}\left(1 - \dfrac{\omega_2}{\omega_1}\right)\dfrac{i}{(\Omega + i\Gamma_N)^2}, & \text{if } \dfrac{k_1}{k_2} < -1\,, \\ 0\,, & \text{otherwise,} \end{cases} \tag{100}$$

where

$$\Omega = \omega_{bc} - \omega_2 - (k_2/k_1)(\omega_{ab} - \omega_1)\,, \tag{101}$$

$$\Gamma_B = \gamma_{bc} + |k_2/k_1|\gamma_{ab}\,, \tag{102}$$

$$\Gamma_N = \begin{cases} \gamma_{ac} + \left(\left|\dfrac{k_2}{k_1}\right| - 1\right)\gamma_{ab}\,, & \text{if } \omega_2 \geqslant \omega_1\,, \\ \left|\dfrac{k_2}{k_1}\right|\gamma_{ac} + \left(1 - \left|\dfrac{k_2}{k_1}\right|\right)\gamma_{bc}\,, & \text{if } \omega_2 \leqslant \omega_1\,. \end{cases} \tag{103}$$

The first Doppler-integral (98) yields the same probe signal as predicted by rate equations (see eq. (65)), *i.e.* a Lorentzian of width Γ_B. The simple hole-burning model can hence give correct results, if the additional signals due to the terms T and D cancel in the velocity average. Equations (99) and (100) predict such a cancellation in the limit of large Doppler widths, if the two laser beams are traveling in the same direction (for a cascade level array! In case of a folded-level array, as in fig. 9*b*) and 9*c*), cancellation occurs for counter-propagating beams). If the two beams are traveling in opposite directions, however, the rate equation results are no longer valid.

If the probe frequency ω_2 is larger than the saturating frequency ω_1, the dynamic Stark effect contribution (100) cancels even for opposing beams, but the two-photon correction (99) contributes to the signal. When added to the hole-burning signal (98), it gives again a Lorentzian line of the same area as before, but with a new line width Γ_N, which can be considerably narrower (see fig. 10*a*)).

If $\omega_2 < \omega_1$, the dynamic Stark effect produces an additional probe signal (100), which is observable even if $n_b^0 = n_a^0$, where the saturating beam cannot produce any hole burning in the atomic velocity distribution. This signal, which is entirely unexpected in the rate equation model, has a characteristic non-Lorentzian line shape, as illustrated in fig. 10*b*), and zero area. It was first predicted and observed by HÄNSCH *et al.* [32, 38].

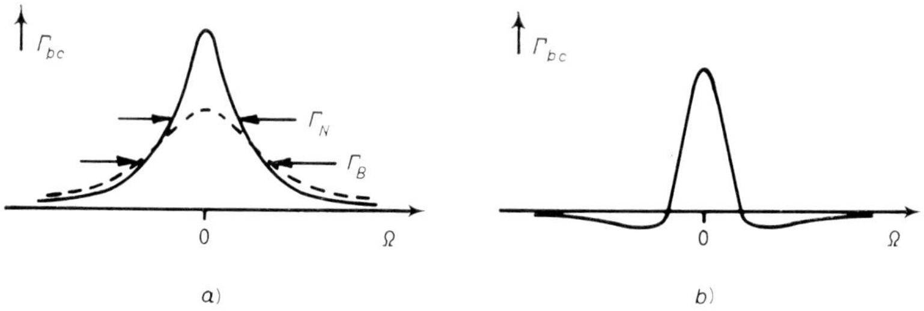

Fig. 10. – Saturation spectra in three-level system, as predicted by the semi-classical model. The probe signal $\Delta\alpha_2$ for a weak saturating beam is plotted for infinite Doppler widths. *a*) Spectra for $n_a > n_b$, $\omega_{bc} > \omega_{ab}$. The dashed curve is expected if saturating beam and probe travel in the same direction (for a cascade level array). It is the same as for pure stepwise excitation. The solid curve corresponds to counter-propagating beams. It is narrower because of coherent two-photon excitation. *b*) Spectrum for $n_a = n_b$ (no hole burning!) and $\omega_{bc} < \omega_{ab}$. The signal is due to the dynamic Stark effect and is observed with counter-propagating beams only.

3. – Saturation spectroscopy experiments.

To illustrate the experimental aspects and the potential of high-resolution saturation spectroscopy, this section will review a number of spectroscopic studies of atoms and simple molecules, and will discuss advantages and limitations of several alternative versions of this nonlinear spectroscopic technique. In accordance with the special interests of the author, the main emphasis will be on experiments in or near the visible region, using widely tunable dye lasers. Many other experiments, using gas lasers of limited tuning range, have been summarized in very good review articles by SHIRLEY [21] and by BETEROV and CHEBOTAEV [31]. An excellent account is also given in a recent monograph by LETOKHOV and CHEBOTAEV [39].

Narrow-band saturation resonances can be observed simply by measuring the total absorption of a standing wave field in a gas sample. Placing the gaseous absorber inside the laser cavity can provide high sensitivity, and the resulting inverted Lamb dips are very useful for the frequency stabilization of gas lasers. For many spectroscopic applications, however, the probe modulation technique illustrated in fig. 2 has a number of important advantages, even though it requires a few more optical components in addition to the laser and the absorption cell, and even though it sacrifices some resolution, because the saturating beam and the probe beam cross at some small finite angle.

The use of a chopped saturating beam and the detection of a synchronous intensity modulation of the probe beam permit one to directly measure the radiation-induced change of the probe absorption $\Delta\alpha_2$, as a given by eq. (26). The saturation spectrum has thus a well-defined zero level, unlike the complicated background slopes of the total intensity, observed by the inverted Lamb-dip method. The attractiveness of this scheme is demonstrated in fig. 11

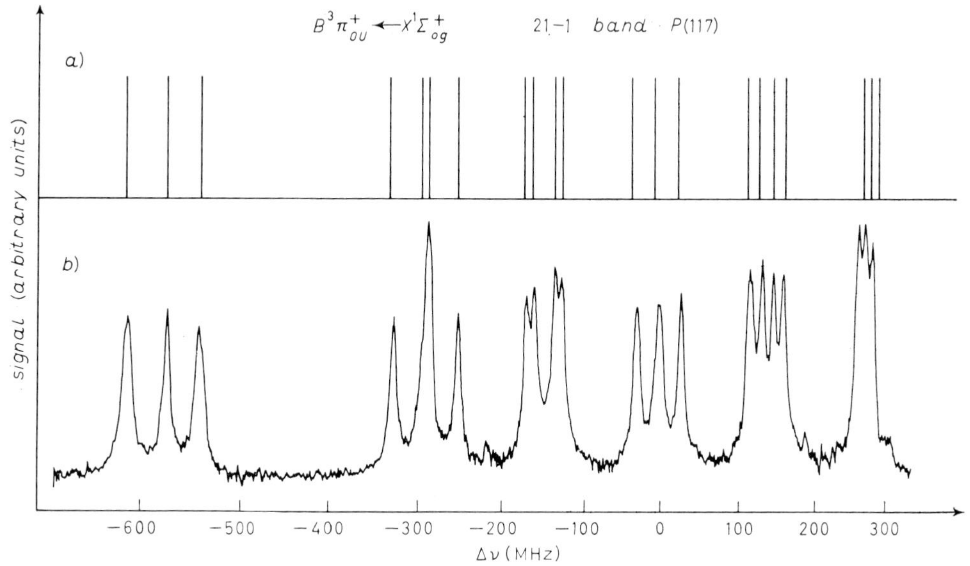

Fig. 11. – Saturation spectrum of $^{127}I_2$ absorption line with theoretical hyperfine structure [14].

with one of the first spectra obtained with this method [14]. An absorption line of I_2 vapor, corresponding to the $P(117)(21\text{-}1)\,B \leftarrow \chi$ transition, was observed with a single-mode Kr^+ laser at 5682 Å, which can be frequency tuned over a 5 GHz wide interval. The saturation spectrum reveals a splitting of this line into 21 hyperfine components, primarily due to the interaction of the nuclear electric-quadrupole moments with the electric-field gradient at the position of the nuclei. The observed line width of 10 MHz corresponds essentially to the (pressure broadened) natural line width. The Doppler width at room temperature is about 600 MHz, for comparison, and completely obscures the hyperfine splitting in conventional high-resolution absorption spectroscopy. Systematic studies of the hyperfine structure of different I_2 lines by saturation spectroscopy have subsequently been reported by LEVENSON and SCHAWLOW [40] and by SOREM *et al.* [41].

The probe modulation technique offers high sensitivity without the need to insert the gas sample inside the delicate cavity of a monochromatic tunable laser. In this way it provides not only experimental convenience, but it also eliminates the dependence of the saturation spectrum on the parameters of the laser medium or the laser cavity. It is even possible to electronically stabilize the laser intensity during scanning.

In addition, this technique can provide a good signal-to-noise ratio even at low laser intensities, where third-order perturbation theory holds, and where a simple hole-burning model can give a quantitative description of the spectra in the limit of large Doppler widths, as discussed in sect. **2**. It also permits one to use a weak probe beam in the presence of a strong saturating beam. As we learned in the preceding section, the signal line shape can be calculated analytically for this situation, whereas experiments with a strong standing wave require cumbersome numerical computations for their quantitative interpretation.

The same method is also well suited for experiments with short-pulse dye lasers, which would offer only poor sensitivity in an inverted Lamb dip experiment because of the limited number of light passes through the intracavity absorber.

The first observation of an atomic-resonance line by high-resolution saturation spectroscopy at Stanford [15] was indeed performed with such a pulsed dye laser, repetitively pumped by a nitrogen laser of only 10 ns pulse width. This dye laser was the first to offer both a narrow line width down to a few MHz and convenient linear and reproducible wavelength tuning, and it has since been widely adopted. Its design principles are described in detail in ref. [42] and subsequent refinements have been published in later papers [43, 44]. In the meantime, single-mode continuous dye lasers have become available, which can operate anywhere in the visible spectrum, and which can substantially exceed the resolution of pulsed dye lasers which is always limited by the uncertainty principle. Pulsed dye lasers remain nonetheless very attractive tools for nonlinear high-resolution spectroscopy, due to their wider easily accessible wavelength range. Their high obtainable peak powers permit one to study transitions of high saturation intensity (29), and also allow an extension of the wavelength range into the ultraviolet and infra-red by sum and difference frequency generation in nonlinear optical materials. In addition, the short pulse length allows time-resolved spectroscopic studies, which are of particular interest for the observation of relaxation processes.

A number of interesting aspects of saturation spectroscopy with a pulsed laser can be illustrated with the early Stanford experiment, studying the yellow Na-*D* resonance lines [15]. A scheme of the experimental set-up is shown in fig. 12. The absorbing Na vapor is contained in a glass cell of 10 cm length, heated by an electric oven to about 110 °C, corresponding to a vapor pressure of about $3 \cdot 10^{-7}$ Torr, or an unsaturated absorption coefficient α_2 of about $0.06\ \text{cm}^{-1}$. The dye laser with external confocal filter interferometer generates pulses of 30 ns length and 7 MHz line width at 80 p.p.s. repetition rate. Attenuators reduce the peak powers of probes and saturating beam to 0.5 and 50 m.w. respectively. The beam diameters are typically 4 mm. Beam splitters and mirrors are used to provide two probe beams in this set-up: one which crosses

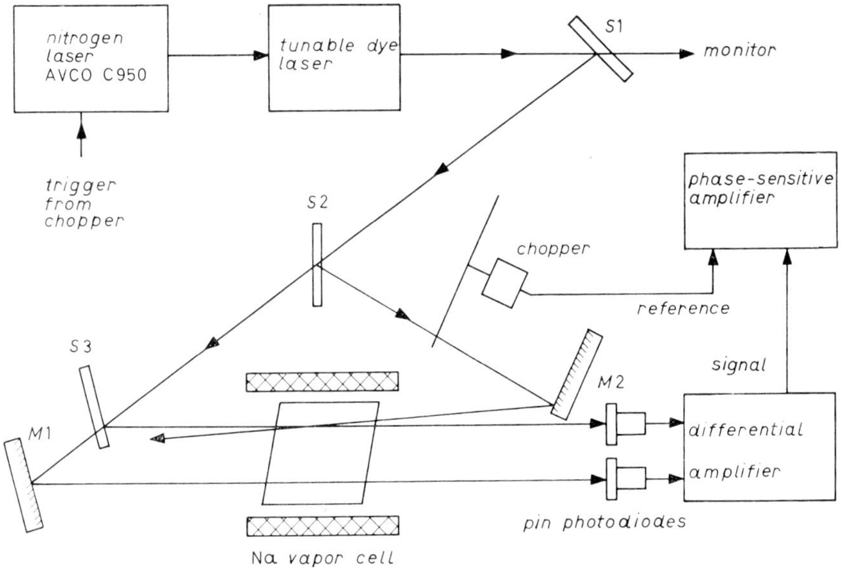

Fig. 12. – Set-up for saturation spectroscopy of Na-D lines with pulsed dye laser [15].

the region bleached by the saturating beam, and a second one which also passes through the absorbing vapor, but does not travel through the bleached region. This second probe serves as a reference in a differential detection scheme which greatly reduces the noise due to random amplitude fluctuations of the pulsed laser. An ordinary lock-in amplifier can be used for the analogue recording of the probe modulation signal, if the short signal pulses of the photodiode detectors are electronically stretched into the millisecond region.

Figure 13 shows the observed saturation spectra of the two D-lines. The theoretical hyperfine structure is shown below for comparison. The hyperfine splitting of the ground state is clearly resolved in both lines, and the splitting of the upper $^2P_{\frac{1}{2}}$-state is resolved in the $D1$-line. These spectra provide instructive examples for saturation spectroscopy. Half-way between the closely spaced $D1$ hyperfine components, which share a common lower level, we recognize additional cross-over lines, as expected from the theory of three-level systems. These resonances do not impair the interpretation of a saturation spectrum, as long as they do not result in undesirable blends of unresolved lines.

The inverted signal located in the middle of the two hyperfine line groups is of a different origin. Here the absorption of the probe beam is actually enhanced by the presence of the bleaching beam. Such an increase can be explained by a velocity-selective optical pumping cycle. The saturating beam selectively excites atoms within a narrow velocity interval from a given hyperfine sublevel of the ground state. These excited atoms decay spontaneously, but only part of them return to their original level. Other atoms decay to the second hyperfine level of the ground state, creating a population bump at the selected velocity, and hence increasing the absorption for the probe which is Doppler shifted into resonance with a transition from this second level.

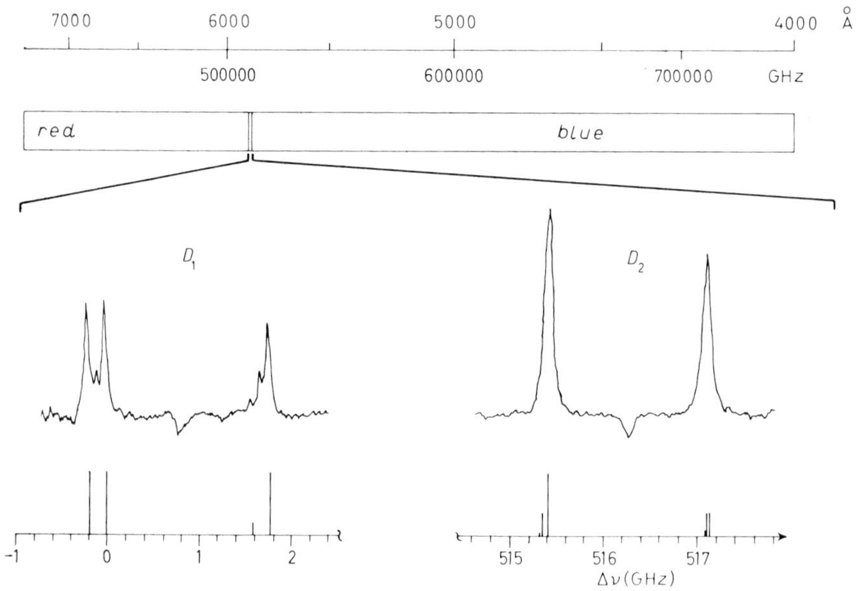

Fig. 13. – Saturation spectra of Na-D lines.

Velocity-selective optical pumping opens an interesting possibility in saturation spectroscopy with a pulsed laser. Since the hyperfine states of the ground level are stable and do not decay by spontaneous emission, the perturbations of the velocity distributions by the saturating light are « permanent » and persist long after the light pulse has passed, until equilibrium is restored by collisions, or until the atoms have moved out of the observed volume. This persistence of the holes and bumps can easily be demonstrated experimentally, if the probe pulse is optically delayed. At low gas pressures, where collisions are negligible, the observed saturation spectrum remains essentially undistorted up to delay times of several microseconds. In the presence of a buffer gas, on the other hand, velocity-changing collisions will tend to restore the Maxwellian velocity distribution, and time-resolved measurements of the saturation spectrum with a delayed probe offer a unique opportunity to observe in detail the kinetics of this return to equilibrium, and to test models of nonequilibrium statistical mechanics. If, for instance, most collisions are

« hard », *i.e.* if they lead to a complete loss of « velocity memory » in a single encounter, then the line width of the saturation resonances will remain narrow with increasing probe delay, but their height will decrease, and a Doppler-broadened pedestal will appear, until finally all atoms have resumed their equilibrium distribution. If, on the other hand, the predominant collision mechanism is small-angle scattering, the lines in the saturation spectrum are expected to broaden due to a diffusion process in velocity space, until they have reached the final equilibrium Doppler width.

Some preliminary experiments were performed with Na vapor in the presence of 3 Torr Ar buffer gas. The lines were found to broaden slowly with increasing delay between saturating and probing light. Since the cross-section for elastic Na-Ar collisions is known, the mean time between two collisions could be estimated to be about 3 ns, and the observations indicated that about 50 such collisions are required to restore thermal equilibrium. A collisional diffusion model seems to provide an adequate description for this particular case.

Except for these initial studies at Stanford, the potential of such collision experiments has so far been left largely unexplored, leaving an interesting field for future research.

BREWER will show in his lectures that velocity-changing collisions can also be observed in a more indirect way via coherent transient effects. Other authors have attempted to extract information about such collisions from the pressure broadening of saturation spectra, observed with continuous lasers under steady-state conditions. This has proven to be very difficult, however, because of competing contributions from quenching collisions or phase interrupting elastic collisions which broaden the natural line width.

Saturation spectroscopy with a delayed probe offers other interesting possibilities. While continuing the experiments with Na vapor, we observed at first a rather unexpected phenomenon. For small probe delays, the saturation resonances become narrower rather than wider with increasing delay time [45]. It was soon confirmed that the same line narrowing occurs also in the absence of buffer gas, thus ruling out any collision effect. Finally we recognized that initially, where probe and saturating beam overlap in time, the observed lines are broadened by the dynamic Stark effect, *i.e.* by the line splitting due to the optical nutations of the atoms in the strong saturating field, as discussed in sect. **2**. A sufficiently large time delay, on the other hand, gives the coherent atomic oscillations time to die out before probing, and hence provides a saturation spectrum which is solely due to population changes, as described by a hole-burning model without complications by atomic coherence effects. This possibility is of particular interest for experiments with short light pulses, for which the steady-state calculations of the preceding section can provide only an approximate description.

After the feasibility of high-resolution saturation spectroscopy with a pulsed dye laser had been demonstrated, we applied the same technique to the red

Balmer line H_α ($n = 2 \div 3$) of atomic hydrogen at 6563 Å [46, 47]. The Balmer α-line is perhaps the most extensively studied of all spectral lines [48]. Many attempts to measure its precise wavelength have been made in order to determine the Rydberg constant, which describes the binding energy between the electron and atomic nucleus, and which is an important cornerstone in the evaluation of other fundamental constants. The relationship between wavelength and rydberg is given by Bohr's formula, corrected for finite nuclear mass, relativistic effects and Lamb shifts.

Wavelength measurements with conventional methods of high-resolution spectroscopy has always encountered difficulties, because the H_α line consists of seven closely spaced fine-structure components (fig. 14*a*)), which appeared blurred and masked by the particularly large Doppler broadening (6000 MHz

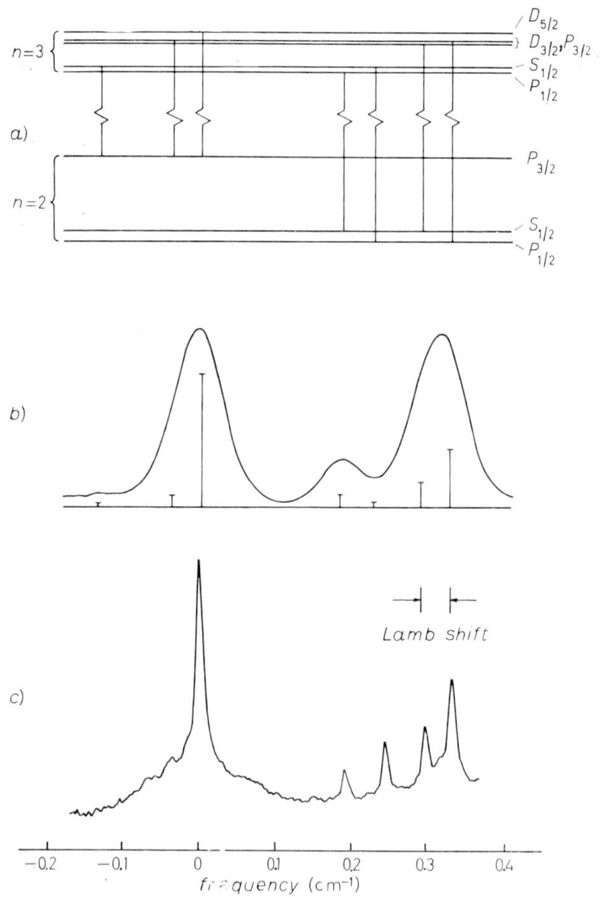

Fig. 14. – Balmer α-line of atomic deuterium: *a*) energy levels with fine-structure transitions, *b*) emission line profile of a cooled deuterium gas discharge and theoretical fine-structure lines with relative transition probabilities ($T = 50$ K), *c*) saturation spectrum with optical resolved Lamb shift.

at room temperature). Doppler broadening can be reduced by operating at cryogenic temperatures, and by the use of the heavier isotope deuterium. An emission line profile, obtained by KIBBLE and co-workers in this way, is shown in fig. 14*b*). But even here it was not possible to resolve single fine-structure components, and it remained necessary to resort to troublesome mathematical deconvolution and fitting procedures.

For comparison, fig. 14*c*) shows a saturation spectrum of this line observed with essentially the same technique as was used in the Na experiments. In order to observe any absorption of the H_α line, the hydrogen atoms were excited to the $n = 2$ state in a simple Wood gas discharge tube at low pressure (0.2 Torr). Stark broadening and level mixing due to electric fields in the plasma could be minimized by observing the saturated absorption in the afterglow, about one microsecond after stalling the d.c. discharge with an electronic switch. This is another example of the versatility offered by a pulsed laser source. The four strongest fine-structure components are clearly resolved in the spectrum, and the Lamb shift could, for the first time, be observed directly in an optical absorption spectrum. The third component is again a cross-over signal due to two coupled transitions with a common lower level. The narrowest line width is about 250 MHz and is primarily determined by the natural line width and by unresolved hyperfine splitting.

Measurements with a delayed probe pulse revealed that the lifetime of the $2P$-states is effectively lengthened by a factor of 1000 owing to resonance trapping of the emitted ultraviolet Lyman-α radiation, and that the velocity distribution of these states is rapidly thermalized by this process in contrast to the metastable $2s$-state.

In a subsequent experiment [47] the wavelengths of the optically resolved Balmer-line components were compared interferometrically with the wavelength of a 6328 Å He-Ne laser, which was electronically frequency stabilized to an absorption line of $^{129}I_2$ vapor inside the cavity, using the inverted Lamb-dip method of saturation spectroscopy. Its wavelength was known in terms of the ^{86}Kr standard to within 1.4 parts in 10^9. A major part of the experimental efforts were devoted to the investigation of possible systematic errors. This measurement yielded a new rydberg value, $R_\infty = 109\,737.3143(10)$, with an almost tenfold improvement in accuracy over the best conventional measurements. High-resolution saturation spectroscopy thus provided the first significant improvement in the precision of this important fundamental constant during the past thirty years. The new rydberg value will be important for the interpretation of other planned or ongoing precision experiments in physics.

FREED and JAVAN [49] were the first to demonstrate, with a CO_2 absorption cell inside a CO_2 laser cavity, that narrow-band saturation signals can also be observed in the intensity of the fluorescent side light, rather than in the absorption of a standing wave field. As discussed in sect. **2**, such a scheme permits high-resolution saturation spectroscopy of much more dilute, optically thin samples, and is of particular interest, if collision effects have to be avoided, or if rare species or weak transitions are to be studied.

SOREM and SCHAWLOW [50] introduced a method of saturation spectroscopy which takes advantage of fluorescence detection but offers a similar sensitivity and convenience as the probe modulation technique. In this technique of « intermodulated fluorescence », a gas sample outside the laser resonator is irradiated with two overlapping counter-propagating laser beams which are chopped at two different frequencies Ω_1 and Ω_2 (see fig. 15). A photomultiplier

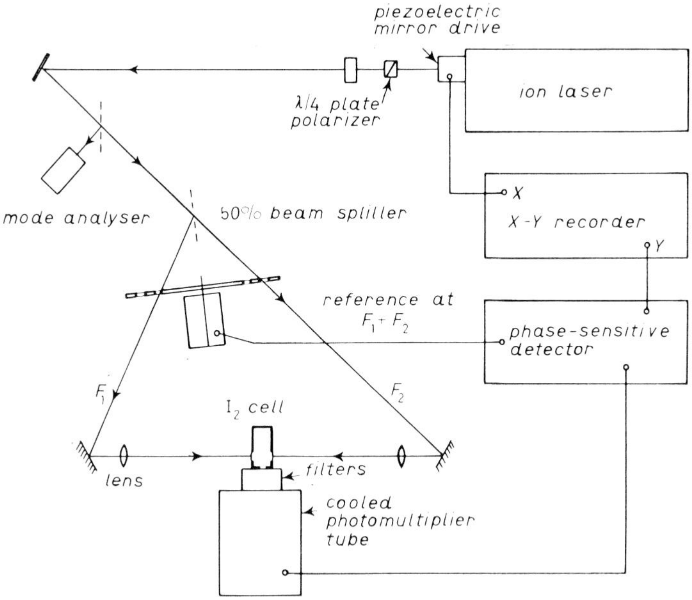

Fig. 15. – Set-up for saturation spectroscopy by intermodulated fluorescence [50].

with phase-sensitive amplifier detects the modulation of the fluorescence at the sum or difference frequency, $\Omega_1 \pm \Omega_2$, which occurs because of the nonlinear mixing of the two intensities in the saturation term (34). A detailed quantitative analysis of this scheme, using rate equations, has been given by SHIMODA [51].

Figure 16 shows a saturation spectrum of the $P(13)$, $R(15)(43\text{-}0)$ lines of $^{127}I_2$ vapor, observed in intermodulated fluorescence with a single mode c.w. Ar ion laser at 5145 Å. The hyperfine structure of the two overlapping lines is clearly resolved. The total fluorescence intensity is displayed on top for comparison. The resonance dips in this broad structure are only barely perceptible. The modulated-fluorescence method permitted measurements down to a vapor pressure of 2.10^{-4} Torr, corresponding to a single-pass absorption of only 0.01 %. The sensitivity was sufficient to detect new cross-over lines due to weakly allowed coupled hyperfine transitions [40].

The same intermodulated-fluorescence technique has also been used by MUIRHEAD *et al.* [52] to study the hyperfine splitting of an absorption line of BO_2, coinciding with the 5145 Å Ar^+ ion laser line.

Since it is possible, under favorable circumstances, to detect single atoms in laser-excited fluorescence, as was recently demonstrated by FAIRBANK *et al.* [53], the method of intermodulated fluorescence should permit the study

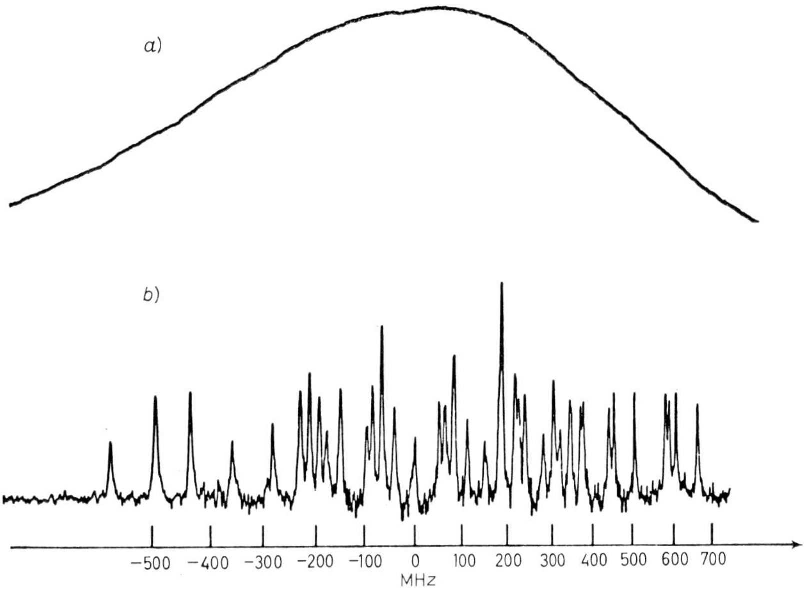

Fig. 16. – Spectra of the $P(13)$, $R(15)(43\text{-}0)$ lines of $^{127}I_2$, excited by a single-mode Ar^+ laser at 5145 Å [50]: *a*) the d.c. fluorescence signal at 10^{-3} Torr as a function of laser frequency; *b*) the intermodulated-fluorescence signal under identical conditions, showing the resolved hyperfine structure.

of extremely dilute fluorescent vapors by high-resolution saturation spectroscopy. Such measurements are of considerable interest, *e.g.* for the study of rare isotopes, or for the spectroscopy of isomeric atoms, *i.e.* atoms whose nucleus is in an excited state as the result of a radioactive decay process, or for the observation of rare exotic atoms such as the hypothetical atoms with (stable) quarks with fractional charge in their nucleus [53].

All the experiments discussed so far detect in effect the radiation-induced changes of the absorption coefficient of a gas. The observation of the fluorescence light is just a convenient means to monitor this quantity. But the change in absorption corresponds to only the imaginary part of the nonlinear susceptibility χ. The real part of χ predicts a concomitant change of the refractive index. This phenomenon of « saturated dispersion » has recently been used by BORDÉ and co-workers [54] in a novel method of saturation spectroscopy.

A sample cell with I_2 vapor is placed inside a Sagnac interferometer, as shown in fig. 17. An Ar laser at 5145 Å is the tunable light source. The laser beam is divided into two parts, which propagate inside the interferometer in opposite directions. An attenuator (At) ensures that one beam is weak inside the sample, whereas the other one is relatively strong. Both beams have equal intensity, however, when they interfere at the output. Two complementary interference signals are monitored by the photomultipliers PM_1 and PM_2. If the refractive

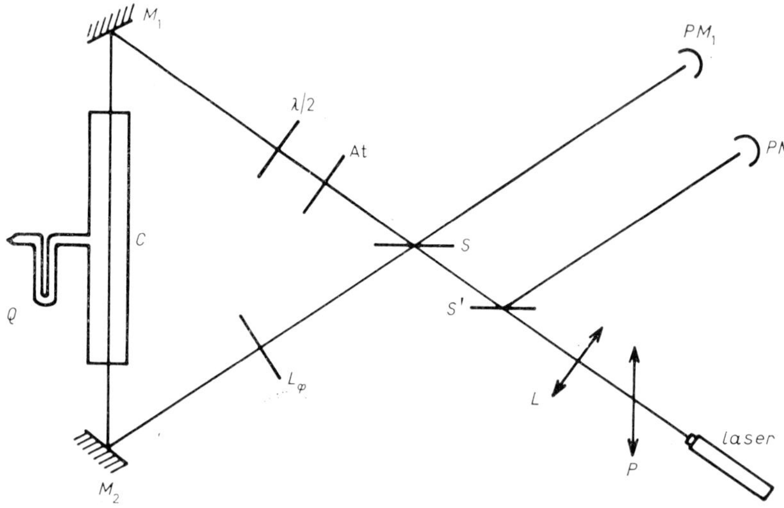

Fig. 17. – Set-up for saturated dispersion spectroscopy [54].

index of the vapor were the same for the two directions of light propagation, the two optical paths would be exactly equal, and the interference conditions would not depend on the laser frequency. For a sensitive detection of any such dependence, it is desirable to introduce an adjustable phase difference between the two waves. This is achieved with a half-wave plate ($\lambda/2$), which gives two orthogonal polarizations of the two waves in the absorption cell, and with a retardation plate L of variable birefringence. Despite these elements, the two beams at the output have the same linear polarization and can thus interfere. The nonlinear saturation results in a difference in the refractive index for the two waves near a resonance, and by proper biasing of the relative phase it is possible to observe a dispersion-shaped saturation signal in the difference of the two multiplier currents, as shown in fig. 18. Such a dispersion-shaped signal is particularly useful for the purpose of laser frequency stabilization, because it can provide the required error signal without any need for a laser frequency modulation. Bordé's group succeeded in stabilizing the frequency of a commercial Ar ion laser within 0.015 MHz by this method.

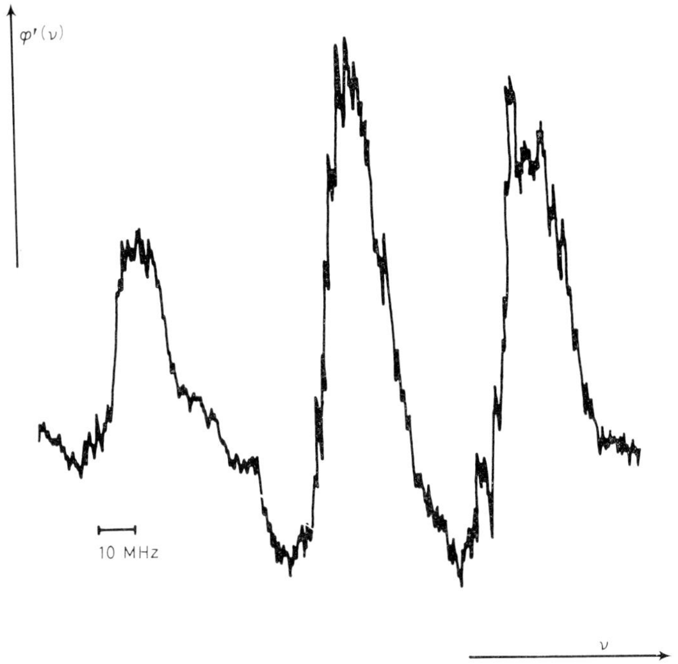

Fig. 18. – Some hyperfine components of the $^{127}I_2$ lines at 5145 Å, observed in saturated dispersion [54].

COUILLAUD and DUCASSE [55] have demonstrated recently that saturated dispersion can also be observed via lenslike and prismlike effects in the saturated gaseous medium, which can cause small deflections of a probe beam.

We conclude our series of experimental examples with some studies [11, 38] which reveal the importance of atomic coherence effects in high-resolution spectroscopy of three-level systems. A folded three-level configuration in neon was studied, consisting of the transitions $2s_2$-$2p_4$ (1.15 μm) and $3s_2$-$2p_4$ (6328 Å), with a common lower level $2p_4$. Figure 19 shows a scheme

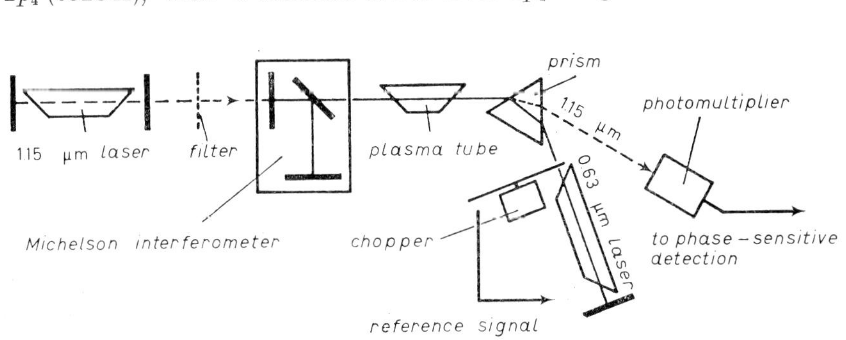

Fig. 19. – Experimental scheme for saturation spectroscopy in a three-level system [38].

of the experiment. The sample cell, a He-Ne discharge tube, is placed inside the cavity of a red single-mode He-Ne laser, which provides the saturating light field in resonance with the $3s_2$-$2p_4$ transition. A second infra-red single-mode He-Ne laser generates a weak probe beam in resonance with the $2s_2$-$2p_4$ transition. The saturating light is chopped, and the resulting intensity modulation of the probe is monitored.

Since there are two saturating traveling-wave components inside the cavity of the red laser, its frequency is kept detuned from the line center so that the

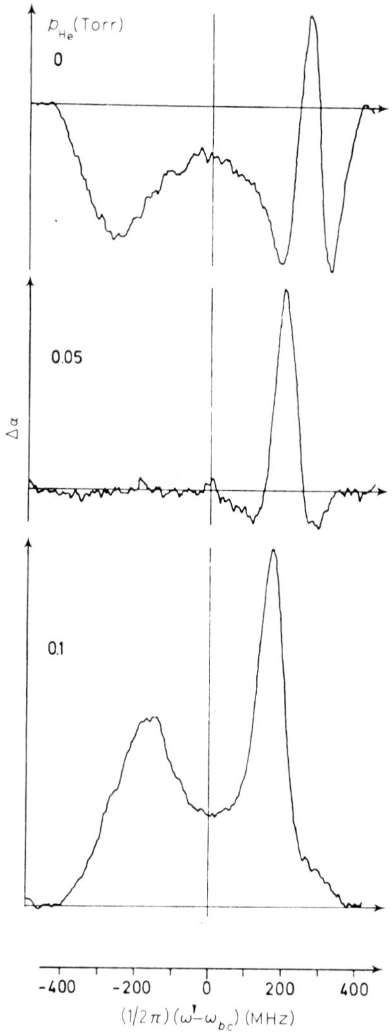

Fig. 20. – Saturation spectra in a (folded) three-level system. The $2s_2$-$2p_4$ transition of Ne ($\lambda = 1.15$ μm) was probed in the presence of a (chopped) standing wave at 0.63 μm ($3s_2$-$2p_4$ of Ne). The spectrum in the middle was obtained with a He-Ne discharge transparent at 0.63 μm [38].

two waves interact with different atoms. Two separate resonances are observed in the saturation spectrum, symmetrically located with respect to the center of the Doppler-broadened probe line and corresponding to a saturating beam traveling in the same and in the opposite direction as the probe. A simple hole-burning model predicts identical line shapes for these two cases. The initial experiments [11] revealed a distinct asymmetry, however. The saturating wave traveling in the same direction as the probe produced a narrower and higher saturation resonance. Such a phenomenon is expected from the contribution of coherent two-photon transitions as given in eq. (99).

In subsequent measurements [38], the background population densities of the three levels were varied within wide limits by changes of the gas mixture and the discharge conditions. Figure 20 shows a series of three saturation spectra, obtained for three different He partial pressures. The population of the $2s_2$-$2p_4$ transition remains always inverted, providing gain on the probe transition. But the population difference $n_{3s_2} - n_{2p_4}$ of the saturated transition changes from positive (top) to zero (center) to negative (bottom), corresponding to gain, complete transparency and absorption, respectively. The center spectrum, recorded for a transparent saturating transition, is remarkable, because it reveals a probe signal in the absence of any hole burning. Only the saturating beam traveling in the same direction as the probe produces such a signal. The saturation spectrum exhibits the characteristic non-Lorentzian shape, which is theoretically expected from the dynamic Stark effect at low intensities (see eq. (100)).

In more recent experiments Toschek and co-workers [56, 57] have extended these measurements to higher saturating intensities, and they were able to observe additional higher-order coherence effects. In particular they observed an actual splitting of the saturation line into two components, separated by the Rabi nutation frequency, which gave a clear demonstration of the dynamic Stark splitting, as is predicted by eq. (90). Other workers have observed dynamic Stark splitting in the spontaneous emission of a Na atomic beam, excited on the Na-D lines with a single-mode c.w. dye laser [58, 59]. A discussion of the interesting implications of the latter experiments is beyond the scope of the present lectures, however [60].

4. – Theory of Doppler-free two-photon spectroscopy.

The perturbation theory of two-photon transitions has been well understood since the thesis of Göppert-Mayer in 1929 [61]. A good review of the general theory, which does not however include any account of the thermal atomic motion and the Dopper effect, may be found in the review articles by Gold [62] and by Worlock [63]. As mentioned in the introduction, the possibility of Doppler-free two-photon spectroscopy of gases was first recognized by CHEBOTAEV and co-workers [16]. A more detailed quantitative discussion was subsequently given by CAGNAC and collaborators [67]. A very good review article on this subject has recently been prepared by BLOEMBERGEN and LEVENSON [64]. Hence the discussions of the present section will be kept rather short.

Let us first try to understand the relationship between two-photon spectroscopy and saturation spectroscopy for the case of a single near-resonant intermediate-energy level. Consider the same physical saturation as discussed in subsect. **2'2**, *i.e.* a gas of three-level atoms, irradiated by two collinear monochromatic laser waves (67). The frequency ω_1 of the first laser is tuned near the resonance frequency ω_{ab} of the transition from the ground state a to the intermediate state b, and the second laser frequency ω_2 is in near resonance with the frequency ω_{bc} of the transition from state b to the final excited state c. If initially all atoms are in the ground state a and if $\omega_1+\omega_2\approx\omega_{ac}$, but $\omega_1\neq\omega_2$, so that neither beam alone has the right frequency for two-photon excitation of the level c, then the total two-quantum excitation rate Γ_{ac} of level c per atom has to be equal to the probe transition rate Γ_{bc} in a saturation spectroscopy experiment, which was calculated in eq. (81) for atoms of a given axial velocity v. If we neglect phase-interrupting elastic collisions, we obtain from eqs. (81) and (88)

$$(104)\qquad \Gamma_{ac}=\Gamma_{bc}=\frac{1}{16}\frac{1}{\hbar^4}|E_1|^2|E_2|^2\frac{\gamma_{ac}}{(\omega_{ac}-(\omega_1+\omega_2)+(k_1+k_2)v)^2+\gamma_{ac}^2}\cdot \cdot\left|\frac{\mu_{ab}\mu_{bc}}{\omega_{ab}-\omega_1+k_1v+i\gamma_{ab}}\right|^2.$$

In our theory of saturation spectroscopy we investigated this transition rate in the immediate neighborhood of the intermediate resonance, where stepwise excitation and coherent two-photon excitation were both important The situation actually becomes quite a bit simpler, if both laser frequencies are tuned away from their respective Doppler-broadened resonance lines, while their sum frequency remains close to ω_{ac}. It is then possible to neglect the Doppler shift $k_1 v$ and the damping $i\gamma_{ab}$ in the off-resonant denominator, and the remaining coherent two-photon excitation rate is a simple Lorentzian function of the sum frequency $\omega_1+\omega_2$ with a line width γ_{ac} and a Doppler shift $(k_1+k_2)v$. Integration over the Maxwellian velocity distribution $f(v)$ (eq. (9)) results in a Doppler width $\Delta(\omega_1+\omega_2)_D=|k_1+k_2|\bar{v}$. If the two laser beams are traveling in the same direction, so that k_1 and k_2 have the same sign, we obtain the full Doppler width, as expected at the sum frequency ω_{ac}. If the two beams are traveling in opposite directions, however, the Doppler width $\big(|k_1|-|k_2|\big)\bar{v}$ can be substantially reduced.

To calculate a more general expression for the two-photon excitation rate, which includes the contributions of many intermediate states, one can use the general third-order nonlinear susceptibility as derived by BLOEMBERGEN.

One obtains in analogy to eq. (104), according to ref. [64], eq. (12),

$$\Gamma_{ac} = \frac{1}{16}\frac{1}{\hbar^4}|E_1|^2|E_2|^2 \frac{\gamma_{ac}}{(\omega_{ac} - (\omega_1 + \omega_2) + (k_1 + k_2)v)^2 + \gamma_{ac}^2} \cdot \left|\sum_n \frac{(\boldsymbol{\mu}_{an}\cdot \boldsymbol{e}_1)(\boldsymbol{\mu}_{nc}\cdot \boldsymbol{e}_2)}{\omega_{an} - \omega_1} + \frac{(\boldsymbol{\mu}_{an}\cdot \boldsymbol{e}_2)(\boldsymbol{\mu}_{nc}\cdot \boldsymbol{e}_1)}{\omega_{an} - \omega_2}\right|^2 . \tag{105}$$

Here, both light fields are permitted to interact with each transition, giving rise to the additional term with denominator $\omega_{an} - \omega_2$. The summation is extended over all intermediate states n. The two light fields can now have arbitrary polarization, described by the unit vectors $\boldsymbol{e}_i$, which enter into the scalar products with the vectors $\boldsymbol{\mu}_{\alpha\beta} = \langle\alpha| - e\boldsymbol{r}|\beta\rangle$. Away from any intermediate resonance, this new result predicts the same line shape and the same possible cancellation of Doppler broadening as the more specialized equation (104).

So far we have excluded the special case $k_1 = -k_2$. Here, the Doppler broadening cancels completely, if one photon is absorbed from each of the beams. But, since $\omega_1 = \omega_2$, it is also possible for an atom to absorb two photons out of the first beam, or two photons out of the second beam alone. The total transition rate is

$$\Gamma_{ac} = \frac{1}{16}\frac{1}{\hbar^4}|E_1|^2|E_2|^2 \cdot \frac{\gamma_{ac}}{(\omega_{ac} - (\omega_1 + \omega_2))^2 + \gamma_{ac}^2}\left|\sum_n \frac{(\boldsymbol{\mu}_{an}\cdot \boldsymbol{e}_1)(\boldsymbol{\mu}_{nc}\cdot \boldsymbol{e}_2) + (\boldsymbol{\mu}_{an}\cdot \boldsymbol{e}_2)(\boldsymbol{\mu}_{nc}\cdot \boldsymbol{e}_1)}{\omega_{an} - \omega_1}\right|^2 + \frac{1}{16}\frac{1}{\hbar^4}|E_1|^4 \frac{\gamma_{ac}}{(\omega_{ac} - 2\omega_1 + 2k_1 v)^2 + \gamma_{ac}^2}\left|\sum_n \frac{(\boldsymbol{\mu}_{an}\cdot \boldsymbol{e}_1)(\boldsymbol{\mu}_{nc}\cdot \boldsymbol{e}_1)}{\omega_{an} - \omega_1}\right|^2 + \frac{1}{16}\frac{1}{\hbar^4}|E_2|^4 \frac{\gamma_{ac}}{(\omega_{ac} - 2\omega_2 + 2k_2 v)^2 + \gamma_{ac}^2}\left|\sum_n \frac{(\boldsymbol{\mu}_{an}\cdot \boldsymbol{e}_2)(\boldsymbol{\mu}_{nc}\cdot \boldsymbol{e}_2)}{\omega_{an} - \omega_2}\right|^2 . \tag{106}$$

The first term in eq. (106) describes the narrow Doppler-free resonance, whereas the other two terms give rise to the Doppler-broadened pedestal. The limits of validity of this expression have been discussed in detail by BLOEMBERGEN and LEVENSON [64].

Second-order Doppler shifts are ignored in eq. (104). They are due to the time dilatation, as predicted by the special theory of relativity (moving atomic « clocks » seem to oscillate more slowly), and are of magnitude $-\frac{1}{2}(v/c)^2\omega_{ac}$. Because of their smallness ($<$ one part in 10^{11} for thermal velocities) such shifts have so far been only rarely observed in optics [65]. A detailed calculation of the expected line broadening and red-shift (for the $1S$-$2S$ two-photon transition in hydrogen) has been given by BAKLANOV and CHEBOTAEV [66].

The expression also ignores light shifts due to the dynamic Stark effect, as predicted by eq. (93) for a simple three-level system. More generally, perturbation theory predicts an energy shift of an atomic level n, induced by a light field of amplitude E and frequency ω [35, 36]:

$$\Delta U_n = \frac{1}{4}\frac{1}{\hbar}|E|^2 \sum_m \left\{ \frac{|\mu_{mn}|^2}{\omega_{mn} - \omega} + \frac{|\mu_{mn}|^2}{\omega_{mn} + \omega} \right\}. \tag{107}$$

The role of such intensity-dependent shifts in two-photon spectroscopy will be investigated in detail in the lectures by BJORKHOLM.

The radiative selection rules for two-photon transitions have been discussed by several authors [63, 67]. Suffice it to mention here that in the electric-dipole approximation the initial and final states must be of the same parity. The selection rules are generally not the same, however, as for stepwise excitation through a resonant intermediate state. Important cancellations occur in the summation over all substates of a given intermediate multiplet, as long as the energy denominator may be taken to be the same for all these states. The resulting rate can be written in a form which depends only on the initial and final state:

$$\Gamma_{ac} = |\langle a|Q|c\rangle|^2, \tag{108}$$

where $Q = aT^2 + bT^0$ can be decomposed into an irreducible tensor operator T^2 of rank 2 and another such operator T^0 of rank 0, and the resulting selection rules can be evaluated by means of the Wigner-Eckart theorem [67].

In the special case of a transition between two atomic S-states, the selection rules require $\Delta F = 0$ and $\Delta m_F = 0$, *i.e.* the atomic angular momentum remains unchanged. The absorption of two circularly polarized photons of the same spin direction is forbidden by conservation of angular momentum, whereas transitions under absorption of two photons of opposing spins are permitted. It is then possible to obtain a narrow resonance without Doppler-broadened pedestal, using excitation with two opposing circularly polarized light beams.

For the interpretation of experiments with short laser pulses, it is interesting to look at the time evolution of the atomic quantum state during two-photon excitation. In the absence of phase-interrupting elastic collisions, and for well-defined initial conditions, the description of the atoms by a density matrix brings no special advantage. A description by a wave function

$$\sum_m C_m(t)|m\rangle \exp[-iU_m t/\hbar] \tag{109}$$

in terms of the amplitudes C_m of the unperturbed energy states $|m\rangle$ may actually be preferable for multilevel atoms, because it entails fewer variables.

Let us consider an atom, initially in its ground state $|a\rangle$, which is irradiated by a monochromatic light pulse of square envelope, starting at time $t=0$ and lasting for a duration τ. The electric field during the pulse in the atomic rest frame is $E(t) = \mathrm{Re}\, E \exp[-i\omega t]$. We want to calculate the development of the atomic state during this pulse. Using standard time-dependent perturbation theory, we set

$$C_m = C_m^{(0)} + C_m^{(1)} + C_m^{(2)} + \dots, \tag{110}$$

where

$$C_m^{(0)} = \delta_{ma} \tag{111}$$

and

$$\left(\frac{\mathrm{d}}{\mathrm{d}t} + \frac{\gamma_m}{2}\right) C_m^{(l+1)} = \frac{1}{i\hbar} \sum_n \langle m|H^I|n\rangle \exp[i\omega_{nm} t]\, C_n^{(l)}\,. \tag{112}$$

As before, $\omega_{nm} = (U_m - U_n)/\hbar$ is the atomic-resonance frequency of the transition n-m. The interaction Hamiltonian for absorption is $H^I = \frac{1}{2}E\mu \exp[-i\omega t]$. Equation (112) is identical to eq. (2.3) of GOLD [62], except for the phenomenological introduction of the decay rate γ_m of level m.

To first order we obtain

$$\dot{C}_m^{(1)} + \frac{\gamma_m}{2} C_m^{(1)} = \frac{1}{i\hbar}\frac{1}{2} E\mu_{am} \exp[i(\omega_{am} - \omega)t]\,. \tag{113}$$

Solution of this linear differential equation yields

$$C_m^{(1)}(t) = (-)\frac{1}{\hbar}\frac{1}{2} E\mu_{am} \frac{\exp[i(\omega_{am} - \omega)t] - \exp[-(\gamma_m/2)t]}{\omega_{am} - \omega - i\gamma_m/2}\,. \tag{114}$$

Away from intermediate resonances, eq. (114) describes small rapid oscillations of the state amplitudes.

Proceeding to the next order, we have to solve

$$\dot{C}_c^{(2)} + \frac{\gamma_c}{2} C_c^{(2)} = \frac{1}{i\hbar} \sum_m \frac{1}{2} E\mu_{mc} \exp[i(\omega_{mc} - \omega)t]\, C_m^{(1)}(t)\,, \tag{115}$$

which yields, if we neglect a small off-resonant term,

$$C_c^{(2)}(t) = \frac{1}{4}\frac{1}{\hbar^2} E^2 \sum_m \frac{\mu_{am}\mu_{mc}}{\omega_{am} - \omega - i\gamma_m/2} \frac{\exp[i(\omega_{ac} - 2\omega)t] - \exp[-(\gamma_c/2)t]}{\omega_{ac} - 2\omega - i\gamma_c/2}\,. \tag{116}$$

The probability of finding the atom in the excited state c at the end of the light pulse is given by $|C_c^{(2)}(\tau)|^2$, and the effective transition « rate » per atom is

$|C_c^{(2)}(t)|^2/\tau$. For short pulses ($\tau \ll |\omega_{ac} - 2\omega + i(\gamma_c/2)|$), the last factor in eq. (116) can be approximated by τ. Just as in the pulsed excitation of a single-photon transition, the excitation probability grows initially with the square of the pulse length, and the excitation rate is proportional to τ, or inversely proportional to the Fourier-transform limited band width of the light pulse. The steady-state expression (105) can be used to calculate the resonant excitation rate for a short pulse, if the line width γ_{ac} is replaced by τ^{-1}. The line shape is described by the absolute square of the last factor in (116), *i.e.* by

$$g(\omega, \tau) = \left| \frac{\exp[i(\omega_{ac} - 2\omega)\tau] - \exp[-(\gamma_c/2)\tau]}{\omega_{ac} - 2\omega - i\gamma_c/2} \right|^2. \tag{117}$$

For long pulses ($\gamma\tau \gg 1$) this is a Lorentzian

$$g(\omega, \tau) = \frac{1}{(\omega_{ac} - 2\omega)^2 + \gamma_{ac}^2} \tag{118}$$

of width $\gamma_{ac} = \gamma_c/2$, just as derived previously for steady-state conditions. For short square pulses ($\gamma_c\tau \ll 1$), on the other hand, the line profile is given by

$$g(\omega, \tau) = \frac{4\sin^2[(\omega_{ac} - 2\omega)\tau/2]}{(\omega_{ac} - 2\omega)^2}, \tag{119}$$

and the line width ($\approx \pi/\tau$) is limited by the laser pulse length rather than by the natural line width.

5. – Doppler-free two-photon spectroscopy experiments.

The large time span of almost four years between the original suggestion of two-photon spectroscopy without Doppler broadening [16] and the first experimental demonstrations [17-19] can perhaps be explained by the fact that virtually all past observations of two-photon absorption in the optical region had been carried out with lasers of peak powers in the megawatt or gigawatt range, and there have simply been no lasers available which would deliver such powers with the required narrow spectral width. CAGNAC and co-workers [67] were the first to point out with a numerical example that the resonant signal enhancement expected in Doppler-free two-photon spectroscopy should actually permit observations in favorable cases with modest powers.

The three initial demonstration experiments [17-19] were all performed in Na vapor. A simplified Na energy level diagram is shown in fig. 21. The excited $5s$ and $4d$ states can be reached from the $3s$ ground state by excitation with two yellow photons, which can easily be generated with an efficient rhodamine 6G dye laser. The $3s$-$3p$ resonance lines are not much more than 100 Å away from the laser wavelength and provide a welcome near-resonant enhancement of the two-photon absorption rate. The excitation can be readily monitored by observing the visible or ultraviolet fluorescence light, emitted in the subsequent spontaneous-emission cascades via the states $3p$ and $4p$.

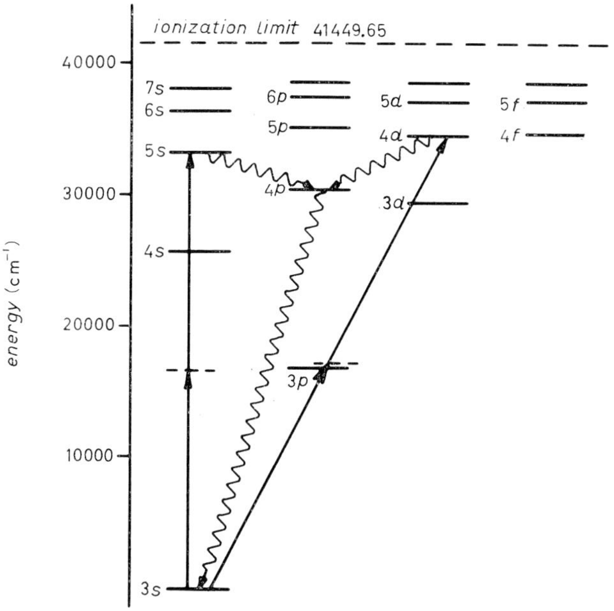

Fig. 21. – Simplified term diagram of Na, illustrating two-photon excitation of the transitions $3s$-$5s$ and $3s$-$4d$.

CAGNAC and co-workers [17] demonstrated that it is indeed possible to obtain line widths narrower than the Doppler width by the new method. They excited the Na $3s$-$5s$ transition with a flash-lamp–pumped dye laser of a few hundred watt peak power at 6022.3 Å. LEVENSON and BLOEMBERGEN [18] studied the same transition with a nitrogen-laser–pumped pulsed dye laser. Owing to the better resolution and better signal-to-noise ratio they were able to make a first quantitative measurement of the $5s$ hyperfine splitting. The extremely high resolution, obtainable by the new technique, has been demonstrated by our group at Stanford [19]. In this experiment, by contrast, the $3s$-$4d$ transition was excited with a single mode c.w. dye laser at 5787.3 Å. The ideal duty cycle and superior resolution of this laser provided easily detectable signals down to powers of a few milliwatt. In an independent experiment PRITCHARD and co-workers [68] studied the same $3s$-$4d$ transition by transverse two-photon excitation of a collimated atomic beam with a c.w. dye laser of 165 mW power.

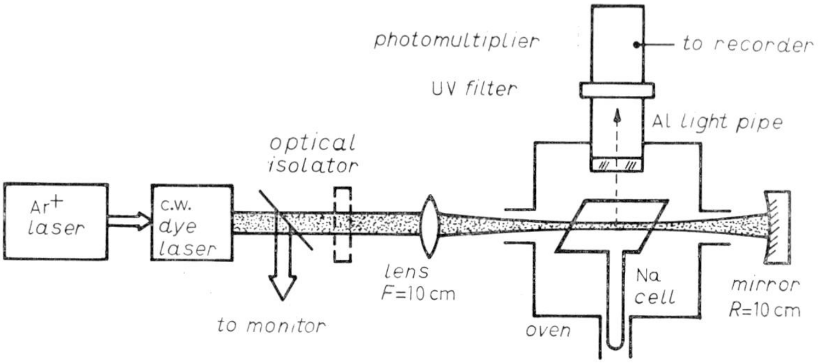

Fig. 22. – Experimental arrangement for Doppler-free two-photon spectroscopy of Na with a c.w. dye laser [19].

A scheme of the very simple set-up for Doppler-free two-photon spectroscopy, used in the c.w. laser experiment at Standford, is shown in fig. 22. The Na vapor ($3 \cdot 10^{-7}$ Torr) is contained in a short heated quartz cell. The output of a commercial single-mode c.w. dye laser, pumped by a 1W argon laser, is focused into the Na cell to a waist diameter of about 0.04 mm. The transmitted light is refocused into the cell by a spherical mirror to provide the required standing wave field. An optical isolator prevents feedback into the laser cavity. A photomultiplier observes the ultraviolet fluorescence light which is emitted by the excited $4d$ atoms if they decay via the $4p$-$3s$ transition at 3302 Å. A UV filter supresses any visible stray light.

Figure 23 shows a two-photon spectrum of the Na $3s$-$4d$ transition obtained with this apparatus by simply scanning the laser frequency across the line and

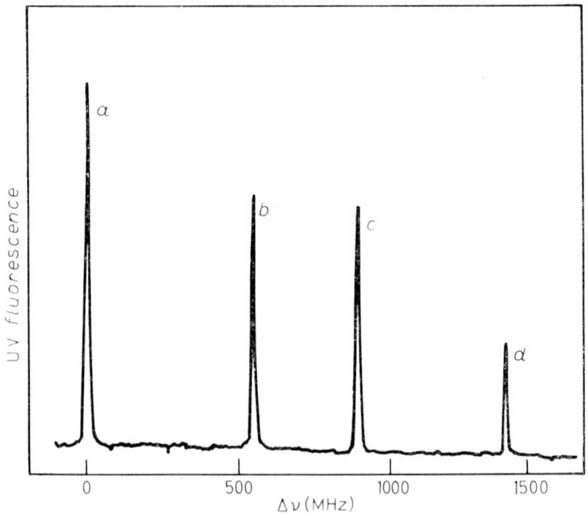

Fig. 23. – Doppler-free two-photon spectrum of Na $3s$-$4d$. The hyperfine splitting of the $3s$ ground state and the fine-structure splitting of the $4d$-state are resolved [19].

plotting the photomultiplier current with a chart recorder. The sharp resonance peaks disappear if the back-reflecting mirror is slightly misaligned so that the two beam waists no longer overlap inside the cell. Then, only the low Doppler-broadened pedestal remains.

The spectrum shows four resolved components, because each state is split into two sublevels. The separation *a-c* and *b-d* corresponds to the (well known) hyperfine splitting of the 3*s*-state ($F = 2, 1$), and the separation *a-b* and *c-d* corresponds to the fine-structure splitting of the 4*d*-state ($J = \frac{5}{2}, \frac{3}{2}$). The hyperfine splitting of the 4*d*-state is too small to be resolved. The line width of these components is less than 15 MHz (in the visible), or less than 1 % of the Doppler width. For comparison the expected natural line width is 3 MHz. A small contribution of about 2 MHz to the observed width originates from the finite transit time of the Na atoms through the small beam waist. The remaining line broadening is ascribed to laser frequency jitter. The signal for the *a*-component at resonance corresponds to about $6 \cdot 10^{10}$ two-photon excitations per second in the observed volume, and is in fair agreement with the estimate using eq. (106). The visible fluorescence from the $4d \to 3p \to 3s$ cascade in this case is actually strong enough to be easily seen with the naked eye.

A two-photon spectrum of the Na 3*s*-5*s* transition, which was recorded in the same way, is displayed in fig. 24. Even though each state is a hyperfine doublet, only two line components are observed, in agreement with the selection rule $\Delta F = 0$ for *S*-*S* transitions. The separation is equal to the difference of the splittings of lower and upper state. CAGNAC as well as LEVENSON demonstrated in their original experiments that the Doppler-broadened background of this line can be completely eliminated by the use of circularly polarized laser light.

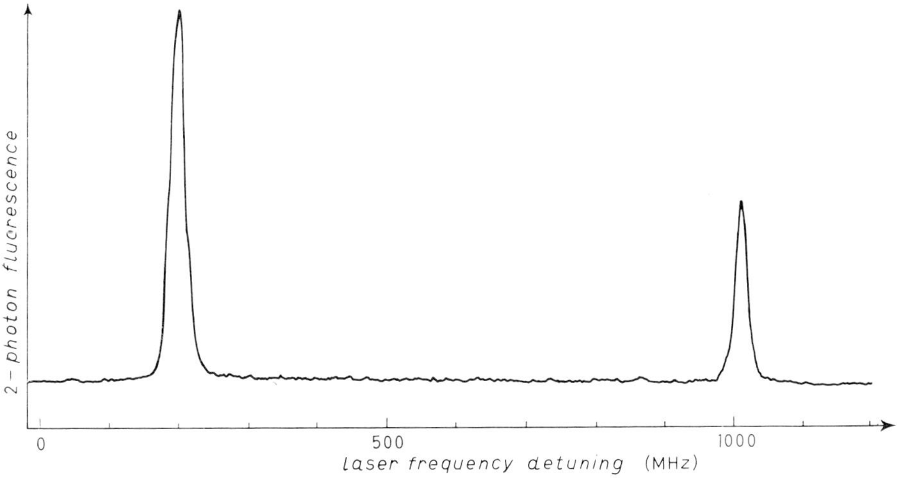

Fig. 24. – Doppler-free two-photon spectrum of Na 3*s*-5*s* with resolved hyperfine splitting [73].

It is quite obvious that such high-resolution two-photon spectra are useful for measuring the spacing between close-lying atomic energy levels. In particular, highly excited states of the same parity as the ground state, which cannot be reached with ordinary single-photon electric-dipole transitions, may be studied with this technique.

Several groups have meanwhile reported measurements of the fine-structure splitting of excited *d*-states and of the hyperfine splitting of excited *s*-states in Na and Rb. A pulsed dye laser [18] was used to gain access to shorter wavelengths than available for a rhodamine 6G c.w. dye laser and a pulsed parametric oscillator was used to reach longer wavelengths [69]. A summary of the experimental results obtained to date has been given in the review by LEVENSON and BLOEMBERGEN [64]. As a particularly beautiful example we shall present in fig. 25 the results obtained recently by KATO and STOICHEFF [70] for the fine-structure intervals of highly excited *nd* Rydberg states in Rb. These workers used a set-up similar to that shown in fig. 22, and they report a strong signal up to $n = 25$, despite the low c.w. dye laser power, and the laser wavelength being more than 1000 Å away from any resonance line.

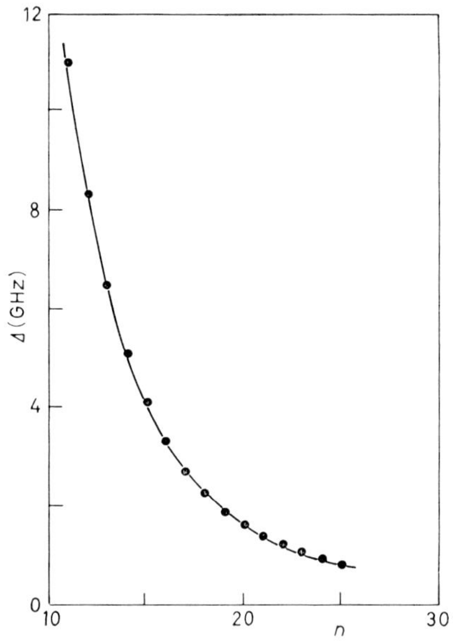

Fig. 25. – Fine-structure intervals of Rb and Rydberg states, measured by Doppler-free two-photon spectroscopy [70].

The accuracy of most reported measurements has not been limited by the spectral resolution, but rather by nonlinearities in the laser tuning characteristic. A considerable improvement should be possible with stabilized calibration interferometers or by the use of multiple dye lasers and heterodyne techniques for direct frequency measurements of optical beats. The potential accuracy should then come close to that offered by level-crossing spectroscopy, optical-r.f. double-resonance spectroscopy or quantum-beat spectroscopy.

Perhaps even more interesting are measurements of isotope shifts. First results for isotope shifts of 5*s*-*ns* and 5*s*-*nd* transitions in Rb have already been reported by ROBERTS and FORTSON [69] and by KATO and STOICHEFF [70]. A recent measurement of the 1*s*-2*s* isotope shift in atomic hydrogen will be discussed in some detail at the end of this section. Doppler-free two-photon spectroscopy not only provides unsurpassed resolution but also permits one to select upper levels with negligible overlap of electron wave function and nucleus. This can greatly facilitate the theoretical interpretation, since only the lower state contributes to the contact interaction.

Another interesting application of the new technique is the study of Zeeman effects. Because the observed volume is typically very small, it is possible

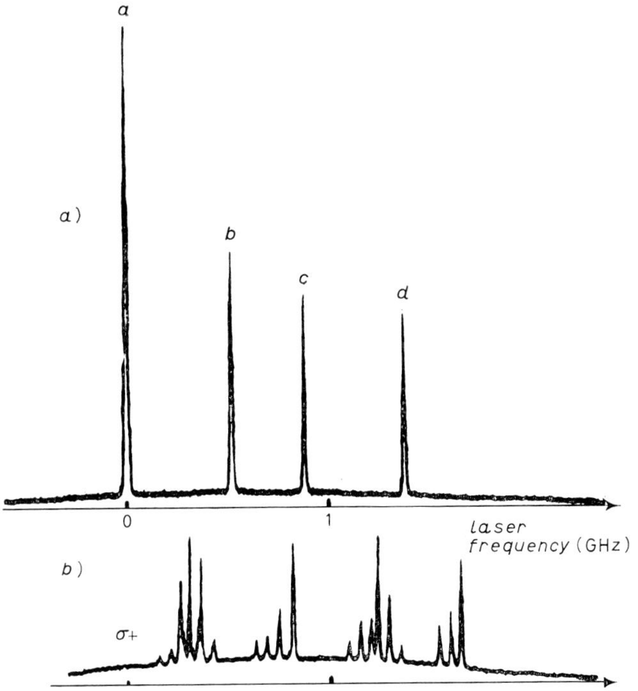

Fig. 26. – Zeeman splitting of the Na 3*s*-4*d* two-photon transition, observed with circulars polarized light [71]: *a*) $H_0=0$, *b*) $H_0=170$ G.

	3s	4d		3s	4d
a:	$F=2\rightarrow J'=\frac{5}{2}$		*c*:	$F=1\rightarrow J'=\frac{5}{2}$	
b:	$F=2\rightarrow J'=\frac{3}{2}$		*d*:	$F=1\rightarrow J'=\frac{3}{2}$	

to observe well-resolved line splittings even for rather inhomogeneous external fields. Complex Zeeman splitting of the Na 3*s*-4*d* spectrum could be observed by simply placing a small horseshoe magnet near the vapor cell. Figure 26 shows the Zeeman splitting of this line, as observed by BIRABEN *et al.* [71]. An axial magnetic field of 170 G was applied in a setup similar to that of fig. 22. The spectrum is simplified by the use of circularly polarized light, which reduces the number of permitted Zeeman transitions. Experiments of this kind should become useful for the measurement of magnetic moments and Landé factors of highly excited states. The same authors have also observed the

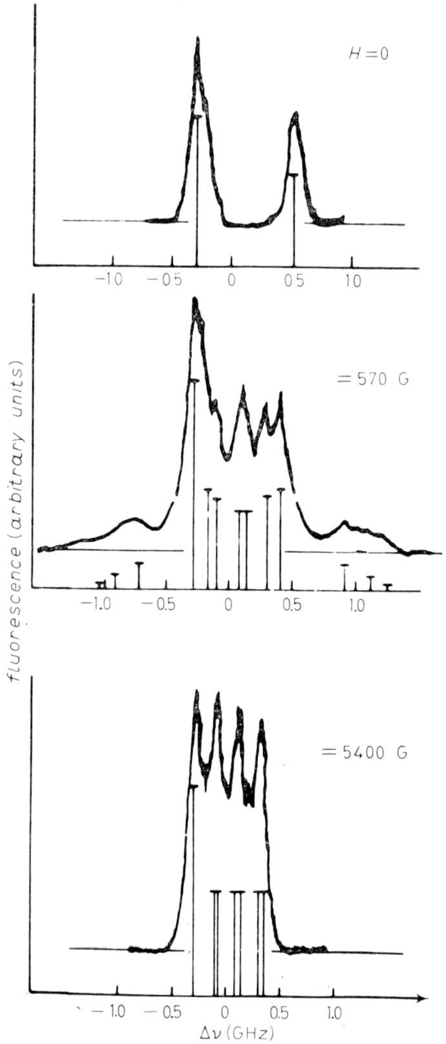

Fig. 27. – Hyperfine splitting of the Na 3*s*-5*s* two-photon transition in zero magnetic field and in magnetic fields of 570 and 5400 G, respectively [72].

Paschen-Back effect at higher fields (10 kG). BLOEMBERGEN *et al.* [72] have studied the Zeeman splitting of the Na $3s$-$5s$ transition, using a nitrogen-laser–pumped pulsed dye laser. Figure 27 gives the line profiles observed at zero magnetic field and at 570 and 5400 G. These spectra are rather remarkable for several reasons. Because of the spherical symmetry of both s-states, the direction of the magnetic field is unimportant. At low fields, both upper and lower states are split in exactly the same way, and the selection rules $\Delta F = 0$ and $\Delta m_F = 0$ permit transitions only between corresponding sublevels. The results is that no splitting whatsoever is observed in this low-field region. At very large fields, electron spin and nuclear spin are decoupled and align separately in the external field. Again, the selection rules permit transitions only between substates of equal m_I and m_s, and the spectrum is field independent. Four lines appear in the spectrum, however, because of the different contact hyperfine interaction in ground state and excited state. This high-field pattern is actually narrower than that in the zero field. A more complicated Zeeman pattern is observed in the intermediate-field region, where transitions with $m_I = -m_s = \pm 1$ are no longer forbidden and give rise to weak satellite lines, in agreement with theoretical predictions. A Zeeman spectrum of this kind, observed by HARVEY [73] with a c.w. dye laser, is shown in fig. 28.

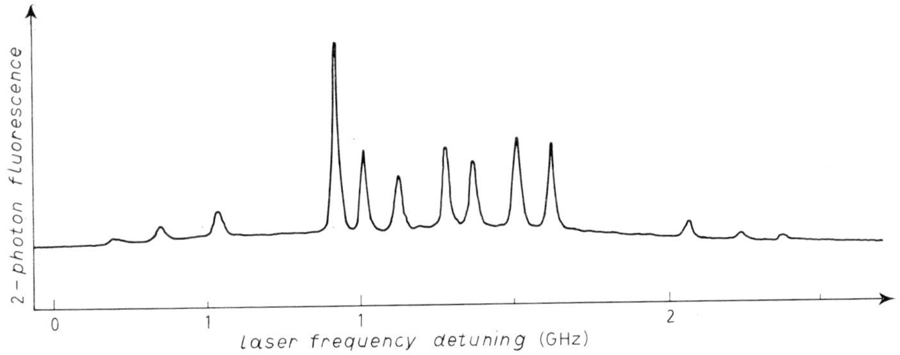

Fig. 28. – Zeeman effect in Na $3s$-$5s$ two-photon spectrum at intermediate field, observed with c.w. dye laser [73].

Doppler-free two-photon spectroscopy can also be used to study the Stark effect of highly excited states. The first such measurements have been reported by HARVEY *et al.* [74], who observed the Na $3s$-$4d$ and $3s$-$5s$ transitions with a c.w. dye laser. A crude, sheetlike atomic beam was used instead of a vapor cell, to facilitate the placement of the Stark-field electrodes, and to reach higher atomic densities without the problems of chemical reactions with the cell walls and windows. The two-photon spectrum of the Na $3s$-$4d$ line in the presence of a transverse electric field of 2.5 kV/cm is shown in fig. 29. The field-free spectrum is given below for comparison. The Stark pattern is considerably simpler than the earlier Zeeman patterns, because only the excited state shows any appreciable splitting, and because sublevels with orientation quantum numbers $\pm m_J$ are degenerate [36]. The spectrum not only reveals a clearly resolved splitting proportional to the square of the electric field but

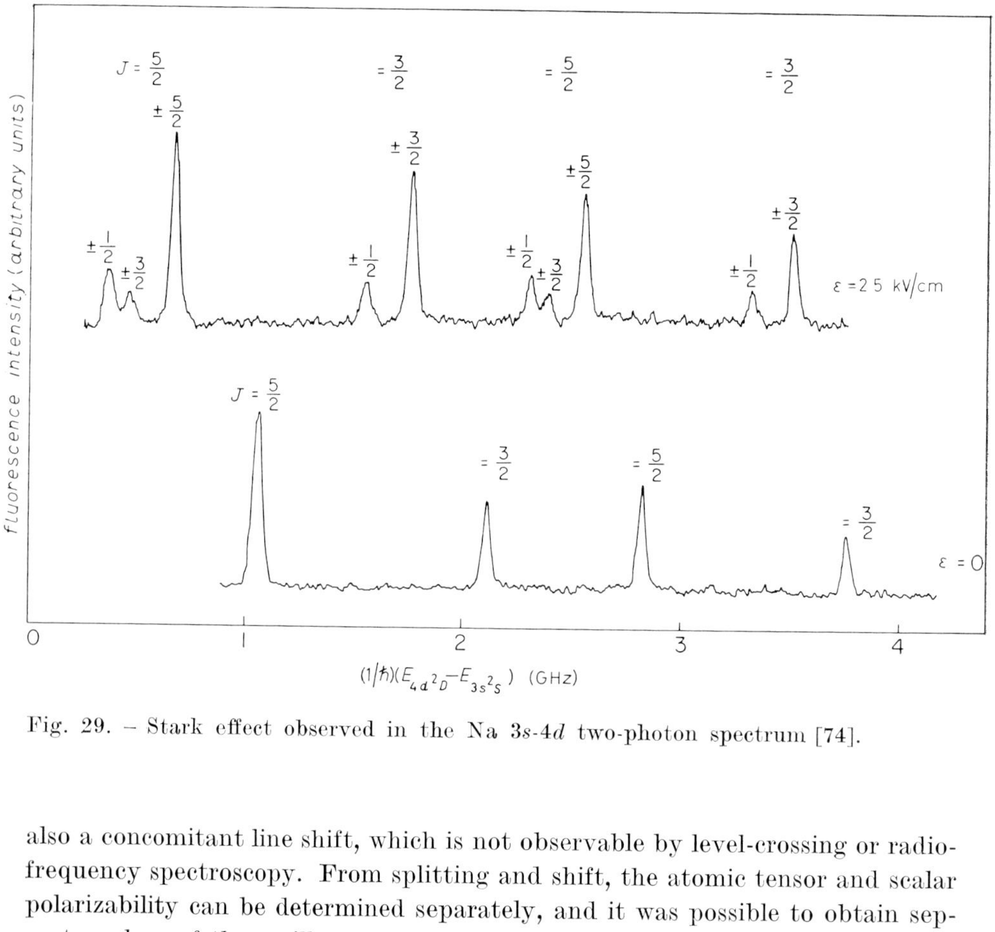

Fig. 29. – Stark effect observed in the Na $3s$-$4d$ two-photon spectrum [74].

also a concomitant line shift, which is not observable by level-crossing or radiofrequency spectroscopy. From splitting and shift, the atomic tensor and scalar polarizability can be determined separately, and it was possible to obtain separate values of the oscillator strengths of the two transitions $4d$-$5p$ and $4d$-$4f$, which are almost exclusively responsible for the $4d$ Stark splitting. In the case of the $3s$-$5s$ transition, no splitting, but only a shift was observed, which revealed the polarizability of the excited $5s$-state.

Other interesting two-photon excitation experiments in Na vapor will be reported in Bjorkholm's lectures, including the observation of a resonant enhancement [75] of up to seven orders of magnitude in the $3s$-$4d$ excitation rate in experiments with two dye lasers, tuned close to the resonance frequencies $3s$-$3p$ and $3p$-$4d$, and an observation of light shifts [76] under these conditions.

BISCHEL and co-workers [77] reported a successful observation of Doppler-free two-photon absorption in a molecular vapor. They studied transitions in methyl fluoride, using two fixed-frequency CO_2 lasers with a few watt of continuous single-mode power near 9.4 μm. Because of the rich rotational-vibrational structure of the molecular spectra it was possible to find near coincidences between two P-branch lines of the CO_2 laser ($P14$ and $P30$) and two coupled molecular transitions of CH_3F belonging to the ν_3 (C-F stretching vibration) bands, as illustrated in fig. 30. The difference between the laser sum frequency

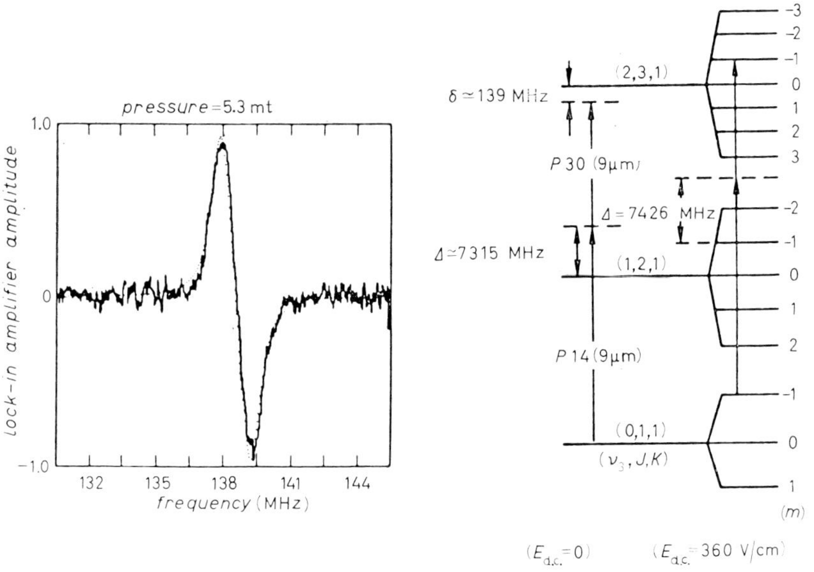

Fig. 30. – Doppler-free two-photon absorption spectroscopy in CH_3F, using two fixed-frequency CO_2 lasers [77].

and the two-photon resonance frequency is only 139 MHz, and it is possible to fine-tune the levels into resonance with a Stark field. The resonance is swept through by applying a ramp voltage between transverse electrodes, and a signal is detected by superimposing a small sinusoidally modulated field to the d.c. Stark field and observing a corresponding modulation in the weak absorption of the ($P30$) CO_2 laser beam, acting as a probe. A lock-in amplifier yields a differentiated two-photon spectrum, as shown in fig. 30. The signal is comfortably strong because of the enhancement by a near-resonant intermediate state. The experimenters were able to measure pressure broadening and pressure shifts.

Studies of the pressure broadening of narrow Doppler-free two-photon resonances should seem rather interesting from a theoretical point of view, because, unlike the case of saturation spectroscopy, these lines are completely insensitive to any change of the atomic velocities in elastic collisions. Any line broadening can be entirely ascribed to an increase of the natural line width, caused by collisional quenching, or by a loss of phase memory during collisional encounters.

We conclude these lectures by discussing some applications of Doppler-free two-photon spectroscopy to atomic hydrogen. Several authors [17, 19, 78]

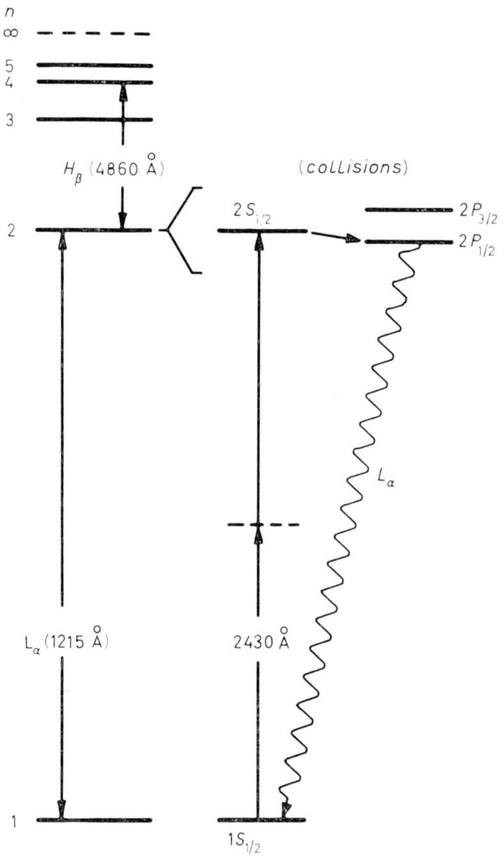

Fig. 31. – Simplified term diagram of atomic hydrogen. Doppler-free two-photon excitation of the $1S$-$2S$ transition promises an extremely narrow natural line width.

have pointed out that it would be particularly interesting to study transitions from the $1S$ ground state of hydrogen to the metastable $2S$-state (fig. 31) by the new technique. Such an experiment requires a laser near 2430 Å, *i.e.* at twice the wavelength of the vacuum ultraviolet Lyman-α resonance line. Since the natural decay mode of the $2S$-state for free atoms is two-photon emission and the lifetime is about $\frac{1}{7}$ s, the natural line width of the $1S$-$2S$ transition is of the order of only a few hertz. Doppler-free spectroscopy of this transition should hence offer an ultimate resolution of about 1 part in 10^{15}, which would surpass even the best resolution achieved with Mössbauer spectroscopy. To reach such a goal, a number of rather formidable technical problems will have to be solved. A tunable ultraviolet laser of extreme monochromaticity has to be constructed. Ways have to be found to ensure a sufficiently long interaction time between atoms and radiation, and to reduce the line broadening by the quadratic Doppler effect which would limit the resolution for hydrogen atoms at room temperature to about 1 part in 10^{11}. It is likely that such an experiment will present an interesting and serious challenge to physicists for many years to come. Even if this ultimate goal cannot be met, high-resolution two-photon spectroscopy in such a simple quantum-mechanical system is still of considerable interest for fundamental physics research, and it has been pointed out repeatedly that an absolute wavelength measurement of the $1S$-$2S$ transition could determine a new accurate value of the Rydberg constant.

In the meantime our research group at Stanford has actually observed Doppler-free two-photon excitation of the $1S$-$2S$ transition in hydrogen and deuterium [79, 80]. A frequency-doubled pulsed dye laser provided the ultraviolet light at 2430 Å, and the excitation was monitored by observing the subsequent collision-induced $2P$-$1S$ fluorescence at the Lyman-α wavelength 1215 Å. The resolution in these initial experiments was limited by the laser to about 2 parts in 10^7 or 2 % of the Doppler width. Nonetheless it was possible to measure the $1S$-$2S$ isotope shift to within 1 part in 10^4, and we also obtained, for the first time, experimental values for the Lamb shift of the $1S$ ground state with better than 2 % accuracy.

Since all hydrogen quantum states are theoretically well known, it is easily possible to calculate the $1S$-$2S$ two-photon excitation rate by numerical evaluation of eq. (106). At resonance and for pulsed excitation one expects a transition rate per atom $\Gamma_{ac} \approx 7 \cdot 10^{-4} I^2/\Delta\omega$, where the light intensity I is measured in W/cm^2, the laser band width $\Delta\omega$ in MHz and Γ in s^{-1}. The situation is actually not so unfavorable as one might expect from the lack of any near-resonant intermediate state, because even very small excitation rates can be detected with high sensitivity by photon counting of the Lyman-α emission.

The dye laser used in the reported experiments consists of a pressure-tuned dye laser oscillator, side pumped by a pulsed nitrogen laser, and equipped with an external confocal filter interferometer, as was used in the earlier saturation spectroscopy experiments. This oscillator is followed by two traveling-wave dye laser amplifier stages, pumped by the same 1MW nitrogen laser. This system generates 10 ns long pulses of $(30 \div 50)$ kW peak power in the visible with a near Fourier-transform limited band width of about 120 MHz at a repetition rate of 17 p.p.s., and offers continuous linear wavelength tuning over ranges up to 5 Å. A detailed description of this particularly versatile tool for high-resolution nonlinear spectroscopy in the visible and ultraviolet region has since been published [44].

A scheme of the $1S$-$2S$ two-photon spectrometer is shown in fig. 32. A lithium formate crystal doubles the dye laser frequency to give ultraviolet pulses of about 300 W peak power. The UV beam is focused to about 0.2 mm diameter inside the observation chamber, and is reflected back into the chamber to produce a standing wave field. The atoms are generated in a d.c. gas discharge in a mixture of H_2 and D_2 at 0.2 Torr, and carried by gas flow and diffusion into the observation chamber. The Lyman-α fluorescence is detected with a solar blind photomultiplier through a MgF_2 window and an interference filter.

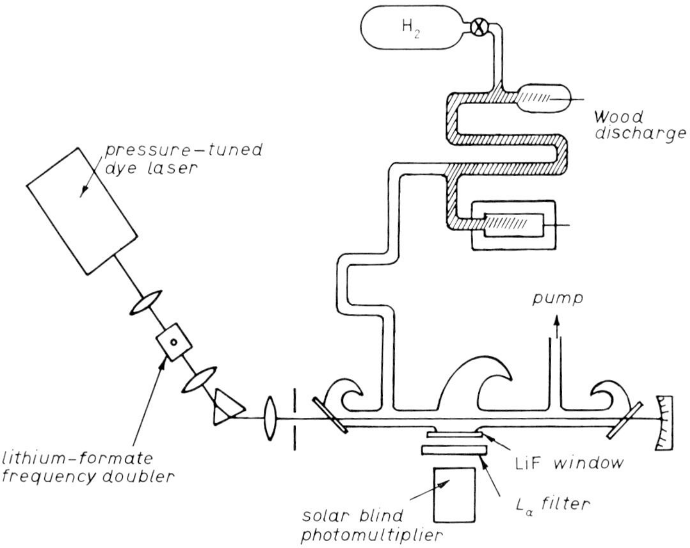

Fig. 32. – Two-photon spectrometer for observation of hydrogen $1S$-$2S$ [79].

Two typical two-photon spectra, obtained by tuning the laser continuously across the isotope shift, are shown in fig. 33. The transmission maxima of a calibrated interferometer have been plotted simultaneously as a frequency marker. The expected hyperfine doublet structure is resolved for hydrogen. The signal at resonance is on the order of 30 registered Lyman-α photons per pulse, *i.e.* about 10^4 times weaker than expected for complete dissociation. This considerable signal loss is ascribed to resonance trapping and quenching of the Lyman-α fluorescence. It appears hence possible to obtain similar signals at much lower gas pressures, where collision effects would be negligible. It would then be necessary, however, to induce $2S$-$2P$ transitions with an external r.f. or magnetic field in order to observe any Lyman-α fluorescence. The a.c. Stark shift or light shift, estimated with eq. (107), is on the order of 1 MHz, *i.e.* quite negligible, and could be further reduced, without loss of resonant signal, by using a laser of lower power and narrower line width.

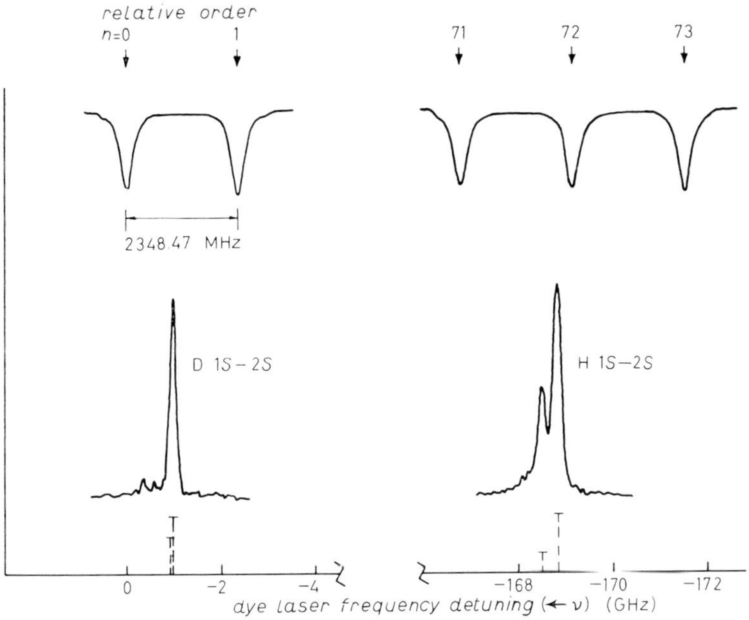

Fig. 33. – Doppler-free two-photon spectra of the $1S$-$2S$ transition in H and D with transmission maxima of calibration interferometer [80].

The $1S$-$2S$ isotope shift has been measured in these experiments to be (670.933 ± 0.056) GHz, in agreement with theoretical predictions. The large shift is primarily caused by the nuclear mass difference, and the accuracy of the present calculations is limited by the uncertainty of the ratio of electron mass m_e to proton mass m_p (1 p.p.m.). Future more precise measurements of this isotope shift should thus become useful to determine a more accurate value of the important ratio m_e/m_p.

Doppler-free two-photon spectroscopy of hydrogen $1S$-$2S$ has also made possible an entirely new approach for measurements of the Lamb shift of the $1S$ ground state. The energy of this state, like that of all hydrogen S-states, is shifted above its Dirac value by quantum electrodynamic corrections, *e.g.* due to vacuum fluctuations of the electromagnetic field. But while the Lamb shifts of excited S-states can be measured with high accuracy by radiofrequency spectroscopy, and provide important tests of low-energy quantum electrodynamics, the largest $1S$ Lamb shift cannot be obtained in this way, because there is no nearby P reference level. The only previous measurement of the Lamb shift of the $1S$-state of deuterium, (7.9 ± 1.1) GHz, has been reported by HERZBERG [81], who used a difficult absolute-wavelength measurement of the Lyman-α line.

Nonlinear laser spectroscopy, however, can provide a closely spaced reference line for the determination of the $1S$ Lamb shift. If the second harmonic of the dye laser is in resonance with the $1S$-$2S$ two-photon transition, the fundamental wavelength (4860 Å) agrees with the Balmer-α line of hydrogen (fig. 31). If Bohr's formula were correct, the $n = 1 \div 2$ interval would be exactly four times larger than the $n = 2 \div 4$ interval. Any actual displacement is due to relativistic and quantum electrodynamic corrections, and is a sensitive indicator for the $1S$ Lamb shift.

To utilize this scheme, a high-resolution saturation spectrum of the Balmer-β line was recorded simultaneously with the $1S$-$2S$ two-photon spectrum. A small fraction of the visible dye laser output was used in a set-up similar to the Balmer-α saturation spectrometer discussed in sect. **3**. Some typical results are shown in fig. 34. The relatively large line widths in the saturation spec-

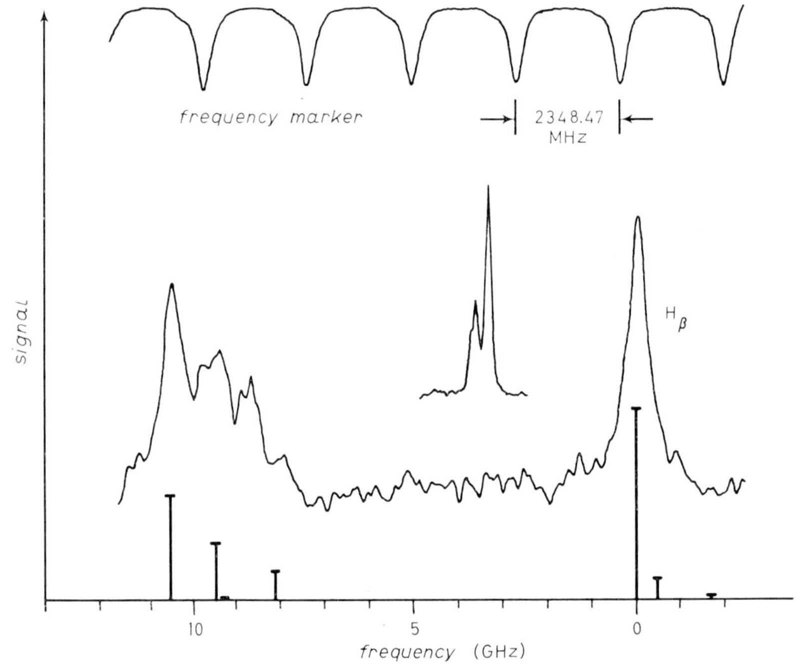

Fig. 34. – Hydrogen $1S$-$2S$ two-photon spectrum with simultaneously recorded saturation spectrum of the Balmer-β line. The $1S$ Lamb shift can be determined from the relative position of these spectra [80].

trum are partly caused by power broadening. The $1S$ Lamb shift was determined by measuring the frequency separation between the $1S$-$2S$ resonance and the strong $2P_{\frac{3}{2}}$-$4D_{\frac{5}{2}}$ component at the right side of the saturation spectrum, and comparing it with the theoretical line separation for a given $1S$ Lamb shift. The incompletely resolved saturation spectrum required a number of systematic corrections. A $1S$ Lamb shift of (8.20 ± 0.10) GHz for hydrogen and of (8.25 ± 0.11) GHz for deuterium was measured in this way, in agreement with present theoretical values.

A considerable improvement in accuracy should be possible, if the resolution of the $1S$-$2S$ two-photon spectrum is improved by reducing the laser line width (a laser power of 1 mW should be sufficient in a c.w. experiment), and if the resolution of the Balmer-β line is improved, *e.g.* by Doppler-free two-photon spectroscopy of the $2S$-$4S$, D transitions. The latter might for instance be accomplished with a GaAs diode laser, whose second harmonic is locked to the visible dye laser frequency. Since neither line width is limited by a short-lived P-state, the ultimate accuracy of such measurements could well exceed that of recent $2S$ Lamb-shift measurements, or even that of present calculations, which are limited by our knowledge of the fine-structure constant and of nuclear-structure effects.

A precise comparison of the energy intervals $1S$-$2S$ and $2S$-$4S$, D within a few parts in 10^9 should also make it possible to experimentally confirm the existence of a theoretically predicted « Dirac shift » [48], which is the same for all fine-structure sublevels of a given principal quantum number and is not accessible to radiofrequency measurements. It describes a relativistic correction due to the nuclear motion, because the Dirac equation of a relativistic two-body problem cannot be strictly solved in terms of a single particle of reduced mass moving in a fixed potential.

Doppler-free two-photon spectroscopy also holds great promise for studies of positronium, *i.e.* the hydrogenlike atoms formed by one electron and one positron, which survive for some $(10 \div 100)$ ns before annihilation into two or three gamma quanta. No nuclear-structure effects are expected for these purely leptonic atoms, and precision spectroscopy of positronium should provide stringent new tests of quantum electrodynamics. It has only recently become possible to observe positronium atoms in their excited $n = 2$ state [82, 83]. Numerous attempts to reach this state by excitation with Lyman-α radiation at 2430 Å have remained unsuccessful. CURRY and SCHAWLOW [84] have shown that free positronium atoms of about 0.3 eV kinetic energy can be produced in vacuum, but the very large Doppler width of the resonance lines seem to preclude any efficient use of laser light for single-photon excitation. Two-photon excitation with two counter-propagating laser beams at 4860 Å should not only provide a rather efficient means to populate the $n = 2$ state but it should also permit a precise measurement of the energy interval 1^3S-2^3S, which would be of considerable interest for the test of relativistic two-body calculations.

These examples have shown that nonlinear high-resolution laser spectroscopy promises many exciting new experiments, even if we restrict our attention to the simplest of all atoms. And there can be little doubt that studies of more complex atoms and molecules by these new techniques will bring many more rewarding results.

REFERENCES

[1] A. C. G. MITCHELL and M. W. ZEMANSKY: *Resonance Radiation and Excited Atoms* (Cambridge, 1961).
[2] A. ABRAGAM: *The Principles of Nuclear Magnetism* (New York, N. Y., 1961).
[3] L. ALLEN and J. H. EBERLY: *Optical Resonance and Two-Level Atoms* (New York, N. Y., 1975).
[4] W. R. BENNETT jr.: *Phys. Rev.*, **126**, 580 (1962).
[5] W. E. LAMB jr.: *Phys. Rev. A*, **134**, 1429 (1964); M. SARGENT, M. O. SCULLY and W. E. LAMB jr.: *Laser Physics* (London, 1974).
[6] A. SZÖKE and A. JAVAN: *Phys. Rev. Lett.*, **10**, 521 (1963).
[7] P. H. LEE and M. L. SKOLNICK: *Appl. Phys. Lett.*, **10**, 521 (1963).
[8] K. SHIMODA and T. SHIMIZU: *Progress in Quantum Electronics*, Vol. **2**, Part 2, edited by J. H. SANDERS and S. STENHOLM (Oxford, 1972).
[9] R. G. BREWER: *Science*, **178**, 247 (1972).
[10] J. L. HALL: *Atomic Physics*, Vol. **3**, edited by S. J. SMITH and G. K. WALTERS (New York, N. Y., 1973).
[11] T. W. HÄNSCH and P. TOSCHEK: *IEEE Journ. Quant. Electr.*, QE-**4**, 467 (1968).
[12] C. BORDÉ: *Compt. Rend.*, **271**, 371 (1970).
[13] P. W. SMITH and T. W. HÄNSCH: *Phys. Rev. Lett.*, **26**, 740 (1971).
[14] T. W. HÄNSCH, M. D. LEVENSON and A. L. SCHAWLOW: *Phys. Rev. Lett.*, **26**, 946 (1971).
[15] T. W. HÄNSCH, I. S. SHAHIN and A. L. SCHAWLOW: *Phys. Rev. Lett.*, **27**, 707 (1971).
[16] L. S. VASILENKO, V. P. CHEBOTAEV and A. V. SHISHAEV: *JETP Lett.*, **12**, 113 (1970).
[17] F. BIRABEN, B. CAGNAC and G. GRYNBERG: *Phys. Rev. Lett.*, **32**, 643 (1974).
[18] M. D. LEVENSON and N. BLOEMBERGEN: *Phys. Rev. Lett.*, **32**, 645 (1974).
[19] T. W. HÄNSCH, K. C. HARVEY, G. MEISEL and A. L. SCHAWLOW: *Opt. Comm.*, **11**, 50 (1974).
[20] C. BORDÉ and J. L. HALL: *Laser Spectroscopy*, edited by R. G. BREWER and A. MOORADIAN (New York, N. Y., 1974), p. 125.
[21] J. H. SHIRLEY: *Phys. Rev. A*, **8**, 347 (1973).
[22] U. FANO: *Rev. Mod. Phys.*, **29**, 74 (1957).
[23] D. H. CLOSE: Scientific Report No. 5, U.S. Gov. Rep. AD 621320 (1965).
[24] C. V. SHANK and S. E. SCHWARTZ: *IEEE Journ. Quant. Electr.*, QE-**4**, 1017 (1968).
[25] E. V. BAKLANOV and V. P. CHEBOTAEV: *Sov. Phys. JETP*, **33**, 300 (1971).
[26] S. HAROCHE and F. HARTMANN: *Phys. Rev. A*, **6**, 1280 (1972).
[27] S. STENHOLM and W. E. LAMB jr.: *Phys. Rev.*, **181**, 618 (1969).
[28] B. J. FELDMAN and M. S. FELD: *Phys. Rev. A*, **1**, 1375 (1970).
[29] P. BERMAN: in *Proceedings of the Second International Conference on Laser Spectroscopy*, edited by S. HAROCHE, J. C. PEBAY-PEYROULA, T. W. HÄNSCH and S. E. HARRIS (Heidelberg, 1975).
[30] B. J. FELDMAN and M. S. FELD: *Phys. Rev. A*, **5**, 899 (1972).
[31] I. M. BETEROV and V. P. CHEBOTAEV: *Progress in Quantum Electronics*, Vol. **3**, Part 1, edited by J. H. SANDERS and S. STENHOLM (Oxford, 1974).
[32] T. W. HÄNSCH and P. TOSCHEK: *Zeits. Phys.*, **236**, 213 (1970).
[33] V. WEISSKOPF: *Zeits. Phys.*, **85**, 451 (1933).

[34] Y. R. SHEN: *Phys. Rev. B*, **9**, 622 (1974).
[35] M. MIZUSHIMA: *Phys. Rev.*, **133**, A 414 (1964).
[36] A. M. BONCH-BRUEVICH and V. A. KHODOVOI: *Sov. Phys. Usp.*, **10**, 637 (1968).
[37] M. S. FELD and A. JAVAN: *Phys. Rev.*, **177**, 540 (1969).
[38] T. W. HÄNSCH, R. KEIL, A. SCHABERT, CH. SCHMELZER and P. TOSCHEK: *Zeits. Phys.*, **226**, 293 (1969).
[39] V. S. LETOKHOV and V. P. CHEBOTAEV: *Principles of Nonlinear Laser Spectroscopy* (Heidelberg, 1976).
[40] M. D. LEVENSON and A. L. SCHAWLOW: *Phys. Rev. A*, **6**, 10 (1972).
[41] M. S. SOREM, T. W. HÄNSCH and A. L. SCHAWLOW: *Chem. Phys. Lett.*, **17**, 300 (1972).
[42] T. W. HÄNSCH: *Appl. Optics*, **11**, 895 (1972).
[43] R. WALLENSTEIN and T. W. HÄNSCH: *Appl. Optics*, **13**, 1625 (1974).
[44] R. WALLENSTEIN and T. W. HÄNSCH: *Opt. Comm.*, **14**, 353 (1975).
[45] T. W. HÄNSCH and I. S. SHAHIN: *Opt. Comm.*, **8**, 312 (1973).
[46] T. W. HÄNSCH, I. S. SHAHIN and A. L. SCHAWLOW: *Nature*, **235**, 63 (1972).
[47] T. W. HÄNSCH, M. H. NAYFEH, S. A. LEE, S. M. CURRY and I. S. SHAHIN: *Phys. Rev. Lett.*, **32**, 1336 (1974).
[48] G. W. SERIES: *Spectrum of Atomic Hydrogen* (Oxford, 1957).
[49] C. FREED and A. JAVAN: *Appl. Phys. Lett.*, **17**, 53 (1970).
[50] M. S. SOREM and A. L. SCHAWLOW: *Opt. Comm.*, **5**, 148 (1972).
[51] K. SHIMODA: *Appl. Phys.*, **1**, 77 (1973).
[52] A. MUIRHEAD, K. V. L. N. SASTRY, R. F. CURL, J. COOK and F. K. TITTEL: *Chem. Phys. Lett.*, **24**, 208 (1974).
[53] W. M. FAIRBANK jr., T. W. HÄNSCH and A. L. SCHAWLOW: *Journ. Opt. Soc. Amer.*, **65**, 199 (1975).
[54] C. BORDÉ, G. CAMY, B. DECOMPS and L. POTTIER: *Colloques Internationaux du C.N.R.S.*, No. 217 (Paris, 1974), p. 231.
[55] B. COUILLAUD and D. DUCASSE: in *Proceedings of the Second International Conference on Laser Spectroscopy*, edited by S. HAROCHE, J. C. PEBAY-PEYROULA, T. W. HÄNSCH and S. E. HARRIS (Heidelberg, 1975).
[56] A. SCHABERT, R. KEIL and P. TOSCHEK: *Opt. Comm.*, **13**, 265 (1975).
[57] A. SCHABERT, R. KEIL and P. TOSCHEK: *Appl. Phys.*, **6**, 181 (1975).
[58] F. SCHUDA, C. R. STROUD jr. and M. HERCHER: *J. Phys. B*, **7**, L198 (1974).
[59] H. WALTHER: in *Proceedings of the Second International Conference on Laser Spectroscopy*, edited by S. HAROCHE, J. C. PEBAY-PEYROULA, T. W. HÄNSCH and S. E. HARRIS (Heidelberg, 1975).
[60] C. COHEN-TANNOUDJI: in *Proceedings of the Second International Conference on Laser Spectroscopy*, edited by S. HAROCHE, J. C. PEBAY-PEYROULA, T. W. HÄNSCH and S. E. HARRIS (Heidelberg, 1975).
[61] M. GÖPPERT-MAYER: *Ann. der Phys.*, **9**, 273 (1931).
[62] A. GOLD: *Rendiconti S.I.F.*, Course XLII (New York, N. Y., 1967).
[63] J. M. WORLOCK: in *Laser Handbook*, edited by F. T. ARECCHI and E. O. SCHULZ-DUBOIS (Amsterdam, 1972).
[64] N. BLOEMBERGEN and M. D. LEVENSON: in *Laser Spectroscopy of Atoms and Molecules*, edited by H. WALTHER (Heidelberg, 1976).
[65] J. J. SNYDER and J. L. HALL: in *Proceedings of the Second International Conference on Laser Spectroscopy*, edited by S. HAROCHE, J. C. PEBAY-PEYROULA, T. W. HÄNSCH and S. E. HARRIS (Heidelberg, 1975).
[66] E. V. BAKLANOV and V. P. CHEBOTAEV: *Sov. Journ. Quant. El.*, **2**, 606 (1975) (in Russian).

[67] B. CAGNAC, G. GRYNBERG and F. BIRABEN: *J. Physique*, **34**, 845 (1973).
[68] D. PRITCHARD, J. APT and T. W. DUCAS: *Phys. Rev. Lett.*, **32**, 641 (1974).
[69] D. E. ROBERTS and E. N. FORTSON: *Opt. Comm.*, **14**, 332 (1975).
[70] Y. KATO and B. P. STOICHEFF: in *Proceedings of the Second International Conference on Laser Spectroscopy*, edited by S. HAROCHE, J. C. PEBAY-PEYROULA, T. W. HÄNSCH and S. E. HARRIS (Heidelberg, 1975).
[71] F. BIRABEN, B. CAGNAC and G. GRYNBERG: *Phys. Lett.*, **48** A, 469 (1974).
[72] N. BLOEMBERGEN, M. D. LEVENSON and M. M. SALOUR: *Phys. Rev. Lett.*, **32**, 867 (1974).
[73] K. C. HARVEY: Ph. D. Thesis, M. L. Report No. 2442, Stanford University (1975).
[74] K. C. HARVEY, R. T. HAWKINS, G. MEISEL and A. L. SCHAWLOW: *Phys. Rev. Lett.*, **34**, 1073 (1975).
[75] J. E. BJORKHOLM and P. F. LIAO: *Phys. Rev. Lett.*, **33**, 128 (1974).
[76] P. F. LIAO and J. E. BJORKHOLM: *Phys. Rev. Lett.*, **34**, 1 (1975).
[77] W. K. BISCHEL, P. J. KELLY and C. K. RHODES: *Phys. Rev. Lett.*, **34**, 300 (1975).
[78] E. V. BAKLANOV and V. P. CHEBOTAEV: *Opt. Comm.*, **12**, 312 (1974).
[79] T. W. HÄNSCH, S. A. LEE, R. WALLENSTEIN and C. WIEMAN: *Phys. Rev. Lett.*, **34**, 307 (1975).
[80] S. A. LEE, R. WALLENSTEIN and T. W. HÄNSCH: *Phys. Rev. Lett.*, submitted for publication (1975).
[81] G. HERZBERG: *Proc. Roy. Soc.*, A **234**, 516 (1956).
[82] K. F. CANTER, A. P. MILLS jr. and S. BERKO: *Phys. Rev. Lett.*, **34**, 177 (1975).
[83] A. P. MILLS jr., S. BERKO and K. F. CANTER: *Phys. Rev. Lett.*, **34**, 1541 (1975).
[84] S. M. CURRY and A. L. SCHAWLOW: *Phys. Lett.*, **37** A, 5 (1971).

Coherent Optical Spectroscopy.

R. G. BREWER

IBM Research Laboratory - San Jose, Cal. 95193, U.S.A.

1. – Introduction.

These lectures review a new branch of optical spectroscopy, what may be called *coherent optical spectroscopy*. Due to the availability of coherent laser light, atomic and molecular quantum states can now be prepared in coherent superposition, in the same way that nuclear-spin systems have been prepared for over 25 years in pulsed nuclear magnetic resonance [1]. Coherently prepared samples of this type exhibit a class of transient phenomena that offer new ways to obtain ultra-high-resolution optical spectra, and they allow the optical spectroscopist to isolate for the first time individual relaxation processes that have remained hidden within the optical line shape. Many of these optical transients have now been realized by using the recently introduced Stark-switching technique [2] and will be reviewed.

We shall treat these coherent transient effects semi-classically using the coupled Schrödinger-Maxwell equations. Our procedure is to first solve the Schrödinger equation, either in density matrix or Bloch form, for a molecular system subject to a pulsed or continuous-wave coherent optical field. This preparative stage generates a sample polarization that can radiate coherent light even after the preparation has passed, as prescribed by Maxwell's equations. Each transient effect corresponds to a particular sequence of one or more preparative steps, followed by periods where the sample freely radiates and where the polarization solutions can be traced from one stage to the next. In the case of a gas, the final polarization must be averaged over the Doppler velocity distribution. As long as the radiated field amplitude is small compared to the applied laser field, as in the present measurements, the laser field can be assumed constant; this is the so-called thin-sample regime. At high optical densities, other coherence phenomena such as self-induced transparency [3] occur.

1'1. *Density matrix equations.* – We consider a molecular two-level quantum configuration where the lower level is labeled 1 and the upper level is 2. In sect. **3**, the corresponding three-level problem is treated. The molecular gas sample encounters a laser field

$$E_x(z, t) = E_0 \cos(\Omega t - kz) \tag{1.1}$$

of frequency Ω and amplitude E_0, polarized along the x-axis and propagating in the z-direction. The time-dependent behavior of the density matrix ϱ is given by the Schrödinger equation of motion

$$i\hbar\dot{\varrho} = [H, \varrho] + \text{relaxation terms}. \tag{1.2}$$

The Hamiltonian

$$H = H_0 + H_I \tag{1.3}$$

contains the free-molecule part H_0 where the eigenenergies are

$$\left\{ \begin{aligned} \langle 1|H_0|1\rangle &= \hbar\omega_1, \\ \langle 2|H_0|2\rangle &= \hbar\omega_2, \end{aligned} \right. \tag{1.4}$$

and the 1-2 level splitting in angular frequency units is

$$\omega_2 - \omega_1 = \omega_{21}. \tag{1.5}$$

The molecule–optical-field interaction term

$$H_I = -\boldsymbol{\mu}\cdot\boldsymbol{E}_x(z, t) \tag{1.6}$$

has the electric-dipole matrix element

$$\langle 1|\mu_x|2\rangle = \mu_{12} \neq 0. \tag{1.7}$$

For this two-level problem, eq. (1.2) becomes

$$\dot{\varrho}_{11} = i\chi(\varrho_{21} - \varrho_{12})\cos(\Omega t - kz) - (\varrho_{11} - \varrho_{11}^0)/T_1, \tag{1.8a}$$

$$\dot{\varrho}_{12} = i\chi(\varrho_{22} - \varrho_{11})\cos(\Omega t - kz) - (i\omega_{12} + 1/T_2)\,\varrho_{12}, \tag{1.8b}$$

where we have introduced the phenomenological decay times T_1, assumed equal for the diagonal elements ϱ_{11} and ϱ_{22}, and T_2 for ϱ_{12}. The source term ϱ_{11}^0 denotes that molecules enter level 1 via relaxation from other quantum states, and we have defined the Rabi-flopping frequency as

$$\chi = \mu_{12} E_0/\hbar \,. \tag{1.9}$$

A similar equation in $\dot{\varrho}_{22}$ can be obtained from (1.8*a*) by the index interchange $1 \leftrightarrow 2$ and an equation in $\dot{\varrho}_{21}$ from (1.8*b*) by taking its complex conjugate.

The rapidly oscillating factors of the off-diagonal elements can be removed with the substitution

$$\varrho_{12} = \tilde{\varrho}_{12} \exp\left[i(\Omega t - kz)\right] \tag{1.10}$$

and by neglecting nonresonant high-frequency terms that oscillate as $\exp\left[2i(\Omega t - kz)\right]$. This procedure, known as the rotating-wave approximation, reduces (1.8) to

$$\frac{\mathrm{d}\varrho_{11}}{\mathrm{d}t} = i\chi(\tilde{\varrho}_{21} - \tilde{\varrho}_{12})/2 - (\varrho_{11} - \varrho_{11}^0)/T_1 \,, \tag{1.11a}$$

$$\left(\frac{\mathrm{d}}{\mathrm{d}t} - i\Delta + 1/T_2\right)\tilde{\varrho}_{12} = i\chi(\varrho_{22} - \varrho_{11})/2 \,, \tag{1.11b}$$

where

$$\Delta = -\Omega + kv_z + \omega_{21} \,. \tag{1.12}$$

We also have recognized that the time derivative

$$\dot{\varrho} = \left(\frac{\partial}{\partial t} + v_z \frac{\partial}{\partial z}\right)\varrho \tag{1.13}$$

includes a second term, because of a molecular velocity component v along the z-axis that results in the Doppler shift kv_z of (1.12). Note that at resonance $\Delta = 0$.

1·2. *Bloch equations.* – It is sometimes convenient to work with combinations of (1.11) and the corresponding equations for $\dot{\varrho}_{22}$ and $\dot{\tilde{\varrho}}_{21}$. For example, we may obtain

$$\dot{\tilde{\varrho}}_{12} + \dot{\tilde{\varrho}}_{21} = -i\Delta(\tilde{\varrho}_{21} - \tilde{\varrho}_{12}) - (\tilde{\varrho}_{12} + \tilde{\varrho}_{21})/T_2 \,, \tag{1.14a}$$

$$i(\dot{\tilde{\varrho}}_{21} - \dot{\tilde{\varrho}}_{12}) - \Delta(\tilde{\varrho}_{12} + \tilde{\varrho}_{21}) + \chi(\varrho_{22} - \varrho_{11}) - i(\tilde{\varrho}_{21} - \tilde{\varrho}_{12})/T_2 \,, \tag{1.14b}$$

$$\dot{\varrho}_{22} - \dot{\varrho}_{11} = -i\chi(\tilde{\varrho}_{21} - \tilde{\varrho}_{12}) - (\varrho_{22} - \varrho_{11})/T_1 + (\varrho_{22}^0 - \varrho_{11}^0)/T_1 \,. \tag{1.14c}$$

These are the Bloch equations and may be written as

$$\dot{u} + \Delta v + u/T_2 = 0 \,, \tag{1.15a}$$

$$\dot{v} - \Delta u - \chi w + v/T_2 = 0 \,, \tag{1.15b}$$

$$\dot{w} + \chi v + (w - w^0)/T_1 = 0 \,, \tag{1.15c}$$

using the variables

(1.16a) $$u = \tilde{\varrho}_{12} + \tilde{\varrho}_{21} \, ,$$

(1.16b) $$v = i(\tilde{\varrho}_{21} - \tilde{\varrho}_{12}) \, ,$$

(1.16c) $$w = \varrho_{22} - \varrho_{11} \, .$$

The set of equations (1.15) can be compactly written as

(1.17) $$\frac{\mathrm{d}\boldsymbol{B}}{\mathrm{d}t} = \boldsymbol{\beta} \times \boldsymbol{B} \, ,$$

where these vectors have components

(1.18) $$\boldsymbol{B} = \boldsymbol{i}u + \boldsymbol{j}v + \boldsymbol{k}w = [u, v, w] \, ,$$

(1.19) $$\boldsymbol{\beta} = \boldsymbol{i}\chi + \boldsymbol{k}\Delta = [-\chi, 0, \Delta] \, ,$$

and we have omitted relaxation, for the moment [4]. Equation (1.17) has the geometric interpretation of a vector of constant length $\boldsymbol{B}$, the Bloch vector, precessing about an effective field $\boldsymbol{\beta}$ (fig. 1). In the nuclear-magnetic-resonance

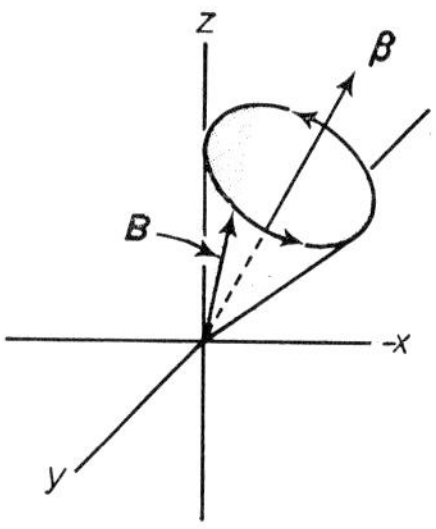

Fig. 1. – The Bloch vector $\boldsymbol{B}$ in its precessional motion about the effective field $\boldsymbol{\beta}$ provides a simple geometrical representation of the time-dependent behaviour of the molecule–optical-field interaction, following directly from Schrödinger's equation. $\mathrm{d}\boldsymbol{B}/\mathrm{d}t = \boldsymbol{\beta} \times \boldsymbol{B}$, $\boldsymbol{B}(t) = [u(t), v(t), w(t)]$, $\boldsymbol{\beta} = [-\chi, 0, \Delta]$.

case formulated by Bloch [5], the spins precess in real space about an effective magnetic field, consisting of a static and an oscillating part. In the electric dipole or optical case considered here, the precession is not in real space, but is only a geometric interpretation of a mathematical result. Relaxation may be included simply in (1.17) if we let $T = T_2 = T_1$, and thus the Bloch vector shrinks with time as $\exp[-t/T]$. In general, the Bloch vector will not only be a function of z and t but of other variables such as molecular velocity $\boldsymbol{v}$, namely $\boldsymbol{B}(\boldsymbol{v}, z, t)$.

The solutions to eq. (1.15), which were discussed initially by TORREY [6], are of the form

$$M(t) = A \exp[-at] + B \exp[-bt] \cos st + C \exp[-bt] \sin st + D, \tag{1.20}$$

where $M(t) = u(t)$, $v(t)$ or $w(t)$. The D-term expresses steady-state behavior and the other terms are transient in nature. The constants cannot be analytically obtained, in general, except for certain specialized cases such as

1) $\Delta = 0$,
2) $T_1 = T_2$,
3) $\chi \gg 1/T_1$ and $\chi \gg 1/T_2$.

1·3. *Polarization and field equations.* – The light wave, eq. (1.1), induces in a sample of molecular density N the polarization

$$P(z, t) = N\mu_{12} \exp[i(\Omega t - kz)]\langle\tilde{\varrho}_{12}\rangle + \text{c.c.} \tag{1.21}$$

The bracket $\langle\,\rangle$ denotes an average over the Maxwellian molecular velocity distribution

$$\langle\tilde{\varrho}_{12}\rangle = \frac{1}{ku\sqrt{\pi}}\int_{-\infty}^{\infty}\tilde{\varrho}_{12} \exp[-(\Delta/ku)^2]\, \mathrm{d}\Delta, \tag{1.22}$$

where Δ is given by (1.12) and u is the r.m.s. velocity, not to be confused with the Bloch-vector component. Note that $\tilde{\varrho}_{12} = (u + iv)/2$. This polarization, in turn, is the origin of a signal field

$$E_s(z, t) = E_{12}(z, t) \exp[i(\Omega t - kz)] + \text{c.c.}, \tag{1.23}$$

which can be calculated from Maxwell's equations

$$\frac{\partial E_{12}}{\partial z} = -2\pi i k N \mu_{12}\langle\tilde{\varrho}_{12}\rangle. \tag{1.24}$$

In eq. (1.24) only the lowest-order terms of the wave equation

$$\frac{\partial^2 E(z, t)}{\partial z^2} - \frac{1}{c^2}\frac{\partial^2 E(z, t)}{\partial t^2} = \frac{4\pi}{c^2}\frac{\partial^2 P(z, t)}{\partial t^2} \tag{1.25}$$

have been retained, the so-called slowly varying envelope approximation.

1·4. *Optical nutation. Simplified case.* – Consider now the transient effect that results when a molecular sample is suddenly exposed to intense resonant laser light. Molecules begin to execute an alternating absorption and re-emission of laser light as they are coherently driven between upper and lower states (fig. 2). The effect has been called « optical nutation » [7] by analogy with Torrey's transient « spin nutation » [6].

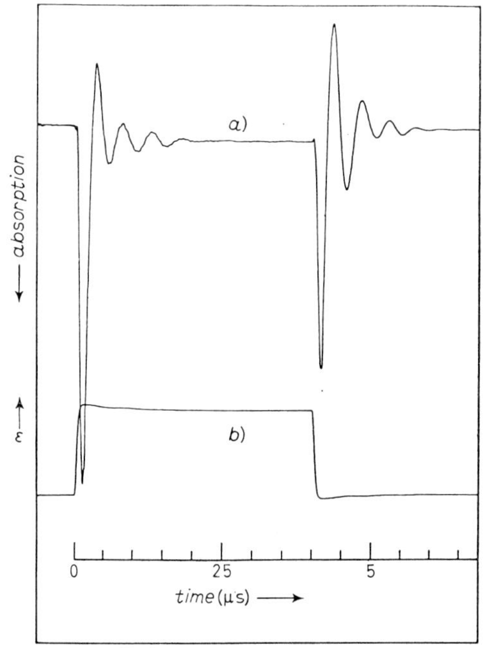

Fig. 2. – The optical nutation effect in $^{13}CH_3F$ following a Stark pulse of amplitude $\varepsilon = 35$ V/cm (from ref. [2]).

To illustrate the solution of the Bloch equations for this effect, we first simplify the problem by omitting the decay terms in (1.15), so that

(1.26a) $$\dot{u} + \Delta v = 0 .$$

(1.26b) $$\dot{v} - \Delta u - \chi w = 0 ,$$

(1.26c) $$\dot{w} + \chi v = 0 .$$

These are easily solved to give in the regime $t > 0$

(1.27a) $$u(t) = \frac{\Delta \chi w(0)}{\beta^2} (\cos \beta t - 1) ,$$

(1.27b) $$v(t) = \frac{\chi w(0)}{\beta} \sin \beta t ,$$

(1.27c) $$w(t) = w(0) \left[1 + \frac{\chi^2}{\beta^2} (\cos \beta t - 1) \right],$$

where the initial conditions at time $t = 0$

$$B(0) = [0,\ 0,\ w(0)] \tag{1.28}$$

have been utilized. Equation (1.27) states that the amplitude of the Bloch vector

$$B(t) = [u^2(t) + v^2(t) + w^2(t)]^{\frac{1}{2}} = w(0) \tag{1.29}$$

is preserved at all times, and that it precesses about the effective field β of (1.19) with a frequency

$$\beta = \sqrt{\Delta^2 + \chi^2}. \tag{1.30}$$

This motion is especially easy to visualize for the case of exact resonance, when $\Delta = 0$, because $u(t) = 0$ and the Bloch vector executes a right-angle rotation about $\beta = \chi$; see fig. 1.

We now apply the results of subsect. **1**·3. According to (1.24), the amplitude of the signal field which exits a sample of length L is

$$E_{12}(L, t) = -2\pi i k N L \mu_{12} \langle \tilde{\varrho}_{12} \rangle\,, \tag{1.31}$$

and the Doppler-averaged density matrix is given by (1.22)

$$\begin{aligned}\langle \tilde{\varrho}_{12} \rangle &= \frac{1}{ku\sqrt{\pi}} \int\limits_{-\infty}^{\infty} \frac{1}{2}(u + iv) \exp[-(\Delta/ku)^2]\,\mathrm{d}\Delta \approx \\ &\approx \frac{i\chi w(0)}{ku\sqrt{\pi}} \exp[-(\Delta_1/ku)^2] \int\limits_0^{\infty} \frac{\sin\sqrt{\Delta^2 + \chi^2}\,t}{\sqrt{\Delta^2 + \chi^2}}\,\mathrm{d}\Delta \approx \\ &\approx \frac{i\sqrt{\pi}}{2ku}\,\chi w(0) \exp[-(\Delta_1/ku)^2] J_0(\chi t)\,.\end{aligned} \tag{1.32}$$

For simplicity, we have assumed that the excitation is centered near the Doppler peak at $\Delta = \Delta_1$, so that the Bloch component $\langle u \rangle \sim 0$. Furthermore, we assume that the band width excited is narrow compared with the Doppler width ku, so that the Gaussian may be taken outside the integral. Combining (1.31) and (1.32), we obtain for the signal field (1.23)

$$E_s(L, t) = \frac{2\pi^{\frac{3}{2}}}{u} (NL)\, \chi w(0) \mu_{12} \exp[-(\Delta_1/ku)^2] J_0(\chi t) \cos(\Omega t - kL)\,. \tag{1.33}$$

The transmitted laser beam thus exhibits a slow oscillation, of frequency $\sim \chi$, that is expressed by the zeroth-order Bessel function $J_0(\chi t)$ and reflects the rate that molecules are driven between lower and upper states. This is essentially the result that is obtained in a more complete treatment where damping is not neglected, the case we will consider next. Figure 2 shows the « nutation » effect which displays the Bessel-function ringing behavior [2].

Nutation with damping. We now give the results of a similar calculation [7, 8] where the damping terms in (1.15) are not neglected. Since the Bloch equations did not yield to an analytical solution in general, we need to assume that $T = T_2 = T_1$. We find in a straightforward manner that for times $t > 0$

$$(1.34a)\qquad u(t) = \exp[-t/T]\left\{u(0) - \Delta\left[v(0) - \frac{\chi w^0/T}{\chi^2+\Delta^2+1/T^2}\right]\frac{\sin\beta t}{\beta} + \right.$$
$$\left. + \Delta\left[\Delta u(0) + \chi w(0) - \frac{\chi w^0/T^2}{\chi^2+\Delta^2+1/T^2}\right]\frac{\cos\beta t - 1}{\beta^2} + \frac{\Delta\chi w^0}{\chi^2+\Delta^2+1/T^2}\right\} -$$
$$- \frac{\Delta\chi w^0}{\chi^2+\Delta^2+1/T^2},$$

$$(1.34b)\qquad v(t) = \exp[-t/T]\left\{\left[v(0) - \frac{\chi w^0/T}{\chi^2+\Delta^2+1/T^2}\right]\cos\beta t + \right.$$
$$\left. + \left[\Delta u(0) + \chi w(0) - \frac{\chi w^0/T^2}{\chi^2+\Delta^2+1/T^2}\right]\frac{\sin\beta t}{\beta}\right\} + \frac{\chi w^0/T}{\chi^2+\Delta^2+1/T^2},$$

$$(1.34c)\qquad w(t) = \exp[-t/T]\left\{w(0) - w^0 - \chi\left[v(0) - \frac{\chi w^0/T}{\chi^2+\Delta^2+1/T^2}\right]\frac{\sin\beta t}{\beta} + \right.$$
$$\left. + \chi\left[\Delta u(0) + \chi w(0) - \frac{\chi w^0/T^2}{\chi^2+\Delta^2+1/T^2}\right]\frac{\cos\beta t - 1}{\beta^2} + \frac{\chi^2 w^0}{\chi^2+\Delta^2+1/T^2}\right\} +$$
$$+ w^0\left(1 - \frac{\chi^2}{\chi^2+\Delta^2+1/T^2}\right).$$

Here, the initial conditions $u(0)$, $v(0)$ and $w(0)$ at $t = 0$ are not assumed to be zero and we recall that $w^0 = \varrho_{22}^0 - \varrho_{11}^0$ is the occupation probability difference in the absence of external radiation. The last term in eqs. (1.34*a*), (1.34*b*) and (1.34*c*) is the steady-state value; it will be of interest in the case of steady-state preparation. The transient solutions will be of interest for the case of pulse preparation, as in an echo experiment.

Proceeding as before, we adopt the initial conditions (1.28) $B(0) = [0, 0, w(0)]$, and find that the Doppler-averaged density matrix element is

$$(1.35)\qquad \langle\tilde{\varrho}_{12}\rangle = \frac{1}{ku\sqrt{\pi}}\int_{-\infty}^{\infty}\frac{1}{2}(u + iv)\exp[-(\Delta/ku)^2]\,\mathrm{d}\Delta \approx$$
$$\approx \frac{i\sqrt{\pi}}{2ku}\chi w(0)\exp[-(\Delta_1/ku)^2]\exp[-t/T]\cdot$$
$$\cdot\left\{J_0(\chi t) + \frac{2w^0/w(0)}{T\sqrt{\chi^2+1/T^2}}\exp[t/T] - \frac{2w^0/w(0)}{\pi T}\int_0^{\infty}\frac{\mathrm{d}\Delta}{\chi^2+\Delta^2+1/T^2}\cdot\right.$$
$$\left.\cdot\left(\cos\sqrt{\Delta^2+\chi^2}\,t + \frac{\sin\sqrt{\Delta^2+\chi^2}\,t}{T\sqrt{\Delta^2+\chi^2}}\right)\right\}.$$

The leading term of (1.35) is the same result as eq. (1.32), except that it is damped by the factor $\exp[-t/T]$. The second term is the steady-state value of $\langle\tilde{\varrho}_{12}\rangle$, and for the present purpose it can be ignored. The remaining two integrals, which cannot be analytically evaluated, can be estimated by replacing the trigonometric functions by unity; they are smaller than the leading term by $\sim 1/\chi T$ and $\sim 1/(\chi T)^2$, respectively; these terms can be usually neglected since $\chi T \sim 10^2$. The nutation signal field then becomes

$$(1.36)\qquad E_s(L,t) = \frac{2\pi^{\frac{3}{2}}}{u}\cdot (NL)\,\chi w(0)\,\mu_{12}\exp[-(\Delta_1/ku)^2]\exp[-t/T]J_0(\chi t)\cos(\Omega t - kL)\,.$$

1.5. *Free-induction decay.* – Let us assume that a molecular sample is resonantly excited under steady-state conditions by a laser beam and that suddenly the excitation is terminated. Under these conditions, the sample will radiate an intense coherent beam of light, which has been called « optical free-induction decay » [9] (FID) by analogy with the NMR effect first seen by HAHN [10]. For the present purpose we shall assume that the preparative stage ends by switching the molecular-transition frequency out of resonance with the laser frequency. The switching mechanism is the Stark effect and will be discussed in the next section.

The preparative steady-state solutions for the period $t \leqslant 0$ can be derived from the Bloch equations (1.15) by setting the time derivatives equal to zero, and yield

$$(1.37)\qquad u(0) = -\Delta\chi w^0/(\chi^2 T_1/T_2 + \Delta^2 + 1/T_2^2)\,,$$

$$(1.38)\qquad v(0) = (\chi w^0/T_2)/(\chi^2 T_1/T_2 + \Delta^2 + 1/T_2^2)\,,$$

$$(1.39)\qquad w(0) = w^0[1 - (\chi^2 T_1/T_2)/(\chi^2 T_1/T_2 + \Delta^2 + 1/T_2^2)]\,.$$

These are the three time-independent terms of (1.34), but without the restriction that $T_1 = T_2$.

At time $t = 0$, we assume that the molecular-transition frequency has been shifted by $\Delta\omega_{21}$ so that

$$(1.40)\qquad \Delta \to \Delta' = \Delta + \Delta\omega_{21}\,,$$

and that the sample is sufficiently far out of resonance with laser light that we may set $\chi = 0$. The Bloch equations are then of the form

$$(1.41a)\qquad \dot{u} + \Delta' v + u/T_2 = 0\,,$$

$$(1.41b)\qquad \dot{v} - \Delta' u + v/T_2 = 0\,,$$

$$(1.41c)\qquad \dot{w} + (w - w^0)/T_1 = 0\,,$$

and the solutions for $t > 0$ are given by

$$u(t) = [u(0) \cos \Delta' t - v(0) \sin \Delta' t] \exp[-t/T_2], \tag{1.42a}$$

$$v(t) = [u(0) \sin \Delta' t + v(0) \cos \Delta' t] \exp[-t/T_2], \tag{1.42b}$$

$$w(t) = w^0 + [w(0) - w^0] \exp[-t/T_1]. \tag{1.42c}$$

To obtain the field radiated we again take the molecular velocity average expressed by (1.22):

$$\begin{aligned}
\langle\tilde{\varrho}_{12}\rangle &= \frac{i}{2\sqrt{\pi}ku} \cdot \\
&\cdot \int_{-\infty}^{\infty} \exp[-(\Delta/ku)^2][u(0)\sin\Delta' t + v(0)\cos\Delta' t]\exp[-t/T_2]\mathrm{d}\Delta \approx \\
&\approx \frac{i}{2\sqrt{\pi}ku} \exp[-(\Delta_1/ku)^2]\,\chi w^0 \exp[-t/T_2]\cos\Delta\omega_{21}t\, \cdot \\
&\cdot \int_{-\infty}^{\infty} \frac{-\Delta\sin\Delta t + (1/T_2)\cos\Delta t}{\chi^2 T_1/T_2 + \Delta^2 + 1/T_2^2}\,\mathrm{d}\Delta \approx \\
&\approx \frac{i\sqrt{\pi}}{2ku}\exp[-(\Delta_1/ku)^2]\chi w^0 \exp[-t/T_2(1+\sqrt{\chi^2 T_1 T_2 + 1})]\, \cdot \\
&\qquad \cdot \left(\frac{1}{\sqrt{\chi^2 T_1 T_2 + 1}} - 1\right)\cos\Delta\omega_{21}t\,.
\end{aligned} \tag{1.43}$$

Here, $\langle u \rangle = 0$ and we have assumed that the Doppler factor can be taken outside the integral. The field amplitude (1.24) associated with the induced polarization is

$$E_{12} = -i(2\pi\Omega/c)\,NL\mu_{12}\langle\tilde{\varrho}_{12}\rangle, \tag{1.44}$$

whereas the total field, including that of the laser, is

$$E_T = (E_{12} + E_0/2)\exp[i(\Omega t - kz)] + \text{c.c.} \tag{1.45}$$

The intensity, therefore, contains a cross-term or beat

$$(E^2)_{\text{beat}} = 2E_{12}E_0 = E_0 Q_{12}(t)\cos\Delta\omega_{12}t, \tag{1.46a}$$

where

$$\begin{aligned}
Q_{12}(t) &= 2\pi^{\frac{3}{2}} NL\mu_{12}^2 E_0 w^0 \left(\frac{1}{\sqrt{\chi^2 T_1 T_2 + 1}} - 1\right)\cdot \\
&\cdot \exp[-(\Delta_1/ku)^2]\exp[-t/T_2(1+\sqrt{\chi^2 T_1 T_2 + 1})]\,.
\end{aligned} \tag{1.46b}$$

The decay behavior of (1.46) has two contributions [8, 9]: 1) a homogeneous part with time constant T_2 and 2) an inhomogeneous part with time constant

$T_2/\sqrt{\chi^2 T_1 T_2 + 1}$ that reflects the velocity band width excited during steady-state preparation. At moderately high laser intensities (a few W/cm^2), the inhomogeneous dephasing can be dominant, the free-induction signal will then decay rapidly and appear modulated at a frequency given by the Stark shift $\Delta\omega_{21}$. The subtle behavior near the time origin is discussed elsewhere [19]. An experimental demonstration of the FID effect is shown in fig. 3.

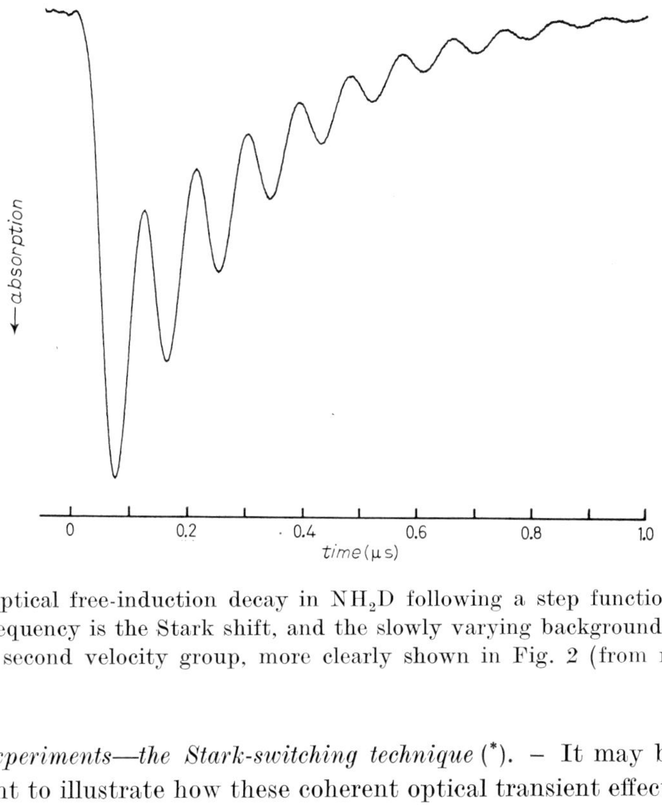

Fig. 3. – Optical free-induction decay in NH_2D following a step function Stark field. The beat frequency is the Stark shift, and the slowly varying background is a nutation signal of a second velocity group, more clearly shown in Fig. 2 (from ref. [9]).

1'6. *Experiments—the Stark-switching technique* (*). – It may be refreshing at this point to illustrate how these coherent optical transient effects have been

(*) *Note added in proofs.* – A new and more universal method which complements the Stark-switching technique of subsect. **1'6** and which extends coherent transient measurements into the visible-ultraviolet region was just reported by R. G. BREWER and A. Z. GENACK: *Phys. Rev. Lett.*, **36**, 959 (1976). In this case, the laser frequency is switched instead of the sample's transition frequency. A stable tunable c.w. dye laser is frequency switched by means of an intracavity electro-optic ADP crystal that is driven by a sequence of low-voltage pulses. The laser frequency follows the refractive-index variations induced in the ADP crystal, and a sample exposed to this radiation exhibits coherent absorption or emission transients. Hence, the experiment is electronically controlled and in such a way that the advantages inherent in the Stark technique are preserved here as well. In addition, with the broad tuning range available in a dye laser, coherent transient phenomena can now be observed with ease in a large number of optical transitions in various atomic, molecular and solid-state systems.

observed in our laboratory at IBM. The underlying idea rests on the fact that gaseous molecules can exhibit exceedingly sharp optical resonances. Laser light, which is essentially monochromatic, will be strongly absorbed only when its frequency closely matches (to one part in 10^8 or 10^9) the molecular-resonance frequency. If the molecular-transition frequency is shifted slightly by some mechanism, the absorption can easily be switched on or off. The switching mechanism is the *Stark effect* and allows the molecular resonance to be *tuned* to or away from the laser frequency by subjecting the molecules to a pulsed d.c. electric field. Instead of applying pulses of laser light, which are difficult to control, the energy level spacing of the molecule is pulsed while the laser's intensity and frequency remain fixed in time. The technique [2] shown in fig. 4 has been demonstrated in molecules in the infra-red spectral region (10 μm) using a stable continuously operating CO_2 laser.

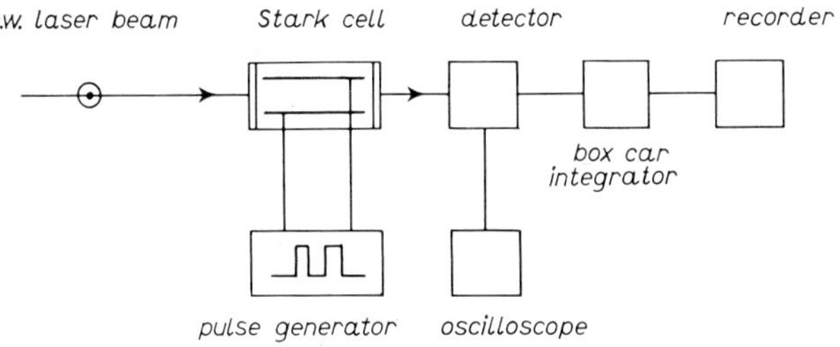

Fig. 4. – Method for observing coherent optical transients following a sequence of Stark pulses (from ref. [2]).

The first-order Stark splitting [11]

$$\Delta w_1 = -\mu \varepsilon MK/J(J+1) \tag{1.47}$$

occurs in symmetric-top molecules such as methyl fluoride, CH_3F, because of the interaction of the molecule's permanent electric-dipole moment μ with a static external electric field ε. The molecular rotational quantum states are designated by J and K, while M labels the space-quantized levels. In fig. 5, the laser photon energy matches the splitting of two of these levels, but if an electric pulse is applied, the upper and lower states will shift by different amounts, so that laser light can no longer be absorbed.

Since the electric pulses (Stark pulses) are generated by conventional electronics, the pulse sequence, amplitude, duration and shape are easily varied—a technique that allows the study of an entire class of coherent optical transient phenomena. This is a much simpler and more versatile method than generating short, very precisely timed laser pulses. Furthermore, since the c.w. laser itself is highly stable, the transient signals derived from it are highly reproducible.

The method may also be applied in other spectral regions, at microwave or visible-ultraviolet frequencies for example, and to atoms or solids at low temperature.

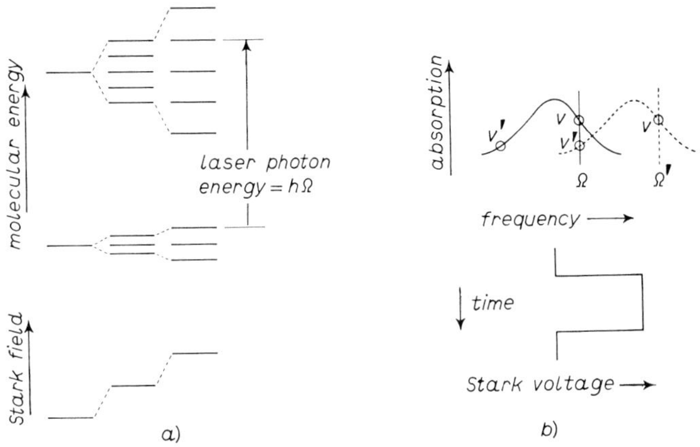

Fig. 5. – The Stark effect is illustrated schematically in *a*) for a molecule such as CH_3F that possesses a permanent dipole moment and, therefore, exhibits a first-order Stark effect. (The level spacings are not drawn to scale as the Stark splittings are generally much smaller than the quantum levels separated by an optical energy.) In *b*) the Stark-switching principle devised by BREWER and SHOEMAKER for observing coherent optical transients is indicated. The sudden application of a d.c. electric field (a pulse) shifts the molecular-level spacing because of the Stark effect and causes the transition frequency to jump from the solid to the dashed curve while the laser frequency Ω remains fixed. The absorption line shape as shown is Doppler broadened. Hence, molecules of velocity v (the component along the laser beam) that were initially excited by laser light are switched out of resonance and spontaneously emit a coherent beam of light at Ω' (fig. 3), the FID signal. The velocity group v' that is suddenly switched into resonance exhibits an optical ringing effect (fig. 2), the nutation effect. When two Stark pulses are applied in succession, the velocity group v' spontaneously emits a photon echo (fig. 9).

The Stark-switching technique has now allowed the observation of optical nutation [2], photon echoes [12] involving two [2] or multiple pulses (stimulated [13] and Carr-Purcell echoes [14]), optical free-induction decay (FID) [9], coherent Raman beats [15], optical adiabatic fast passage [16, 17], the optical analogue of spin locking [18], FID interference pulses [19] and quantum beats [20]. With the exception of perhaps Raman beats, all of these effects are the optical analogues of spin transients that have been studied by nuclear magnetic resonance (NMR) over the past 25 years [1, 6, 10, 21, 22].

Let us review, by referring to fig. 5, how a few of these coherent transients may be observed when a two-level system is suddenly subjected to a Stark pulse. Initially, molecules having a longitudinal velocity v are excited in steady state by laser light of frequency Ω, thereby preparing the transition levels in coherent superposition. When a pulse appears, this velocity group is no longer in resonance, but, because of its preparation, it will freely radiate an intense coherent infra-red beam—the free-induction decay signal (FID). At the same time, a second velocity group v' may be switched into resonance and will alternately absorb and emit laser radiation. This is the optical nutation effect. When the pulse terminates, the group v is suddenly excited and it too begins to nutate, while the second group v' now emits a FID signal. If two pulses are applied, it is possible to observe a photon echo, the optical analogue of a spin echo. The optical nutation, FID and echo effects are illustrated in fig. 2, 3 and 9, respectively.

Transient light signals that are emitted by molecules switched out of resonance propagate in the forward direction because of the preparative step ($\boldsymbol{k}$-vector conservation), and are monitored together with the transmitted laser beam by a photodetector. Heterodyne detection is possible, as in the FID and photon echo experiments, since the emission signal is Stark shifted from the laser frequency; heterodyne detection increases the emission signal amplitude 1000-fold and, in the infra-red, enhances detection sensitivity. In fact, incoherent infra-red emission is many orders of magnitude weaker and is rarely observed.

Most of the experiments discussed here involve an infra-red vibration-rotation transition of $^{13}CH_3F$. The transition is a fundamental ν_3-band $R(4)$-line, $(J, K) = (4, 3) \rightarrow (5, 3)$, that overlaps the $P(32)$ CO_2 laser line at 1035.474 cm^{-1} (9.66 μm). The CO_2 laser, which is described elsewhere [23], is free running and possesses high amplitude and frequency stability. The output is ~ 1 W and the beam is expanded by a Galilean telescope to ~ 1 cm diameter to lengthen the molecule-optical interaction time. Furthermore, the light is linearly polarized, permitting $\Delta M = 0$ selection rules, when its orientation is parallel to the d.c. Stark field, or $\Delta M = \pm 1$ transitions, when it is perpendicular to the Stark field.

In the above discussion of FID we have considered coherent emission in a collection of two-level systems that are not degenerate. A new effect [19] arises, however, when the transition levels are degenerate during the steady-state preparative stage. The emission that results after a step function Stark field is applied is no longer a simple decay but instead appears as a train of sharp pulses regularly spaced in time due to a repetitive interference. The effect is shown in fig. 6*a*) for $^{13}CH_3F$, where the selection rule is $\Delta M = 0$. This situation arises because an entire set of infra-red transitions within the Stark split manifold are initially prepared. This emission, which beats with the laser, produces a heterodyne beat spectrum consisting of a set of regularly spaced frequencies that is the Fourier transform of the slowly decaying pulse progression observed. By means of a spectrum analyzer, the time behavior of fig. 6*a*) can be Fourier transformed into the spectrum given in fig. 6*b*).

We thus demonstrate what is the optical analogue of the well-known NMR method of high-resolution pulse Fourier spectroscopy [24]. In this simple example, the regular frequency interval between spectral lines is due to the first-order Stark shift, but, with other molecules, hyperfine structure and other spectral features might be investigated, where very high resolution is required. In fig. 6*b*) the line width is ~ 0.5 MHz, considerably narrower than the 66 MHz Doppler width. The usual advantages of pulse Fourier spectroscopy apply,

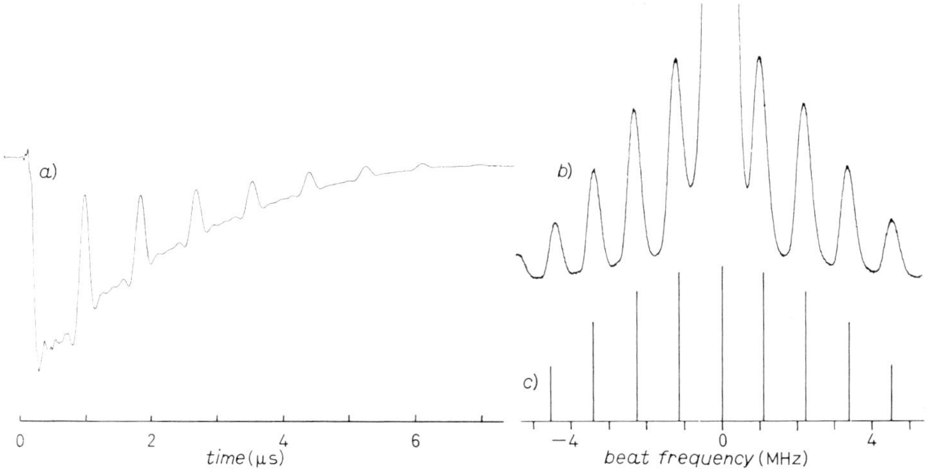

Fig. 6. – *a*) Free-induction decay interference pulses in $^{13}CH_3F$, *b*) the corresponding beat spectrum or its Fourier transform, and *c*) the predicted spectrum (from ref. [19]).

namely 1) an enhanced signal-to-noise ratio compared to steady-state techniques because of averaging repetitively produced signals, 2) rapid data acquisition and 3) the possibility of monitoring *spectra* as a function of decay time in a variety of coherent transient experiments.

These interference pulses superficially resemble mode locking of a laser, because several evenly spaced frequency components are involved. However, in this case the regular frequency spacing is inherent in the transitions involved and no interaction between them is needed. The detailed behavior of the pulse train agrees, in fact, with a FID theory [19] that assumes the transitions to be uncoupled. Using eq. (1.46*a*), we can show that, for $\Delta M = 0$ selection rules, the photodetector will monitor cross-terms in the optical intensity of the form

$$(E^2)_{\text{beat}} = \sum_{[1,2]} E_0 Q_{12}(t) \cos \Delta\omega_{12} t = E_0 Q \frac{\sin(9\delta\varepsilon t/2)}{\sin(\delta\varepsilon t/2)}, \tag{1.48}$$

where $\delta\varepsilon$ is the frequency interval between two neighboring lines of fig. 6*b*) due to a d.c. Stark field ε and the number 9 signifies the number of transitions involved. For simplicity, we have taken all transition matrix elements to be equal. Equation (1.48) shows the characteristics displayed in fig. 6*a*), *i.e.* a pulse interval given by $T = 2\pi/\delta\varepsilon$ and a pulse width $\Delta t = 2\pi/(9\delta\varepsilon) = T/9$, so that their ratio $T/\Delta T = 9$ is just the number of transitions contributing. The more rigorous theory gives $T/\Delta T = 6.25$. The time scale becomes compressed as the number of transitions, the Stark tuning rate δ or the Stark field ε increases. It is tempting to consider the effect of higher Stark fields. At the present time, pulse widths of ~ 10 ns have been realized with a modest Stark field of ~ 350 V/cm. At a field of $\sim 100\,000$ V/cm, coherent infra-red pulses of ~ 50 ps might be achieved under suitable conditions.

2. – Photon echoes and molecular collisions [25].

2·1. *Conventional echo theory.* – The inhomogeneous dephasing discussed in the FID effect is a reversible phenomenon. Coherent electromagnetic energy stored momentarily in a sample can be partially recovered if the dephasing process is reversed in time. The spin echo [21], which set a precedent for this class of phenomena, provides such a mechanism. It was discovered by HAHN, and the echo concept has been extended to the optical region by ABELLA, KURNIT and HARTMANN [12].

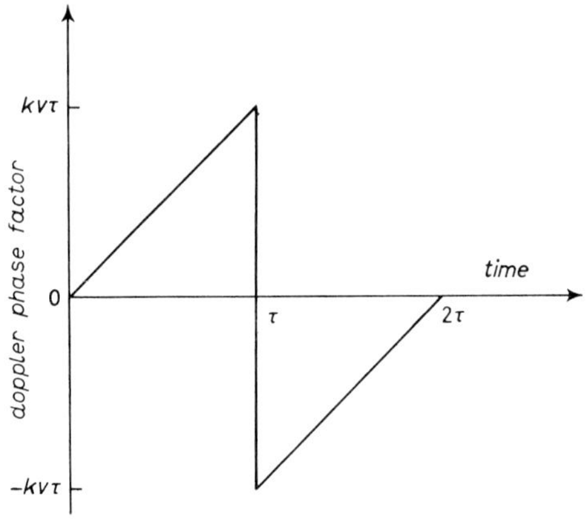

Fig. 7. – The relative Doppler phase of molecules with longitudinal velocity v during a two-pulse sequence. The molecules are all in phase because of a $\pi/2$ pulse at $t = 0$. The phase then advances but reverses sign at $t = \tau$ by application of a π pulse. At $t = 2\tau$, the molecular induced dipoles have all rephased again and an echo signal is produced (from ref. [25]).

To understand the dephasing-rephasing echo cycle in its simplest terms, consider fig. 7. This shows the relative Doppler phase $\boldsymbol{k}\cdot\boldsymbol{v}t$ of a gaseous molecule moving with velocity $\boldsymbol{v}$, where $\boldsymbol{k}$ is the radiation propagation vector and t is the elapsed time after excitation by an optical pulse.

Assume another pulse applied to the sample at time τ can reverse the sign of the phase from $\boldsymbol{k}\cdot\boldsymbol{v}t$ to $-\boldsymbol{k}\cdot\boldsymbol{v}t$. Since the phase continues to increase by an amount $\boldsymbol{k}\cdot\boldsymbol{v}\,\Delta t$ in a time Δt, we see that, at time $t=2\tau$, the initial phase at time $t=0$ will be reproduced. Different velocity groups behave in the same way, although they traverse different phase trajectories, and, therefore, all velocity groups will come back into phase at $t=2\tau$. When this happens, the macroscopic sample can coherently radiate a spontaneous burst of light—the photon echo.

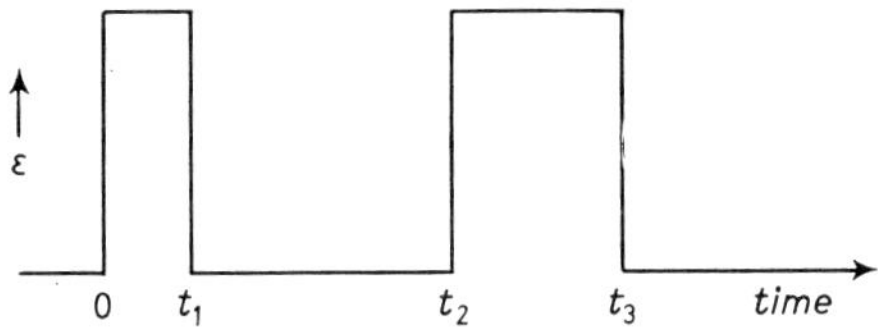

Fig. 8. – Echo pulse sequence showing the Stark-field amplitude *vs.* time. The two pulses occur over the intervals $t=0$ to t_1, and $t=t_2$ to t_3.

We shall first present the conventional echo theory [21] for the Stark-pulse sequence shown in fig. 8. A comparison of these results with the geometrical Bloch-vector description [4] will then be made. The role of elastic and inelastic molecular collisions in photon echo theory will then be treated and compared with recent experiments.

Assume that the molecular gas sample is switched into resonance with the laser beam during the Stark-pulse intervals $0\div t_1$ and $t_2\div t_3$ of fig. 8 and is out of resonance at other times. The tuning parameter (1.12) in the relevant time regimes is

$$0<t<t_1 \text{ and } t_2<t<t_3:\ \Delta\,, \tag{2.1a}$$

$$t_1<t<t_2 \text{ and } t_3<t\quad:\ \Delta'=\Delta+\Delta\omega_{21}\,, \tag{2.1b}$$

where $\Delta\omega_{21}$ is the Stark shift.

$0<t<t_1$. For a sufficiently brief pulse, $t_1\ll T_2$, T_1, relaxation terms in the Bloch equations (1.15) can be dropped. Furthermore, the w^0-term does not contribute to echo formation and can be removed, so that (1.15) reduces to

$$\dot{u}+\Delta v=0\,, \tag{2.2a}$$

$$\dot{v}-\Delta u-\chi w=0\,, \tag{2.2b}$$

$$\dot{w}+\chi v=0\,. \tag{2.2c}$$

The solutions, subject to the initial conditions

$$B(0)=[0,\,0,\,w(0)]\,, \tag{2.3}$$

are given by the nutation results (1.34) when we set $T=\infty$ and $w^0=0$. This procedure yields

$$u(t_1)=\frac{\Delta\chi w(0)}{\beta^2}(\cos\theta_{10}-1)\,, \tag{2.4a}$$

$$v(t_1)=\frac{\chi w(0)}{\beta}\sin\theta_{10}\,, \tag{2.4b}$$

$$w(t_1)=w(0)+\frac{\chi^2 w(0)}{\beta^2}(\cos\theta_{10}-1)\,, \tag{2.4c}$$

where the pulse areas are

$$\begin{cases}\theta_{10}=\sqrt{\chi^2+\Delta^2}\,t_1\,,\\ \theta_{32}=\sqrt{\chi^2+\Delta^2}\,(t_3-t_2)\,,\end{cases} \tag{2.5}$$

and we recall that

$$\beta=\sqrt{\chi^2+\Delta^2}.$$

$t_1<t<t_2$. Over this interval, it is assumed that the sample is sufficiently far out of resonance that $\chi\sim 0$, and the Bloch equations become

$$\dot{u}+\Delta' v+u/T_2=0\,, \tag{2.6a}$$

$$\dot{v}-\Delta' u+v/T_2=0\,, \tag{2.6b}$$

$$\dot{w}+w/T_1=0\,. \tag{2.6c}$$

The w^0-term has been omitted in (2.6c) again. These are the same equations used to describe FID, and the solutions, evaluated at $t=t_2$, are the same as (1.42):

$$u(t_2)=[u(t_1)\cos\Delta'(t_2-t_1)-v(t_1)\sin\Delta'(t_2-t_1)]\exp[-(t_2-t_1)/T_2]\,, \tag{2.7a}$$

$$v(t_2)=[u(t_1)\sin\Delta'(t_2-t_1)+v(t_1)\cos\Delta'(t_2-t_1)]\exp[-(t_2-t_1)/T_2]\,, \tag{2.7b}$$

$$w(t_2)=w(t_1)\exp[-(t_2-t_1)/T_1]\,, \tag{2.7c}$$

but subject to the initial conditions

$$B(t_1)=[u(t_1),\,v(t_1),\,w(t_1)]\,. \tag{2.8}$$

$t_2<t<t_3$. For the second pulse, the solutions of (2.2) are again given by the nutation results (1.34) where $T=\infty$, $w^0=0$ and the initial condition is

$$B(t_2)=[u(t_2),\,v(t_2),\,w(t_2)]\,. \tag{2.9}$$

We find at $t = t_3$ that

(2.10a) $$u(t_3) = u(t_2) - \frac{\Delta}{\beta} v(t_2) \sin\theta_{32} - \frac{2\Delta}{\beta^2} [\Delta u(t_2) + \chi w(t_2)] \sin^2(\theta_{32}/2)\,,$$

(2.10b) $$v(t_3) = v(t_2) \cos\theta_{32} + \frac{1}{\beta} [\Delta u(t_2) + \chi w(t_2)] \sin\theta_{32}\,,$$

(2.10c) $$w(t_3) = w(t_2) - \frac{\chi}{\beta} v(t_2) \sin\theta_{32} - \frac{2\chi}{\beta^2} [\Delta u(t_2) + \chi w(t_2)] \sin^2(\theta_{32}/2)\,.$$

$t_3 < t$. For the period following the second pulse, the FID eqs. (2.7) yield

(2.11a) $$u(t) = [u(t_3) \cos \Delta'(t - t_3) - v(t_3) \sin \Delta'(t - t_3)] \exp[-(t - t_3)/T_2]\,,$$

(2.11b) $$v(t) = [u(t_3) \sin \Delta'(t - t_3) + v(t_3) \cos \Delta'(t - t_3)] \exp[-(t - t_3)/T_2]\,,$$

(2.11c) $$w(t) = w(t_3) \exp[-(t - t_3)/T_1]\,.$$

The earlier solutions (2.10) and (2.7) when inserted in (2.11) result in terms that contain as a factor $\cos \Delta'(t - 2\tau)$, where

(2.12) $$2\tau = t_3 + t_2 - t_1\,.$$

It is clear that these terms are responsible for echo formation at time

(2.13) $$t = 2\tau\,,$$

since the Doppler phase vanishes. The other terms such as $\sin \Delta'(t - 2\tau)$ are zero at $t = 2\tau$ or do not rephase and need not be considered. Another point is that the echo terms of $u(t)$ are an odd function in Δ, so that, if the optical excitation selects molecules near the Doppler peak, we may consider the Doppler-averaged quantity to be

(2.14) $$\langle u_c \rangle \sim 0\,.$$

The subscript c designates that only rephasing or echo terms are included. For this case, the polarization (1.21) and the echo field (1.23) that it radiates at time $t = 2\tau$ are due solely to the v_c-component of (2.11b). It is of the form

(2.15) $$\langle v_c(t) \rangle = -\chi^3 w(0) \exp[-t/T_2] \cos \Delta\omega_{21}(t - 2\tau) \cdot \cdot \left\langle \frac{1}{\beta^3} \sin\theta_{10} \sin^2(\theta_{32}/2) \cos \Delta(t - 2\tau) \right\rangle.$$

Keeping in mind that $\langle \tilde{\varrho}_{12} \rangle = \frac{1}{2} \langle u + iv \rangle$, we find that the echo field amplitude

$$E_{12}(L, t) = -2\pi i k N L \mu_{12} \langle \tilde{\varrho}_{12}(t) \rangle$$

and the laser field (1.1) add to give (1.45). As in FID, the intensity contains a beat term

$$(2.16)\qquad [E_c^2(t)]_{\text{beat}} = 2E_{12}(t)E_0 = 2\pi\hbar kNL\chi^4 w(0)\exp[-t/T_2]\cos\Delta\omega_{21}(t-2\tau)\cdot \cdot\left\langle\frac{1}{\beta^3}\sin\theta_{10}\sin^2(\theta_{32}/2)\cos\Delta(t-2\tau)\right\rangle,$$

where the beat frequency is the Stark shift $\Delta\omega_{21}$. The echo signal (*) reaches a maximum value at time $t = 2\tau$ and (2.16) becomes

$$(2.17)\qquad [E_c^2(t=2\tau)]_{\text{beat}} = 2\pi\hbar kNL\chi^4 w(0)\exp[-t/T_2]\left\langle\frac{1}{\beta^3}\sin\theta_{10}\sin^2(\theta_{32}/2)\right\rangle.$$

The envelope function of the echo signal therefore decays with a time constant T_2 that is independent of the more rapid inhomogeneous Doppler dephasing encountered in FID. Equation (2.17) also indicates that the signal will be largest when the pulse areas are $\theta_{10} = \pi/2$ and $\theta_{32} = \pi$.

An example [2] of the photon echo effect is shown in fig. 9 for $^{13}CH_3F$. The echo is shown as the third optical pulse in the upper trace; the first two pulses

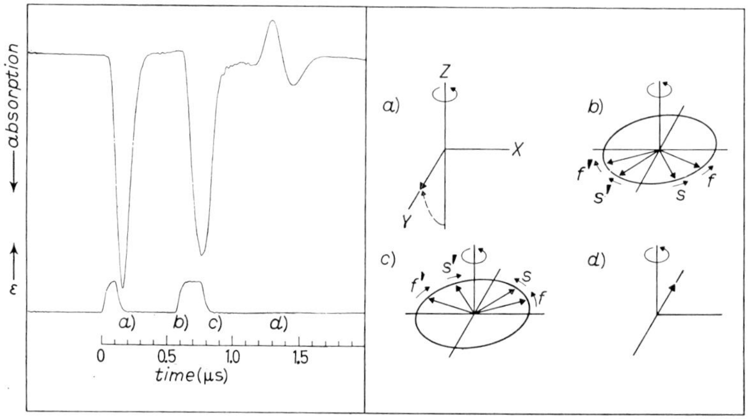

Fig. 9. – Left: A photon echo in $^{13}CH_3F$ is shown as the third optical pulse in the upper trace; the first two optical pulses are nutation signals that accompany the two Stark pulses shown on the lower trace (ref. [2]). The various times labeled *a*) through *d*) of the pulse sequence are described on the right in terms of the Bloch vector's precessional motion.

(*) *Note added in proofs.* – The echo Doppler integral of (2.17) has been analytically evaluated for the limiting cases where either a large or a small fraction of the Doppler line width is prepared. See A. SCHENZLE, S. GROSSMAN and R. G. BREWER: *Phys. Rev. A*, 1891 (1976).

are manifestations of nutation behavior (preparation) that accompany the two Stark pulses shown on the lower trace. The various times labeled *a*) through *d*) of the pulse sequence are described on the right in terms of the Bloch vector's precessional motion. The dephasing-rephasing cycle of the various molecular-velocity groups occurs in the following way. For the case of exact resonance, optical excitation during the first Stark pulse causes the Bloch vector to precess about the X-axis. At the end of the pulse *a*), the Bloch vector stops, in this case in the (X-Y)-plane. Over the time interval *a*)-*b*), the vector decomposes into a set of components that precess about the Z-axis at different speeds, fast (f) and slow (s). Each component vector corresponds to one molecular-velocity group emitting at its own unique frequency because of the Doppler shift. The second optical excitation in the interval *b*)-*c*) causes the « pancake » of vectors to again rotate about the X-axis, thereby placing the slow vectors ahead of the fast ones. Following the second pulse, these vectors continue to rotate in the same direction about Z, as before. At *d*) the fast vectors catch up with the slow ones, and the macroscopic sample now with its induced dipoles all in phase emits a coherent burst of light, the photon echo. Note that the Bloch vector at this stage is reversed in direction from that at *a*). Therefore, while the vectors rephase, they do not retrace their behavior in time.

Many of the features predicted by (2.16) and (2.17) appear to be verified by our experiments at IBM. For example, as fig. 9 illustrates, the echo is observed as a heterodyne beat signal. In addition, the echo signal is observed to be proportional to the number of molecules N, in agreement with (2.17), and, hence, the echo intensity varies as N^2. The signal amplitude also goes through a maximum value when the first and second pulse areas are approximately $\pi/2$ and π, respectively.

However, there is one serious disagreement with the observations and that is in the decay behavior. The echo amplitude for different pulse delay times does not follow the simple exponential decay $\exp[-t/T_2]$ of (2.17) but instead behaves as in the lower curve of fig. 10 [14]. The limiting time dependence is in fact of the form

$$\left\{\begin{array}{ll}\exp[-Kt^3] & \text{for short times},\\ \exp[-\Gamma t] & \text{for long times}.\end{array}\right. \tag{2.18}$$

The reason for the discrepancy with (2.17) is that the traditional echo theory does not allow for elastic collisions between molecular pairs where the linear velocity exhibits a diffusive character through small-angle scattering. The transition frequency of the radiating sample is no longer pure, therefore, but displays a band width, corresponding to the velocity spread produced by elastic collisions. The velocity packets diverge in their phase relationship, causing the echo amplitude to diminish with a characteristic time dependence that distinguishes the velocity diffusion mechanism from other dephasing processes. An echo theory that includes elastic collisions is presented in the next section.

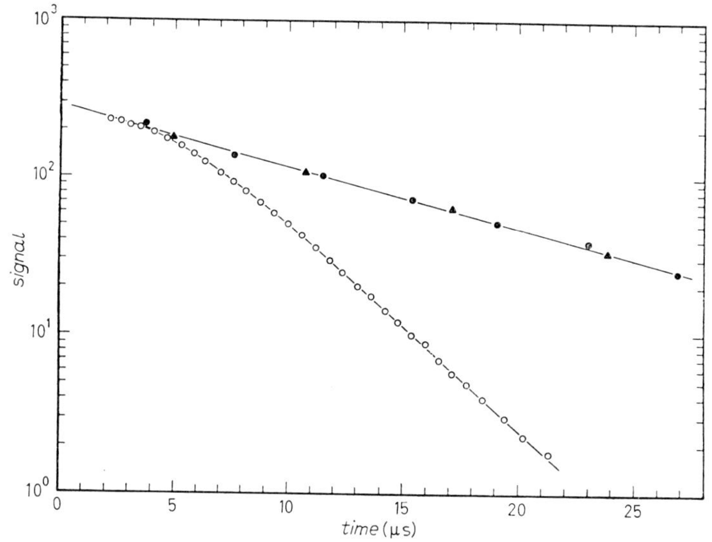

Fig. 10. – Decay behaviour of two coherent optical transients in $^{13}CH_3F$. The upper curve is obtained either by multiple-pulse photone choes (Carr-Purcell echoes, •) or by the delayed nutation technique, ▲, and measures the decay time T_1. The lower curve is due to the two-pulse photon echo, ○; its deviation from the upper curve is due to elastic molecular collisions (from ref. [14, 25]).

2'2. *The effect of elastic collisions.* – To allow for an elastic-collision dephasing mechanism in echo formation, we introduce two new terms in the Bloch equations:

$$\dot{u}(\boldsymbol{v}, t) = -\Delta v(\boldsymbol{v}, t) - (\Gamma_1 + \Gamma)\, u(\boldsymbol{v}, t) + \int \mathrm{d}^3\boldsymbol{v}'\, W(\boldsymbol{v}' \to \boldsymbol{v})\, u(\boldsymbol{v}', t)\,, \tag{2.19a}$$

$$\dot{v}(\boldsymbol{v}, t) = \Delta u(\boldsymbol{v}, t) + \chi w(\boldsymbol{v}, t) - (\Gamma_1 + \Gamma)\, v(\boldsymbol{v}, t) + \int \mathrm{d}^3\boldsymbol{v}'\, W(\boldsymbol{v}' \to \boldsymbol{v})\, v(\boldsymbol{v}', t)\,, \tag{2.19b}$$

$$\dot{w}(\boldsymbol{v}, t) = -\chi v(\boldsymbol{v}, t) - (\Gamma_1 + \Gamma)\, w(\boldsymbol{v}, t) + \int \mathrm{d}^3\boldsymbol{v}'\, W(\boldsymbol{v}' \to \boldsymbol{v})\, w(\boldsymbol{v}', t)\,, \tag{2.19c}$$

which follow from a Boltzmann or quantum-mechanical transport equation [25-27]. (Note that the Bloch-vector component v is not to be confused with molecular velocity $\boldsymbol{v}$.) In contrast to subsect. **2'1**, the Bloch vector $\boldsymbol{B}(\boldsymbol{v}, t)$ for an N-molecule sample is now weighted by the thermal equilibrium velocity distribution $G(\boldsymbol{v})$, where

$$\boldsymbol{B}(\boldsymbol{v}, t) = \boldsymbol{B}'(\boldsymbol{v}, t) \exp[-\Gamma_1 t] = \widetilde{\boldsymbol{B}}(\boldsymbol{v}, t)\, N G(\boldsymbol{v}) \exp[-\Gamma_1 t]\,, \tag{2.20}$$

$\Gamma_1 = 1/T_1$ and the $\exp[-\Gamma_1 t]$ factor will permit us to remove the Γ_1 terms in (2.19). The quantity $-\int d^3\boldsymbol{v}'\, W(\boldsymbol{v}\to\boldsymbol{v}')\boldsymbol{B}(\boldsymbol{v},t) = -\Gamma\boldsymbol{B}(\boldsymbol{v},t)$ expresses the reduction of $\boldsymbol{B}(\boldsymbol{v},t)$ due to elastic collisions associated with the collision kernel $W(\boldsymbol{v}\to\boldsymbol{v}')$, that bring molecules from velocity $\boldsymbol{v}$ to $\boldsymbol{v}'$. The factor Γ is the total rate of elastic collisions. The second term $\int d^3 v'\, W(\boldsymbol{v}'\to\boldsymbol{v})\boldsymbol{B}(v',t)$ represents the inverse process and gives the increase in $\boldsymbol{B}(\boldsymbol{v},t)$ due to elastic collisions that change the velocity from $\boldsymbol{v}'$ to $\boldsymbol{v}$.

In (2.19) we have neglected the possibility of phase-interrupting collisions [26, 27], so that $T_2 = T_1$. That is, for our present purpose, an elastic collision is state independent. This simplification is suggested by our current experiments in the infra-red, although it may not be valid in the visible or ultraviolet region. As before, the source term w^0 is dropped in (2.19*c*), since unprepared molecules that enter the transition states do not contribute to echo formation. We also neglect any spatial variation of $\boldsymbol{B}$ over the laser beam geometry.

To carry out the calculation, a collision kernel $W(\boldsymbol{v}'\to\boldsymbol{v})$ must be selected. *Since the functional time dependence of the results will be independent of the choice of kernel*, we have adopted the Brownian-motion kernel of KEILSON and STORER [28]:

$$W(\boldsymbol{v}'\to\boldsymbol{v}) = \Gamma(\pi\Delta u^2)^{-\frac{3}{2}} \exp[-(\boldsymbol{v}-\alpha\boldsymbol{v}')^2/\Delta u^2], \tag{2.21}$$

where α is a constant close to but less than unity. Equation (2.21) and detailed balancing

$$W(\boldsymbol{v}'\to\boldsymbol{v})\, G(\boldsymbol{v}') = G(\boldsymbol{v})\, W(\boldsymbol{v}\to\boldsymbol{v}') \tag{2.22}$$

impose the constraint that

$$\Delta u^2 = (1-\alpha^2)\, u^2 \approx 2(1-\alpha)\, u^2, \tag{2.23}$$

where Δu is root two times the r.m.s. change in velocity per collision, while u is the most probable speed of the *thermal equilibrium distribution*. This kernel is mathematically simple to treat and the assumption that $\alpha\sim 1$, which corresponds to having very small changes in velocity per collision, will be supported by the experimental data.

Simplified case. To illustrate the physics of the problem, we consider a simplified case where the pulses are sufficiently short that $|\Delta t_1| \ll 1$ and $|\Delta(t_3-t_2)| \ll 1$, while, at the same time, the optical-field strength is large enough that $\chi t_1 \sim 1$ and $\chi(t_3-t_2)\sim 1$. Physically, this corresponds to assuming that the pulses uniformly excite the entire thermal distribution of molecules or that they are all on resonance ($\Delta = 0$).

The solutions are now traced for each period as in the previous treatment.

$0 < t < t_1$. Equations (2.19) become

$$\dot{u}'(\boldsymbol{v}, t) = 0\,, \tag{2.24a}$$

$$\dot{v}'(\boldsymbol{v}, t) = \chi w'(\boldsymbol{v}, t)\,, \tag{2.24b}$$

$$\dot{w}'(\boldsymbol{v}, t) = -\chi v'(\boldsymbol{v}, t)\,, \tag{2.24c}$$

where the system is assumed to be on resonance and relaxation for short pulses can be ignored. The solution at $t = t_1$, subject to the initial condition

$$B'(\boldsymbol{v}, 0) = [0, 0, w'(0)]\,, \tag{2.25}$$

is the same as eq. (2.4) when we set $\Delta = 0$, namely

$$u'(\boldsymbol{v}, t_1) = u'(\boldsymbol{v}, 0) = 0\,, \tag{2.26a}$$

$$v'(\boldsymbol{v}, t_1) = w'(\boldsymbol{v}, 0) \sin \chi t_1 = -NG(\boldsymbol{v}) \sin \chi t_1\,, \tag{2.26b}$$

$$w'(\boldsymbol{v}, t_1) = w'(\boldsymbol{v}, 0) \cos \chi t_1 = -NG(\boldsymbol{v}) \cos \chi t_1\,. \tag{2.26c}$$

These results may be further simplified by choosing t_1 such that $\chi t_1 = \pi/2$ to yield

$$u'(\boldsymbol{v}, t_1) = 0\,, \tag{2.27a}$$

$$v'(\boldsymbol{v}, t_1) = -NG(\boldsymbol{v})\,, \tag{2.27b}$$

$$w'(\boldsymbol{v}, t_1) = 0\,. \tag{2.27c}$$

$t_1 < t < t_2$. In this interval, χ is effectively zero and (2.19) becomes

$$\dot{\tilde{u}}'(\boldsymbol{v}, t) = -\Delta' \tilde{v}'(\boldsymbol{v}, t) - \Gamma \tilde{u}'(\boldsymbol{v}, t) + \int \mathrm{d}^3 v'\, W(\boldsymbol{v} \to \boldsymbol{v}')\, \tilde{u}'(\boldsymbol{v}', t)\,, \tag{2.28a}$$

$$\dot{\tilde{v}}'(\boldsymbol{v}, t) = \Delta' \tilde{u}'(\boldsymbol{v}, t) - \Gamma \tilde{v}'(\boldsymbol{v}, t) + \int \mathrm{d}^3 v'\, W(\boldsymbol{v} \to \boldsymbol{v}')\, \tilde{v}'(\boldsymbol{v}', t)\,, \tag{2.28b}$$

$$\dot{\tilde{w}}'(\boldsymbol{v}, t) = -\Gamma \tilde{w}'(\boldsymbol{v}, t) + \int \mathrm{d}^3 v'\, W(\boldsymbol{v} \to \boldsymbol{v}')\, \tilde{w}'(\boldsymbol{v}', t)\,. \tag{2.28c}$$

Due to the Stark shift $\Delta\omega_{21}$

$$\Delta'(\boldsymbol{v}) = \Delta(\boldsymbol{v}) + \Delta\omega_{21}\,.$$

Here we have incorporated (2.20), thereby removing the Γ_1 term in (2.19). In addition, the detailed balancing relation (2.22) has been utilized in the integral of (2.19), so that the factor $NG(v)$ common to every term of this equation can be removed.

From the assumed thermal equilibrium condition, implied by (2.22), the solution of (2.28*c*) is simply

$$\tilde{w}'(\boldsymbol{v}, t) = \tilde{w}'(\boldsymbol{v}, t_1)\,, \tag{2.29a}$$

and from (2.27c)

$$w'(\boldsymbol{v}, t_2) = 0 . \tag{2.29b}$$

That is, the $\pi/2$ pulse equalizes the population in levels 1 and 2, and thereafter it remains constant according to our simplifying assumption.

Equations (2.28a) and (2.28b) are solved by combining them as in (2.7), *i.e.*

$$\dot{\tilde{u}}'(\boldsymbol{v}, t) - i\dot{\tilde{v}}'(\boldsymbol{v}, t) = -\big(i\Delta'(\boldsymbol{v}) + \Gamma\big)\big(\tilde{u}'(\boldsymbol{v}, t) - i\tilde{v}'(\boldsymbol{v}, t)\big) + \\ + \int \mathrm{d}^3 v'\, W(\boldsymbol{v} \to \boldsymbol{v}')\big(\tilde{u}'(\boldsymbol{v}', t) - i\tilde{v}'(\boldsymbol{v}', t)\big) . \tag{2.30}$$

In terms of the density matrix element $\tilde{\varrho}_{21} = \frac{1}{2}(u - iv)$, eq. (2.30) reduces to

$$\frac{\mathrm{d}}{\mathrm{d}t}\big(\tilde{\varrho}'_{21}(\boldsymbol{v}, t)\exp[i\Delta'(v)t]\big) = -\Gamma\tilde{\varrho}'_{21}(\boldsymbol{v}, t)\exp[i\Delta'(v)t] + \\ + \int \mathrm{d}^3 v'\, W(\boldsymbol{v} \to \boldsymbol{v}')\,\tilde{\varrho}'_{21}(\boldsymbol{v}', t)\exp[i\Delta'(\boldsymbol{v})t] . \tag{2.31}$$

We, therefore, define the new variable

$$\tilde{\varrho}'_{21}(\boldsymbol{v}, t, t_1) = \tilde{\varrho}'_{21}(\boldsymbol{v}, t)\exp[i\Delta'(\boldsymbol{v})(t - t_1)] , \tag{2.32}$$

which obeys the equation

$$\dot{\tilde{\varrho}}'_{21}(\boldsymbol{v}, t, t_1) = -\Gamma\tilde{\varrho}'_{21}(\boldsymbol{v}, t, t_1) + \int \mathrm{d}^3 v'\, W(\boldsymbol{v} \to \boldsymbol{v}')\,\tilde{\varrho}'_{21}(\boldsymbol{v}', t, t_1) \cdot \\ \cdot \exp[i\boldsymbol{k}\cdot(\boldsymbol{v} - \boldsymbol{v}')(t - t_1)] , \tag{2.33}$$

and is subject to the initial condition

$$\tilde{\varrho}'_{21}(\boldsymbol{v}, t_1, t_1) = \tfrac{1}{2}[\tilde{u}'(\boldsymbol{v}, t_1) - i\tilde{v}'(\boldsymbol{v}, t_1)] = i/2 . \tag{2.34}$$

Note that the $\exp[-i\Delta' t_1]$ factor in (2.32) is arbitrary and is added for convenience, as will be apparent later.

We now try a solution of the form

$$\tilde{\varrho}'_{21}(\boldsymbol{v}, t, t_1) = \frac{i}{2}\exp[i\boldsymbol{x}(t, t_1)\cdot\boldsymbol{v} + y(t, t_1)] , \tag{2.35}$$

where the variables $\boldsymbol{x}$ and y satisfy initial conditions

$$\boldsymbol{x}(t_1, t_1) = 0 , \qquad y(t_1, t_1) = 0 . \tag{2.36}$$

If we substitute (2.35) into (2.33), there results

$$i\dot{\boldsymbol{x}}(t, t_1)\cdot\boldsymbol{v} + \dot{y}(t, t_1) = -\Gamma + \int \mathrm{d}^3 v'\, W(\boldsymbol{v} \to \boldsymbol{v}')\exp\big[i[\boldsymbol{k}(t - t_1) - \boldsymbol{x}](\boldsymbol{v} - \boldsymbol{v}')\big] = \\ = -\Gamma + \Gamma\exp\big[i[\boldsymbol{k}(t - t_1) - \boldsymbol{x}]\cdot\boldsymbol{v}(1 - \alpha)\big]\exp\big[-[\boldsymbol{k}(t - t_1) - \boldsymbol{x}]^2\Delta u^2/4\big] , \tag{2.37}$$

where the collision kernel (2.21) has been used. The second term on the right will contribute only if

$$[\boldsymbol{k}(t-t_1)-\boldsymbol{x}]^2\Delta u^2 \lesssim 1\,. \tag{2.38}$$

In that case, if we use $1-\alpha=\Delta u^2/2u^2$, the leading exponential of (2.37) is

$$[\boldsymbol{k}(t-t_1)-\boldsymbol{x}]\cdot\boldsymbol{v}(1-\alpha)=[\boldsymbol{k}(t-t_1)-\boldsymbol{x}]\cdot\boldsymbol{v}\Delta u^2/2u^2 \lesssim 1\,, \tag{2.39}$$

and this term may be expanded to give

$$i\dot{\boldsymbol{x}}\cdot\boldsymbol{v}+\dot{y}\approx -\Gamma+\Gamma\big(1+i(1-\alpha)[\boldsymbol{k}(t-t_1)-\boldsymbol{x}]\cdot\boldsymbol{v}\big)\cdot \cdot\exp\big[-[\boldsymbol{k}(t-t_1)-\boldsymbol{x}]^2\Delta u^2/4\big]\,. \tag{2.40}$$

This equation immediately yields two equations:

$$\dot{\boldsymbol{x}}(t,t_1)=\Gamma(1-\alpha)[\boldsymbol{k}(t-t_1)-\boldsymbol{x}]\exp\big[-[\boldsymbol{k}(t-t_1)-\boldsymbol{x}]^2\Delta u^2/4\big]\,, \tag{2.41a}$$

$$\dot{y}(t,t_1)=-\Gamma+\Gamma\exp\big[-[\boldsymbol{k}(t-t_1)-\boldsymbol{x}]^2\Delta u^2/4\big]\,, \tag{2.41b}$$

which may be integrated to find $x(t,t_1)$ and $y(t,t_1)$. If the exponent is taken to be very small, previous derivations of the spin echo problem result where the starting point is the Fokker-Planck rather than the transport equation [29-31].

It is possible, however, to find an analytic solution to eqs. (2.41) for a case of practical interest. Since echo signals are only observable for pulse separations τ such that $\Gamma\tau \lesssim 1$ and since $1-\alpha \ll 1$ for all t of interest, we have

$$\Gamma(1-\alpha)\,t=\Gamma t\Delta u^2/2u \ll 1\,. \tag{2.42}$$

Applied to (2.41a), this shows that the thermal equilibrium velocity distribution is approached in a time $1/\Gamma(1-\alpha)$ that is orders of magnitude longer than the experimental observation time $1/\Gamma$ given by (2.41b).

To solve (2.41) in this limit, we further make the reasonable assumption (to be justified) that

$$|\boldsymbol{x}| \ll |\boldsymbol{k}|(t-t_1)\,, \tag{2.43}$$

and (2.41) becomes

$$\dot{\boldsymbol{x}}(t,t_1)\approx\Gamma(1-\alpha)\,\boldsymbol{k}(t-t_1)\exp\big[-[\boldsymbol{k}\Delta u(t-t_1)]^2/4\big]\,, \tag{2.44a}$$

$$\dot{y}(t,t_1)\approx -\Gamma+\Gamma\exp\big[-[\boldsymbol{k}\Delta u(t-t_1)]^2/4\big]\,, \tag{2.44b}$$

leading to the immediate solution at $t = t_2$

$$(2.45a)\qquad \boldsymbol{x}(t_2, t_1) = 2\Gamma(1-\alpha)\,\boldsymbol{k}/(k\Delta u)^2\big(1 - \exp\big[-[\boldsymbol{k}\Delta u(t_2 - t_1)]^2/4\big]\big)\,,$$

$$(2.45b)\qquad y(t_2, t_1) = -\Gamma(t_2 - t_1) + (2\Gamma/k\Delta u)\int\limits_0^{k\Delta u(t_2-t_1)/2} \exp[-\eta^2]\,\mathrm{d}\eta\,.$$

Equation (2.45a) is seen to be consistent with (2.43).

Recalling that

$$(2.46a)\qquad \tilde{\varrho}'_{21}(\boldsymbol{v}, t_2, t_1) = \tilde{\varrho}'_{21}(\boldsymbol{v}, t_2)\exp[i\Delta'(v)(t_2 - t_1)]\,,$$

$$(2.46b)\qquad = \frac{i}{2}\exp[i\boldsymbol{x}(t_2, t_1)\cdot\boldsymbol{v} + y(t_2, t_1)]\,,$$

$$(2.46c)\qquad \tilde{\varrho}'_{21}(\boldsymbol{v}, t_2) = \frac{i}{2}\exp[-i\Delta'(v)(t_2 - t_1)]\exp[i\boldsymbol{x}(t_2, t_1)\cdot\boldsymbol{v} + y(t_2, t_1)]\,,$$

we obtain

$$(2.47a)\qquad u'_{21}(\boldsymbol{v}, t_2) = \frac{i}{2}NG(\boldsymbol{v})\exp[-i\Delta'(\boldsymbol{v})(t_2 - t_1)]\cdot \cdot\exp[i\boldsymbol{x}(t_2, t_1)\cdot\boldsymbol{v} + y(t_2, t_1)] + \text{c.c.}\,,$$

$$(2.47b)\qquad v'_{21}(\boldsymbol{v}, t_2) = -\frac{N}{2}G(\boldsymbol{v})\exp[-i\Delta'(\boldsymbol{v})(t_2 - t_1)]\cdot \cdot\exp[i\boldsymbol{x}(t_2, t_1)\cdot\boldsymbol{v} + y(t_2, t_1)] + \text{c.c}\,.$$

$t_2 < t < t_3$. Equations (2.24) can be solved again to give

$$(2.48a)\qquad u'(\boldsymbol{v}, t_3) = u'(\boldsymbol{v}, t_2)\,,$$

$$(2.48b)\qquad v'(\boldsymbol{v}, t_3) = v'(\boldsymbol{v}, t_2)\cos\chi(t_3 - t_2) + w'(\boldsymbol{v}, t_2)\sin\chi(t_3 - t_2)\,,$$

$$(2.48c)\qquad w'(\boldsymbol{v}, t_3) = w'(\boldsymbol{v}, t_2) - v'(\boldsymbol{v}, t_2)\sin\chi(t_3 - t_2) - 2w'(\boldsymbol{v}, t_2)\sin^2\chi(t_3 - t_2)/2\,.$$

For a pulse where $\chi(t_3 - t_2) = \pi$

$$(2.49a)\qquad u'(\boldsymbol{v}, t_3) = u'(\boldsymbol{v}, t_2)\,,$$

$$(2.49b)\qquad v'(\boldsymbol{v}, t_3) = -v'(\boldsymbol{v}, t_2)\,,$$

$$(2.49c)\qquad w'(\boldsymbol{v}, t_3) = w'(\boldsymbol{v}, t_2) = 0\,.$$

$t_3 < t$. The procedure is the same as in the $t_1 < t < t_2$ region. Note first that

$$(2.50)\qquad w'(\boldsymbol{v}, t) = 0\,.$$

We seek a solution for $\tilde{\varrho}'_{21}(\boldsymbol{v}, t, t_3+t_2-t_1)$ which is defined by (2.32), satisfies the differential equation (2.33) and is subject to the initial condition

$$(2.50') \qquad \begin{aligned} \tilde{\varrho}'_{21}(\boldsymbol{v}, t_3, t_3+t_2-t_1) &= \tilde{\varrho}'_{21}(\boldsymbol{v}, t_3)\exp[i\Delta'(\boldsymbol{v})(-t_2+t_1)] = \\ &= \tfrac{1}{2}[u(\boldsymbol{v}, t_3)-iv(\boldsymbol{v}, t_3)]\exp[-i\Delta'(\boldsymbol{v})(t_2-t_1)] = \\ &= \tfrac{1}{2}[u(\boldsymbol{v}, t_2)+iv(\boldsymbol{v}, t_2)]\exp[-i\Delta'(\boldsymbol{v})(t_2-t_1)] = \tilde{\varrho}'_{21}(\boldsymbol{v}, t_2, t_1)^* = \\ &= -\frac{i}{2}\exp[-i\boldsymbol{x}(t_2, t_1)\cdot\boldsymbol{v}+y(t_2, t_1)]\,. \end{aligned}$$

The above makes use of eqs. (2.49) and (2.46*b*). We see that the effect of a π pulse is simply to convert $\tilde{\varrho}'_{21}(\boldsymbol{v}, t_2, t_1)$ to its complex conjugate and that the seemingly arbitrary choice of the argument $t_3+t_2-t_1$ makes this relation evident.

We attempt a solution of the form

$$(2.51) \qquad \tilde{\varrho}'_{21}(\boldsymbol{v}, t, 2\tau) = -\frac{i}{2}\exp[i\boldsymbol{x}(t, 2\tau)\cdot\boldsymbol{v}+y(t, 2\tau)],$$

where the mean pulse delay time

$$\tau = (t_3+t_2-t_1)/2\,.$$

Equation (2.51) is subject to the initial conditions

$$(2.52a) \qquad \boldsymbol{x}(t_3, 2\tau) = -\,\boldsymbol{x}(t_2, t_1)\,,$$

$$(2.52b) \qquad y(t_3, 2\tau) = y(t_2, t_1)\,,$$

and satisfies (2.33)

$$(2.53) \qquad \begin{aligned} \dot{\tilde{\varrho}}'_{21}(\boldsymbol{v}, t, 2\tau) = -\,\Gamma\tilde{\varrho}'_{21}(\boldsymbol{v}, t, 2\tau) + \int \mathrm{d}^3 v'\, W(\boldsymbol{v}\to\boldsymbol{v}')\,\cdot \\ \cdot\,\tilde{\varrho}_{21}(\boldsymbol{v}', t, 2\tau)\exp[i\boldsymbol{k}\cdot(\boldsymbol{v}-\boldsymbol{v}')(t-2\tau)]\,. \end{aligned}$$

Essentially the same solution for x and y are found as before, although the lower limit of the time integral is t_3 instead of t_1. The result is

$$(2.54a) \qquad \boldsymbol{x}(t, 2\tau) = -2\boldsymbol{x}(t_2, t_1) + \frac{2\Gamma(1-\alpha)\,\boldsymbol{k}}{(k\,\Delta u)^2}\Big[1-\exp[-[k\,\Delta u(t-2\tau)]^2/4]\Big]\,,$$

$$(2.54b) \qquad y(t, 2\tau) = 2y(t_2, t_1) - \Gamma(t-2\tau) + \frac{2\Gamma}{k\,\Delta u}\int\limits_0^{k\Delta u(t-2\tau)/2}\exp[-\eta^2]\,\mathrm{d}\eta\,.$$

Using (2.54), (2.51) and (2.32), we find that

$$(2.55)\qquad \tilde{\varrho}_{21}(\boldsymbol{v}, t) = -\frac{i}{2} N G(\boldsymbol{v}) \exp[-\Gamma_1 t]\cdot \cdot \exp[i\boldsymbol{x}(t, 2\tau)\cdot \boldsymbol{v} + y(t, 2\tau)] \exp[-i\Delta'(\boldsymbol{v})(t-2\tau)] .$$

To obtain the signal field

$$(2.56)\qquad E_{21} = i(2\pi\Omega/c)\, L\mu_{12} \langle \tilde{\varrho}_{21}(\boldsymbol{v}, t)\rangle ,$$

we require the Doppler-averaged quantity

$$(2.57)\qquad \begin{cases} \langle \tilde{\varrho}_{21}(\boldsymbol{v}, t)\rangle = -\frac{i}{2} N \exp[-\Gamma_1 t] \int\limits_{-\infty}^{\infty} \frac{\mathrm{d}v}{\sqrt{\pi} u} \exp[-(v/u)^2]\cdot \\ \qquad \cdot \exp[i\boldsymbol{x}(t, 2\tau)\cdot \boldsymbol{v} + y(t, 2\tau)] \exp[-i\Delta'(\boldsymbol{v})(t-2\tau)] , \\ \qquad = -\frac{i}{2} N \exp[-\Gamma_1 t] \exp[y(t, 2\tau)]\cdot \\ \qquad \cdot \cos(-\Omega + \omega_{21} + \Delta\omega_{21})(t-2\tau) \exp\big[-[\boldsymbol{x} - \boldsymbol{k}(t-2\tau)]^2 u^2/4\big] . \end{cases}$$

Since the signal and laser fields, (2.56) and (1.1), add to give

$$E_T = (E_{21} + E_0/2) \exp[i(\Omega t - kz)] + \text{c.c.} ,$$

the beat term in the intensity E_T^2 is

$$(2.58)\qquad [E^2(t)]_{\text{beat}} = 2E_{21}(t)\, E_0 = = (2\pi\hbar k) N L \chi \exp[-\Gamma_1 t] \exp[y(t, 2\tau)] \cos(-\Omega + \omega_{21} + \Delta\omega_{21})(t - 2\tau)\cdot \cdot \exp\big[-[\boldsymbol{x} - \boldsymbol{k}(t - 2\tau)]^2 u^2/4\big] .$$

Equation (2.58) shows that the echo signal exhibits a beat of low frequency $(-\Omega + \omega_{21} + \Delta\omega_{21})$, which is essentially the Stark shift $\Delta\omega_{21}$ since $-\Omega + \omega_{21} \sim 0$. Evaluating the echo beat signal (2.58) at $t = 2\tau$ yields

$$(2.59a)\qquad E_c^2(t - 2\tau) - 2\pi\hbar k N L \chi \exp[\quad \Gamma_1 t] \exp[2y(t_2, t_1)] \exp\big[-[\boldsymbol{x}(t_2, t_1)\, u]^2 -$$

$$= 2\pi\hbar k N L \chi \exp\Big[-\Gamma_1 t - \Gamma t + (4\Gamma/k\Delta u)\int\limits_0^{k\Delta u t/4} \exp[-\eta^2 \mathrm{d}\eta] -$$

$$(2.59b)\qquad - [2\Gamma(1-\alpha)\, \boldsymbol{k}/(k\Delta u)^2]^2 \big[1 - \exp\big[-[\boldsymbol{k}\Delta u(t_2 - t_1)]^2/4\big]\big]^2 u^2\Big| .$$

The last term in the exponent is quite small and may be neglected since

$$(2.60)\qquad \Gamma/ku \ll 1 ,$$

so that the maximum absolute value of echo amplitude normalized to unity and denoted by $\bar{E}_c^2(t=2\tau)$ is

$$\bar{E}_c^2(t=2\tau)=\exp\left[-\Gamma_1 t-\Gamma t+(4\Gamma/k\Delta u)\int_0^{k\Delta ut/4}\exp[-\eta^2]\,d\eta\right]. \tag{2.61}$$

The nature of the solution (2.61) depends on the value of the r.m.s. collisional change in the Doppler phase $k\Delta u\tau$. The asymptotic limits are

$$k\Delta u\tau\ll 1,\qquad \bar{E}_c^2(t=2\tau)\sim\exp[-\Gamma_1 t-\Gamma t^3(k\Delta u)^2/48], \tag{2.62a}$$

$$k\Delta u\tau\gg 1,\qquad \bar{E}_c^2(t=2\tau)\sim\exp[-\Gamma_1 t-\Gamma t+2\pi^{\frac{1}{2}}\Gamma/(k\Delta u)]. \tag{2.62b}$$

$\exp[-\Gamma_1 t]$ represents the decrease in echo amplitude due to population decay and is easily understood. The other terms in (2.62) represent the effects of velocity-changing collisions. Additional physical insight into these terms may be obtained by looking at a mathematically nonrigorous picture of echo formation.

The maximum echo signal arises when the net Doppler phase factor

$$\exp[i\boldsymbol{k}\cdot\boldsymbol{v}(t_2-t_1)-i\boldsymbol{k}\cdot\boldsymbol{v}(t-t_3)] \tag{2.63}$$

goes to unity—*i.e.* the dephasing-rephasing process of echo formation has been accomplished. Collisions will produce a change in the Doppler phase factor which roughly goes as

$$H=\langle\exp[i\boldsymbol{k}\cdot\Delta\boldsymbol{v}\tau]\rangle, \tag{2.64}$$

where $\langle\ \rangle$ represents a collision average. Let Δu be $\sqrt{2}$ times the r.m.s. change in velocity per collision. If $k\Delta u\tau\gg 1$, any collision produces destructive phase interference, so that the only term which survives in eq. (2.64) will be the one in which no collision occurs during the time $t=2\tau$. Since the associated probability is $\exp[-\Gamma t]$, one finds

$$k\Delta u\tau\gg 1,\qquad H\sim\exp[-\Gamma t], \tag{2.65a}$$

corresponding to (2.62*b*).

On the other hand, if $k\Delta ut\ll 1$, each collision produces only a small phase change such that

$$H\approx 1-\frac{k^2}{2}\langle(\Delta v)^2\rangle\,\tau^2,$$

where we have assumed $\langle\Delta\boldsymbol{v}\rangle=0$ for simplicity. The quantity $\langle(\Delta v^2)\rangle$ will be equal to (number of collisions in time $t=\Gamma t)\times[\langle\Delta v^2\rangle$ for one collision $=\Delta u^2/2]$, so that

$$k\Delta u\tau\ll 1,\qquad H\sim 1-\tfrac{1}{4}k^2\Delta u^2\tau^2\Gamma t\approx\exp[-\Gamma t^3(k\Delta u)^2/16]. \tag{2.65b}$$

Except for numerical factors, eqs. (2.65) agree with the asymptotic forms (2.62) of the transport equation solution. Thus, the decrease in echo amplitude due to velocity-changing collisions results simply because the collisions destroy the perfect Doppler-phase cancellation which would have occurred had the collisions been absent.

The approximations used to obtain (2.61) require further discussion. First, note that, by expanding the exponential in (2.37) and carrying through the calculation as above, it is easy to derive the following formula for the echo amplitude valid for any type Brownian-motion collision kernel $W(\boldsymbol{v}' \to \boldsymbol{v})$ when $k\Delta u\tau \ll 1$:

$$(2.66) \qquad \bar{E}_c(t = 2\tau) = \exp\Big[-\Gamma_1 t + 2\sum_{m=1}^{\infty}[(-1)^m(\tau)^{2m+1}/(2m)!]\cdot \cdot \int d^3 v W(\boldsymbol{v} \to \boldsymbol{v}')[\boldsymbol{k}\cdot(\boldsymbol{v} - \boldsymbol{v}')]^{2m}\Big].$$

In addition, eq. (2.65*a*) will still be valid for $k\Delta u\tau \gg 1$, so that eqs. (2.66) and (2.65*a*) generalize our results to any Brownian-motion kernel. This argument substantiates our claim that the t^3- and t-dependence associated with velocity-changing collisions are independent of the choice of collision kernel.

Previous theories of echo formation employed a Fokker-Planck equation (FPE) rather than a transport equation to treat collisions [21, 29-31]. For times $\Gamma t(1-\alpha) \ll 1$, these theories give the limiting form (2.62*a*) for the echo amplitude rather than the entire eq. (2.61). It has been shown that the transition from the transport equation to the FPE is valid only if a single collision produces a negligible change in the distribution function or density matrix elements [32]. For the echo problem, the effect of a single collision is negligible only if $k\Delta u\tau \ll 1$. Thus, one can expect the FPE treatment to fail (as it does) when $k\Delta u\tau \gtrsim 1$.

In the actual experiment, it is possible to use field strengths such that the optimal conditions $\chi t_1 \approx \pi/2$ and $\chi(t_3 - t_2) \approx \pi$ are maintained. Due to power broadening, a monochromatic source excites a frequency band width of order χ in a system of two-level molecules. For the echo problem under consideration, the velocity band width that may be excited is of order

$$u_0 = \chi/k\,.$$

In the simple model above, we assumed that u_0 was much greater than the thermal speed u, so that the field interacted with the entire thermal distribution of molecules. Moreover, since $u_0 = \chi/k \gg u$, the power broadening was so great that the excitation spectrum was constant over the entire thermal distribution of velocities.

Experimentally, $u_0 \approx 10^3$ cm/s, while $u \approx 10^4$ cm/s, indicating that only a fraction of the molecules will interact with the laser field. The excitation spectrum for the molecules will *not* be constant, but, instead, will have a width of order u_0. This, in turn, implies that, if the field represents a π pulse for one velocity subgroup, it will not be a π pulse for another velocity subgroup. Thus, the simplification of assuming $\pi/2$ and π pulses for the entire sample cannot be maintained.

The fact that only a fraction of the molecules is excited by the laser means that collisions, in addition to producing the changes in the Doppler phase factors discussed above, will also cause the excited velocity distribution of molecules to decay towards equilibrium. However, if velocity diffusion is slow, $\Gamma t\Delta u^2/u_0^2 \ll 1$, as it is in our experiment, changes in the velocity distribution can be neglected and eq. (2.61) is not affected [32]. If collisions did produce large Δu jumps such that $\Delta u > u_0$, these collisions would remove molecules from the excitation band width and would be reflected in an effective increase in the population loss rate Γ_1; however, this situation does not prevail according to our experiments.

The fact that there is an excitation spectrum rather than $\pi/2$ and π pulses for the entire sample will lead to a modification of the shape of the echo signal, but not any change in its functional time dependence on pulse separation. In other words, the *relative* contribution of each velocity subgroup to the echo amplitude is unchanged as the pulse separation is varied. Thus, eq. (2.61) can still be used to describe the dependence of echo amplitude on pulse separation.

Our observations, as indicated in fig. 10, are in agreement with (2.61). The observed functionality in time is precisely what is predicted by (2.62), namely

$$(2.67)\qquad \begin{cases} \exp[-Kt^3] & \text{for short times}, \\ \exp[-\Gamma t] & \text{for long times}. \end{cases}$$

The agreement with theory constitutes primary evidence that elastic molecular collisions involving small changes in velocity play a crucial role in photon echo measurements. In fig. 10, the departure of the echo decay (lower curve) from the T_1 decay (upper curve) is due solely to the velocity-changing collision mechanism. In other measurements [25, 33, 34], not described here, the T_1 decay channel results from inelastic collisions that do not exhibit velocity diffusion. From the observed decay constants K and Γ of (2.67), the $^{13}CH_3F$-$^{13}CH_3F$ elastic-collision parameters are

$$(2.68)\qquad \begin{cases} \sigma = 430\ \text{Å}^2, \\ \Delta u = 85\ \text{cm/s}, \end{cases}$$

where the total elastic-collision cross-section is defined by $\sigma = \Gamma/(Nu\sqrt{2})$ and from (2.62*a*) $\Delta u = (1/k)\sqrt{48K/\Gamma}$. We note the smallness of the characteristic velocity jump per collision Δu and the large cross-section which must be associated with the long-range permanent dipole-dipole forces. From this value of Δu and (2.23), we see that $1-\alpha = 2.6\cdot 10^{-6}$, so that the velocity jumps are closely clustered about the initial velocity, as required in a weak-collision model. Furthermore, it follows that an arbitrary initial velocity approaches a thermal equilibrium distribution as $\exp[-\beta t]$ in a time $1/\beta = 1/[\Gamma(1-\alpha)] = 5$ s, which contrasts greatly with the time scale of an echo experiment $1/\Gamma \sim 15$ μs at 1 mTorr pressure.

3. – Coherent two-photon processes [35].

In sect. **1** and **2** we treated coherent transient phenomena that can occur in a two-level quantum system. Here, we discuss a class of coherent two-photon [36] processes that can occur in a three-level quantum system under transient or steady-state conditions. The problem arises also in nuclear magnetic and quadrupole resonance [37] and has emerged recently at optical frequencies, as in the phenomenon of coherent Raman beats [35, 38, 39], two-photon absorption with oppositely directed beams [40, 41] and self-induced transparency [42].

In the case of Raman beats, observed recently by SHOEMAKER and BREWER [38], three molecular levels are prepared initially in superposition

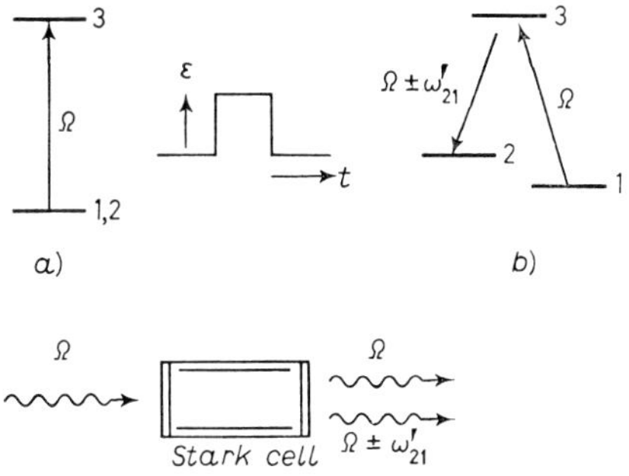

Fig. 11. – Molecular energy level diagrams illustrate *a*) coherent preparation of a three-level quantum system during a Stark pulse when the levels 1 and 2 are degenerate ($\omega_{21} = 0$) and *b*) transient forward scattering following the pulse because of the coherent Raman-beat effect. A second frequency component $\Omega + \omega_{21}$, not shown in *b*), is also emitted (from ref. [35]).

by means of continuous laser radiation. Two of the levels are degenerate and connect optically with a third level during a resonant steady-state preparative phase (see fig. 11*a*)). Upon application of a d.c. Stark field, the molecular-level degeneracy is removed and coherent forward Raman scattering then occurs in the presence of the same laser field during the nonresonant condition. The two beams, laser and Raman light, strike a photodetector, where they produce a coherent beat at a frequency that corresponds to the level splitting between initial and final states in this two-photon process (see fig. 12). The Raman-beat effect has been analysed in terms of a three-level density matrix perturbation treatment, both during the *steady-state preparation* and the subsequent transient Raman scattering which follows.

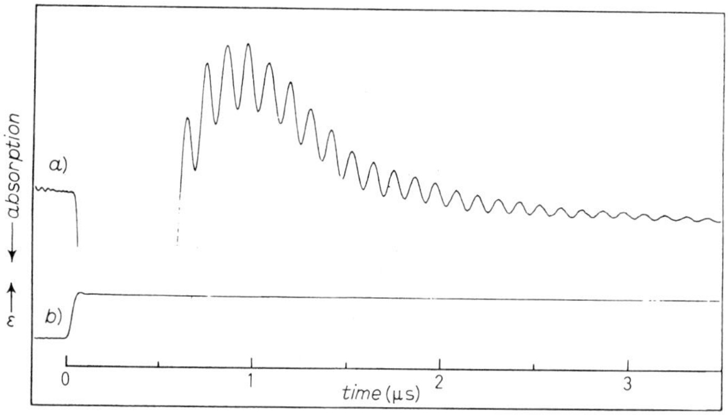

Fig. 12. – *a*) Coherent Raman beat signal in $^{13}CH_3F$ following (*b*)) a step function Stark pulse ((0 ÷ 46) V/cm). The modulation or beat frequency corresponds to the level splitting between initial and final states in the two-photon process. The slow decay is independent of Doppler dephasing or elastic collisions (from ref. [38]).

In this section, we treat the Raman-beat effect and also the related problem of two-photon absorption of oppositely directed beams for the case of *pulse preparation*. In contrast to previous attempts, an exact pulse solution is attainable for certain experimentally interesting situations. An exact solution is necessary because the relevant density matrix elements swing through large excursions in time and perturbation techniques become invalid. We will show that Bloch-like equations of motion for the three-level problem can be derived to yield a solution quite analogous to the standard two-level case. The transient regime which follows the pulse preparation is still well approximated by a perturbation calculation. Exact steady-state solutions for the Raman effect and two-photon absorption are included also, applicable to continuous-wave spectroscopy or for transient experiments requiring an initial preparation of quantum states. In general, the form of the solutions is the same for the Raman and two-photon absorption problems, either for transient or steady-state conditions.

3'1. *Raman beats.*

3'1.1. Pulse preparation: basic equations. We consider the Raman-beat case first using the molecular-level configuration of fig. 11. The two-photon absorption problem is fundamentally the same and will be handled later. The molecular gas sample is subject to a laser field

$$E_x(z, t) = \tfrac{1}{2} E_0(\exp[i(\Omega t - kz)] + \text{c.c.}) \tag{3.1}$$

of frequency Ω and amplitude E_0, polarized along the x-axis and propagating in the z-direction. Normally levels 1 and 2 are split by a frequency ω_{12} due to a d.c. Stark bias field ε applied along the y-axis. However, when a Stark pulse of amplitude $-\varepsilon$ appears, levels 1 and 2 become degenerate and the laser field excites the transitions 1-3 and 2-3, placing all three levels in superposition (fig. 11*a*)). In so far as the Raman-beat effect is concerned, the preparation of states is contained in the off-diagonal density matrix element ϱ_{12}, which persists after the pulse is removed and gives rise to the coherent transient Raman scattering that follows (fig. 11*b*)).

Our objective, therefore, is to obtain a closed-form solution for $\varrho_{12}(t_w)$ as a result of a Stark pulse of arbitrary duration t_w. This exact solution constitutes the initial condition for transient Raman scattering for times $t > t_w$. For this latter period, a perturbation treatment developed in ref. [39] suffices and will be referred to again.

The time-dependent behavior of the density matrix ϱ is given by

$$i\hbar\dot{\varrho} = [H, \varrho]\,, \tag{3.2}$$

where we neglect damping terms, assuming the pulse to be sufficiently short compared with actual molecular decay times. Molecules interact with the laser field, eq. (3.1), through the dipole interaction part

$$H_I = -\boldsymbol{\mu}\cdot\boldsymbol{E}(z, t) \tag{3.3}$$

of the total Hamiltonian H.

The electric-dipole matrix elements are given by

$$\begin{cases} \langle 1|\mu|3\rangle \equiv \mu_{13} \neq 0\,, \\ \langle 2|\mu|3\rangle \equiv \mu_{23} \neq 0\,, \\ \langle 1|\mu|2\rangle = 0\,. \end{cases} \tag{3.4}$$

The time dependence arising from molecular motion along the z-direction is introduced by defining (1.13)

$$\dot{\varrho} = \left(\frac{\partial}{\partial t} + v\frac{\partial}{\partial z}\right)\varrho\,. \tag{3.5}$$

We remove the rapidly oscillating factors of the off-diagonal elements by using

$$\varrho_{13} = \tilde{\varrho}_{13} \exp[i(\Omega t - kz)]\,, \tag{3.6a}$$

$$\varrho_{23} = \tilde{\varrho}_{23} \exp[i(\Omega t - kz)]\,, \tag{3.6b}$$

and adopt the definitions

$$\omega_{ij} = (E_i - E_j)/\hbar\,,$$
$$\alpha = \mu_{13} E_0/2\hbar\,,$$
$$\beta = \mu_{23} E_0/2\hbar\,,$$
$$\Delta = \Omega - kv - \omega_{31} = \Omega - kv - \omega_{32}\,.$$

The equations of motion in the rotating-wave approximation then become

(3.7*a*) $$\dot{\tilde{\varrho}}_{13} + i\Delta\tilde{\varrho}_{13} = i\alpha(\varrho_{33} - \varrho_{11}) - i\beta\varrho_{12}\,,$$

(3.7*b*) $$\dot{\tilde{\varrho}}_{23} + i\Delta\tilde{\varrho}_{23} = i\beta(\varrho_{33} - \varrho_{22}) - i\alpha\varrho_{21}\,,$$

(3.7*c*) $$\dot{\varrho}_{12} + i\omega_{12}\varrho_{12} = i\alpha\tilde{\varrho}_{32} - i\beta\tilde{\varrho}_{13}\,,$$

(3.7*d*) $$\dot{\varrho}_{11} = i\alpha(\tilde{\varrho}_{31} - \tilde{\varrho}_{13})\,,$$

(3.7*e*) $$\dot{\varrho}_{22} = i\beta(\tilde{\varrho}_{32} - \tilde{\varrho}_{23})\,,$$

(3.7*f*) $$\dot{\varrho}_{33} = i\alpha(\tilde{\varrho}_{13} - \tilde{\varrho}_{31}) + i\beta(\tilde{\varrho}_{23} - \tilde{\varrho}_{32})\,.$$

An exact solution of eqs. (3.7) is easily obtained for the Raman pulse preparation case when $\omega_{12} = 0$, a condition that has been easily satisfied in previous experiments [38].

3·1.2. Three-level Bloch equations. Pairs of equations in (3.7) may be combined to give Bloch-like functions u, v and w, where the matrix elements are

(3.8*a*) $$u_{ij} = \tilde{\varrho}_{ij} + \tilde{\varrho}_{ji}\,,$$

(3.8*b*) $$iv_{ij} = \tilde{\varrho}_{ij} - \tilde{\varrho}_{ji}\,,$$

(3.8*c*) $$w_{ij} = \tilde{\varrho}_{ii} - \tilde{\varrho}_{jj}\,.$$

The condition $\omega_{12} = 0$ is imposed for the Raman problem, and we obtain the following set of equations:

(3.9*a*) $$\dot{u}_{13} - v_{13}\Delta - \beta v_{12} = 0\,,$$

(3.9*b*) $$\dot{u}_{23} - v_{23}\Delta + \alpha v_{12} = 0\,,$$

(3.9*c*) $$\dot{v}_{13} + u_{13}\Delta + 2\alpha w_{13} + \beta u_{12} = 0\,,$$

(3.9*d*) $$\dot{v}_{23} + u_{23}\Delta + 2\beta w_{23} + \alpha u_{12} = 0\,,$$

(3.9*e*) $$\dot{w}_{13} - 2\alpha v_{13} - \beta v_{23} = 0\,,$$

(3.9*f*) $$\dot{w}_{23} - 2\beta v_{23} - \alpha v_{13} = 0\,,$$

(3.9g) $$\dot{u}_{12} - \alpha v_{23} - \beta v_{13} = 0\,,$$

(3.9h) $$\dot{v}_{12} - \alpha u_{23} + \beta u_{13} = 0\,,$$

(3.9i) $$\dot{w}_{12} - \alpha v_{13} + \beta v_{23} = 0\,.$$

Equations (3.9) can now be symmetrized in the following way. First, (3.9a) and (3.9b) are multiplied respectively by the dimensionless quantities $\alpha_1 = \alpha/\varepsilon$ and $\beta_1 = \beta/\varepsilon$, and then added together. The parameter

$$\varepsilon = 2\sqrt{\alpha^2 + \beta^2}$$

evolves from the treatment below. Similarly, (3.9c) is multiplied by α_1 and (3.9d) by β_1 and the two equations added. In the resulting equations we identify terms of the form

(3.10a) $$U = \alpha_1 u_{13} + \beta_1 u_{23}\,,$$

(3.10b) $$V = \alpha_1 v_{13} + \beta_1 v_{23}\,,$$

(3.10c) $$W = (\alpha_1^2 w_{13} + \beta_1^2 w_{23} + \alpha_1 \beta_1 u_{12})/\sqrt{\alpha_1^2 + \beta_1^2}\,.$$

After substitution of expressions for $\dot{w}_{13}$, $\dot{w}_{23}$ and $\dot{u}_{12}$ from eqs. (3.9) into $\dot{W}$, the particular identity

$$\dot{W} = \varepsilon V$$

results. (This ε is not to be confused with the d.c. Stark field.) A set of reduced Bloch equations may now be written as

(3.11) $$\begin{cases} \dot{U} - \Delta V = 0\,, \\ \dot{V} + \Delta U + \varepsilon W = 0\,, \\ \dot{W} - \varepsilon V = 0\,. \end{cases}$$

For constant ε, the well-known solutions [7] to eq. (3.11) are given by

(3.12a) $$U(t) = K\Delta[\cos\gamma t - 1]\,,$$

(3.12b) $$V(t) = -K\gamma\sin\gamma t\,,$$

(3.12c) $$W(t) = W(0) + K\varepsilon[\cos\gamma t - 1]$$

with

$$\gamma = \sqrt{\Delta^2 + \varepsilon^2}\,,$$
$$K = \varepsilon W(0)/(\Delta^2 + \varepsilon^2)\,.$$

When the Stark pulse is applied at $t = 0$, the initial conditions are $W(t) = W(0)$ and $U(0) = V(0) = 0$.

Note that the quantity γ is an effective precession frequency in the frame of reference rotating at frequency Ω, and γt is the angle through which the three-level Bloch vector turns in a time t.

3·1.3. Pulse solution of ϱ_{12}. The next step is to derive solutions for $u_{12}(t)$ and $v_{12}(t)$ using (3.9)-(3.12), since the desired quantity over the pulse interval $0 \leqslant t \leqslant t_w$ is

$$\varrho_{12}(t) = \tfrac{1}{2}[u_{12}(t) + iv_{12}(t)]\,. \tag{3.13}$$

These equations ultimately reduce to

$$\dddot{v}_{12} + [\Delta^2 + 2(\alpha^2 + \beta^2)]\ddot{v}_{12} + (\alpha^2 + \beta^2)^2 v_{12} = 0\,. \tag{3.14}$$

With the initial conditions

$$v_{12}(0) = \dot{v}_{12}(0) = \ddot{v}_{12}(0) = 0\,,$$
$$\dddot{v}_{12}(0) = 2\Delta\alpha\beta w_{12}(0)\,,$$

we find from (3.14) that

$$\begin{cases} v_{12}(t) = \dfrac{\alpha\beta w_{12}(0)}{\delta}\left[\dfrac{\sin(\delta - \Delta/2)t}{\delta - \Delta/2} - \dfrac{\sin(\delta + \Delta/2)t}{\delta + \Delta/2}\right], \\ \delta = (\alpha^2 + \beta^2 + \Delta^2/4)^{\frac{1}{2}}\,. \end{cases} \tag{3.15}$$

Note that $w_{12}(0) = \varrho_{11}(0) - \varrho_{22}(0)$ is the occupation probability difference at time $t = 0$ preceding the pulse. The quantity

$$u_{12}(t) = \frac{8\alpha\beta K}{\varepsilon}(\cos\gamma t - 1) - \frac{\alpha\beta}{\delta}\frac{\alpha^2 - \beta^2}{\alpha^2 + \beta^2} w_{12}(0)\left[\frac{\cos(\delta - \Delta/2)t - 1}{\delta - \Delta/2} + \frac{\cos(\delta + \Delta/2)t - 1}{\delta + \Delta/2}\right] \tag{3.16}$$

derives from (3.9*g*), where $v_{23} = \varepsilon(V - \alpha_1 v_{13})/\beta$, V is given by (3.12*b*), and

$$v_{13} = \frac{-\alpha\beta^2 w_{12}(0)}{\delta(\alpha^2 + \beta^2)}[\sin(\delta - \Delta/2)t + \sin(\delta + \Delta/2)t] - \frac{\alpha K\gamma\varepsilon}{\alpha^2 + \beta^2}\sin\gamma t\,. \tag{3.17}$$

Equation (3.17) follows by taking the time derivative of (3.9*h*), and then inserting into it expressions for $\dot{u}_{23}$ and $\dot{u}_{13}$ from (3.9*b*) and (3.9*a*) and for v_{23} from (3.10*b*).

Equations (3.15) and (3.16) combine finally to give $\varrho_{12}(t_w)$ of eq. (3.13), evaluated at $t = t_w$ at the end of a pulse of duration t_w (*).

3'1.4. Steady-state preparation. An exact expression for ϱ_{12} under steady-state conditions can also be derived and is given by eq. (3.34*c*). It reduces to the perturbation result [39]

$$\varrho_{12}(\text{s.s.}) = \frac{2(\tau_2/T_2)\alpha\beta(\varrho^0_{33} - \varrho^0_{22})}{(\omega - \omega_0)^2 + 1/T_2^2} \tag{3.18}$$

when the levels 1 and 2 are degenerate during preparation. For this case, $\omega_0 = \omega_{31} = \omega_{32}$, and the thermal difference in state occupation probability is $\varrho^0_{33} - \varrho^0_{22} = \varrho^0_{33} - \varrho^0_{11}$ in the absence of an external field. Equation (3.18) is included here to allow an easy comparison of the steady-state case with that of pulse preparation, eqs. (3.13), (3.15) and (3.16).

3'1.5. Transient decay. Previously, a perturbation solution was derived for the heterodyne beat signal following preparation under steady-state conditions. The result is given in eq. (35) of ref. [39]. In the steady-state situation described by eq. (3.18), ϱ_{12} is real, but, in the present pulse solution for $\varrho_{12}(t_w)$, real and imaginary parts of (3.13) contribute to the beat signal, so that a slight generalization of our earlier treatment is required (see two-photon subsect. **3**'2.2 for additional details).

We find, for the period $t \geqslant t_w$ following the pulse, that the velocity-averaged beat amplitude of the optical flux in terms of $\varrho_{12}(t_w)$ is

$$(E_T E_T^*)^b = -4\pi NL\frac{\hbar\Omega}{c}\alpha\beta\exp[-t/\tau_2]\cdot \left\langle [v_{12}(t_w)\cos\omega'_{12}t + u_{12}(t_w)\sin\omega'_{12}t]\times\left(\frac{1}{\Delta'_{13}} - \frac{1}{\Delta'_{23}}\right)\right\rangle_{\text{vel}}. \tag{3.19}$$

E_T is the total optical field and levels 1 and 2 are no longer degenerate due to the Stark bias field, so that $\omega'_{31} = \omega_{31} + \Delta\omega_{31}$ and $\omega'_{32} = \omega_{32} + \Delta\omega_{32}$, and

$$\omega'_{12} \neq 0,$$

$$\Delta'_{13} = \Omega - kv_z - \omega'_{31},$$

$$\Delta'_{23} = \Omega - kv_z - \omega'_{32}.$$

(*) *Note added in proofs.* – Essentially the same result can be obtained more simply by using wave functions instead of the density matrix. The author thanks M. Sargent III for bringing this point to his attention. See M. Sargent III and P. Horwitz: *Phys. Rev. A*, 1962 (1976), for a discussion of the three-level wave function solution and its limitations.

The quantity N is the molecular density, L is the sample length and τ_2 is the (Raman) dephasing time associated with ϱ_{12}.

For steady-state preparation, we see that eq. (3.19) reduces to eq. (35) of ref. [39]:

$$(3.20)\qquad (E_T E_T^*)^b_{\text{s.s.}} = -8\pi NL\frac{\hbar\Omega}{c}\cdot \cdot\alpha\beta\exp[-t/\tau_2]\left\langle \varrho_{12}(\text{s.s.})\sin\omega'_{12}t\left(\frac{1}{\Delta'_{13}}-\frac{1}{\Delta'_{23}}\right)\right\rangle_{\text{vel}},$$

because $\varrho_{12}(\text{s.s.}) = u_{12}(\text{s.s.})/2$ is real and $v_{12}(\text{s.s.}) = 0$ according to (3.18). (Equation (3.20) corrects the error in sign of the Δ'_{23} term in ref. [39].)

For pulse preparation, considerable simplification of (3.19) results when there is a zero difference in occupation probability $w_{12}(0) \sim 0$. This condition corresponds to the experiments of SHOEMAKER and BREWER [38], so that (3.15) and (3.16) become

$$(3.21)\qquad \begin{cases} v_{12}(t_w) \sim 0\,, \\ u_{12}(t_w) \sim \dfrac{8\alpha\beta W(0)}{\Delta^2 + 4(\alpha^2+\beta^2)}(\cos\gamma t_w - 1)\,. \end{cases}$$

Note that $W(0) = [\alpha_1^2 w_{13}(0) + \beta_1^2 w_{23}(0)]/\sqrt{\alpha_1^2+\beta_1^2}$ is large compared with $w_{12}(0)$. The effect of a Stark pulse is now easily visualized as causing a precession of an effective Bloch vector through the angle γt_w during the pulse interval t_w. The maximum beat amplitude in (3.19) will occur for a $\pi/2$ pulse for the resonance case, *i.e.*

$$(3.22)\qquad \gamma t_w = 2\sqrt{\alpha^2+\beta^2}\,t_w = \pi/2\,.$$

If the level structure has more than three levels, so that different three-level combinations are possible, each corresponding to a different $\alpha^2+\beta^2$ term in (3.21), then varying the pulse width t_w will give a series of maxima as the various $\pi/2$ conditions are achieved. Since (3.19) is velocity dependent through the terms Δ_{13}, Δ'_{13}, Δ_{23} and Δ'_{23}, it must be Doppler averaged (*).

(*) *Note added in proofs.* – The Doppler integral of (3.19) is not in general analytic. However, the case of steady-state preparation can be treated and is given here, since the integration was not given explicitly by BREWER and HAHN [39]. We begin at an earlier stage than (3.19) by using the Raman-beat terms of eq. (18) of [39]:

$$(\text{A}.1)\qquad \varrho^b_{13}(t) \sim \frac{\exp[-ikz_0](\exp[i(\omega+|\omega_{12}|)t - t/\tau_2] - \exp[i|\omega_{13}|t - t/T_2])}{[i(\omega-|\omega_{23}|) + (1/T_2 - 1/\tau_2)][(\omega-\omega_0)^2 + 1/T_2^2]}\,.$$

The Raman-beat signal $(E_T E_T^*)^b$ is proportional to $\varrho^b_{13}(t)$ and to $\varrho^b_{23}(t)$, which can be obtained from $\varrho^b_{13}(t)$ by the index interchange $2 \leftrightarrow 3$. The quantity $z_0 = z - v_z t$ and

3·2. *Two-photon absorption and emission.*

3·2.1. Pulse preparation. We will show that the exact solutions obtained for pulse preparation of Raman beats also apply to pulsed two-photon absorption when the two light beams propagate collinearly in opposite directions. The transient two-photon emission which follows the pulse is treated in a manner analogous to the Raman problem. In addition, exact steady-state solutions for two-photon absorption (preparation) are presented in subsect. 3·2.3 in connection with recent experiments. The appropriate molecular-level structure is shown in fig. 13, where the sample is excited by two c.w. collinear optical beams

$$E_x(z, t) = E_1 \cos(\Omega_1 t - kz) + E_2 \cos(\Omega_2 t + kz) \tag{3.23}$$

---

the laser frequency $\omega' = \omega + kv_z$. The Doppler integral

$$\langle \varrho_{13}^b(t) \rangle = \frac{1}{ku\sqrt{\pi}} \int_{-\infty}^{\infty} \exp[-(\Delta/ku)^2]\, \varrho_{13}^b(t)\, d\Delta$$

reduces in the limit where a small fraction of the Doppler width ku is excited, permitting the Gaussian to be factored from the integral, to the form

$$\langle \varrho_{13}^b(t) \rangle \sim \frac{T_2 \exp[i(\omega' + |\omega_{12}|)t - t/T_2]}{2/T_2 - 1/\tau_2 - i(|\omega_{23}| - \omega_0)}\,. \tag{A.2}$$

Here, we have dropped the uninteresting term $\exp[i(|\omega_{13}| + \omega' - \omega_0)t - 2t/T_2]$ and set the Gaussian equal to unity. This result, which is obtained by contour integration, is independent of the sign of $2/T_2 - 1/\tau_2$ and does not diverge when $|\omega_{23}| - \omega_0 = 0$. Had we integrated eq. (19) of [39] or the final result eq. (35) of [39] instead, we would have omitted the $\exp[i|\omega_{13}|t - t/T_2]$ of $\varrho_{13}^b(t)$ in (A.1) and obtained the result

$$\langle \varrho_{13}^b(t) \rangle \sim \frac{T_2 \exp[i(\omega' + |\omega_{12}|)t - t/T_2]}{2/T_2 - 1/\tau_2 - i(|\omega_{23}| - \omega_0)} \cdot \left(1 + \begin{cases} 0 & \text{for } \dfrac{1}{T_2} - \dfrac{1}{\tau_2} > 0 \\[2ex] \dfrac{-2/T_2}{i(\omega_0 - |\omega_{23}|) - 1/\tau_2} & \text{for } \dfrac{1}{T_2} - \dfrac{1}{\tau_2} < 0 \end{cases}\right). \tag{A.3}$$

We see, however, that (A.2) and (A.3) do not different significantly, since

$$\frac{-2/T_2}{i(\omega_0 - |\omega_{23}|) - 1/\tau_2} \ll 1\,.$$

This is because $\omega_0 - |\omega_{23}| \sim |\omega_{12}| \gg 1/\tau_2,\ 1/T_2$, *i.e.* the Raman-beat period must be smaller than the decay time, otherwise beats would never be observed. Hence, to the degree of approximation involved in this perurbation calculation, either (A.2) or (A.3) is acceptable. Note that eq. (3.19) also neglects the $\exp[i|\omega_{13}|t - t/T_2]$ term of $\varrho_{13}^b(t)$ and the corresponding term for $\varrho_{23}^b(t)$. The author appreciates a communication received from M. S. FELD, who recently has obtained results similar to (A.2) and who essentially has pointed out to me the difference between (A.2) and (A.3).

polarized in the x-direction and propagating along the z-axis in opposite directions. The frequency Ω_1 is assumed to be in near resonance with the 3-1 transition and Ω_2 is in near resonance with the 2-3 transition. The two frequencies are sufficiently different that each beam excites only one transition rather than both transitions simultaneously. Yet it can be assumed that $\Omega_1 - \Omega_2$ is sufficiently small, say in the microwave region, that we can neglect in (3.23) the

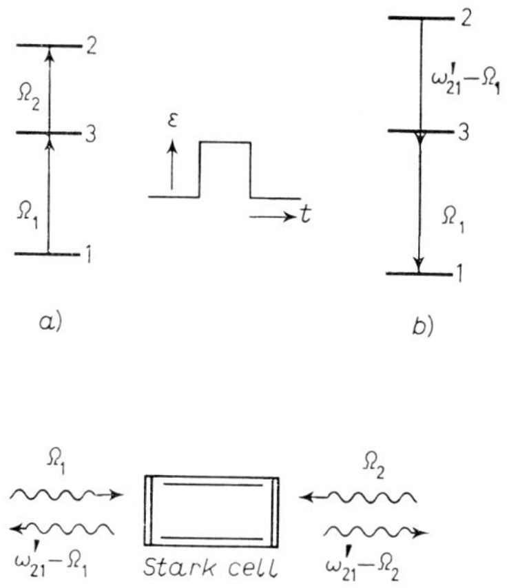

Fig. 13. – Molecular energy level diagrams illustrate *a*) coherent preparation of a three-level quantum system by two oppositely directed beams (Ω_1 and Ω_2); a Stark pulse shifts the levels so that the condition $\Omega_1 + \Omega_2 = \omega_{21}$ is satisfied; and *b*) transient two-photon emission (backward scattering) following the pulse. A second frequency component $\omega_{21} - \Omega_2$, not shown in *b*), is also emitted (from ref. [35]).

difference in propagation vectors $k_i = \Omega_i/c$ ($i = 1, 2$). We also retain the optical selection rules expressed by eq. (3.4). The effect of a Stark pulse, directed along the y-axis, is to shift the molecular levels into resonance for a two-photon transition where the frequency condition $\Omega_1 + \Omega_2 = \omega_{21}$ is satisfied (fig. 13*a*)). This process places the three levels in coherent superposition and constitutes the preparative stage over the pulse duration. As in the Raman beat effect, memory of the preparation is contained in the off-diagonal element ϱ_{12} and gives rise to a transient two-photon emission following the pulse (fig. 13*b*)). Two light beams are emitted in *opposite directions* along the z-axis where each beam exhibits a *beat*, due to the Stark shift, that is analogous to the Raman beat.

The appropriate density matrix equations are of the form

$$\dot{\tilde{\varrho}}_{13} + i\tilde{\varrho}_{13}(\Delta - i/T_2) = i\alpha(\varrho_{33} - \varrho_{11}) - i\beta\tilde{\varrho}_{12}\,, \tag{3.24a}$$

$$\dot{\tilde{\varrho}}_{23} - i\tilde{\varrho}_{23}(\Delta' + i/T_2) = i\beta(\varrho_{33} - \varrho_{22}) - i\alpha\tilde{\varrho}_{21}\,, \tag{3.24b}$$

$$\dot{\tilde{\varrho}}_{12} + i\tilde{\varrho}_{12}(\Delta + \Delta' - i/\tau_2) = i\alpha\tilde{\varrho}_{32} - i\beta\tilde{\varrho}_{13}\,, \tag{3.24c}$$

$$(3.24d)\qquad \dot{\varrho}_{11} = i\alpha(\tilde{\varrho}_{31} - \tilde{\varrho}_{13}) - (\varrho_{11} - \varrho^0_{11})/T_1\,,$$

$$(3.24e)\qquad \dot{\varrho}_{22} = i\beta(\tilde{\varrho}_{32} - \tilde{\varrho}_{23}) - (\varrho_{22} - \varrho^0_{22})/T_1\,,$$

$$(3.24f)\qquad \dot{\varrho}_{33} = i\alpha(\tilde{\varrho}_{13} - \tilde{\varrho}_{31}) + i\beta(\tilde{\varrho}_{23} - \tilde{\varrho}_{32}) - (\varrho_{33} - \varrho^0_{33})/T_1\,,$$

where now

$$(3.24g)\qquad \Delta = \Omega_1 - kv - \omega_{31}\,,$$

$$(3.24h)\qquad \Delta' = \Omega_2 + kv - \omega_{23}\,,$$

rapidly oscillating terms are removed with the definitions

$$(3.25a)\qquad \varrho_{13} = \tilde{\varrho}_{13} \exp\,[i(\Omega_1 t - kz)]\,,$$

$$(3.25b)\qquad \varrho_{23} = \tilde{\varrho}_{23} \exp\,[-i(\Omega_2 t + kz)\,,$$

$$(3.25c)\qquad \varrho_{12} = \tilde{\varrho}_{12} \exp\,[i(\Omega_1 + \Omega_2)t]\,,$$

and antiresonant terms have been dropped. The quantity ϱ^0_{ii} represents the thermal equilibrium population of level i. The decay time T_1 is assumed to be the same for all diagonal elements, and T_2 is taken to be the same for ϱ_{13} and ϱ_{23}, but we designate a different decay time τ_2 for ϱ_{12}. For Stark pulses shorter than the decay times T_1 and T_2, the relaxation and source terms in (3.24) can be neglected, but they will be of importance in the steady-state calculation of subsect. **3**.2.3. We notice that, if the two-photon resonance condition

$$(3.26)\qquad \Delta + \Delta' = \Omega_1 + \Omega_2 - \omega_{21} = 0$$

is satisfied, eqs. (3.24*a*)-(3.24*f*) are identical to eqs. (3.7) for the pulse Raman case when

$$(3.27)\qquad \omega_{12} = 0\,.$$

In this situation, the pulse Raman solutions for ϱ_{12} (Raman) (eqs. (3.13), (3.15) and (3.16)) will be valid also for the $\tilde{\varrho}_{12}$ (two photon) case.

The condition (3.26) is seen to be no more restrictive than that of (3.27). For eq. (3.27) to be satisfied, the d.c. Stark field must be zero during the pulse and this condition is easily achieved in practice. On the other hand, for (3.26) to be valid it is only necessary that ω_{21} match the sum of the two laser frequencies and this can also be accomplished with a Stark pulse of appropriate magnitude.

3.2.2. *Transient decay.* For the period following pulse excitation, $t > t_w$, transient two-photon emission will occur. In addition to the two fields of eq. (3.23) which are always present, two new fields will be generated. These are seen in the linearized form of Maxwell's equations:

$$\text{(3.28}a\text{)} \qquad \frac{\partial E_{23}(z,t)}{\partial z} = -2\pi i k_2 N \mu_{23} \langle \tilde{\varrho}_{23}(z,t)\rangle_{\text{vel}}\,,$$

$$\text{(3.28}b\text{)} \qquad \frac{\partial E_{13}(z,t)}{\partial z} = -2\pi i k_1 N \mu_{13} \langle \tilde{\varrho}_{13}(z,t)\rangle_{\text{vel}}\,,$$

where $\langle\,\rangle_{\text{vel}}$ denotes a velocity summation over the Doppler distribution. The total field then becomes

$$\text{(3.29)} \qquad E_T = (E_{13} + E_1/2)\exp[i(\Omega_1 t - kz)] + \\ + (E_{23} + E_2/2)\exp[-i(\Omega_2 t + kz)] + \text{c.c.}$$

The matrix elements $\tilde{\varrho}_{13}$ and $\tilde{\varrho}_{23}$ are obtained from eqs. (3.24a)-(3.24c) in a perturbation treatment following ref. [39]. This assumes that the right-hand side of (3.24c) can be neglected, giving

$$\text{(3.30)} \qquad \tilde{\varrho}_{12}(t) = \tilde{\varrho}_{12}(t_w)\exp[-i(\Omega_1 + \Omega_2 + \omega'_{12})t - t/\tau_2]\,,$$

where ω'_{12} is the level splitting for times $t > t_w$.

Considering only the two-photon (beat) terms of (3.24a) and (3.24b), it follows that

$$\text{(3.31}a\text{)} \qquad \tilde{\varrho}^{\,b}_{23}(t) = -\frac{\alpha\tilde{\varrho}_{21}(t_w)\exp[i(\Omega_1 + \Omega_2 + \omega'_{12})t - t/\tau_2]}{\Omega_1 - kv_z + \omega'_{13}}\,,$$

$$\text{(3.31}b\text{)} \qquad \tilde{\varrho}^{\,b}_{13}(t) = -\frac{\beta\tilde{\varrho}_{12}(t_w)\exp[-i(\Omega_1 + \Omega_2 + \omega'_{12})t - t/\tau_2]}{-\Omega_2 - kv_z + \omega'_{23}}\,.$$

Note that the pulse preparation is contained in $\varrho_{12}(t_w)$ as given by eqs. (3.13), (3.15) and (3.16).

The absolute square of (3.29) yields beat terms in the intensity:

$$\text{(3.32)} \qquad (E_T E_T^*)^b = -4\pi i N L\langle \alpha(\hbar\Omega_1/c)(\tilde{\varrho}^{\,b}_{13} - \tilde{\varrho}^{\,b}_{31}) + \beta(\hbar\Omega_2/c)(\tilde{\varrho}^{\,b}_{23} - \tilde{\varrho}^{\,b}_{32})\rangle_{\text{vel}} = \\ = -4\pi N L\alpha\beta\exp[-t/\tau_2]\Big\langle[-v_{12}(t_w)\cos(\Omega_1 + \Omega_2 + \omega'_{12})t + \\ + u_{12}(t_w)\sin(\Omega_1 + \Omega_2 + \omega'_{12})t]\left(\frac{\hbar\Omega_2/c}{\Omega_1 - kv + \omega'_{13}} + \frac{\hbar\Omega_1/c}{\Omega_2 + kv - \omega'_{23}}\right)\Big\rangle_{\text{vel}}\,,$$

where the in and out of phase components oscillate at a beat frequency $\Omega_1 + \Omega_2 + \omega'_{12}$. Since $\Omega_1 + \Omega_2 + \omega_{12} = 0$ during the Stark pulse, we see that, after the pulse, the beat $\Omega_1 + \Omega_2 + \omega'_{12} = \omega'_{12} - \omega_{12}$ is simply the Stark shift. As in the Raman case, eq. (3.19), the Doppler average must be, in general, handled numerically.

Examination of the frequency content of (3.29), using (3.28) and (3.31), shows that there are two possibilities for two-photon emission (as there are in Raman emission). In the first case, a photon of frequency $\omega'_{21} - \Omega_1$ is emitted in the $-z$-direction and one of frequency Ω_1 in the $+z$-direction, the direction of emission being contained in the k-vectors of (3.29). In the second case, a photon of frequency Ω_2 is emitted in the $-z$-direction and one of frequency $\omega'_{21} - \Omega_2$ in the $+z$-direction. Therefore, a photodetector that samples either the $+z$ or the $-z$ propagating beams will detect both two-photon processes simultaneously. The relative contributions of each process are weighted by the off-resonant tuning behavior expressed in the last two terms of (3.32). Classically speaking, we may say that the coherently prepared sample modulates at a frequency ω'_{21} the Ω_1 beam to produce an oppositely directed sideband at $\omega'_{21} - \Omega_1$; similarly, the sample also modulates the Ω_2 beam to generate an oppositely directed sideband at $\omega'_{21} - \Omega_2$.

It is interesting to speculate on whether or not the beat signal (3.32) would be observed if the two fields E_1 and E_2 of (3.23) prepared the sample, but only one of them, say E_1, was allowed to probe the sample afterwards. In such a situation, the total field becomes

$$E_T = (E_{13} + E_1/2) \exp[i(\Omega_1 t - kz)] + E_{23} \exp[-i(\Omega_2 t + kz)] + \text{c.c.} \tag{3.33a}$$

instead of (3.29), and the intensity associated with the beat signal is

$$(E_T E_T^*)^b = E_1(E_{13} + E_{13}^*) , \tag{3.33b}$$

where $E_1 = E_1^*$. The beat term $E_1(E_{13} + E_{13}^*)$, which appears in (3.32) through the term containing $\Omega_2 + kv - \omega'_{23}$, represents one of the two possible two-photon processes, while the second process contained in the term $\Omega_1 - kv + \omega'_{13}$ of (3.32) is absent. We see, therefore, that the two fields E_1 and E_2 are not only needed in the preparation stage but in the detection process as well.

Note that $u_{12}(t_w)$ and $v_{12}(t_w)$ in (3.32) do not simplify as in (3.21) for the Raman case because $w_{12}(0) \neq 0$. Nevertheless, most of the features discussed here for two photon-emission apply to the Raman-beat problem. A comparison of the Raman-beat expression eq. (3.19) with the two-photon expression (3.32) shows indeed that they are of the same form. A unique feature of transient two-photon emission, due to the nature of the preparation, is that the two waves are emitted collinearly in opposite directions. If, on the other hand, the beams of frequency Ω_1 and Ω_2 propagated in the same direction, the two photons would also be emitted in the same direction, but then the Doppler shifts would not cancel and the decay times would be severely shortened because of the Doppler-dephasing effect. In the case of coherent Raman beats, only forward scattering is possible, where Doppler dephasing does not appear and the incident beam is modulated at frequency ω'_{21} corresponding to the $1 \rightarrow 2$ and $2 \rightarrow 1$ transitions.

We might add that, when the E_1 and E_2 fields travel in the same direction, Doppler dephasing could be eliminated in a two-pulse photon echo experiment, in a manner analogous to the two-level photon echo problem.

Another property of eqs. (3.19) and (3.32) is that the two-photon emission and Raman-beat effects decay as $\exp[-t/\tau_2]$, independent of molecular velocity. This result is expected in lowest-order perturbation theory, because the first-order shift vanishes for these collinear two-photon processes. All of the velocity packets prepared contribute to the beat signal. It follows that elastic collisions should have no effect on the decay time for either two-photon emission of oppositely directed beams or for Raman beats, because the superposition of states remains unaffected. Raman-beat experiments [18, 25] verify this conclusion for molecular vibration-rotation transitions. Whether or not two-photon electronic transitions behave differently because of the influence of phase-interrupting elastic collisions on state superposition is not known at present. However, power-dependent frequency shifts can occur in higher-order solutions, as discussed in the next section, and then these two-photon processes are no longer velocity independent.

3·2.3. Steady-state preparation. Several experiments [41] have been recently reported involving steady-state two-photon absorption of two oppositely propagating beams, following the results of an earlier perturbation calculation [40]. We include here an exact treatment of this problem, which reveals significant power-dependent line broadening and frequency shift that are not evident in the perturbation treatment of VASILENKO *et al.* [40]. However, the exact calculation given in this section does verify our previous estimate [39] of the power-dependent shift and broadening for the Raman-beat effect. In fact, the steady-state two-photon and Raman solutions are again of the same form, as in the transient case.

We consider the two-photon absorption problem first and then the Raman case, and continue to use the optical arrangement and level structure of fig. 13, where laser radiation of frequency Ω_1 is in near resonance with the 1-3 transition, while Ω_2 is in near resonance with the 2-3 transition, but $\Omega_1 \neq \Omega_2$. The nine equations, given by the off-diagonal elements eqs. (3.24*a*)-(3.24*c*) and their complex conjugates and by the diagonal elements eqs. (3.24*d*)-(3.24*f*), can then be simultaneously solved in a straightforward manner by setting the time derivatives equal to zero.

The solutions are

$$w_{23} = \frac{Dw_{23}^0 - Pw_{12}^0}{T_1(MP - DQ)}, \tag{3.34a}$$

$$w_{12} = \frac{Qw_{12}^0 - Mw_{23}^0}{T_1(MP - DQ)}, \tag{3.34b}$$

(3.34c) $$\tilde{\varrho}_{12} = a\left(\frac{w_{23}}{\Delta' - i/T_2} + \frac{w_{12} + w_{23}}{\Delta - i/T_2}\right),$$

(3.34d) $$\tilde{\varrho}_{23} = \frac{\alpha\tilde{\varrho}_{21} + \beta w_{23}}{\Delta' + i/T_2},$$

(3.34e) $$\tilde{\varrho}_{13} = \frac{-\beta\tilde{\varrho}_{12} - \alpha(w_{12} + w_{23})}{\Delta - i/T_2},$$

where $w^0_{ij} \equiv \varrho^0_{ii} - \varrho^0_{jj}$ is the occupation probability difference of levels i and j in the absence of an external field, and we have defined the following quantities:

(3.35a) $$a = \alpha\beta\Big/\left[(\Delta + \Delta' - i/\tau_2) - \frac{\alpha^2}{\Delta' - i/T_2} - \frac{\beta^2}{\Delta - i/T_2}\right],$$

(3.35b) $$M = a^* b^* e + abe^* - \frac{2\alpha^2/T_2}{\Delta^2 + 1/T_2^2} + \frac{2\beta^2/T_2}{\Delta'^2 + 1/T_2^2},$$

(3.35c) $$Q = a^* b^* j + abj^* - \left(\frac{2\alpha^2/T_2}{\Delta^2 + 1/T_2^2} + \frac{4\beta^2/T_2}{\Delta'^2 + 1/T_2^2} + 1/T_1\right),$$

(3.35d) $$e = i\alpha\beta\left(\frac{-1}{\Delta + i/T_2} + \frac{1}{\Delta' + i/T_2}\right),$$

(3.35e) $$b = \frac{1}{\Delta - i/T_2} + \frac{1}{\Delta' - i/T_2},$$

(3.35f) $$D = -\left(\frac{2\alpha^2/T_2}{\Delta^2 + 1/T_2^2} + 1/T_1\right) + \frac{a^* e}{\Delta + i/T_2} + \frac{ae^*}{\Delta - i/T_2},$$

(3.35g) $$j = -i\alpha\beta\left(\frac{2}{\Delta' + i/T_2} + \frac{1}{\Delta + i/T_2}\right),$$

(3.35h) $$P = -\frac{2\alpha^2/T_2}{\Delta^2 + 1/T_2^2} + \frac{a^* j}{\Delta + i/T_2} + \frac{aj^*}{\Delta - i/T_2}.$$

It will be seen that eqs. (3.34) considerably simplify for the level configuration of fig. 13, because the thermal occupation probability $w^0_{23} \approx 0$, whereas $w^0_{12} \neq 0$.

The set of solutions (3.34) also apply to the corresponding steady-state Raman process of fig. 11 when the off-resonance tuning terms are defined as

$$\Delta = \Omega - kv - \omega_{31},$$
$$-\Delta' = \Omega - kv - \omega_{32},$$

in place of eqs. (3.24g) and (3.24h). However, now $w^0_{23} \neq 0$, $w^0_{12} \approx 0$ and the Raman-tuning behavior is given by $\Delta + \Delta' = \omega_{12}$ in contrast to the two-photon case, eq. (3.26), $\Delta + \Delta' = \Omega_1 + \Omega_2 - \omega_{21}$.

To be specific, we return to the two-photon absorption solution which reduces to the expected two-level solution when we set one of the two external fields equal to zero. For example, when $\beta = 0$, eq. (3.34g) becomes

$$\tilde{\varrho}_{13} = -\frac{\alpha w_{13}^0}{\Delta - i/T_2}\left(1 - \frac{4\alpha^2 T_1/T_2}{\Delta^2 + 1/T_2^2 + 4\alpha^2 T_1/T_2}\right), \tag{3.36}$$

which is the result anticipated.

Equation (3.34e) also reduces to the first-order two-photon perturbation solution [40, 43]

$$\tilde{\varrho}_{23}^{(1)} = \frac{\alpha^2 \beta w_{13}^0}{\Delta' + i/T_2}\,\frac{1}{\Delta + i/T_2}\left[\frac{1}{\Delta + \Delta' + i/\tau_2} - \frac{2T_1/T_2}{\Delta - i/T_2}\right] \tag{3.37}$$

when one of the external fields is much weaker than the other, corresponding to the condition $\beta \ll \alpha$. The velocity average of (3.37)

$$\langle \tilde{\varrho}_{23}^{(1)} \rangle_{\text{vel}} = \frac{1}{\sqrt{\pi}u}\int_{-\infty}^{\infty} \exp[-v^2/u^2]\,\tilde{\varrho}_{23}^{(1)}\,dv \simeq \tag{3.38}$$

$$\simeq \frac{2\sqrt{\pi}}{ku}\,\alpha^2 \beta w_{13}^0 \exp[-v_1^2/u^2]\,\frac{\tau_2}{\Delta + \Delta' + i/\tau_2}$$

can be carried out by contour integration under the assumption that the excited velocity group v_1 has a much narrower frequency band width kv_1 than the Doppler width ku, so that the Gaussian may be factored from the integral. For simplicity, we have also assumed that $T_1 = T_2 = \tau_2$. The polarization associated with (3.38) generates according to (3.28a) a field E_{23} that propagates in the same direction as the E_2 laser beam, as in the transient case. A detector viewing this light monitors a cross-term in the total intensity E_T of eq. (3.29):

$$|(E_T E_T^*)|^b = 16\pi^{\frac{3}{2}} NL\alpha^2\beta^2\,\frac{\exp[-v_1^2/u^2]}{ku}\,(\hbar\Omega_2/c)\,w_{13}^0\,\frac{1}{(\Delta + \Delta')^2 + 1/\tau_2^2}, \tag{3.39}$$

which displays the expected two-photon resonance tuning behavior having a Doppler-free line width $1/\tau_2$ that is independent of the intermediate level and of elastic collisions. A second detector viewing light from the opposite direction would show essentially the same behavior as (3.39) due to the $\langle \tilde{\varrho}_{13}^{(1)} \rangle$ source term.

When the exact forms of $\tilde{\varrho}_{23}$ and $\tilde{\varrho}_{13}$ in eq. (3.34) are to be used in obtaining $(E_T E_T^*)^b$ from (3.28) and (3.29), a numerical evaluation is required as the Doppler average cannot be analytically performed. However, it is still possible to investigate the two-photon resonance denominator or tuning behavior through

(3.35*a*), namely

$$a=\alpha\beta\Big/\left[(\Delta+\Delta'-i/\tau_2)-\frac{\alpha^2}{\Delta'-i/T_2}-\frac{\beta^2}{\Delta-i/T_2}\right]=$$

$$=\alpha\beta\Big/\left\{\Delta\left(1-\frac{\beta^2}{\Delta^2+1/T_2^2}\right)+\Delta'\left(1-\frac{\alpha^2}{\Delta'^2+1/T_2^2}\right)-\right.$$

$$\left.-i\left[1/\tau_2+(1/T_2)\left(\frac{\beta^2}{\Delta^2+1/T_2^2}+\frac{\alpha^2}{\Delta'^2+1/T_2^2}\right)\right]\right\}.$$

This exhibits an effective line width

$$1/T_2(\text{eff})=1/\tau_2+(1/T_2)\left(\frac{\alpha^2}{\Delta'^2+1/T_2^2}+\frac{\beta^2}{\Delta^2+1/T_2^2}\right) \tag{3.40}$$

and a frequency shift

$$\nu=-\left(\frac{\beta^2\Delta}{\Delta^2+1/T_2}+\frac{\alpha^2\Delta'}{\Delta'^2+1/T_2^2}\right) \tag{3.41}$$

that are power dependent, and thus corrects the perturbation result (3.39). Furthermore, we see that the two-photon reasonance condition is no longer independent of velocity and elastic collisions. It is interesting that the exact expressions (3.40) and (3.41) agree with an earlier estimate [39] of the power-dependent line broadening and frequency shift in the case of Raman beats.

When the frequency shift (3.41) is not excessively large, so that $\Delta+\Delta'\sim 0$, its magnitude

$$\nu\sim\frac{\Delta(\alpha^2-\beta^2)}{\Delta^2+1/T_2^2} \tag{3.42}$$

can be readily estimated. For instance, in the case $\alpha=\beta$, the frequency shift vanishes. More typically, the condition $\alpha\gg\beta$ (or $\beta\gg\alpha$) prevails, and then the shift

$$\nu\sim\Delta\alpha^2/(\Delta^2+1/T_2^2) \tag{3.43}$$

can be quite large, where the direction of the shift is determined by the sign of Δ. For example, when $\Delta\gg 1/T_2$, $\nu\sim\alpha^2/\Delta$, which may be orders of magnitude larger than the homogeneous line width $1/T_2$. Large power-dependent shifts of this kind have been observed in recent two-photon absorption measurements [44].

3·2.4. Summary. We have shown that the coherent Raman-beat effect with a single beam [38, 39] and the two-photon absorption of oppositely directed laser beams [40, 41] contain identical mechanisms for yielding Doppler-free spectra. The difference between the two cases is only one of level spacing.

In the Raman problem, two of the levels are nearly degenerate and optically connect with a third level. In the two-photon problem, the third level is intermediate in eigenenergy with respect to the other two where the selection rules are the same as in the Raman case. Thus far, the Raman-beat effect has been seen only in the transient regime with Stark switching [38], whereas two-photon absorption has been observed only under steady-state conditions [41]. However, our solutions apply to both transient and steady-state cases for these two problems. We predict that transient two-photon emission should be observable in backward scattering and will exhibit a beat oscillation that is analogous to the Raman beat seen in forward scattering. Exact steady-state solutions for the Raman and two-photon problems reveal power-dependent frequency shifts and line broadening that must be considered in precision spectroscopy.

REFERENCES

[1] A. Abragam: *The Principles of Nuclear Magnetism* (Oxford, 1961).
[2] R. G. Brewer and R. L. Shoemaker: *Phys. Rev. Lett.*, **27**, 631 (1971).
[3] S. L. McCall and E. L. Hahn: *Phys. Rev. Lett.*, **18**, 908 (1967); *Phys. Rev.*, **183**, 457 (1969); *Phys. Rev. A*, **2**, 861 (1970).
[4] R. P. Feynman, F. L. Vernon and R. W. Hellwarth: *Journ. Appl. Phys.*, **28**, 49 (1957).
[5] F. Bloch: *Phys. Rev.*, **70**, 460 (1946).
[6] H. C. Torrey: *Phys. Rev.*, **76**, 1059 (1949).
[7] G. L. Tang and B. D. Silverman: *Physics of Quantum Electronics*, edited by P. Kelley, B. Lax and P. E. Tannenwald (New York, N. Y., 1966), p. 280.
[8] F. A. Hopf, R. F. Shea and M. O. Scully: *Phys. Rev. A*, **7**, 2105 (1973).
[9] R. G. Brewer and R. L. Shoemaker: *Phys. Rev. A*, **6**, 2001 (1972).
[10] E. L. Hahn: *Phys. Rev.*, **77**, 297 (1950).
[11] C. H. Townes and A. L. Schawlow: *Microwave Spectroscopy* (New York, N. Y., 1955), p. 248.
[12] N. A. Kurnit, I. D. Abella and S. R. Hartmann: *Phys. Rev. Lett.*, **13**, 567 (1964); *Phys. Rev.*, **141**, 391 (1966).
[13] R. G. Brewer: *Very high resolution spectroscopy*, in *Proceedings of the Rank Prize Fund Symposium*, edited by R. A. Smith (London, 1976), p. 127.
[14] J. Schmidt, P. R. Berman and R. G. Brewer: *Phys. Rev. Lett.*, **31**, 1103 (1973).
[15] R. L. Shoemaker and R. G. Brewer: *Phys. Rev. Lett.*, **28**, 1430 (1972); R. G. Brewer and E. L. Hahn: *Phys. Rev. A*, **8**, 464 (1973); **11**, 1641 (1975).
[16] M. M. T. Loy: *Phys. Rev. Lett.*, **32**, 814 (1974).
[17] J. M. Levy and R. G. Brewer: unpublished.
[18] J. Schmidt and R. G. Brewer: unpublished.
[19] K. L. Foster, S. Stenholm and R. G. Brewer: *Phys. Rev. A*, **10**, 2318 (1974).
[20] R. L. Shoemaker and F. A. Hopf: *Phys. Rev. Lett.*, **33**, 1527 (1974).
[21] E. L. Hahn: *Phys. Rev.*, **80**, 580 (1950).
[22] H. Y. Carr and E. M. Purcell: *Phys. Rev.*, **94**, 630 (1954).
[23] C. Freed: *IEEE Journ. Quantum Electron.*, **4**, 404 (1968); **3**, 203 (1967).

[24] W. A. ANDERSON, R. FREEMAN and H. HILL: *Pure and Applied Chemistry*, Vol. **32** (London, 1972), p. 27.
[25] This section is abstracted from the article of P. R. BERMAN, J. M. LEVY and R. G. BREWER: *Phys. Rev. A*, **11**, 1668 (1975).
[26] P. R. BERMAN: *Phys. Rev. A*, **5**, 927 (1972); **6**, 2157 (1972).
[27] P. R. BERMAN and W. E. LAMB jr.: *Phys. Rev. A*, **2**, 2435 (1972).
[28] J. KEILSON and J. E. STORER: *Quart. Appl. Math.*, **10**, 243 (1952).
[29] B. HERZOG and E. L. HAHN: *Phys. Rev.*, **103**, 148 (1956).
[30] M. SCULLY, M. J. STEPHEN and D. C. BURNHAM: *Phys. Rev.*, **171**, 213 (1968).
[31] H. C. TORREY: *Phys. Rev.*, **104**, 563 (1956).
[32] P. R. BERMAN: *Phys. Rev. A*, **9**, 2170 (1974).
[33] R. L. SHOEMAKER, S. STENHOLM and R. G. BREWER: *Phys. Rev. A*, **10**, 2037 (1974).
[34] T. OKA: in *Advances in Atomic and Molecular Physics*, edited by D. R. BATES, Vol. **9** (New York, N. Y., 1973), p. 127.
[35] This section is abstracted from the article of R. G. BREWER and E. L. HAHN: *Phys. Rev. A*, **11**, 1641 (1975).
[36] M. GÖPPERT-MAYER: *Ann. der Phys.*, **9**, 273 (1931).
[37] M. BLOOM, E. L. HAHN and B. HERZOG: *Phys. Rev.*, **97**, 1699 (1955); P. A. FEDDERS and E. Y. C. LU: *Phys. Rev. B*, **8**, 5156 (1973).
[38] R. L. SHOEMAKER and R. G. BREWER: *Phys. Rev. Lett.*, **28**, 1430 (1972).
[39] R. G. BREWER and E. L. HAHN: *Phys. Rev. A*, **8**, 464 (1973); erratum, *Phys. Rev. A*, **9**, 1479 (1974).
[40] L. S. VASILENKO, V. P. CHEBOTAEV and A. V. SHISHAEV: *Pis'ma Žurn. Ėksp. Teor. Fiz.*, **12**, 161 (1970) (English translation: *JETP Lett.*, **12**, 113 (1970)); I. M. BETEROV, YU. A. MATYUGIN and V. P. CHEBOTAEV: *Sov. Phys. JETP*, **37**, 756 (1973).
[41] D. PRITCHARD, J. APT and T. W. DUCAS: *Phys. Rev. Lett.*, **32**, 641 (1974); F. BIRABEN, B. CAGNAC and C. GRYNBERG: *Phys. Rev. Lett.*, **32**, 643 (1974); M. D. LEVENSON and N. BLOEMBERGEN: *Phys. Rev. Lett.*, **32**, 645 (1974); T. W. HÄNSCH, K. C. HARVEY, G. MEISEL and A. L. SCHAWLOW: *Opt. Comm.*, **11**, 50 (1974); N. BLOEMBERGEN, M. D. LEVENSON and M. M. SALOUR: *Phys. Rev. Lett.*, **32**, 867 (1974); J. E. BJORKHOLM and P. F. LIAO: *Phys. Rev. Lett.*, **33**, 128 (1974); T. W. HÄNSCH, S. A. LEE, R. WALLENSTEIN and C. WIEMAN: *Phys. Rev. Lett.*, **34**, 307 (1975); W. K. BISCHEL, P. J. KELLY and C. K. RHODES: *Phys. Rev. Lett.*, **34**, 300 (1975).
[42] N. TAN-NO, K. YOKOTO and H. INABA: *Phys. Rev. Lett.*, **29**, 1211 (1972).
[43] M. S. FELD and A. JAVAN: *Phys. Rev.*, **177**, 540 (1969).
[44] P. F. LIAO and J. E. BJORKHOLM: *Phys. Rev. Lett.*, **34**, 1 (1975).

Two-Photon Spectroscopy Using Counter-Propagating Laser Beams and the a.c. Stark Effect.

J. E. BJORKHOLM

Bell Telephone Laboratories - Holmdel, N. J. 07733

During the last year and one-half there has been a great deal of interest in experimental techniques which allow Doppler broadening to be eliminated or greatly reduced in two-photon spectroscopy. In these techniques atoms are caused to absorb one photon from each of two laser beams of equal, or nearly equal, frequencies which propagate in opposite directions. The basic ideas involved were pointed out in 1970 [1], however it was not until early 1974 that the first experimental demonstrations were carried out [2-4]. Since that time the counter-propagating beams techniques have been widely used to carry out high-resolution spectroscopy. A listing of such work, which is probably not all-inclusive, is given by ref. [5-14].

In this paper I review recent work carried out by my colleague P. F. LIAO and myself relevant to two-photon spectroscopy using counter-propagating laser beams.

1. – Two-photon absorption using counter-propagating beams.

The total or partial elimination of Doppler effects for a two-photon transition is easily understood. Suppose a gas of atoms is subjected to radiation at the frequencies ν_1 and ν_2 traveling in the $+z$ and $-z$ directions respectively. The situation for a typical atom, whose component of velocity along the z-direction is v, is illustrated in fig. 1. In the rest frame of this atom the two incident waves are Doppler shifted and appear at the frequencies

$$\nu_1' = \nu_1 \left(1 - \frac{v}{c}\right) \quad \text{and} \quad \nu_2' = \nu_2 \left(1 + \frac{v}{c}\right).$$

The sum of the two apparent frequencies is

$$\nu_s(v) \equiv \nu_1' + \nu_2' = \nu_1 + \nu_2 + \frac{v}{c}(\nu_2 - \nu_1). \tag{1}$$

This sum depends on v unless $\nu_1 = \nu_2$, in which case all velocity groups see the same apparent sum frequency. In this special case all the atoms of the Maxwellian thermal distribution will be simultaneously resonant when $2\nu_1 = \nu_3$, where $h\nu_3$ is the energy of the two-photon transition. In the more general case $\nu_1 \neq \nu_2$, only those atoms for which $\nu_s(v)$ lies within several homogeneous line widths of ν_3 are active in two-photon transitions. Consequently there can be a large reduction in the effective two-photon absorption cross-section resulting from the decreased utilization of available atoms.

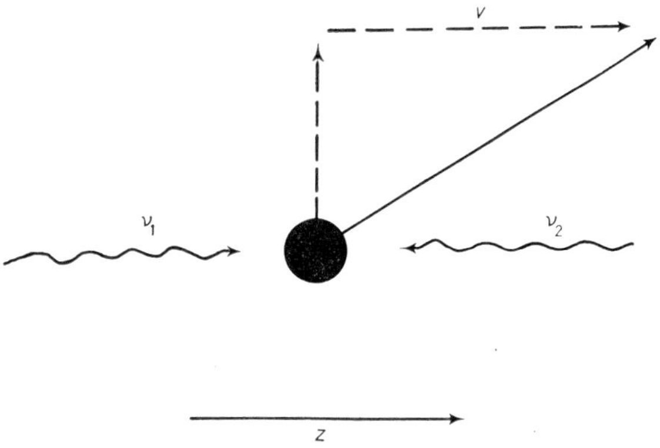

Fig. 1. – Schematic diagram of an atom having a z-component of velocity v interacting with oppositely propagating light beams having the frequencies ν_1 and ν_2.

For purposes of spectroscopy or for the determination of the frequency stability requirements of $\nu_1 + \nu_2$ in potential applications of two-photon absorption, the line width of the two-photon transition is an important quantity. If $\nu_1 = \nu_2$, eq. (1) shows no dependence on velocity v and hence the line width is given by the homogeneous line width (which, in the absence of collision broadening, is the natural line width). If $\nu_1 \neq \nu_2$ some residual Doppler broadening is present, but, as seen in eq. (1), it is reduced by the factor $(\nu_2 - \nu_1)/\nu_3$ with respect to normal Doppler broadening.

Given the simple energy level diagram shown in fig. 2, let us now calculate the two-photon transition rate from the ground state g to the excited state f for the atoms in the gas. We follow the analysis given in ref. [15] (*) and start by considering the transition rate for the velocity group of atoms having a

(*) The results in this article are correct and differ slightly from those previously given by us in ref. [15] and in *Laser Spectroscopy: Proceedings of the Second International Conference, Megéve, June 23-27, 1975*, edited by S. HAROCHE *et al.* (Berlin, 1975), p. 176. In the previous work we incorrectly used $\Delta\nu = 1/\pi\tau$ instead of the correct expression $\Delta\nu = 1/2\pi\tau$ used here, as the relationship between the natural line width and the natural lifetime of an atomic energy level.

z-component of velocity equal to v. We assume that $\nu_2 > \nu_1$ and that $2\nu_2$ and $2\nu_1$ are at least several Doppler widths off resonance for the two-photon transition. The latter restriction means that two-photon transitions cannot occur by absorption of two photons from the same light wave. For linearly polarized light the transition rate of an atom is given by [16]

$$W(\nu_1, \nu_2, v) = AI_1I_2 \left| \sum_j \langle f|z|j\rangle \langle j|z|g\rangle \left\{ \frac{1}{\nu_j - \nu_1'} + \frac{1}{\nu_j - \nu_2'} \right\} \right|^2 \cdot g(\nu_s) , \tag{2}$$

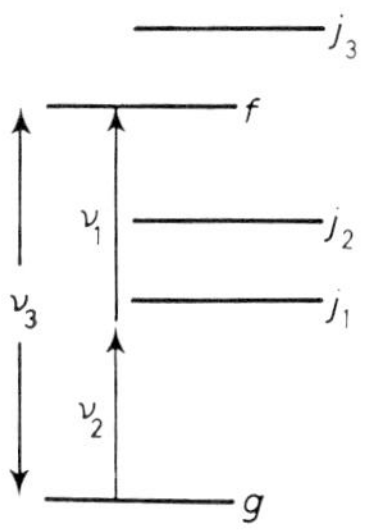

Fig. 2. – Representative atomic energy level diagram. Light at the frequencies ν_1 and ν_2 induces the atom to make two-photon transitions from the ground state g to the excited state f (having the same parity as g). The states g and f are separated in energy by $h\nu_3$, and $\nu_1 + \nu_2 = \nu_3$. The states j are various intermediate states.

where ν_1', ν_2' and ν_s are all functions of v and the summation is carried out over all intermediate states of energy $h\nu_j$. We explicitly assume that $\nu_j - \nu_1'$ and $\nu_j - \nu_2'$ are large compared to the homogeneous width of the j-th intermediate state so that natural damping of the intermediate states can be ignored. The light intensities at the frequencies ν_1 and ν_2 are I_1 and I_2 respectively, and A is a constant. The function $g(\nu_s)$ is the normalized line shape function for the state f:

$$g(\nu_s) = \frac{1}{4\pi^2\tau_2[(\nu_s(v) - \nu_3)^2 + (1/4\pi\tau_2)^2]} , \tag{3}$$

where $1/2\pi\tau_2 = \Delta\nu_H$ is the homogeneous line width (FWHM). The parameter τ_2 includes contributions due to the natural lifetime and both inelastic and phase-interrupting elastic collisions.

In order to simplify the calculation, but without unduly restricting the applicability of the result, we shall assume that τ_2 does not depend on v and that both $\nu_j - \nu_1'(v)$ and $\nu_j - \nu_2'(v)$ are large compared to both the homogeneous and Doppler widths of the j-th intermediate state. As a result, eq. (2) can be rewritten as

$$W(\nu_1, \nu_2, v) = B(\nu_1, \nu_2)\, g[\nu_s(v)]\, I_1 I_2 , \tag{4}$$

where

$$B(\nu_1, \nu_2) = A \left| \sum_j \langle f|z|j\rangle \langle j|z|g\rangle \left\{ \frac{1}{\nu_j - \nu_1} + \frac{1}{\nu_j - \nu_2} \right\} \right|^2 . \tag{5}$$

The density $n(v)$ of atoms belonging to the v-th velocity group is given by the Maxwellian thermal distribution and the number of $g \to f$ transitions made by these atoms per unit time and volume is

$$\frac{\mathrm{d}n(v)}{\mathrm{d}t} = W(\nu_1, \nu_2, v) n(v) . \tag{6}$$

The total transition rate per unit volume is obtained by combining eqs. (3)-(6) and integrating over all velocity groups (over all values of v). The peak transition rate occurs for $\nu_1 + \nu_2 = \nu_3$ and is given by

$$\frac{\mathrm{d}n}{\mathrm{d}t} = n_0 4\tau_2 B(\nu_1, \nu_2) I_1 I_2 \eta , \tag{7}$$

where

$$\eta = \sqrt{\pi}\, y \operatorname{Re} w(iy) \tag{8}$$

and

$$y = \sqrt{\ln 2}\, \frac{\Delta\nu_H}{\Delta\nu_{\mathrm{D}}} \frac{\nu_3}{\nu_2 - \nu_1} . \tag{9}$$

In these equations n_0 is the density of atoms in the vapor, $\Delta\nu_{\mathrm{D}}$ is the normal Doppler width (FWHM) of the two-photon transition, and $w(z)$ is the error function for a complex argument [17]. In essence η is the fraction of atoms in the Doppler-broadened line that can simultaneously interact with the light. For $y \gg 1$ (as for the case $\nu_1 = \nu_2$), η goes to 1; for $y \ll 1$, η goes to $\sqrt{\pi}\, y$. Note that y is proportional to $\Delta\nu_H$ divided by the residual Doppler broadening $((\nu_2 - \nu_1)/\nu_3)\, \Delta\nu_{\mathrm{D}}$.

The line shape of the transition is given by

$$\alpha_2(\nu_r) = \sqrt{\pi}\, y \operatorname{Re} w\{(4\pi\tau_2\nu_r + i) y\} , \tag{10}$$

where $\nu_r = \nu_3 - \nu_1 - \nu_2$. This has the form of a Lorentzian line having a line width (FWHM) of $\Delta\nu_H$ convolved with a Gaussian line having a line width equal to the residual Doppler broadening. Thus for $y \gg 1$, $\alpha_2(\nu_r)$ is a Lorentzian line shape function with a line width of $\Delta\nu_H$. For $y \ll 1$, $\alpha_2(\nu_r)$ has the familiar form of a Gaussian line with Lorentzian tails and the line width is $((\nu_2 - \nu_1)/\nu_3)\Delta\nu_{\mathrm{D}}$, the residual Doppler broadening.

The preceding results show that as ν_1 and ν_2 are made more and more unequal the line width of the two-photon transition increases and the fraction of atoms which can simultaneously interact with the light decreases. On the other hand, eq. (7) shows that the total transition rate is proportional to $B(\nu_1, \nu_2)\,\eta$ and B can be made very large by choosing ν_1 or ν_2 to be nearly equal to one of the ν_j's. In this way the total transition rate can be greatly increased even though ν_1 and ν_2 are very different. This is termed resonant enhancement of the two-photon absorption.

2. – Resonant enhancement.

We have used two single-mode, c.w. dye lasers operating at different frequencies to study resonant enhancement of two-photon absorption in sodium vapor [6]. Enhancements of over 7 orders of magnitude and dramatic destructive interference effects due to neighboring intermediate states were clearly observed.

In these experiments we observed two-photon transitions from the $3S$ ground state of sodium to the $4D$ excited state. The dominant intermediate states for these transitions are the $3P$ levels, as shown in fig. 3. The basic

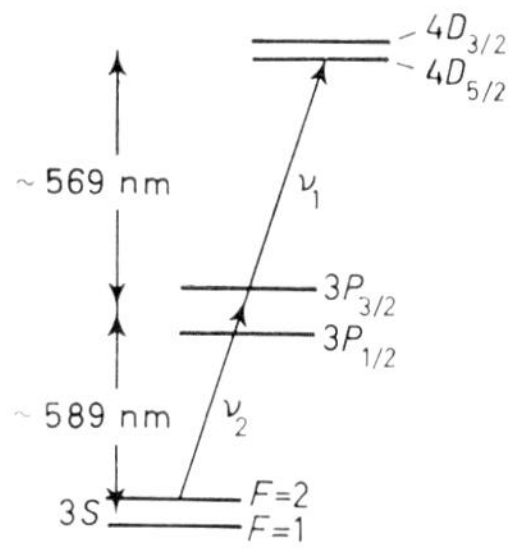

Fig. 3. – The energy levels of atomic sodium relevant to the two-photon absorption experiments described here.

experimental set-up was straightforward and is shown in fig. 4. The light from two single-axial-mode, c.w. dye lasers propagated in opposite directions through a pyrex cell containing sodium vapor. The minimum beam radius in the vapor was about 40 μm and typical laser powers were 50 mW, so that maximum light intensities were about 10^3 W/cm². The vapor pressure was of the order of 10^{-5} Torr. Two-photon transitions to the $4D$ levels were monitored by using a 1P28 photomultiplier to detect the 330 nm fluorescence ($4P$-$3S$ transition) resulting from the decay from the $4D$ levels.

In making an experimental run, one laser was operated at a fixed frequency ν_2, while the other laser was operated at the frequency ν_1, which was electronically tuned across the two-photon absorption lines. High-resolution spectra such as that

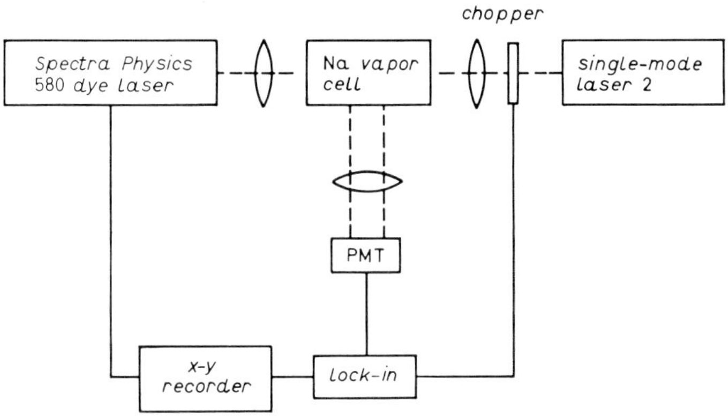

Fig. 4. – Schematic diagram of the experimental set-up. The horizontal axis of the x-y recorder was driven by the electronic frequency scanning circuit of the Model 580 dye laser.

shown in fig. 5 were obtained. Using eq. (10), we find that the line width calculated for the conditions of fig. 5 is 30 MHz; the experimentally observed line width of 60 MHz is larger because of the frequency jitter of the two lasers. The normal Doppler width of the two-photon transition is $\Delta\nu_D = 3.4$ GHz; clearly the four transitions could not be resolved without use of the opposed-beam technique.

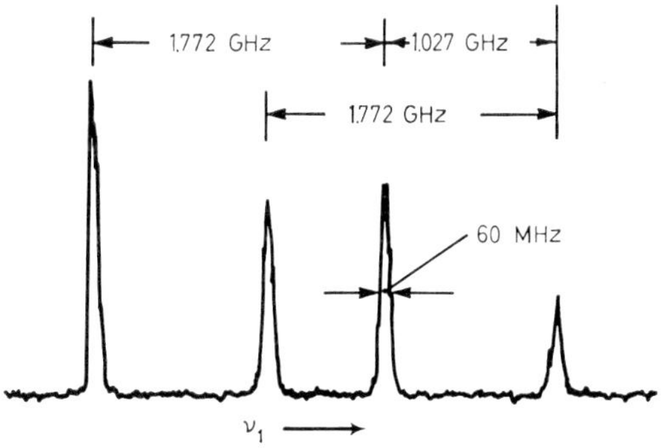

Fig. 5. – The excitation spectrum for two-photon excitation of the $4D$ level with $\lambda_2 = 5835$ Å. As we proceed in the direction of high frequency, the peaks correspond to the following transitions: $3S(F=2)\to 4D_{\frac{5}{2}}$, $3S(F=2)\to 4D_{\frac{3}{2}}$, $3S(F=1)\to 4D_{\frac{5}{2}}$, $3S(F=1)\to 4D_{\frac{3}{2}}$. The splittings of 1.772 and 1.027 GHz correspond to the ground-state hyperfine splitting and our measurement of the $4D$ fine-structure splitting, respectively.

To demonstrate resonant enhancement, a series of runs giving curves like that of fig. 5 were carried out, each with a different λ_2, the wavelength of the fixed-frequency laser. Of course, λ_1 was also varied so that $\nu_1 + \nu_2 = \nu_3$. In fig. 6 are plotted the normalized strengths of the $3S(F=2)\rightarrow 4D_{\frac{5}{2}}$ and $3S(F=2)\rightarrow 4D_{\frac{3}{2}}$ transitions as a function of λ_2. These normalized strengths were obtained by dividing the peak line strengths of each transition by the product of the incident laser powers. Both transitions show strong resonant enhancement as the laser frequency ν_2 approaches the frequency of the relevant $3P$ intermediate states ($B(\nu_1, \nu_2)$ becomes large as $\nu_{3P}-\nu_2$ goes to zero). For the

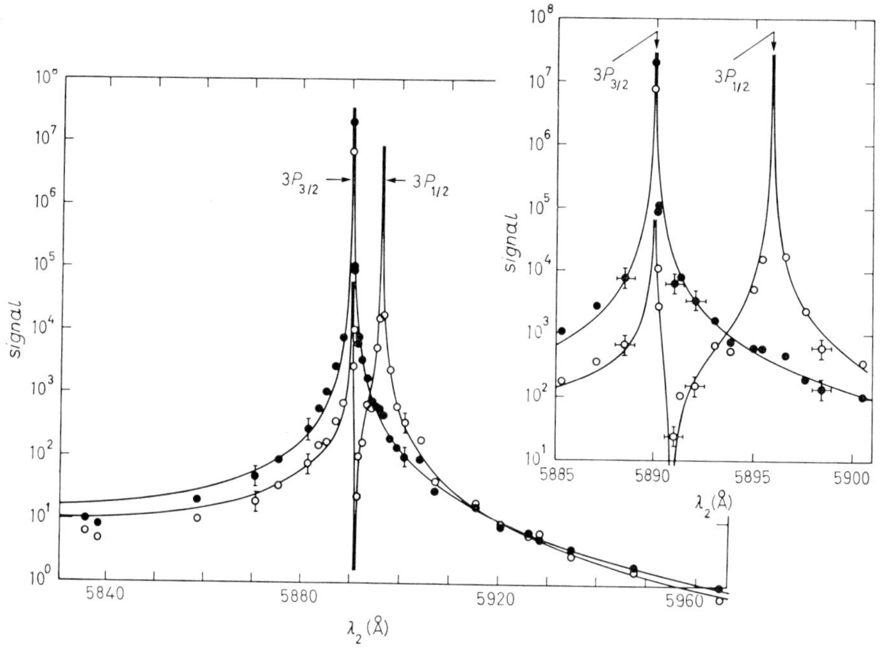

Fig. 6. – Normalized two-photon transition rates for the $3S(F=2)\rightarrow 4D_{\frac{5}{2}}$ (•) and $3S(F=2)\rightarrow 4D_{\frac{3}{2}}$ (○) transitions as a function of the wavelength of the fixed-frequency laser λ_2. (Note that $\nu_1=\nu_2$ corresponds to $\lambda_2=5787$ Å.) The points are experimental and the curves are theoretical. The inset shows the behaviour in the region from 5885 Å to 5900 Å with an expanded horizontal axis.

$3S(F=2)\rightarrow 4D_{\frac{5}{2}}$ transition only the $3P_{\frac{3}{2}}$ level acts as an intermediate state since the $3P_{\frac{1}{2}}\rightarrow 4D_{\frac{5}{2}}$ dipole matrix element $\langle 3P_{\frac{1}{2}}|z|4D_{\frac{5}{2}}\rangle$ is zero (because $\Delta J=2$). For the $3S(F=2)\rightarrow 4D_{\frac{3}{2}}$ transition both the $3P_{\frac{3}{2}}$ and $3P_{\frac{1}{2}}$ levels at $\lambda_2=5890$ Å and 5896 Å respectively are intermediate states. The normalized signal strength varies over a range of 10^7 as λ_2 is varied from 5830 Å to 5970 Å (the corresponding λ_1 varies from 5758 Å to 5629 Å). Also notice that the signal strength for the $3S(F=2)\rightarrow 4D_{\frac{3}{2}}$ transition drops rapidly on the long-wavelength side of the $3P_{\frac{3}{2}}$ level and has a minimum only 1.00 Å from its maximum at the $3P_{\frac{3}{2}}$ level. This most interesting behavior is due to destructive interference between the contributions of the two intermediate states for wavelengths lying between them. To be specific, in this region the resonant denominators $\nu_{3P_{\frac{3}{2}}} - \nu_2$ and $\nu_{3P_{\frac{1}{2}}} - \nu_2$ which appear in eq. (5) have opposite signs. The curves in fig. 6 give the theoretically calculated behavior for $B(\nu_1, \nu_2)\eta$, where the sum over intermediate states in eq. (5) has been restricted to the $3P$ levels. The parameters for the $3S$-$4D$ transition are $\nu_1 + \nu_2 = \nu_3 = 1.036 \cdot 10^{15}$ Hz and $\tau_2 = 52$ ns, the natural lifetime of the $4D$ level. Over the tuning range for λ_2, η varied from 0.38 to 0.08 (the calculated line width varied from 30 to 100 MHz). The relative amplitudes of all matrix elements were determined using angular-momentum relations [18]. Except for an overall normalization factor, there are no adjustable parameters in the calculation. The fit between experiment and theory is very satisfying. The slight discrepancy at the shortest wavelengths occurs because the frequency jitter of the lasers contributes appreciably to broadening of the experimental lines, resulting in a reduction of the peak signal. Taking this into account removes the discrepancy.

Resonant enhancement can make two-photon absorption a very strong process. For example, for the $3S$-$4D$ transition in sodium, the absorption cross-section for light at ν_1 induced by light at ν_2 is approximately $5 \cdot 10^{-14} I_2$ cm^2 when $\nu_{3P} - \nu_2 = 3$ GHz. This can easily be made comparable to the single-photon absorption cross-section per atom $(\lambda^2/4\pi)$ of the strongly allowed $3S$-$3P$ resonance transition which is about $3 \cdot 10^{-10}$ cm^2. Consequently it is not surprising that fluorescence caused by two-photon absorption was brightly visible in our experiments.

Multiphoton processes are usually calculated in the electric-dipole approximation, as we have done here. Another aspect of resonant enhancement bearing on this approximation has recently been discussed in a theoretical paper by others [19]. It is shown that contributions to multiphoton transition rates due to higher-order (forbidden) transitions, such as electric-quadrupole transitions, can in some cases be larger than the contributions from the electric-dipole transitions. For this to occur, the forbidden transitions must be resonantly enhanced while at the same time all electric-dipole transitions must be far from resonance. Such situations can occur and an experimental observation of electric-quadrupole transitions for such a case has recently been made in the three-photon ionization of atomic sodium [20]. It is also possible that the slight « bulge » that we observed around $\lambda_2 = 5896$ Å for the $3S(F=2) \to 4D_{\frac{5}{2}}$ two-photon transition rate, as shown in fig. 6, is due to a similar effect.

3. – A.c. Stark effect and two-photon spectroscopy.

Long-lived excited states are accessible via two-photon absorption; consequently it should be possible to achieve extremely narrow line widths using Doppler-free two-photon spectroscopy. Thus the technique can, in principle, be used for ultra-precise measurements. Caution must be used in carrying out such measurements, however, since the light used to induce the two-photon transitions can also cause the atomic energy levels to shift. These shifts are known as the a.c. Stark effect [21] and they are due to virtual transitions between atomic levels caused by nonresonant light. Because two-photon transitions generally involve virtual transitions to an intermediate atomic state, such level shifts are *intrinsic* to two-photon processes (and to other multiphoton processes also) (*).

Previous observations of the a.c. Stark effect in optical transitions [22] have required high-power lasers since the transitions were Doppler broadened. Recently we have shown, however, that a.c. Stark shifts of several hundred MHz can easily be observed using low-power c.w. dye lasers and Doppler-free techniques [10]. This demonstrates that, when carrying out high-resolution two-photon spectroscopy (even with low-power lasers), it is important to account for the possibility of relatively large energy level shifts caused by the a.c. Stark effect.

Optically induced shifts of atomic energy levels is a subject about which there has been considerable interest. These energy level shifts have probably been most extensively studied in optical double-resonance experiments in which the radiation that optically pumped the atomic vapor also produced « light shifts » [23]. Such shifts were observed as small changes (several hundred Hz) in the microwave or r.f. transition frequencies between various hyperfine components of the atomic ground state. Earlier studies of level shifts produced by electromagnetic radiation considered the shifts produced by radio frequency [24] and microwave radiation [25].

The shift of an atomic level n, induced by the optical field $\mathcal{E} = \mathrm{Re}\, \mathcal{E}_0 \exp[-i(kz - \omega t)]$, is given by perturbation theory as [21, 26]

$$\Delta E_n = h\, \delta\nu_n = \frac{1}{4} \sum_m \left\{ \frac{|\boldsymbol{P}_{mn} \cdot \mathcal{E}_0|^2}{E_n - E_m - \hbar\omega} + \frac{|\boldsymbol{P}_{mn} \cdot \mathcal{E}_0|^2}{E_n - E_m + \hbar\omega} \right\}. \tag{11}$$

The summation is taken over all unperturbed atomic states $|m\rangle$ having energy E_m, and $\boldsymbol{P}_{mn}$ is the electric-dipole matrix element between states m and n. In the above we have neglected natural damping of the states so that eq. (11) is applicable only as long as $1/h$ times the energy denominator is large compared to the natural line width of the levels. Another restriction is that δE_n must be small compared to the energy denominators. If this is not the case the level shift as a function of light intensity tends to saturate and, in the limit, δE_n becomes linearly proportional to $\mathcal{E}_0$ [21].

For simplicity we now restrict our attention to the case of the two-level

(*) Theoretical aspects of level shifts in two-photon absorption have been considered by B. Cagnac, G. Grynberg and F. Biraben: *J. Physique*, **34**, 845 (1973); P. L. Kelley, H. Kildal and H. R. Schlossberg: *Chem. Phys. Lett.*, **27**, 62 (1974).

atom shown in fig. 7. For this case eq. (11) reduces to

$$\delta\nu = \frac{1}{4}\frac{|\boldsymbol{P}_{12}\cdot\mathscr{E}_0|^2}{h^2(\nu_0-\nu)}\,, \tag{12}$$

where $h\nu_0$ is the energy separation of the unperturbed levels and ν is the frequency of the applied light. The two levels experience equal, but opposite, shifts. If $\nu < \nu_0$, then the energy separation of the perturbed levels is greater than that of the unperturbed levels; for $\nu > \nu_0$, the opposite is the case.

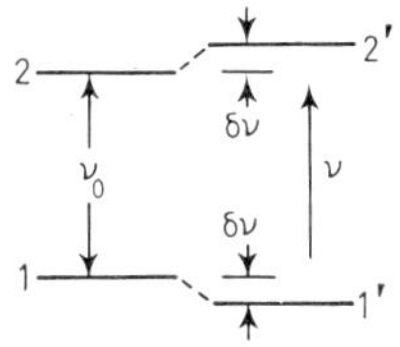

Fig. 7. – Energy level shift for the case $\nu_0 > \nu$, where ν_0 is the energy separation of the unperturbed energy levels 1 and 2 and ν is the frequency of the applied light. The levels 1′ and 2′ are the perturbed levels which have been shifted by $h\,\delta\nu$, where $\delta\nu$ is given by eq. (12).

A physical understanding of eq. (12) can be obtained by considering the need to conserve energy for long periods of time. Consider two systems. One is the unperturbed atom having energy levels 1 and 2 separated in energy by $h\nu_0$. The other system is a linearly polarized radiation field composed of N photons of energy $h\nu$. In the absence of an interaction between the two systems the atom is in its ground state 1 and the total energy of the two systems is $E_1 + Nh\nu$. Now let the two systems interact. The light induces the atom to make virtual transitions between levels 1 and 2 at a rate W. A simple rate equation analysis shows that the atom spends the fraction $f = W\tau/(2W\tau + 1)$ of its time in level 2, where τ is the natural lifetime of level 2. When the atom makes a virtual transition from level 1 to level 2, gaining the energy $h\nu_0$, a photon of energy $h\nu$ is removed from the system. Thus the combined system gains an amount of energy given by $h(\nu_0 - \nu)$. This energy imbalance can be tolerated for times of the order of $1/(\nu_0 - \nu)$ according to the uncertainty principle. However, for energy to be conserved on the average a shift of energy levels is required. Assume that the levels are shifted by equal, but opposite, amounts as shown in fig. 7. Then set the energy of the two systems in their noninteracting states equal to the sum of their energies when interacting. We obtain

$$E_1 + Nh\nu = (1-f)(E_1 - h\delta\nu + Nh\nu) + f(E_1 + h\nu_0 + h\,\delta\nu + (N-1)\,h\nu)\,,$$

which yields

$$\delta\nu = \frac{f}{1-2f}(\nu_0 - \nu)\,. \tag{13}$$

To solve for $\delta\nu$ we use the standard expressions [27]

$$W = \frac{\lambda^2 I}{8\pi h\nu\tau}\cdot\frac{1}{4\pi^2\tau[(\nu_0-\nu)^2+(1/4\pi\tau)^2]}\,,$$

$$I = \frac{c}{8\pi}\mathscr{E}_0^2$$

and

$$\frac{1}{\tau} = \frac{32\pi^3\nu^3 P_{12}^2}{\hbar c^3}\,.$$

The result is

$$\delta\nu = \frac{P_{12}^2\mathscr{E}_0^2}{4h^2}\,\frac{\nu_0-\nu}{(\nu_0-\nu)^2+(1/4\pi\tau)^2}\,, \tag{14}$$

which agrees with eq. (12) for the case $|\nu_0-\nu| \gg 1/4\pi\tau$. This result indicates that the origin of the level shifts are the energy-nonconserving virtual transitions.

As described in ref. [10], we observed the a.c. Stark effect by extending our studies of $3S \to 4D$ two-photon transitions in sodium vapor (which were discussed in sect. **2**). The basic experimental set-up was similar to that shown in fig. 4. As in the previous work, data were taken with the frequency ν_2 held constant while the frequency ν_1 of the other laser was swept repetitively through the various two-photon resonances. In this experiment, however, the lasers were tuned such that either the $3P_{\frac{3}{2}}$ or the $3P_{\frac{1}{2}}$ intermediate state was nearly resonant ($\lambda_2 \sim 589$ nm and $\lambda_1 \sim 569$ nm). The resulting large resonant enhancement gave strong two-photon signals and allowed us to eliminate the chopper, lock-in amplifier and x-y recorder shown in fig. 4. Instead the signals were directly displayed on an oscilloscope whose horizontal deflection was proportional to changes in ν_1. Energy level shifts of the $3S$ ground state and the $4D$ excited state were unambiguously observed as absolute shifts in the positions (in the ν_1 sweep) of the two-photon signals as the laser intensities were changed. The resonance denominators in eq. (11) assured that the shifts of the $3S$ state were primarily induced by the light at 589 nm while the $4D$ state was shifted primarily by the 569 nm light. By independently adjusting the intensities of the two beams, we could operate such that essentially only the $3S$ level or only the $4D$ level was being shifted, not both.

Since the level shift is proportional to the light intensity, spatial nonuniformities of the intensity within the region of observation cause the two-photon absorption lines to be broadened as well as shifted, and the broadening can obscure the shift. This problem was overcome by focusing one beam (ν_1) to a spot size much smaller than that of the other beam (ν_2) and by appropriately masking the region observed. Thus shifts of the $3S$ level caused by the light at $\lambda_2 \approx 589$ nm were observed as clean shifts, free of spatial effects. An example of such data is shown in fig. 8*a*) where the $3S$ level is shifted 745 MHz by 26 mW of 589 nm light; the shift of the $4D$ level caused by 569 nm light was negligible. Likewise, fig. 8*b*) shows the shift of the $4D$ level caused by 30 mW of the sharply focused 569 nm light in the absence of a $3S$ level shift. In this case the shift is observed as severe line broadening and distortion because of the spatial nonuniformities. For both these cases $\Delta\nu$, which equals the $3S$-$3P_{\frac{3}{2}}$ transition frequency minus ν_2, was 4 GHz.

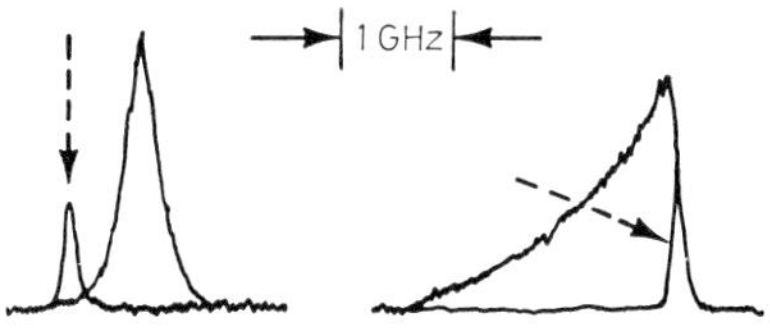

Fig. 8. – Oscillographs of the $3S(F=2)\rightarrow 4D_{\frac{5}{2}}$ two-photon absorption line. The frequency ν_1 increases to the right. Each contains two lines, one obtained with both lasers at low (<3 mW) power (shown by arrows) and the other obtained by increasing *a*) only the 589 nm power to 26 mW, and *b*) only the 569 nm power to 30 mW.

It is apparent from fig. 8*a*) that, in addition to the level shift, there is line broadening. This broadening is *not* due to spatial nonuniformities of the 589 nm beam. Instead it occurs because of Doppler effects; different atomic velocity groups experience different level shifts since the applied light has a different apparent frequency for each velocity group. For the case $\delta\nu \ll \Delta\nu$, this effect is easily incorporated into the analysis of sect. **1** by including the level shift in eq. (1). The result is that the line shape retains the form given by eq. (10), but the peak is shifted by the amount $\delta\nu_0$ (the level shift of the zero-velocity group) and the line width is proportional to

$$\frac{1}{\nu_1+\nu_2}\left[(\nu_2-\nu_1)+\nu_2\frac{\delta\nu_0}{\Delta\nu}\right]$$

(shift caused by light at ν_2). The second term in the line width accounts for the broadening due to the level shift.

The dependence given by eq. (12) for the level shift upon the light intensity I and the mistuning $\Delta\nu$ was verified for the $3S(F=2)$ level, under restrictions related to those explained earlier. The first restriction was that $\Delta\nu$ be larger than twice the Doppler width of the $3S$-$3P_{\frac{3}{2}}$ transition ($3P_{\frac{3}{2}}$ was near resonance) so that $\Delta\nu$ was many times the natural line width for all velocity groups having an appreciable population. The second restriction was $\delta\nu \ll \Delta\nu$. Under these conditions we found $\delta\nu \approx 1.4 \cdot 10^{15} I/\Delta\nu$, where I is in units of W/cm^2. The measured numerical factor agrees well with theory. We also verified that shifts of the $3S$ levels due to near resonance with the $3P_{\frac{1}{2}}$ level are one-half as large as those due to near resonance with the $3P_{\frac{3}{2}}$ level, as expected from an evaluation of the relevant matrix elements. When the $3P_{\frac{1}{2}}$ level was nearly resonant there were no shifts of the $4D_{\frac{5}{2}}$ level produced by the 569 nm light, whereas the shifts for the $4D_{\frac{3}{2}}$ level were very strong. This behavior results because the dipole matrix element connecting the $3P_{\frac{1}{2}}$ state with the $4D_{\frac{5}{2}}$ state is zero. Finally we verified that large level shifts can exist even in the absence of saturation of the two-photon transition.

The size of the atomic energy level shifts is basically related to the two-photon transition rate (notice the similarity of eqs. (2) and (11)). Specifically, for the special case in which only one intermediate state is of importance (as for the experiments described in this section), the product of the level shift of the ground state with the level shift of the excited state is proportional to the two-photon transition rate. This shows that the energy level shifts will be large whenever the two-photon transition rate is large, *regardless of the amount of resonant enhancement*. Also notice that, if one level shift is large and the other is small, it is possible to have a large optically induced shift of the two-photon transition frequency even though the two-photon transition rate is small. Thus, as we observed, large shifts of the two-photon resonance frequency can occur in the absence of saturation of the two-photon transition.

In closing, we point out that in addition to being an effect that must be accounted for, the a.c. Stark effect may also find useful applications. For instance, the intensity dependence of the shifts provides an accurate method for measuring transition dipole moments. In fact, such an experiment has very recently been performed on the ammonia molecule, NH_3, using two infra-red lasers, resonant enhancement and opposed-beam techniques [28]. In addition to the dipole-moment measurement, a previously unmeasured energy level was accurately determined. To do this it was necessary to make measurements at several intensity levels and then extrapolate to the case of zero a.c. Stark shift.

It seems safe to speculate that the recent surge of interest in two-photon spectroscopy which has been spurred by the development of the opposed-beam techniques will give rise to many new and interesting experiments in the near future.

REFERENCES

[1] L. S. Vasilenko, V. P. Chebotaev and A. V. Shishaev: *Pis'ma Žurn. Ėksp. Teor. Fiz.*, **12**, 161 (1970 (English translation: *JETP Lett.*, **12**, 113 (1970)).
[2] F. Biraben, B. Cagnac and G. Grynberg: *Phys. Rev. Lett.*, **32**, 643 (1974).

[3] M. D. LEVENSON and N. BLOEMBERGEN: *Phys. Rev. Lett.*, **32**, 645 (1974).
[4] T. W. HÄNSCH, K. C. HARVEY, G. MEISEL and A. L. SCHAWLOW: *Opt. Comm.*, **11**, 50 (1974).
[5] N. BLOEMBERGEN, M. D. LEVENSON and M. M. SALOUR: *Phys. Rev. Lett.*, **32**, 867 (1974).
[6] J. E. BJORKHOLM and P. F. LIAO: *Phys. Rev. Lett.*, **33**, 128 (1974).
[7] F. BIRABEN, B. CAGNAC and G. GRYNBERG: *Compt. Rend.*, **279** B, 51 (1974).
[8] F. BIRABEN, B. CAGNAC and G. GRYNBERG: *Phys. Lett.*, **48** A, 469 (1974).
[9] F. BIRABEN, B. CAGNAC and G. GRYNBERG: *Phys. Lett.*, **49** A, 71 (1974).
[10] P. F. LIAO and J. E. BJORKHOLM: *Phys. Rev. Lett.*, **34**, 1 (1975).
[11] T. W. HÄNSCH, S. A. LEE, R. WALLENSTEIN and C. WIEMAN: *Phys. Rev. Lett.*, **34**, 307 (1975).
[12] W. K. BISCHEL, P. J. KELLY and C. K. RHODES: *Phys. Rev. Lett.*, **34**, 300 (1975).
[13] F. BIRABEN, B. CAGNAC and G. GRYNBERG: *Compt. Rend.*, **280** B, 235 (1975).
[14] K. C. HARVEY, R. T. HAWKINS, G. MEISEL and A. L. SCHAWLOW: *Phys. Rev. Lett.*, **34**, 1073 (1975).
[15] J. E. BJORKHOLM and P. F. LIAO: *IEEE Journ. Quantum Electron.*, QE-**10**, 906 (1974).
[16] For example, see J. M. WORLOCK: in *Laser Handbook*, edited by F. T. ARECCHI and E. O. SCHULZ-DUBOIS (Amsterdam, 1972), p. 1323.
[17] W. GAUTSCHI: in *Handbook of Mathematical Functions*, edited by M. ABRAMOWITZ and I. A. STEGUN (Washington, D. C., 1965).
[18] See E. U. CONDON and G. H. SHORTLEY: *The Theory of Atomic Spectra* (London, 1967).
[19] P. LAMBROPOULOS, G. DOOLEN and S. P. ROUNTREE: *Phys. Rev. Lett.*, **34**, 636 (1975).
[20] M. LAMBROPOULOS, S. E. MOODY, S. J. SMITH and W. C. LINEBERGER: *Phys. Rev. Lett.*, **35**, 159 (1975).
[21] For a comprehensive review of a.c. Stark effect theory and early experiments see A. M. BONCH-BRUEVICH and V. A. KHODOVOI: *Usp. Fiz. Nauk*, **93**, 71 (1967) (English translation: *Sov. Phys. Usp.*, **10**, 637 (1968)).
[22] A. M. BONCH-BRUEVICH, N. N. KOSTIN, V. A. KHODOVOI and V. V. KHROMOV: *Žurn. Èksp. Teor. Fiz.*, **56**, 144 (1969) (English translation: *Sov. Phys. JETP*, **29**, 82 (1969)); P. PLATZ: *Appl. Phys. Lett.*, **14**, 168 (1969); B. DUBREUIL, P. RANSON and J. CHAPELLE: *Phys. Lett.*, **42** A, 323 (1972).
[23] For a recent review see W. HAPPER: in *Progress in Quantum Electronics*, Vol. **1**, Part 2 (Oxford, 1971), p. 51.
[24] F. BLOCH and A. SIEGERT: *Phys. Rev.*, **57**, 522 (1940).
[25] S. H. AUTLER and C. H. TOWNES: *Phys. Rev.*, **100**, 703 (1955).
[26] M. MIZUSHIMA: *Phys. Rev.*, **133**, A 414 (1964).
[27] For example, see, A. YARIV: *Quantum Electronics*, Chap. 13 (New York, N. Y., 1967).
[28] W. K. BISCHEL, P. J. KELLY and C. K. RHODES: Talk 7.2 presented at the *Conference on Laser Engineering and Applications, May 28-30, 1975, Washington, D. C.*

The Nonlinear Optics of Autoionizing Resonances.

J. A. ARMSTRONG and J. J. WYNNE

IBM T. J. Watson Research Center - Yorktown Heights, N. Y. 10598, U.S.A.

1. – Introduction.

The main aim of these lectures is to discuss the importance of the so-called autoionizing resonances in determining the nonlinear optical properties of atoms and simple molecules in the vacuum ultraviolet. These resonances give a highly structured character to certain parts of the continuum absorption spectra of many-electron atoms. See fig. 1. These « structured continua » show up strongly not only in single-photon absorption [1] and dielectronic recombination cross-sections, but also in 4-wave mixing susceptibilities [2] and in multiphoton ionization cross-sections.

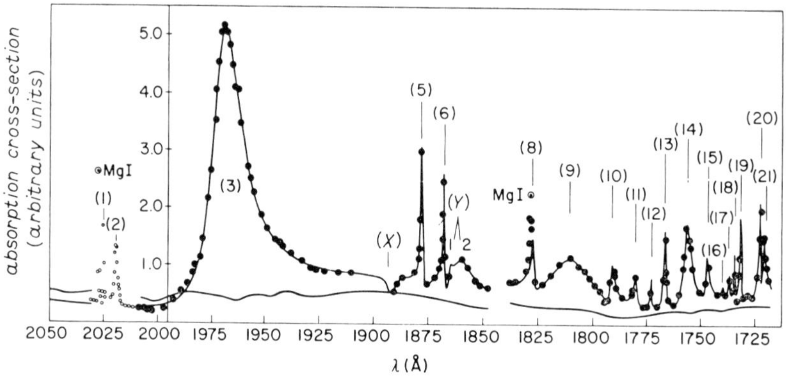

Fig. 1. – The absorption spectrum of atomic strontium in the vacuum UV autoionizing region. The ionization limit is off to the left of the figure at about 2180 Å. See ref. [1].

The nature and origin of these autoionizing (AI) states is, briefly, as follows. The ionization limit E_I of an atom is the energy required to remove the least tightly bound electron. For example, the ground configuration of SrI is $5s^2$, ${}^1S_0$ and the first ionization limit occurs as the limit of the series $5s^2\,{}^1S_0 - 5snp\,{}^1P_1^0$, *i.e.* as the limit of the excitation of one of the 5s-electrons into np-states with unchanged spin direction. This limit occurs [3] at 45932 cm^{-1}, or about 5.69 eV. If one gives the electron in the np-state still more energy, it breaks free of the atom, but still has p-type angular wave functions and spin antiparallel to the remaining 5s-electrons. These continuum states are denoted $5s\varepsilon p\,^1P_1^0$, where the ε denotes a free electron with continuous positive energy.

But it *may* happen that the energy required to excite both 5s-electrons to nominally bound, one-electron states is greater than the energy required to remove one 5s-electron altogether. For example, if both 5s-electrons are promoted to 4d-states, the configuration $4d^2\,^3P$ has levels which are bound, although loosely, at around 44600 cm^{-1}. But exciting the $5s^2$-configuration to $4d4f\,^1P^0$ requires 53546 cm^{-1} of energy, which is well above the ionization limit at 45932 cm^{-1}.

Note that the ionization limit corresponds to $5s\varepsilon p\,^1P_1^0$ with $\varepsilon = 0$; so with $\varepsilon = (53546 - 45932)$ cm^{-1}, we clearly have a degeneracy between $4d4f\,^1P_1^0$ and $5s\varepsilon p\,^1P_1^0$. This degeneracy between a nominally bound configuration (*e.g.*, $4d4f$) and a continuum of the same J is « lifted » by the configuration interaction, as was shown in an important set of papers by FANO [4, 5]. This interaction mixes the wave functions of, for example, $4d4f$ and a band of the $5s\varepsilon p$, and produces new continuum states which constitute the autoionizing resonance.

Fano's calculation will be outlined in some detail in sect. **2**. In sect. **3** we show how the autoionizing states play a role in resonantly enhanced 4-wave mixing experiments, and we calculate the nonlinear susceptibility $\chi^{(3)}$ for the process $2\nu_1 + \nu_2 = \nu_{\text{VUV}}$, where ν_1 and ν_2 are tunable laser frequencies, and $2\nu_1$ corresponds to a two-photon allowed transition. Section **4** gives a discussion of experimental work on autoionizing resonances as seen by 4-wave mixing, as well as a brief discussion of the process of resonantly enhanced photoionization as it applies to autoionizing levels; and, finally, we make concluding remarks about further areas in which AI states may be expected to play an important role.

2. – Fano's theory of AI states.

Fano's treatment of autoionization involves two distinct phases: the first phase, comprising ref. [4-6], deals essentially with isolated interaction between nominally bound, highly excited states and the continuum. In the second phase, ref. [7, 8], multichannel quantum defect theory (MQDT) has been applied to analyze whole Rydberg-type series of AI lines. For the present purpose, however, we will concentrate on the earlier and simpler theory, and in particular on the basic paper, ref. [4].

Since our aim is to introduce Fano theory to the reader, we will go through parts of the 1961 paper in detail, using Fano's notation and commenting on various points which may be unfamiliar to the nonlinear optician.

The « difficulties » of the theory of AI resonances stem mainly from the general unfamiliarity of perturbation theory as applied to overlapping discrete and continuous states.

We begin with the notation. Let the Hamiltonian be denoted by $\mathscr{H}$. The bound states of $\mathscr{H}$ are φ_j, with discrete energy eigenvalues, and the unbound states of $\mathscr{H}$ are ψ_E with continuous energy eigenvalues. The states φ_j are orthonormalized in the usual fashion, and we have

$$\langle\varphi_j|\mathscr{H}|\varphi_i\rangle = E_i\,\delta_{ij}\,. \tag{1a}$$

However, the unbound states ψ_E cannot be normalized in the usual sense, and one must choose one of a number of possible normalization schemes. The choice is not important so long as it is applied consistently throughout the calculation. Commonly used schemes are *a*) normalization in a box, *b*) Weyl's scheme involving wave packets [9] and *c*) delta-function normalization [10]. FANO chose to use scheme *c*), so that the analogue of eq. (1*a*) is

$$\langle\psi_{E''}|\mathscr{H}|\psi_{E'}\rangle = E'\,\delta(E''-E')\,. \tag{1b}$$

Moreover, the states are such that $\langle\psi_E|\varphi_j\rangle = 0$, and $\langle\psi_E|\psi_{E'}\rangle = \delta(E-E')$.

However, because of degeneracy between at least one φ_k and some $\psi_{E'}$, the matrix of $\mathscr{H}$ is not diagonal, and we wish to diagonalize that part of the matrix generated by the state φ_k and $\psi_{E'}$; the off-diagonal matrix element is

$$\langle\varphi_k|\mathscr{H}|\psi_E\rangle = V_{k,E}\,. \tag{1c}$$

As regards normalization and dimensions, from (1*b*) and from the property of the δ-function that $\int\delta(E)\,\mathrm{d}E = 1$, we see that the continuum wave functions have dimension $(\text{energy})^{-\frac{1}{2}}$ relative to the discrete wave functions. Therefore, $V_{kE'}$ has the dimension $(\text{energy})^{\frac{1}{2}}$, so that $|V_{kE}|^2$ will be an energy.

The diagonalization will produce a new state at each energy E, Ψ_E, which is a linear combination of φ_k and the ψ_E, as follows:

$$\Psi_E = a(E)\,\varphi + \int b_{E'}(E)\,\psi_{E'}\,\mathrm{d}E'\,. \tag{2}$$

(For the present we suppress the label on the state φ.). The new nondegenerate state Ψ_E at energy E *is an unbound state*, but its wave function near the atom reflects a contribution from the « bare » discrete state φ_k (*).

(*) Note that the integrand in (2) does not have any explicit factor for a density of states; that factor may be thought of as included in the expansion coefficient $b_{E'}$.

Apply the Hamiltonian operator to Ψ_E and project the resulting state in turn on φ and on $\psi_{E''}$. One finds

$$aE_\varphi + \int b_{E'} V^*_{E'} \mathrm{d}E' = aE , \tag{3a}$$

$$aV_{E''} + b_{E''}(E'' - E) = 0 . \tag{3b}$$

One cannot simply solve (3*b*) for $b_{E''}$ and blindly substitute in (3*a*), since E'' equals E during the integration in (3*a*), and hence the naive solution of (3*b*), *viz.* $b_{E''} = -aV_{E''}/(E_E - E)$, involves dividing by zero. In the section of his book [10] dealing with the properties of the delta-function, DIRAC points out that, since $\int x\delta(x)\,\mathrm{d}x = 0$, from an equation of the form $A(x) = B(x)$, if x can equal zero, one cannot conclude $A/x = B/x$, but rather $A/x = B/x + z\delta(x)$, where z is any quantity continuous at $x = 0$.

Thus, the solution of (3*b*) for $b_{E''}$ is

$$b_{E''} = aV^*_{E''}\left[\frac{1}{E - E''} + z(E)\delta(E'' - E)\right] . \tag{4}$$

The quantity $z(E)$ will be determined in what follows, and the integral over $(E - E')^{-1}$ is to be taken as the principal part. Put (4) in (3*a*), obtaining

$$aE_\varphi + a\int |V_{E''}|^2 \left\{\frac{1}{E - E''} + z(E)\delta(E'' - E)\right\} \mathrm{d}E'' = aE . \tag{5}$$

Since a is in general $\neq 0$, eq. (5) determines $z(E)$, *viz.*

$$z(E) = \left\{E - \left(E_\varphi + \mathrm{P}\int \frac{|V_{E''}|^2}{E - E''}\,\mathrm{d}E''\right)\right\} / |V_E|^2 . \tag{6}$$

The principal-part integral in (6) might be though of as shift in energy of the « bare » bound state φ due to the interaction with the continuum. However, this is somewhat misleading, since the proper eigenfunctions Ψ_E are *unbound* for all E, including the energy

$$E_s \equiv E_\varphi + \mathrm{P}\int \frac{|V_{E''}|^2\,\mathrm{d}E'}{E - E'} \equiv E_\varphi + F(E) .$$

This shifted energy E_s does, however, play an important role, as it is the energy which determines the « center » of the autoionizing feature in the continuum absorption; $F(E)$ is assumed to be constant over any one AI feature.

The quantity $z(E)$ can be interpreted as a phase shift by an argument which examines the particular superposition of continuum states occurring in the definition of Ψ_E (eq. (2)). It is well known [9] that the continuum wave functions can be chosen to have the asymptotic form

$$\psi_E \sim \sin\left(\frac{\sqrt{2mE}}{\hbar}\, r + \delta\right).$$

If such an expression is used, along with the $b_{E''}$ from (4) and (6), to calculate the quantity $b_{E'}\psi_{E'}\,\mathrm{d}E'$, which occurs in Ψ_E, we find

$$\text{(7)}\qquad \int b_{E'}\psi_{E'}\,\mathrm{d}E' = \int \mathrm{d}E'\, aV^*_{E'}\left\{\frac{1}{E-E'} + z(E)\delta(E'-E)\right\}\sin\left(\frac{\sqrt{2mE'}}{\hbar}\, r + \delta\right) =$$
$$= aV^*_E z(E)\psi_E + a\cdot \mathrm{P}\int \frac{V^*_{E'}\sin\left(r\sqrt{2mE'}/\hbar + \delta\right)}{E-E'}\,\mathrm{d}E'\,.$$

In the evaluation of the principal-part integral we assume $V^*_{E'}$ to be a slowly enough varying function of E', so that it can be replaced by V^*_E and taken out of the integral, which then can be evaluated to give

$$-\pi\cos\delta\cos\sqrt{\frac{2mE}{\hbar^2}}\, r + 2\sin\delta\left[\mathrm{Ci}\left(\sqrt{\frac{2mE}{\hbar^2}}\right) r\cos\sqrt{\frac{2mE}{\hbar^2}}\, r + \right.$$
$$\left. + \,\mathrm{Si}\left(\sqrt{\frac{2mE}{\hbar^2}}\, r\right)\sin\sqrt{\frac{2mE}{\hbar^2}}\, r\right].$$

The functions $\mathrm{Ci}(x)$ and $\mathrm{Si}(x)$ have, for large arguments (*i.e.* in the asymptotic region of interest to us), the limits 0 and $\pi/2$, respectively, so that we have finally

$$\text{(8)}\qquad \int b_{E'}\psi_{E'}\,\mathrm{d}E' \sim aV^*_E z(E)\sin\left(\sqrt{\frac{2mE}{\hbar^2}}\, r + \delta\right) - \pi a V^*_E\cos\left(\sqrt{\frac{2mE}{\hbar^2}}\, r + \delta\right).$$

Now note that, if we define an angle Δ such that $\mathrm{tg}\,\Delta = -\pi/z(E)$, the right-hand side of the asymptotic expression in (8) may be written as

$$\sqrt{\pi^2 + z^2}\, aV^*_E\sin\left(\sqrt{\frac{2mE}{\hbar^2}}\, r + \delta + \Delta\right) =$$
$$= aV^*_E\sqrt{\pi^2+z^2}\left(\cos\Delta\sin\left(\sqrt{\frac{2mE}{\hbar^2}}\, r + \delta\right) + \sin\Delta\cos\left(\sqrt{\frac{2mE}{\hbar^2}}\, r + \delta\right)\right).$$

This justifies the interpretation of Δ $(= -\mathrm{tg}^{-1}\pi/z(e))$ as a phase shift in the continuum wave function caused by its interaction with the bare bound state.

We have still to determine the quantity a which occurs in the expression for the new eigenfunction Ψ_E. It is determined from the requirement that

$$\text{(9)}\qquad \langle\Psi_E|\Psi_{\bar{E}}\rangle = \delta(E-\bar{E})\,.$$

Using eqs. (2) and (4) in (9) and recalling that φ and the ψ_E are orthogonal, we have

$$\langle \Psi_E | \Psi_{\bar{E}} \rangle = a^*(\bar{E})a(E) + \int b^*_{E'}(\bar{E}) b_{E'}(E)\, dE' = \tag{10}$$

$$= a^*(\bar{E})a(E)\left\{1 + \int dE' |V_{E'}|^2 \left[\frac{z(E)\delta(E'-E)}{\bar{E}-E'} + \frac{z(\bar{E})\delta(E'-\bar{E})}{E-E'} + \right.\right.$$

$$\left.\left. + z(E)z(\bar{E})\delta(E'-\bar{E})\delta(E'-E) + \frac{1}{(\bar{E}-E')(E-E')}\right]\right\}.$$

To evaluate the integral in (10) is a *tour de force* in the use of δ-functions. In addition to the following well-known relations [10]:

$$\delta(\bar{E}-E')\,\delta(E-E') = \delta(\bar{E}-E)\,\delta\big(E' - \tfrac{1}{2}(E+\bar{E})\big)\,, \tag{11a}$$

$$\delta(E-E')f(E') = \delta(E-E')f(E)\,, \tag{11b}$$

one needs the result, derived by FANO in appendix A of the 1961 paper, that the integration over the double pole in (10) is properly treated as follows:

$$\frac{1}{(\bar{E}-E')}\,\frac{1}{(E-E')} = \frac{1}{\bar{E}-E'}\left\{\frac{1}{E-E'} - \frac{1}{\bar{E}-E'}\right\} + \tag{11c}$$

$$+ \pi^2\delta(\bar{E}-E')\,\delta\left[E' - \frac{1}{2}(\bar{E}+E)\right].$$

Using relations (11), one can easily put eq. (10) in the form

$$\langle \Psi_E | \Psi_{\bar{E}} \rangle = |a(E)|^2\,[z^2(E)+\pi^2]\,|V_E|^2\,\delta(\bar{E}-E) = \delta(\bar{E}-E)\,, \tag{12}$$

from which the normalizing value of $|a|^2$ is obviously

$$|a|^2 = \frac{1}{|V_E|^2(\pi^2+z^2)} = \frac{|V_E|^2}{[E-E_\varphi-F(E)]^2+\pi^2|V_E|^4}\,. \tag{13a}$$

The coefficient $a(E)$ can be written in terms of the phase shift $\Delta = -\operatorname{tg}^{-1}\pi/z$

$$a = (\sin\Delta)/\pi V_E\,. \tag{13b}$$

So, finally, the continuum wave function of energy E in the neighborhood of an autoionizing feature can be written

$$\Psi_E = \frac{\sin\Delta}{\pi V_E}\,\varphi + \frac{\sin\Delta}{\pi V_E}\,P\!\int \frac{V_{E'}\psi_{E'}}{E-E'}\,dE' - (\cos\Delta)\psi_E - \tag{14a}$$

$$= \frac{\sin\Delta}{\pi V_E}\left\{\varphi + P\!\int \frac{V_{E'}\psi_{E'}\,dE'}{E-E'}\right\} - (\cos\Delta)\psi_E \equiv \tag{14b}$$

$$\equiv \frac{\sin\Delta}{\pi V_E}\,\Phi - (\cos\Delta)\psi_E\,. \tag{14c}$$

The quantity in braces in (14*b*) has been denoted Φ and is called by FANO the « modified discrete state ». But it should be borne in mind that the modified discrete state Φ is an *unbound* state; and the radial-part wave function Φ may bear little or no resemblance to that of state φ.

We found in discussing the phase shift Δ that the principal-part integral in Φ can be readily evaluated under the assumption that the $\psi_{E'}$ have the asymptotic character of standing waves, $\sin(kr+\delta)$. Hence, well out from the nucleus, Φ has the general form

$$\Phi \sim \varphi + \varkappa \cos(kr+\delta)\,, \tag{15}$$

where $\varkappa$ is independent of r, is given by the normalization of $\psi_{E'}$, and contains the angular dependence of $\psi_{E'}$. Clearly, there can be interferences between the two parts of Φ, and the character of φ may not be reflected at all faithfully in the modified discrete state. This may show up, for example, in that the transition matrix element between the ground state and Φ may be very different from what it would be between the ground state and the bare discrete state φ; we will return to this point when discussing the meaning of Fano's q-parameter.

Consider now the transition matrix element of an operator T (which may be the dipole-moment operator) between the ground state φ_g and an AI state whose wave function is given by eq. (14*c*):

$$\langle\Psi_E|T|\varphi_g\rangle = \frac{\sin\Delta}{\pi V_E^*}\langle\Phi|T|\varphi_g\rangle - \cos\Delta\langle\psi_E|T|\varphi_g\rangle\,. \tag{16}$$

As a function of the continuum energy E, the only *rapid* variation of $\langle\Psi_E|T|\varphi_g\rangle$ comes from the phase factors $\sin\Delta$ and $\cos\Delta$. Since these are given by

$$\sin\Delta = \pi|V_E|^2/\sqrt{\pi^2|V_E|^4 + (E_\varphi + F(E) - E)^2} \tag{17a}$$

and

$$\cos\Delta = (E_\varphi + F(E) - E)/\sqrt{\pi^2|V_E|^4 + (E_\varphi + F(E) - E)^2}\,, \tag{17b}$$

we see that $\sin\Delta$ varies from 0 through 1 to 0, and $\cos\Delta$ varies from 1 to 0 to -1, as E goes from below resonance to above resonance. Resonance is defined as $E = E_\varphi + F(E)$; the width of the resonance is given by $\pi|V_E|^2$. Because of the above rapid variations in Δ as E goes through resonance, the transition matrix element (16) *always* goes through zero near the resonance. Physically, this means that the two transition dipoles, $((\sin\Delta)/\pi V_E^*)\langle\Phi|T|\varphi_g\rangle$ and $(\cos\Delta)\langle\psi_E|T|\varphi_g\rangle$, will always be equal in strength and 180° out of phase at some energy close to resonance. This gives rise in continuum photoabsorption spectra to the characteristic « windows » or low-absorption regions associated with AI resonances.

The transition probability associated with the operator T between states φ_g and Ψ_E is, from (16),

$$|\langle\Psi_E|T|\varphi_g\rangle|^2 = |\langle\psi_E|T|\varphi_g\rangle|^2|q\sin\Delta - \cos\Delta|^2\,, \tag{18}$$

where we have introduced Fano's q-parameter

$$q \equiv \frac{\langle\Phi|T|\varphi_g\rangle}{\pi V_E^*\langle\psi_E|T|\varphi_g\rangle}\,. \tag{19}$$

From (18) we can form the ratio between the transition probability to the mixed state Ψ_E and that to the bare continuum state ψ_E, *viz.*

$$\frac{|\langle\Psi_E|T|\varphi_g\rangle|^2}{|\langle\psi_E|T|\varphi_g\rangle|^2} = \frac{(q+\varepsilon)^2}{1+\varepsilon^2}\,, \tag{20}$$

where the quantity ε is defined as

$$\varepsilon = \frac{E - E_\varphi - F(E)}{\pi|V_E|^2} \tag{21}$$

(see eqs. (17*a*), (17*b*)).

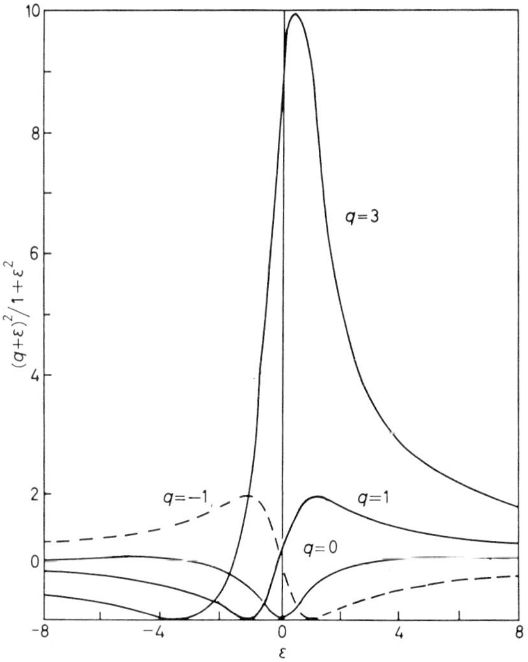

Fig. 2. – The various Fano-Beutler profiles for the absorption in an autoionizing feature. See ref. [4].

Formula (20) gives the so-called Fano-Beutler profile for the photoabsorption cross-section near an AI resonance. The various shapes which occur, depending on the value of q, are shown in fig. 2. Note that the cross-section goes to zero at $\varepsilon = -q$. For large q, the line is sharp and nearly Lorentzian. For $q = 0$, the absorption cross-section has an « antiresonance », or window. The condition for $q = 0$ corresponds to $\langle\Phi|T|\varphi_g\rangle = 0$; from eq. (15) we see that

$$\langle\Phi|T|\varphi_g\rangle \sim \langle\varphi_k|T|\varphi_g\rangle + \varkappa\langle\cos(kr+\delta)|T|\varphi_g\rangle\,, \tag{22}$$

and that, if $q = 0$, it means that the projection of the ground state onto Φ is composed of two equal but opposite contributions, one from the bare discrete state and the other from the admixed continuum state.

3. – Autoionizing resonances in 4-wave parametric interactions.

In this section we describe how the 1961 Fano theory of AI resonances can be extended to explain recent experiments in which AI states were studied by the generation of tunable vacuum UV radiation [2]. The experiment consists of mixing two tunable dye laser beams (of frequencies ν_1 and ν_2) in a cell containing a metal vapor, *e.g.* Sr [11]. The experimental set-up is shown in fig. 3.

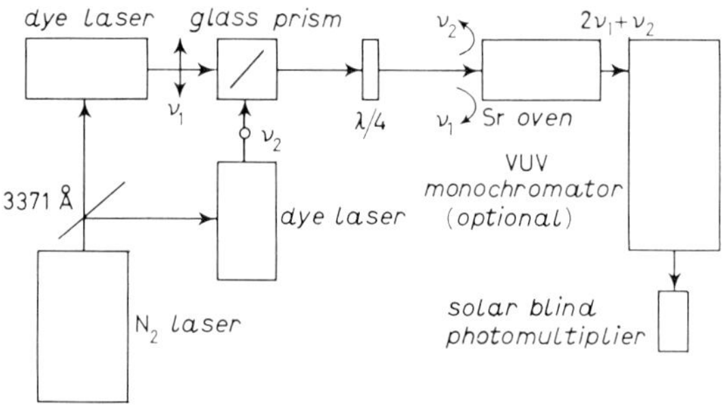

Fig. 3. – The experimental set-up for VUV generation by 4-wave parametric interactions. The ($\lambda/4$)-plate is used to suppress third-harmonic generation. See ref. [11].

The nonlinear process of interest is the one in which the combination $2\nu_1 + \nu_2 = \nu_{\text{VUV}}$ is generated, with $2\nu_1$ chosen to resonate a two-photon allowed transition [12]. This results in a large enhancement of the nonlinear susceptibility. The second laser ν_2 is then tuned, and scans ν_{VUV}. When ν_{VUV} is tuned through an autoionizing resonance, there is a dramatic increase in the VUV signal, and a line shape of generation is observed which is to be associated with the autoionizing resonance transition. Such an output spectrum is shown in fig. 4 for the case of Sr; compare with fig. 1.

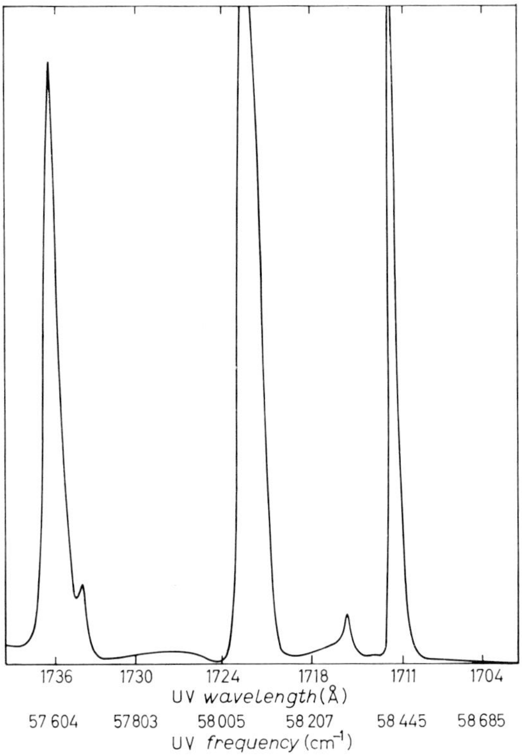

Fig. 4. – The 4-wave parametric generation spectrum of strontium in the vacuum UV; cf. fig. 1.

This was explained in ref. [2] as follows. Because of the two-photon resonant enhancement, the nonlinear susceptibility $\chi^{(3)}$ for the process $2\nu_1 + \nu_2 = \nu_{\text{VUV}}$ may be approximated by [13]

$$\chi^{(3)}(\nu_{\text{VUV}}) \propto \sum_{\substack{j,k \\ \text{only}}} P_{gk} P_{kj'} P_{j'j} P_{jg} / (\nu_{kg} - \nu_{\text{VUV}})(\nu_{j'g} - 2\nu_1)(\nu_{jg} - \nu_1) \,. \tag{23}$$

This may be rewritten as

$$\chi^{(3)} \propto \frac{1}{\nu_{j'g} - 2\nu_1} \sum_j \frac{P_{j'j} P_{jg}}{\nu_{jg} - \nu_1} \sum_k \frac{P_{gk} P_{kj'}}{\nu_{kg} - \nu_{\text{VUV}}} \,. \tag{24}$$

Here P is the electric-dipole operator, g labels the ground state, j and j' label discrete, excited states (with j' being the two-photon state), and k labels the states of the continuum, the states we have called Ψ_E in sect. **2**. The factors in (24) require some comment. If the spectral width of the first dye laser $\Delta\nu_1$ is such that $\Delta\nu_1 \ll \Gamma_{j'}$, where $\Gamma_{j'}$ is the line width of the two-photon excited state, then

$1/(\nu_{j'g} - 2\nu_1)$ should be set equal to $i\Gamma_j^{-1}$; if, as was the case in ref. [2], $2\Delta\nu_1 \gg \Gamma_{j''}$, then an approximate treatment of the implied integration over all frequency components of ν_1 is simply to replace $\nu_{j'g} - 2\nu_1$ by $i(2\Delta\nu_1)^{-1}$. If $\Delta\nu_1$ and $\Gamma_{j'}$ are comparable, a more careful treatment is required.

The second factor in (24), the sum over the states j, may be replaced for present purposes by $N \equiv [P_{j'j}P_{jg}/(\nu_{jg} - \nu_1)]_{\text{effective}}$. This is because a) we are not varying ν_1 and b) we are not trying to calculate the magnitude of $\chi^{(3)}$ but only its rapid variation near the AI state.

Hence we have

$$\chi^{(3)} \propto \frac{N}{2\,\Delta\nu_1} \sum_k \frac{P_{gk}\,P_{kj'}}{\nu_{kg} - \nu_{\text{VUV}}}\,. \tag{25}$$

The quantity P_{gk} is given by (see eq. (18))

$$P_{gk} = \langle\varphi_g|P|\psi_E\rangle(q_g \sin\Delta - \cos\Delta)\,, \tag{26a}$$

and $P_{kj'}$ is given by

$$P_{kj'} = \langle\psi_E|P|\varphi_{j'}\rangle(q_{j'} \sin\Delta - \cos\Delta)\,. \tag{26b}$$

Observe that Δ is the same in each case, being a property of the AI state only; moreover, we have introduced two q's, one describing the connection between the AI state and the ground state, the other describing the connection between the AI state and the two-photon state. Explicitly, we have set

$$q_g = \frac{\langle\Phi|P|\varphi_g\rangle}{\pi V_E^*\langle\psi_E|P|\varphi_g\rangle} \tag{27a}$$

and

$$q_{j'} = \frac{\langle\Phi|P|\varphi_{j'}\rangle}{\pi V_E^*\langle\psi_E|P|\varphi_{j'}\rangle}\,. \tag{27b}$$

Now, returning to eq. (25), replacing the sum over states k by the integral over energy E, we find

$$\chi^{(3)} \propto \frac{N}{2\,\Delta\nu_1}\langle\varphi_g|P|\psi_E\rangle\,\langle\psi_E|P|\varphi_{j'}\rangle \int \frac{q_g\,q_{j'}\sin^2\Delta + \cos^2\Delta - (q_g + q_{j'})\sin\Delta\cos\Delta}{E - h_{\text{VUV}}}\,\mathrm{d}E \propto$$

$$\propto \frac{N}{2\,\Delta\nu_1}\int P_{g\psi_E}P_{\psi_E j'}\left\{\frac{(q_g q_{j'} - 1)\sin^2\Delta - (q_g + q_{j'})\sin\Delta\cos\Delta + 1}{E - h\nu_{\text{VUV}}}\right\}\mathrm{d}E\,.$$

In evaluating the above integral we have to be careful about the singularity at $E = h\nu_{\text{VUV}}$. This is properly done [14] by recalling that, although the AI states Ψ_E are eigenstates of $\mathscr{H}$, they do in fact decay due to processes we have not taken into account, and hence E should be replaced by $E - i\gamma$, where γ is some small, E-independent broadening of the level. With this substitution, and the fact that $\sin \Delta = 1\sqrt{1+\varepsilon^2}$ from eq. (17), we find that

$$\chi^{(3)} \propto \frac{N}{2\,\Delta\nu_1}\int \frac{P_{g\psi_E}P_{\psi_{Ej'}}}{E-h\nu_{\mathrm{VUV}}-i\gamma}\left\{\frac{q_g q_{j'}-1}{1+[(E-E_s)/\pi V_E^2]^2}+\right.$$
$$\left.+\frac{q_g+q_{j'}}{1+[(E-E_s)/\pi V_E^2]^2}\,\frac{E-E_s}{\pi V_E^2}+1\right\}\mathrm{d}E\,.$$

Recall that $E_s = E_\varphi + F(E)$ is the energy of the shifted, modified discrete state Φ; let us measure E from E_s in units of $\pi|V_E|^2$, the line width. That is, let $y=(E-E_s)/\pi|V_E|^2$, and call

$$x=(E_s-h\nu_{\mathrm{VUV}})/\pi|V_E|^2\,,\qquad \gamma'\equiv\gamma/\pi|V_E|^2\,,$$

then

$$\chi^{(3)} \propto \frac{N}{2\,\Delta\nu_1}\int_{-\infty}^{\infty}\frac{P_{g\psi_E}P_{\psi_{Ej'}}}{y+x-i\gamma'}\left\{\frac{q_g q_j-1+(q_g+q_{j'})y+1+y^2}{1+y^2}\right\}\mathrm{d}y\,. \tag{28}$$

The integral is done easily by contour integration. The denominator in the braces has factors $(i+y)(-i+y)$. Hence, in the lower half-plane the only pole is at $y=-i$. We use the following contour:

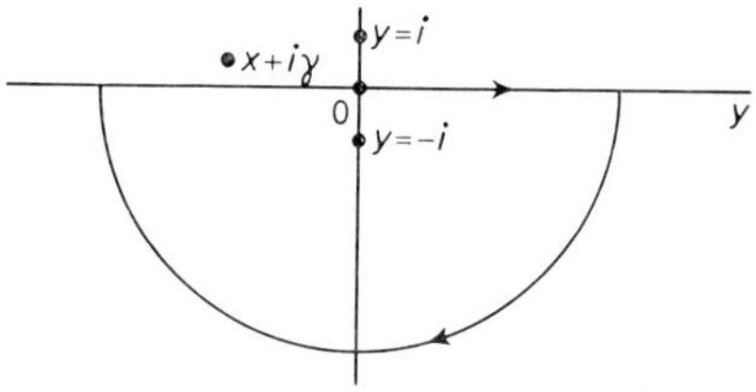

The contribution from the clockwise semi-circle is $-i\pi\big((N/2\,\Delta\nu_1)\,P_{g\psi_E}P_{\psi_{Ej'}}\big)$; the residue times $-2\pi i$ is

$$\frac{\pi N}{2\,\Delta\nu_1}\,\frac{P_{g\psi_E}P_{\psi_{Ej}}[(q_g q_{j'}-1)-i(q_g+q_{j'})]}{x-i(1+\gamma)}\,,$$

so we have finally that

$$\chi^{(3)} \propto \frac{\pi N}{2\,\Delta\nu_1}\,P_{g\psi_E}P_{\psi_{Ej'}}\left\{\frac{(q_a q_{j'}-1)-i(q_g+q_{j'})}{x-i(1+\gamma)}+i\right\}. \tag{29}$$

We can set $\gamma=0$ in eq. (29), since $\gamma \ll \pi|V_E|^2$; then expressed in terms of real and imaginary parts, we have (omitting for clarity the common constant factors)

$$\mathrm{Re}\,\chi^{(3)} \propto \frac{q_g+q_{j'}+x(q_g q_{j'}-1)}{1+x^2}\,, \tag{30a}$$

$$\mathrm{Im}\,\chi^{(3)} \propto \frac{(x-q_g)(x-q_{j'})}{1+x^2}\,. \tag{30b}$$

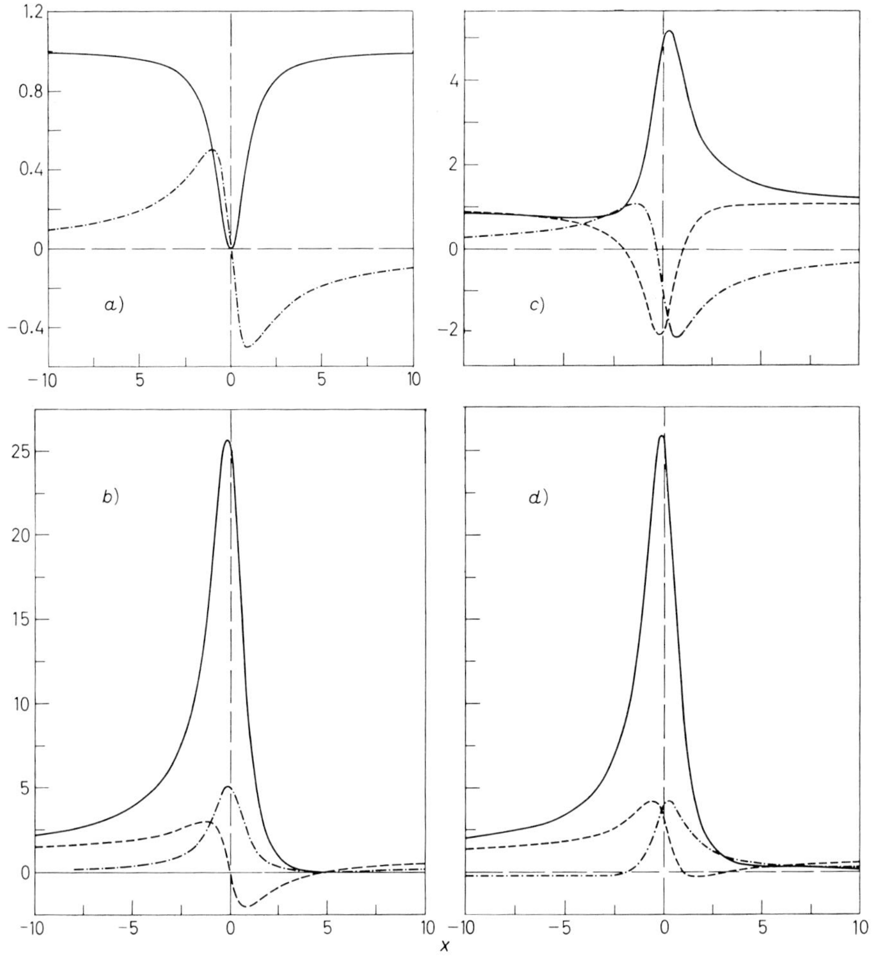

Fig. 5. – The real and imaginary parts of $\chi^{(3)}$, along with $|\chi^{(3)}|^2$ for several pairs of q's. $x = (E_s - h\nu)/\pi V_E^2$. *a*) ——— χ^2, Im χ; — — — Re χ; $q_1 = q_2 = 0$. *b*) ——— χ^2, —·—·— Re χ, — — — Im χ, $q_1 = 0$, $q_2 = 5$. *c*) ——— χ, —·—·— Re χ, — — — Im χ, $q_1 = -2$, $q_2 = 1$. *d*) ——— χ, —·—·— Re χ, — — — Im χ, $q_1 = 1$, $q_2 = 3$.

In ref. [2], only the expression for Re $\chi^{(3)}$ was derived because of failure to take into account the small imaginary part of the eigenvalue E. The real and imaginary parts of $\chi^{(3)}$ are shown in fig. 5*a*)-*d*) for several pairs of q values, along with $|\chi|^2$.

A few comments about the autoionizing 4-wave susceptibility (30) are in order.

The fact that $\chi^{(3)}$ is complex does not imply that there is loss at $\nu_{\text{VUV}} = 2\nu_1 + \nu_2$, but rather that the nonlinear polarization at $2\nu_1 + \nu_2$ has shifted phase with respect to the phase of the driving nonlinear field $E^2_{\nu_1}(z, t)\, E_{\nu_2}(z, t)$.

This phase shift is

$$\eta = \operatorname{tg}^{-1}\frac{(x-q_g)(x-q_{j'})}{q_g+q_{j'}+x(q_g q_{j'}-1)}\,. \tag{31}$$

Far from the AI resonance $\operatorname{Re}\chi^{(3)}=0$, $\operatorname{Im}\chi^{(3)}=$ constant, and $\eta=(\pi/2)\cdot\operatorname{sgn}(q_g q_{j'}-1)$. Physically this reflects the fact that the nonlinear susceptibility of the *smooth* continuum is purely imaginary and constant (apart from the slow variation of $P_{g\psi_E}$ and of $P_{\psi_E j'}$) in energy.

The expression for $|\chi^{(3)}|^2$ is readily found from eq. (29) to be [14] (with $\gamma=0$)

$$|\chi^{(3)}|^2 \propto \frac{q_g^2 q_{j'}^2+[q_g+q_{j'}-x]^2}{1+x^2}\,. \tag{32}$$

This form differs from that of ref. [14] because of our different convention for x. For large x, $|\chi^{(3)}|^2$ becomes constant at a value *roughly* $q_g^2 q_{j'}^2$ times smaller than at the peak of the line (this is for large q_i). The shape of $|\chi^{(3)}|^2$ depends on both q's; if q_g is known from linear measurements, observation of the 4-wave generation spectrum allows one to infer the value of $q_{j'}$. This information is not readily available otherwise.

Clearly, we can use the preceding theoretical expressions, in connection with 4-wave mixing experiments, to study the autoionizing states themselves. This is a nonlinear, vacuum UV spectroscopy which can be done without a grating (*), with little or no vacuum, and with a resolution equal to $\sim 2\Delta\nu_1+\Delta\nu_2$, which can easily be better than that of all but the largest and costliest gratings.

But, in addition to their spectroscopy, the AI states are of interest in the generation of ever shorter wavelengths by multiple-wave mixing experiments. The AI states drastically rearrange the oscillator strength of the continuum, and make it feasible to do, *e.g.*, 6-wave mixing experiments, using resonant enhancement in the continuum. This will be discussed further in sect. **5**.

4. – Experimental studies.

In ref. [2], ARMSTRONG and WYNNE described the following experimental observations, made on 4-wave VUV generation in strontium vapor. The AI state k under study was one belonging to the $4d4f$-configuration at 1867 Å. In the first instance the two-photon state $|j'\rangle$ was chosen to be $5p^2\,{}^1D_2$, and in the second instance $|j'\rangle$ was $5s5d\,{}^1D_2$. In each case ν_2 was then tuned so that $2\nu_1+\nu_2$ scanned through the $4d4f$ resonance, giving rise to a strong variation

(*) Third-harmonic signals $(3\nu_1, 3\nu_2)$ are suppressed by circularly polarizing both laser beams.

in generated VUV. The experimental results are shown in fig. 6. The observation is that the spectral shape of the generated output at 1867 Å depends markedly on the two-photon intermediate state $|j'\rangle$. Since the spectrum is to be understood in terms of eq. (32), we infer that the $q_{j'}$ connecting $5p^2\,{}^1D_2$ to $4d4f\,{}^1P_1$ must be substantially different from the $q_{j'}$ connecting $5s5d\,{}^1D_2$ to $4d4f\,{}^1P_1$. In fact, since the q_g connecting $5s^2\,{}^1S_0$ to $4d4f\,{}^1P_1$ is known from

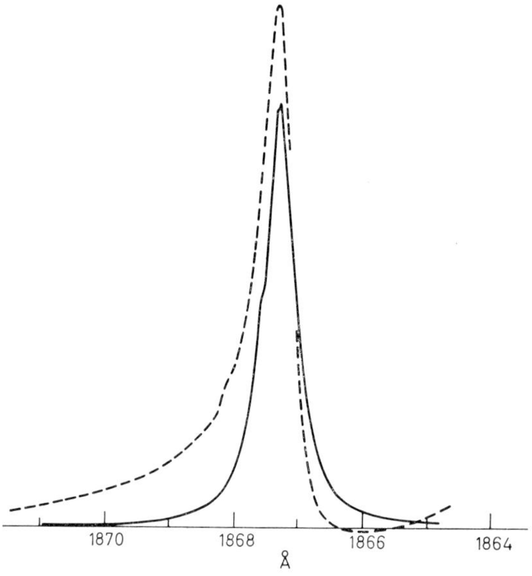

Fig. 6. – Experimental dependence of generated spectrum at an AI line on the nature of the 2-photon intermediate state. See ref. [2]. $4d4f$ AI resonance, 2-photon intermediate, – – – $5p^2$, ——— $5s5d$.

ref. [1] to be -3.5 (*), a careful fit of the data in fig. 6 to eq. (32) will produce different values for the different $q_{j'}$. Figures 7*a*) and *b*) are taken from ref. [14]; they show that the experimental generation spectra are well described by the 2-q formula with $q_{5p^2-4d4f} = -0.6$ and $q_{5s5d-4d4f} = 2.1$. As mentioned previously, the slow fall-off of the line shape in the wings is due to the nonlinear susceptibility of the « pure » continuum. Each of the fits produced a line width in good agreement with the result of ref. [1] obtained from linear absorption.

There are several precautions which were taken to ensure that the observed spectrum was determined only by $|\chi|^2$ and not by other factors. First, the phase matching of $2\nu_1 + \nu_2 = \nu_{\text{VUV}}$ by buffer gas [12] (usually Xe) was

(*) The sign of a q value from ref. [1] must be changed in order to be in agreement with Fano's definition.

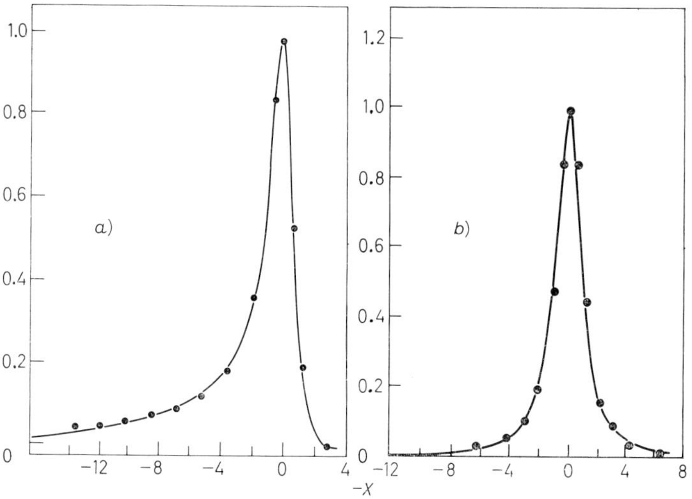

Fig. 7. – Comparison of experimental and theoretical line shapes for 4-wave parametric generation in the vacuum UV. See ref. [2, 14]. —— experiment, • theory. *a*) $q_g = -3.5$, $q_{j'} = -0.6$, $5p^2$; *b*) $q_g = -3.5$, $q_{j''} = 2.1$, $5s5d$.

purposely adjusted *away* from optimum, so that it would not be sensitive either to the slight variation in the index of refraction of Sr vapor at the AI feature itself or to the change of index at ν_2 as that frequency was tuned. Second, the input beams were brought to a focus just before the output window of the cell, so that absorption of the generated VUV light would not be a complication. In addition, since the absorption coefficient occurs exponentially in the expression for the generated light, observations were made at the lowest possible pressure consistent with a good signal-to-noise ratio.

Three-photon ionization. Another related experiment, carried out by WYNNE and SOROKIN, was to replace to detector of ν_{VUV} by an electron collector, which detected the electrons released by the decay of the AI state. As before, $2\nu_1$ was adjusted to resonate a two-photon allowed intermediate state, and ν_2 was then tuned to scan $2\nu_1 + \nu_2$ through the AI resonance while the electron current was monitored. This is a lower-order process, involving only three waves and a *real transition* of the atom from its ground state to the AI state. Moreover, since the system does not return to its ground state, the selection rules are different. That is, one can study AI states whose J value is different by 3 from that of the ground state. For the case of strontium this implies that F-states (or, more accurately, states with $J = 3$) can be investigated; these states are not accessible either by conventional absorption spectroscopy or by the 4-wave parametric processes described earlier in these lectures.

The matrix element for the real transition, assuming two-photon resonant enhancement as before, will be of the form

$$M \propto \frac{1}{\nu_{j'g} - 2\nu_1} \left\{ \sum_j \frac{P_{j'j} P_{jg}}{\nu_{jg} - \nu_1} \right\} P_{kj'} . \tag{33}$$

As before, the only term which varies rapidly in energy near the AI feature is $P_{kj'}$, which is given by (26*b*) and (27*b*) as before. It therefore becomes possible to see the Fano-Beutler resonance profile associated with a one-photon transition to the AI state from the $|j'\rangle$-state, even though the latter is not actually populated to a significant degree. The $q_{j'}$ determined in this fashion should be the same as that deduced from the 4-wave process in cases where the AI feature is accessible by both techniques.

Further work along several directions is needed in the nonlinear spectroscopy of AI resonances.

1) Systematic experimental studies of the $q_{j'}$ for Mg, Ca, Sr and Ba should be done.

2) The experiments done to date have measured $|\chi^{(3)}|^2$ rather than $\chi^{(3)}$ itself. It is possible to measure Re $\chi^{(3)}$ and Im $\chi^{(3)}$ separately and exhibit their particular resonant line shapes. A technique analogous to heterodyne detection would be as follows. The two dye laser beams ν_1, ν_2 of fig. 3 could be passed first through a cell containing a nonresonant nonlinear medium (some other atomic vapor, for example). There would be generated a strong optical field at $2\nu_1 + \nu_2$, say $F(2\nu_1 + \nu_2)$. The field F, along with the fields at ν_1 and ν_2, would then go into the cell containing the resonant medium (*e.g.*, Sr). The field at $2\nu_1 + \nu_2$ in this second medium, say $f_{\text{res}}(2\nu_1 + \nu_2)$, will be phase coherent with F. Their sum $F + f$ is detected as $(F + f)(F^* + f^*) \simeq |F|^2 + 2\,\text{Re}\,Ff$ if $|F| \gg |f|$. $|F|^2$ is nonresonant, so the signal 2 Re (Ff) is what varies as ν_2 is tuned through the AI resonance. If a variable phase delay at $2\nu_1 + \nu_2$ (Xe gas at adjustable pressure, for instance) is interposed between the two cells, 2 Re (Ff) may be made proportional either to Re $\chi^{(3)}$ or to Im $\chi^{(3)}$. Analogous techniques have been used on other nonlinear susceptibilities (see ref. [15]).

3) The multichannel quantum defect method could be applied to the calculation of $\chi^{(3)}$ for a whole Rydberg series and the adjoining series of AI resonances of the same J and parity. Some work is already progressing along these lines [16].

4) The basic paper of FANO on AI lines [4] treated, in addition to the case of an isolated line interacting with a single continuum, the more complex cases of *a*) several discrete lines interacting with a single continuum and *b*) a single line interacting with several continua. Case *a*) above can readily be applied to the calculation of 4-wave generation line shapes when several AI resonances are close to each other.

5) Since the AI lines provide structure in the continuum, they may be used for resonant enhancement of the $\chi^{(5)}$ susceptibility, which would govern 6-wave parametric interactions and which would allow generation of light still deeper into the VUV.

6) Further and systematic studies of AI lines and $q_{j'}$ values by means of the resonantly enhanced multiphoton ionization technique should be carried out.

REFERENCES

[1] W. R. S. GARTON, G. L. GRASDALEN, W. H. PARKINSON and E. M. REEVES: *J. Phys. B*, **1**, 114 (1968).

[2] J. A. ARMSTRONG and J. J. WYNNE: *Phys. Rev. Lett.*, **33**, 1183 (1974).

[3] W. R. S. GARTON and K. CODLING: *J. Phys. B*, **1**, 106 (1968).

[4] U. FANO: *Phys. Rev.*, **124**, 1866 (1961).

[5] U. FANO and J. W. COOPER: *Phys. Rev.*, **137**, A 1364 (1963).

[6] U. FANO: *Nuovo Cimento*, **12**, 156 (1935).

[7] U. FANO: *Phys. Rev. A*, **2**, 353 (1970).

[8] C.-M. LEE and K. T. LU: *Phys. Rev. A*, **8**, 1241 (1973).

[9] H. A. BETHE and E. SALPETER: *Quantum Mechanics of One- and Two-Electron Atoms* (New York, N. Y., 1957).

[10] P. A. M. DIRAC: *The Principles of Quantum Mechanics*, 4th Ed. (Oxford, 1958).

[11] R. T. HODGSON, P. P. SOROKIN and J. J. WYNNE: *Phys. Rev. Lett.*, **32**, 343 (1974).

[12] R. B. MILES and S. F. HARRIS: *IEEE Journ. Quantum Electron.*, **9**, 470 (1973).

[13] J. A. ARMSTRONG, N. BLOEMBERGEN, J. DUCUING and P. S. PERSHAN: *Phys. Rev.*, **127**, 1918 (1962).

[14] L. ARMSTRONG jr. and B. L. BEERS: *Phys. Rev. Lett.*, **34**, 1290 (1975).

[15] R. K. CHANG, J. DUCUING and N. BLOEMBERGEN: *Phys. Rev. Lett.*, **15**, 6 (1965). Also J. J. WYNNE: *Phys. Rev.*, **178**, 1295 (1969).

[16] U. FANO and K. T. LU: private communication.

Some Fundamental Aspects of Nonlinear Optics.

Y. R. SHEN

Department of Physics, University of California
Inorganic Materials Research Division, Lawrence Berkeley Laboratory - Berkeley, Cal. 94720

We discuss here a number of selected topics related to the fundamentals of nonlinear optics. First, in sect. **1**, we give a brief description on the theory of nonlinear optical susceptibilities. Emphasis is on the basic properties, on the recently developed bond theory and on the dispersion of nonlinear susceptibilities. Then, in sect. **2**, we discuss the coupled-wave approach. We show that the transverse resonant excitation in a material can be treated as a material excitational wave, and most nonlinear-optics problems can be understood as results of coupling between e.m. waves or between e.m. and material excitational waves. A number of examples of current interest are presented. These include sum and difference frequency generation, Raman scattering by coherent excitation, stimulated Raman and stimulated polariton scattering, four-wave mixing and vibrational relaxation measurements. Finally, in sect. **3**, we consider the physical distinction between laser emission and resonant stimulated Raman scattering in an optically pumped three-level system. A similar problem on the distinction of hot luminescence and resonant Raman scattering has recently attracted much attention. We show that the hot-laser action is directly connected with the population inversion, while the resonant stimulated Raman scattering is connected with transverse excitation. The problem is, however, more complicated for a multilevel system.

1. – Nonlinear susceptibilities.

The theoretical description of any optical phenomenon always begins with the Maxwell equations or the resulting wave equations

$$(1)\qquad \begin{cases} \left[\nabla\times(\nabla\times)+\dfrac{1}{C^2}\dfrac{\partial^2}{\partial t^2}\right]\boldsymbol{E}(\boldsymbol{r},t)=-\dfrac{4\pi}{C^2}\dfrac{\partial^2}{\partial t^2}\boldsymbol{P}(\boldsymbol{r},t)\,,\\[2ex] \nabla\cdot\boldsymbol{E}=-4\pi\nabla\cdot\boldsymbol{p}\,, \end{cases}$$

where $\boldsymbol{P}(\boldsymbol{r}, t)$ is the electric-dipole polarization. More generally, $\boldsymbol{P}$ should be replaced by a generalized electric polarization $\boldsymbol{\pi}$ which contains not only $\boldsymbol{P}$ but also all the multipole contributions, including the conduction current density $\boldsymbol{J}_{\text{cond}}$, the dipole magnetization $\boldsymbol{M}$, the electric-quadrupole polarization $\mathbf{Q}$, etc. [1-3]:

$$\partial\boldsymbol{\pi}/\partial t = \boldsymbol{J}_{\text{cond}} + \partial P/\partial t - C\nabla\times\boldsymbol{M} + \partial\nabla\cdot\mathbf{Q}/\partial t + \dots . \tag{2}$$

In these lectures, for the sake of convenience in presentation, we shall assume $\boldsymbol{\pi} = \boldsymbol{P}$ unless othewise specified.

As a response to the external field, $\boldsymbol{P}$ should of course be a function of $\boldsymbol{E}$. While $\boldsymbol{P}$ must be linear in $\boldsymbol{E}$ in the weak-field limit, it is in general a nonlinear function of $\boldsymbol{E}$. In principle, a nonlinear optical effect is fully described if $\boldsymbol{P}$ as a nonlinear function of $\boldsymbol{E}$ is known and the solution of eq. (1) can be found. Unfortunately, this is usually not the case. Theories of nonlinear optics deal mainly with the problem of how reasonable approximations can be made to describe $\boldsymbol{P}(\boldsymbol{r}, t)$ properly and to solve eq. (1) accordingly.

Formally, $\boldsymbol{P}(\boldsymbol{r}, t)$ as a function of $\boldsymbol{E}(\boldsymbol{r}, t)$ obeys an equation of motion, which can be derived quantum mechanically from [2, 4]:

$$\begin{cases} \boldsymbol{P}(\boldsymbol{r}, t) = \mathrm{Tr}\,[\varrho(\boldsymbol{r}, t)\boldsymbol{p}]\,, \\ i\hbar\dfrac{\partial\varrho}{\partial t} = [\mathscr{H}, \varrho] + i\hbar\left(\dfrac{\partial\varrho}{\partial t}\right)_{\text{damping}}, \end{cases} \tag{3}$$

where ϱ is the density matrix operator, $\boldsymbol{p}$ is the electric-dipole operator, and $\mathscr{H}$ is the total Hamiltonian. However, such an equation for $\boldsymbol{P}$ is usually too complicated to be useful. In some case, a simple equation for $\boldsymbol{P}$ does exist. For example, for an effective two-level system, $\boldsymbol{P}$ obeys the Bloch equation, which, together with eq. (1), properly describes most of the observed transient coherent optical phenomena [5].

Here we shall limit our discussion to steady-state problems. They often involve interaction of several monochromatic or quasi-monochromatic field components in a nonlinear medium. It is then convenient to decompose $\boldsymbol{E}(\boldsymbol{r}, t)$ and $\boldsymbol{P}(\boldsymbol{r}, t)$ into Fourier components:

$$\begin{cases} \boldsymbol{E}(\boldsymbol{r}, t) = \sum\limits_i \boldsymbol{E}(\omega_i)\,, \\ \boldsymbol{P}(\boldsymbol{r}, t) = \sum\limits_i \boldsymbol{P}(\omega_i)\,, \end{cases} \tag{4}$$

where $\boldsymbol{E}(\omega_i) = \boldsymbol{\mathscr{E}}(\omega_i)\exp[i\boldsymbol{k}_i\cdot\boldsymbol{r} - i\omega t]$, and $\boldsymbol{P}(\omega_i)$ can be written as

$$\boldsymbol{P}(\omega_i) = \boldsymbol{\chi}^{(1)}(\omega_i)\cdot\boldsymbol{E}(\omega_i) + \sum_{j,k}\boldsymbol{\chi}^{(2)}(\omega_i = \omega_j + \omega_k)\,\boldsymbol{E}(\omega_j)\,\boldsymbol{E}(\omega_k) + \\ + \sum_{j,k,l}\boldsymbol{\chi}^{(3)}(\omega_i = \omega_j + \omega_k + \omega_l)\,\boldsymbol{E}(\omega_j)\,\boldsymbol{E}(\omega_k)\,\boldsymbol{E}(\omega_l) + \dots . \tag{5}$$

The tensors $\chi^{(n)}$ in the above equation are known as the n-th–order susceptibility tensors. They are independent of the wave vectors $\boldsymbol{k}_i$'s in the electric-dipole approximation. The so-called n-th–order nonlinear optical effect is then governed by $\chi^{(n)}$. Clearly, the importance of $\chi^{(n)}$ in nonlinear optics is the same as the importance of $\chi^{(1)}$ (or linear dielectric constant $\boldsymbol{\varepsilon}$) in linear optics. For this reason, there has always been strong research interest in nonlinear susceptibilities of materials.

How do we theoretically find $\chi^{(n)}$ for a given medium? As an n-th–rank tensor, $\chi^{(n)}$ has 3^n elements, but not all of them are independent because of symmetry of the medium. We can therefore, divide the problem into two parts. First, from the symmetry of the given medium, we can find the independent nonvanishing elements of $\chi^{(n)}$, and then, from a microscopic theory, the magnitudes of these independent elements.

The first part is straightforward. For a given medium with a certain class of symmetry, there is a set of symmetry operations such as rotations, inversions, etc., which, operating on the medium, leave the properties of the medium unchanged. Let S be one of the symmetry operations. Then, under the transformation of S, the tensor elements of $\chi^{(n)}$ should remain unchanged. We have, for example,

$$(\mathbf{S}^{\dagger}\cdot\hat{\imath})\cdot\chi^{(2)}:(\mathbf{S}\cdot\hat{\jmath})(\mathbf{S}\cdot\hat{k}) = \hat{\imath}\cdot\chi^{(2)}:\hat{\jmath}\hat{k}\,. \tag{6}$$

There is one such equation for every symmetry operation. These equations will make some elements of $\chi^{(n)}$ vanish and some depend on others. As an example, eq. (6) leads immediately to the well-known result that a medium with inversion symmetry has zero $\chi^{(2)}$. The reduced forms of $\chi^{(2)}$ and $\chi^{(3)}$ for various classes of materials can be found in ref. [6].

The microscopic expression for $\chi^{(n)}$ is also not difficult to find, at least in principle. It comes directly from an n-th–order perturbation calculation, *e.g.* from eq. (3) by iteratively solving the equation of motion for ϱ to obtain the n-th–order term $\varrho^{(n)}$.

As an illustration, we derive here the microscopic expression for $\chi^{(2)}(\omega_i = \omega_j + \omega_k)$ [4]. We shall consider only the electronic contribution.

We can expand $\varrho(t)$ into a series of ascending order:

$$\varrho(t) = \varrho^{(0)} + \varrho^{(1)} + \varrho^{(2)} + \dots\,, \tag{7}$$

where $\varrho^{(0)}$ is the density matrix at equilibrium in the absence of the field, $\varrho^{(1)}$ is linear in $\boldsymbol{E}$, $\varrho^{(2)}$ is quadratic in $\boldsymbol{E}$, etc. For each term in eq. (7), we can also decompose it into Fourier components:

$$\varrho^{(n)}(t) = \sum_j \varrho^{(n)}(\omega_j) \tag{8}$$

with $\partial \varrho^{(n)}(\omega_j)/\partial t = -i\omega_j\, \varrho^{(n)}(\omega_j)$. We are interested in finding $\varrho^{(2)}(\omega_i = \omega_j + \omega_k) \propto \propto \boldsymbol{E}(\omega_j)\boldsymbol{E}(\omega_k)$, from which we can obtain the second-order nonlinear polarization

$$(9)\qquad \boldsymbol{P}^{(2)}(\omega_i = \omega_j + \omega_k) = \\ = \boldsymbol{\chi}^{(2)}(\omega_i = \omega_j + \omega_k) : \boldsymbol{E}(\omega_j)\boldsymbol{E}(\omega_k) = N\, \mathrm{Tr}\,[\varrho^{(2)}(\omega_i = \omega_j + \omega_k)\boldsymbol{p}]\ ,$$

and hence $\boldsymbol{\chi}^{(2)}(\omega_i = \omega_j + \omega_k)$.

With the electric-dipole approximation, the total Hamiltonian in eq. (3) takes the form

$$(10)\qquad \begin{cases} \mathscr{H} = \mathscr{H}_0 + \mathscr{H}^{(1)}_{\text{int}}\,, \\ \mathscr{H}_0|n\rangle = E_n|n\rangle\,, \qquad E_n - E_m = \hbar\omega_{nm}\,, \\ \mathscr{H}^{(1)}_{\text{int}} = -\boldsymbol{p}\cdot\boldsymbol{E} + \text{complex conjugate}\,, \end{cases}$$

where $\mathscr{H}_0$ is the unperturbed Hamiltonian of the material system, $|n\rangle$ is the corresponding eigenstate and E_n is the eigenenergy. The equation of motion for ϱ can then be decomposed into a set of equations

$$(11)\qquad i\hbar\frac{\partial\varrho^{(n)}}{\partial t} = [\mathscr{H}_0, \varrho^{(n)}] + [\mathscr{H}^{(1)}_{\text{int}}, \varrho^{(n-1)}] + i\hbar\left(\frac{\partial\varrho^{(n)}}{\partial t}\right)_{\text{damping}}.$$

We usually assume [4]

$$(12)\qquad \begin{cases} \left(\dfrac{\partial}{\partial t}\langle n|\varrho|n\rangle\right)_{\text{damping}} = \sum\limits_{n'}[W_{nn'}\langle n'|\varrho|n'\rangle - W_{n'n}\langle n|\varrho|n\rangle]\,, \\ \left(\dfrac{\partial}{\partial t}\langle n|\varrho|n'\rangle\right)_{\text{damping}} = -\Gamma_{nn'}\langle n|\varrho|n'\rangle \qquad \text{for } n \neq n'\,, \end{cases}$$

where $W_{nn'}$ is the transition probability from $|n'\rangle$ to $|n\rangle$ due to a random perturbation and $\Gamma_{nn'}$ is a damping coefficient. With the help of eqs. (10) and (12), the solution of eq. (11) can be easily obtained for a certain Fourier component of $\varrho^{(n)}$. We find, for example,

$$(13)\qquad \begin{cases} \langle n|\varrho^{(1)}(\omega_j)|n'\rangle = \dfrac{\langle n|-\boldsymbol{p}\cdot\boldsymbol{E}(\omega_j)|n'\rangle}{\omega_j - \omega_{nn'} + i\Gamma_{nn'}}(\varrho^{(0)}_{n'} - \varrho^{(0)}_n)\,, \\ \langle n|\varrho^{(2)}(\omega_i = \omega_j + \omega_k)|n'\rangle = \\ \qquad = \dfrac{\langle n|\{[-\boldsymbol{p}\cdot\boldsymbol{E}(\omega_j), \varrho^{(1)}(\omega_k)] + [-\boldsymbol{p}\cdot\boldsymbol{E}(\omega_k), \varrho^{(1)}(\omega_j)]\}|n'\rangle}{\hbar(\omega_i - \omega_{nn'} + i\Gamma_{nn'})}\,, \end{cases}$$

where $\varrho^{(0)}_n \equiv \langle n|\varrho^{(0)}|n\rangle$ is the equilibrium population in $|n\rangle$. Then, from eq. (9), we obtain

$$\chi^{(2)}(\omega_i=\omega_j+\omega_k)=\frac{1}{\hbar^2 g}\sum_{n,n'}\sum\left\{\frac{\langle g|\boldsymbol{p}|n\rangle\langle n'|\boldsymbol{p}|g\rangle\langle n|\boldsymbol{p}|n'\rangle}{(\omega_i-\omega_{ng}+i\Gamma_{ng})(\omega_j-\omega_{n'g}+i\Gamma_{n'g})}+\right.$$
$$+\frac{\langle g|\boldsymbol{p}|n\rangle\langle n|\boldsymbol{p}|n'\rangle\langle n'|\boldsymbol{p}|g\rangle}{(\omega_i-\omega_{ng}+i\Gamma_{ng})(\omega_k-\omega_{n'g}+i\Gamma_{n'g})}+\frac{\langle n'|\boldsymbol{p}|g\rangle\langle n|\boldsymbol{p}|n'\rangle\langle g|\boldsymbol{p}|n\rangle}{(\omega_i-\omega_{gn'}+i\Gamma_{gn'})(\omega_k-\omega_{gn}+i\Gamma_{gn})}+$$
$$+\frac{\langle n'|\boldsymbol{p}|g\rangle\langle g|\boldsymbol{p}|n\rangle\langle n|\boldsymbol{p}|n'\rangle}{(\omega_i-\omega_{gn'}+i\Gamma_{gn'})(\omega_j-\omega_{gn}+i\Gamma_{gn})}-\frac{\langle n'|\boldsymbol{p}|n\rangle\langle g|\boldsymbol{p}|n'\rangle\langle n|\boldsymbol{p}|g\rangle}{(\omega_i-\omega_{nn'}+i\Gamma_{nn'})(\omega_j-\omega_{gn'}+i\Gamma_{gn'})}-$$
$$-\frac{\langle n'|\boldsymbol{p}|n\rangle\langle g|\boldsymbol{p}|n'\rangle\langle n|\boldsymbol{p}|g\rangle}{(\omega_i-\omega_{nn'}+i\Gamma_{nn'})(\omega_k-\omega_{ng}+i\Gamma_{ng})}-\frac{\langle n'|\boldsymbol{p}|n\rangle\langle n|\boldsymbol{p}|g\rangle\langle g|\boldsymbol{p}|n'\rangle}{(\omega_i-\omega_{nn'}+i\Gamma_{nn'})(\omega_k-\omega_{gn'}+i\Gamma_{gn'})}-$$
$$\left.-\frac{\langle n'|\boldsymbol{p}|n\rangle\langle n|\boldsymbol{p}|g\rangle\langle g|\boldsymbol{p}|n'\rangle}{(\omega_i-\omega_{nn'}+i\Gamma_{nn'})(\omega_j-\omega_{ng}+i\Gamma_{ng})}\right\}\varrho_g^{(0)}. \tag{14}$$

When all frequencies are far away from resonance, so that the Γ's in the denominators can be neglected, eq. (14) leads immediately to the permutation symmetry relation [7]

$$\chi^{(2)*}_{lmn}(\omega_i=\omega_j+\omega_k)=\chi^{(2)}_{mnl}(\omega_j=-\omega_k+\omega_i)=\chi^{(2)}_{nlm}(\omega_k=\omega_i-\omega_j)\,. \tag{15}$$

The above result is valid for any material system as long as appropriate eigenstates of the system are used. While localized electronic states may be used for gases, liquids, molecular solids, and low-lying states of ionic solids, band electronic states should be used for covalent and even weakly covalent solids. In the case of condensed matter, since the applied field is different from the local field, one should also incorporate in $\chi^{(n)}$ a local-field correction factor $L^{(n)}$. The expression for $L^{(n)}$ is often complicated [3, 8]. For covalent solids, even the theory of local-field correction is still in a primitive stage.

In order to calculate the magnitude of $\chi^{(n)}$, we must know both the eigenenergies and the eigen wave functions of the material system. This information is usually not available except for simple atomic systems. Therefore, for most practical cases, a microscopic expression for $\chi^{(n)}$ in a form similar to eq. (14) is almost useless. We must use appropriate approximations to simplify the calculation. For example, one simplifying assumption often used is to replace the frequency denominator in every term of $\chi^{(n)}$ by an averaged one. The approximation is presumably fair when all frequencies are far away from resonance. Then, by the closure property of the eigenstates, the matrix element part can be written in terms of moments of the ground-state charge distribution [9, 10]. The problem is thus reduced to finding an accurate ground-state wave function [10, 11].

Recently, the bond model [12] has been used to calculate $\chi^{(2)}$ and $\chi^{(3)}$ for solids in the low-frequency limit with apparent success [13, 14]. The bond theory assumes that the induced polarization in a crystal is the vectorial sum of the induced dipoles on all bonds connecting the atoms in a unit volume, and that identical bonds in different solids have the same properties. We can, therefore, write

$$\boldsymbol{\chi}^{(n)} = \sum_i \boldsymbol{\beta}_i^{(n)}\,, \tag{16}$$

where $\boldsymbol{\beta}_i^{(n)}$ is the n-th–order polarizability tensor for the i-th bond and the summation is over all the bonds in a unit volume. In order to calculate $\boldsymbol{\chi}^{(n)}$, we only need to find $\boldsymbol{\beta}^{(n)}$ for different types of bonds. We begin by first finding $\boldsymbol{\beta}^{(1)}$.

Consider a semiconductor or insulator of cubic symmetry composed of identical bonds. We expect

$$\beta_{\parallel}^{(1)} = g\beta_{\perp}^{(1)} = G\chi_{ll}^{(1)}\,, \tag{17}$$

where $\beta_{\parallel}^{(1)}$ and $\beta_{\perp}^{(1)}$ are linear polarizabilities parallel and perpendicular to the bond, respectively, with $\beta_{\perp}^{(1)}/\beta_{\parallel}^{(1)} = \varkappa$, and g and G are geometric factors depending on the crystal structure. At sufficiently low temperatures, $\chi_{ll}^{(1)}$ is given by the well-known expression [15]

$$\chi_{ll}^{(1)} = \frac{1}{\hbar}\sum_g \left\{\sum_n \frac{|\langle n|p_l|g\rangle|^2}{\omega_{ng}^2 - \omega^2} 2\omega_{ng}\right\}, \tag{18}$$

which can be easily derived from eq. (13). If we assume $\omega^2 \ll \omega_{ng}^2$ and approximate $\omega_{ng}^2 - \omega^2$ in eq. (18) by an average $\bar{\omega}_{ng}^2 - \omega^2$, then, by the Thomas-Reiche-Kuhn sum rule [15], $\chi_{ll}^{(1)}(\omega)$ becomes

$$\chi_{ll}^{(1)}(\omega) = \Omega_p^2/4\pi(\bar{\omega}_{ng}^2 - \omega^2)\,, \tag{19}$$

where Ω_p is the plasma frequency of the valence electrons. Equation (19) has been more rigorously derived by PENN in the limiting case $\omega \to 0$ [16].

How do we find $\bar{\omega}_{ng}$? According to the bond theory of PHILLIPS, the average energy gap $\bar{E}_g \equiv \hbar\bar{\omega}_{ng}$ is given by [12, 17]

$$\bar{E}_g^2 = E_h^2 + C^2\,, \tag{20}$$

where the homopolar gap E_h and the heteropolar gap C have the expressions

$$E_h^{-2} \simeq a d^{2s}\,, \qquad C \simeq b(Z_A/r_A - Z_B/r_B)\exp[-k_s d/2]\,. \tag{21}$$

Here, a, b and s are constants, Z_A and Z_B are the valences and r_A and r_B are the covalent radii of the A and B atoms forming the bond, $d = r_A + r_B$ is the bond length and $\exp[-k_s d/2]$ is the Thomas-Fermi screening factor. Equation (20) can be seen more physically from the molecular-orbital theory [18]. There are two eigenstates for the bond electrons, a lower bonding state and an upper antibonding state. The energy difference between the two states is $\bar{E}_g$. For a homopolar bond ($A = B$), the bond electrons see a symmetric potential with respect to the bond center, and we have $\bar{E}_g = E_h$. For a heteropolar bond ($A \neq B$), the bond electrons see an asymmetric potential, and $\bar{E}_g^2 = E_h^2 + C^2$ with C proportional to the antisymmetric part of the potential. In fig. 1 we

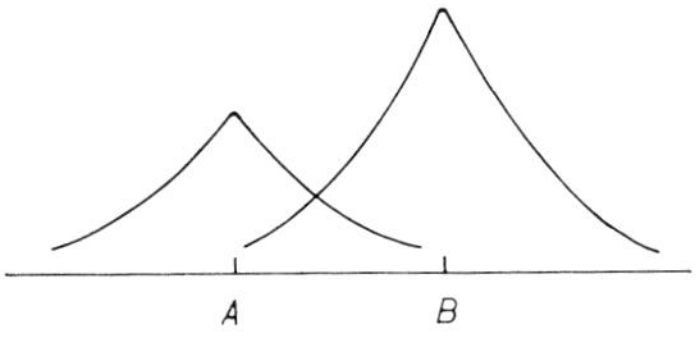

Fig. 1. – Sketch of bonding electron wave function along a heteropolar bond.

sketch the wave function of the bonding state along a heteropolar bond. We notice that there is a charge transfer from the less electronegative atom to the more electronegative atom. From the simple molecular-orbital theory, the amount of charge transfer Q is related to the heteropolar gap C by

$$Q = -2eC/\bar{E}_g \,. \tag{22}$$

Figure 1 also shows that there is a bond charge cloud between the two atoms. According to PHILLIPS [17], the magnitude of the bond charge is

$$q = -2e\bar{E}_g^2/(\bar{E}_g^2 + \hbar^2 \Omega_p^2) \,. \tag{23}$$

LEVINE [13] has suggested that the bond charge may be considered as a point charge sitting at a distance r_A and r_B, respectively, from the atoms A and B.

We have seen how to calculate $\chi_{ll}^{(1)}$ from eq. (19) and hence $\boldsymbol{\beta}^{(1)}$. What happens when the bond is subject to another low-frequency field $\boldsymbol{E}(\omega')$? Clearly, $\boldsymbol{\beta}^{(1)}$ should now depend on $\boldsymbol{E}(\omega')$. The dependence is through $\bar{\omega}_{ng}$ or $\bar{E}_g$ in eq. (19). We do not expect the bond length d to vary with $\boldsymbol{E}(\omega')$. Therefore, eq. (21) suggests that, at least in the first-order approximation, E_h is independent of $\boldsymbol{E}(\omega')$. On the other hand, the bond charge position, or the amount of charge transfer, or both, can depend on $\boldsymbol{E}(\omega')$, and hence C is a strong function of $\boldsymbol{E}(\omega')$. Knowing $\boldsymbol{\beta}^{(1)}$ as a function of $\boldsymbol{E}(\omega')$, we can then find $\boldsymbol{\beta}^{(n)}$ from the relation

$$\boldsymbol{\beta}^{(n)} = \partial^{(n-1)}\boldsymbol{\beta}^{(1)}/[\partial \boldsymbol{E}(\omega')]^{n-1} = \{\partial^{(n-2)}/[\partial \boldsymbol{E}(\omega')]^{n-2}\}\{(\partial\boldsymbol{\beta}^{(1)}/\partial C)[\partial C/\partial \boldsymbol{E}(\omega')]\} \,. \tag{24}$$

Two models have been used to calculate $\partial C/\partial \boldsymbol{E}(\omega')$. The bond charge model [13] assumes that the field $\boldsymbol{E}(\omega')$ induces only a change $\Delta \boldsymbol{r}$ in the bond charge position. We then have $\partial C/\partial \boldsymbol{E}(\omega') = (\partial C/\partial r_A)\cdot(\partial r_A/\partial \boldsymbol{E}) + (\partial C/\partial r_B)\cdot$ $\cdot(\partial r_B/\partial \boldsymbol{E})$, where $\partial r_A/\partial \boldsymbol{E}(\omega')$ and $\partial r_B/\partial \boldsymbol{E}(\omega')$ can be obtained from the relation

$\boldsymbol{\beta}^{(1)}\cdot \boldsymbol{E}(\omega')=q\Delta\boldsymbol{r}$. Usually, the calculation is simplified by assuming that the field component $E_\perp(\omega')$ perpendicular to the bond has little effect on the bond properties, so that $\partial C/\partial E_\perp(\omega')=0$. The charge transfer model [14] assumes that the field $\boldsymbol{E}(\omega')$ induces only a change ΔQ in the amount of charge transfer. We have $\partial C/\partial \boldsymbol{E}(\omega')=(\partial C/\partial Q)(\partial Q/\partial \boldsymbol{E})$, where $\partial Q/\partial \boldsymbol{E}(\omega')$ can be obtained from $\boldsymbol{\beta}^{(1)}\cdot \boldsymbol{E}(\omega')=(\Delta Q)\boldsymbol{d}$, by assuming again that $E_1(\omega')$ is not effective in inducing the change. Thus, all the quantities in eq. (24) for $\boldsymbol{\beta}^{(n)}$ are easily calculable.

LEVINE [13] has used the bond charge model to calculate $\chi^{(2)}$ and $\chi^{(3)}$ for a large number of crystals and found good agreement with the measured values. TANG and co-workers [14], on the other hand, have used the charge transfer model to calculate $\chi^{(2)}$ for a number of semiconductors. Their results also agree well with the measured values. Then, an immediate question one would ask is which model is more realistic.

Actually, both models are only crude approximations of the real situation. The empirical pseudopotential calculation [19] shows that the distribution of the valence electrons around the bond is rather broad with its peak situated at the center of the bond. An example is shown in fig. 2. When a d.c. field is

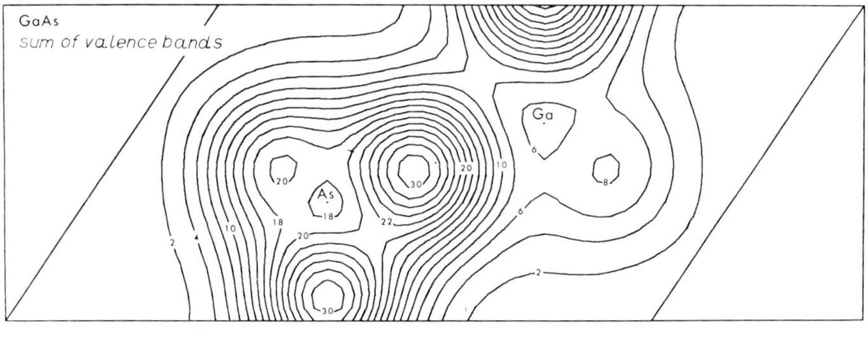

Fig. 2. – Contour map of valence electron density distribution (in units of e per primitive cell) for GaAs in the $(1, -1, 0)$-plane. (Taken from ref. [19].)

applied along the bond, this broad charge distribution becomes only slightly more asymmetric with its peak essentially unshifted [20]. This is sketched in fig. 3 for the charge distribution along a bond in Si and GaAs. Presumably, in the bond charge model, the field-induced shift in the bond charge position refers to the shift in the center of gravity of the valence charge distribution. In the charge transfer model, the field-induced charge transfer may refer to the redistribution of valence charges rather than to net charge transferred from one ion core to the other.

The above discussion tells us that large nonlinear bond polarizabilities are essential for large nonlinear susceptibilities. For large $\chi^{(2)}$, we should in addition require the crystal structure to be as asymmetric as possible, so that the summation in eq. (16) has as little cancellation as possible from different $\boldsymbol{\beta}_i^{(2)}$.

The calculations discussed above are, of course, only valid in the low-frequency limit. The approximations break down when the optical frequencies

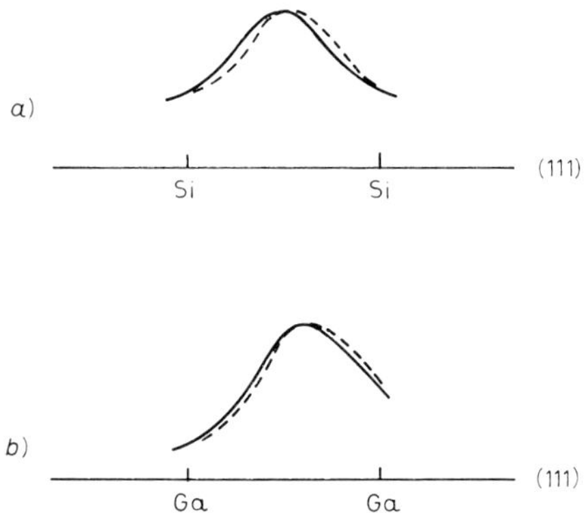

Fig. 3. – Sketch of the charge distribution along a bond in *a*) Si and *b*) GaAs. Solid and dashed curves refer to cases with and without an external field, respectively.

are near or within the band absorption region. Because of resonant enhancement, those transitions with transition frequencies closer to the optical frequencies contribute much more to $\chi^{(n)}$. In order to calculate $\chi^{(n)}$ and its dispersion in these cases, we must use the full microscopic expression for $\chi^{(n)}$. These calculations have been made for the nonlinear susceptibilities $\chi^{(2)}(2\omega)$ of zincblende semiconductors, using eq. (14) ($\omega_j=\omega_k$) with different degrees of approximation.

Assuming constant matrix elements independent of the band states, Chang *et al.* [21] rearranged the expression of $\chi^{(2)}_{xyz}(2\omega)$ into the form

$$\chi^{(2)}_{xyz}(2\omega)=A\int_{\mathrm{BZ}}\mathrm{d}^3k\sum_{c,v}\left[\frac{Q^{(1)}_{cv}(\boldsymbol{k})}{\omega^2-\omega^2_{cv}(\boldsymbol{k})}+\frac{Q^{(2)}_{cv}(\boldsymbol{k})}{4\omega^2-\omega^2_{cv}(\boldsymbol{k})}\right], \tag{25}$$

where A is a constant. If $Q^{(1),(2)}_{cv}(\boldsymbol{k})$ are assumed to be independent of $\boldsymbol{k}$, then $\chi^{(2)}_{xyz}(2\omega)$ could be written as a linear combination of $\chi^{(1)}(\omega)$ and $\chi^{(1)}(2\omega)$. The structure in $\chi^{(1)}(\omega)$ and $\chi^{(1)}(2\omega)$ would then be reflected in $\chi^{(2)}_{xyz}(2\omega)$ *vs.* ω. However, in eq. (14) with $\omega_j=\omega_k$, we can have two types of resonances, *e.g.* single resonances with $\omega=\omega_{cv}(\boldsymbol{k})$ or $2\omega=\omega_{cv}(\boldsymbol{k})$ and double resonances with both $\omega=\omega_{cv}(\boldsymbol{k})$ and $2\omega=\omega_{c'v}(\boldsymbol{k})$ at the same $\boldsymbol{k}$. At double resonances, $Q^{(1)}_{cv}$ or $Q^{(2)}_{cv}$ in eq. (25) should have a singularity, and hence they can no longer be independent of $\boldsymbol{k}$.

Bell [22] used a simplified three-band model to calculate $\chi^{(2)}_{xyz}(2\omega)$. He also assumed constant matrix elements and anticipated the structure in $\chi^{(2)}_{xyz}(2\omega)$ to arise from critical-point transitions at Γ and along Λ in the Brillouin zone.

It seems that his calculation has given too much weight to the transitions at Γ, where the density of states is known to be small.

Recently, FONG and SHEN [23] have calculated $\chi^{(2)}_{xyz}(2\omega)$ from eq. (14), using the wave functions and energies of the band states obtained from the empirical pseudopotential method. Such a method has been very successful in reproducing the observed $\chi^{(1)}(\omega)$ for zincblende semiconductors [24], and, presumably, it will also give fairly accurate results for $\chi^{(2)}(2\omega)$. The calculation of FONG and SHEN has included the contributions from 4 lowest conduction bands and 4 highest valence bands. It was found that the $\boldsymbol{k}$-dependent dipole matrix elements could significantly change the dispersion of $\chi^{(2)}(2\omega)$. An example is shown in fig. 4 for InSb. The shoulder at 1.6 eV and the peak at 1.8 eV in the solid curve arise from double resonances in the ΓKX-plane near Γ and in the ΓLK-plane near Γ, respectively. Comparison with experimental results [25] shows fair agreement in the positions of the structure in $\chi^{(2)}(2\omega)$, while the shapes of the structure may look different.

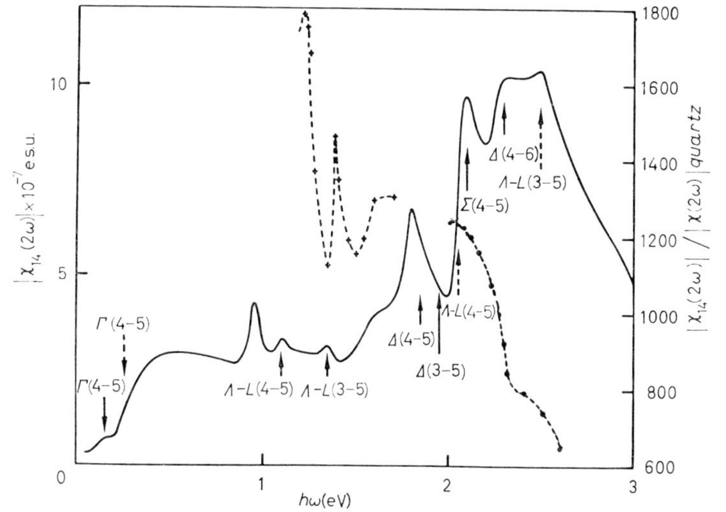

Fig. 4. – Comparison of $|\chi^{(2)}_{14}(2\omega)|$ of InSb calculated by means of an empirical pseudopotential method with available experimental data. (Taken from ref. [23].) ——— theory, — — + — — exp. CHANG *et al.*, — — • — — exp. 77K (Shen's group).

The discrepancy between theory and experiment is, at least partially, due to experimental difficulties. Aside from the inherently worse accuracy in nonlinear optical experiments, the high-intensity laser beam necessary for second-harmonic generation may also appreciably heat up the sample and change the sample characteristics. This would be the case when the laser intensity is not far from the surface damage threshold of the sample. The discrepancy can, of course, also be due to inaccuracy in the numerical calculations of the matrix elements.

2. – Coupling between electromagnetic and material excitational waves.

We now assume that $\chi^{(2)}(\omega_i = \omega_j + \omega_k)$ is known for a given medium. To find the corresponding second-order nonlinear optical effects, we must solve eq. (1). This is usually done by decomposing it into a set of three coupled equations for the three Fourier components at ω_i, ω_j and ω_k, with $\boldsymbol{P}(\omega_i) = \chi^{(1)}(\omega_i)\cdot \cdot\boldsymbol{E}(\omega_i) + \chi^{(2)}(\omega_i = \omega_j + \omega_k):\boldsymbol{E}(\omega_j)\boldsymbol{E}(\omega_k)$. The solution is well known [3, 7], and we will not repeat it here. However, such a procedure is strictly correct only when all frequencies are in the transparent region. If one of the frequencies is in the absorbing region, the problem becomes somewhat different. The medium is now being excited, and this leads to the following questions.

First, how do we describe the material excitation and its coupling to the electromagnetic waves? How do we take care of the transient response of the material excitation? Second, while $\chi^{(2)}$ of eq. (14) with its complex denominators governs the second-order nonlinear effects, does it also take into account two-photon absorption, or should the absorption be described by a separate coefficient? Third, is sum or difference frequency generation a higher-order nonlinear effect than two-photon absorption, or *vice versa*?

In this section, we shall try to answer these questions by considering a special case where ω_i $(= \omega_j + \omega_k)$ is near resonance with a discrete transition from $\langle g|$ to $\langle f|$ with transition frequency ω_{fg}. We shall apply the results to illustrate a number of interesting physical problems, *e.g.* sum frequency generation and two-photon absorption near a discrete resonance, tunable far–infra-red generation, Raman scattering by coherent excitation, stimulated light scattering, stimulated polariton scattering, four-wave mixing and vibrational relaxation measurements.

In fig. 5 we show schematically the simultaneous one-photon and two-photon resonant excitation from $\langle g|$ to $\langle f|$. For such a case, we expect the po-

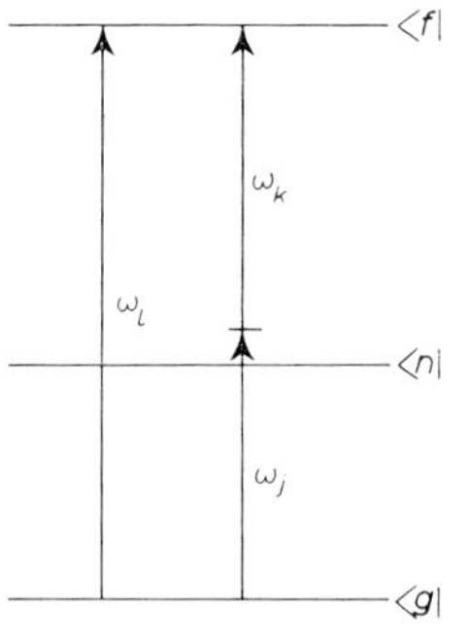

Fig. 5. – Sketch showing one-photon and two-photon absorption processes in a three-level system.

larization $\boldsymbol{P}$ to have a resonant part and a nonresonant part:

$$(26)\qquad \begin{cases} \boldsymbol{P}(\omega_i) = \boldsymbol{P}_R(\omega_i) + \boldsymbol{P}_{\text{NR}}(\omega_i) = N\,\text{Tr}\,[\varrho(\omega_i)\boldsymbol{p}]\,, \\ \boldsymbol{P}_R(\omega_i) = N\,\text{Tr}\,[\varrho_R(\omega_i)\boldsymbol{p}]\,, \\ \boldsymbol{P}_{\text{NR}}(\omega_i) = \chi^{(1)}_{\text{NR}}\cdot\boldsymbol{E}(\omega_i) + \chi^{(2)}_{\text{NR}}(\omega_i = \omega_j + \omega_k):\boldsymbol{E}(\omega_j)\,\boldsymbol{E}(\omega_k) + \dots\,. \end{cases}$$

The complete expression for $\boldsymbol{P}(\omega_i)$ or $\varrho(\omega_i)$ can be derived, in principle, from the density matrix method as sketched in the previous section. The resonant part, however, can be most easily obtained by using an effective interaction Hamiltonian $\mathscr{H}'_{\text{eff}}$ for the resonant excitation [26]:

$$(27)\qquad \begin{cases} \mathscr{H}'_{\text{eff}} = -\,\boldsymbol{p}\cdot\boldsymbol{E}(\omega_i) - \mathbf{q}:\boldsymbol{E}(\omega_j)\,\boldsymbol{E}(\omega_k) + \text{complex conjugate}\,, \\ \mathbf{q}:\boldsymbol{E}(\omega_j)\,\boldsymbol{E}(\omega_k) = \sum\limits_n \left[\dfrac{\boldsymbol{p}\cdot\boldsymbol{E}(\omega_j)|n\rangle\langle n|\boldsymbol{p}\cdot\boldsymbol{E}(\omega_k)}{\omega_k - \omega_{ng}} + \dfrac{\boldsymbol{p}\cdot\boldsymbol{E}(\omega_k)|n\rangle\langle n|\boldsymbol{p}\cdot\boldsymbol{E}(\omega_j)}{\omega_j - \omega_{ng}}\right], \end{cases}$$

where $|n\rangle$ is an intermediate state in fig. 5. We then have, from the equation of motion for the density matrix,

$$(28)\qquad \hbar\left(i\frac{\partial}{\partial t} - \omega_{fg} + i\Gamma_{fg}\right)\langle f|\varrho_R(\omega_i)|g\rangle = \\ = -\,[\langle f|\boldsymbol{p}|g\rangle\cdot\boldsymbol{E}(\omega_i) + \langle f|\mathbf{q}|g\rangle\,\boldsymbol{E}(\omega_j)\boldsymbol{E}(\omega_R)](\varrho_g - \varrho_f)\,,$$

where ϱ_g and ϱ_f are the populations in $|g\rangle$ and $|f\rangle$, respectively. Clearly, $\langle f|\varrho_R(\omega_i)|g\rangle$ is the only resonant matrix element of $\varrho_R(\omega_i)$ as ω_i approaches ω_{fg}. For the sake of simplicity we shall assume in the following all fields to be linearly polarized and use the notations

$$(29)\qquad \begin{cases} \psi(\omega_i) \equiv \langle f|\varrho_R(\omega_i)|g\rangle\,, \\ A_{fg} \equiv \langle f|\boldsymbol{p}\cdot\hat{e}(\omega_i)|g\rangle\,, \\ B_{fg} \equiv \langle f|\mathbf{q}:\hat{e}(\omega_j)\,\hat{e}(\omega_k)|g\rangle\,. \end{cases}$$

From eq. (24) we obtain

$$(30)\qquad P_R(\omega_i) = NA^*_{fg}\,\psi(\omega_i)\,.$$

The polarizations at other frequencies can be derived by using the average coupling energy, whose resonant part is

$$(31)\qquad \langle\mathscr{H}'_{\text{eff}}\rangle = -\,\langle p\rangle^* E(\omega_i) - \langle q\rangle^* E(\omega_j)\,E(\omega_k) + \text{complex conjugate} = \\ = -\,A_{fg}\,\psi^*(\omega_i)\,E(\omega_i) - B_{fg}\,\psi^*(\omega_i)\,E(\omega_j)\,E(\omega_k) + \text{c.c.}$$

We find

$$(32)\qquad P_R(\omega_j) = -N\,\partial\langle\mathscr{H}'_{\text{eff}}\rangle/\partial H^*(\omega_j) = NB^*_{fg}\,\psi(\omega_i)\,E^*(\omega_k)$$

and a similar expression for $P_R(\omega_k)$. The nonresonant part is similar to that for a transparent medium [3].

As shown here, $\psi(\omega_i)$ corresponds to an off-diagonal density matrix element. In terms of magnetic-resonance language, it is a transverse excitation [27]. If the driving fields are coherent, then ψ is a coherent excitational wave. Its phase memory decays with a transverse relaxation time $T_2 = 1/\Gamma_{fg}$. We should also expect in addition a longitudinal excitation. This corresponds to a population change $\Delta\varrho$ in $\langle g|$ and $\langle f|$:

$$(33)\qquad \Delta\varrho = \varrho_f - \varrho_f^{(0)} = \varrho_g^{(0)} - \varrho_g\,,$$

where $\varrho_f^{(0)}$ and $\varrho_g^{(0)}$ are the populations at thermal equilibrium. The rate equation for $\Delta\varrho$ can be obtained from simple physical argument:

$$(34)\qquad \left(\frac{\partial}{\partial t}+\frac{1}{T_1}\right)\Delta\varrho = \frac{1}{2\omega_i}\sum_{\lambda=i,j,k}\left[\frac{\partial P^*(\omega_\lambda)}{\partial t}E(\omega_\lambda) + \text{complex conjugate}\right] =$$
$$= \frac{i}{2}N[A_{fg}\psi^*(\omega_i)E(\omega_i) - A^*_{fg}\psi(\omega_i)E^*(\omega_i) + $$
$$+ B_{fg}\psi^*(\omega_i)E(\omega_j)E(\omega_k) - B^*_{fg}\psi(\omega_i)E^*(\omega_j)\,E^*(\omega_k)]\,.$$

As seen in the equation, $\Delta\varrho$ has a longitudinal relaxation time T_1.

We can summarize the above discussion as follows: under one-photon and two-photon resonant excitation the medium is excited both « transversely » and « longitudinally ». The transverse excitation ψ obeys the wave equation in eq. (29), or

$$(35)\qquad \hbar\left[i\frac{\partial}{\partial t} - \omega_{fg}(\boldsymbol{k}) + i\Gamma_{fg}\right]\psi(\omega_i) = -\,[A_{fg}E(\omega_i) + B_{fg}E(\omega_j)E(\omega_k)](\varrho_g - \varrho_f)\,,$$

where ω_{fg} is, in general, a function of the wave vector $\boldsymbol{k}$ of the excitational wave ψ. The longitudinal excitation $\Delta\varrho$ obeys eq. (34). These resonant excitations should affect the propagation of the electromagnetic waves at ω_i, ω_j and ω_k through the wave equations

$$(36)\qquad \left\{\begin{aligned}
&\left[\nabla\times(\nabla\times) + \frac{\varepsilon_i}{c^2}\frac{\partial^2}{\partial t^2}\right]E(\omega_i) = -\frac{4\pi}{c^2}\frac{\partial^2}{\partial t^2}[NA^*_{fg}\psi(\omega_i) + \chi^{(2)}_{\text{NR}}E(\omega_j)E(\omega_k)]\,,\\
&\left[\nabla\times(\nabla\times) + \frac{\varepsilon_j}{c^2}\frac{\partial^2}{\partial t^2}\right]E(\omega_j) = -\frac{4\pi}{c^2}\frac{\partial^2}{\partial t^2}[B^*_{fg}\psi(\omega_i)E^*(\omega_k) + \chi^{(2)*}_{\text{NR}}E(\omega_i)E^*(\omega_k)],\\
&\left[\nabla\times(\nabla\times) + \frac{\varepsilon_k}{c^2}\frac{\partial^2}{\partial t^2}\right]E(\omega_k) = -\frac{4\pi}{c^2}\frac{\partial^2}{\partial t^2}[B^*_{fg}\psi(\omega_i)E^*(\omega_j) + \chi^{(2)*}_{\text{NR}}E(\omega_i)E^*(\omega_j)].
\end{aligned}\right.$$

We have neglected here terms of $\chi^{(3)}_{NR}|E|^2 E$. In the following discussion, they will be inserted when necessary. Equations (34)-(36) form a set of coupled equations. They govern all the lower-order nonlinear optical phenomena created by the three nonlinearly coupled e.m. waves in the resonant medium.

Equations (34) and (35) describe the time-dependent response of the material excitations to the applied fields. In the steady-state case we can replace $\partial\psi(\omega_i)/\partial t$ by $-i\omega_i\psi(\omega_i)$. The dispersion of $\psi(\omega_i)$ for $k \leqslant 10^5\ \text{cm}^{-1}$ is often negligible except in special cases. Then we have

$$\psi(\omega_i) = -\frac{[A_{fg}E(\omega_i) + B_{fg}E(\omega_j)E(\omega_k)](\varrho_g - \varrho_f)}{\hbar(\omega_i - \omega_{fg} + i\Gamma_{fg})}\,. \tag{37}$$

We can, therefore, eliminate ψ in the coupled wave equations. Substitution of eq. (37) into eq. (36) gives

$$\left\{\begin{aligned}
&\left[\nabla\times(\nabla\times) - \frac{\omega_i^2}{c^2}(\varepsilon_i)_{\text{eff}}\right]E(\omega_i) = \frac{4\pi\omega_i^2}{c^2}(\chi_i^{(2)})_{\text{eff}}E(\omega_j)E(\omega_k)\,,\\
&\left[\nabla\times(\nabla\times) - \frac{\omega_j^2}{c^2}(\varepsilon_j)_{\text{eff}}\right]E(\omega_j) = \frac{4\pi\omega_j^2}{c^2}(\chi_j^{(2)})_{\text{eff}}E(\omega_i)E^*(\omega_k)\,,\\
&\left[\nabla\times(\nabla\times) - \frac{\omega_k^2}{c^2}(\varepsilon_k)_{\text{eff}}\right]E(\omega_k) = \frac{4\pi\omega_k^2}{c^2}(\chi_k^{(2)})_{\text{eff}}E(\omega_i)E^*(\omega_j)\,.
\end{aligned}\right. \tag{38}$$

We have defined

$$\left\{\begin{aligned}
&(\varepsilon_i)_{\text{eff}} = \varepsilon_i - 4\pi N|A_{fg}|^2(\varrho_g - \varrho_f)/D\,,\\
&(\varepsilon_j)_{\text{eff}} = \varepsilon_j - 4\pi N|B_{fg}|^2|E(\omega_k)|^2(\varrho_g - \varrho_f)/D\,,\\
&(\varepsilon_k)_{\text{eff}} = \varepsilon_k - 4\pi N|B_{fg}|^2|E(\omega_j)|^2(\varrho_g - \varrho_f)/D\,,\\
&(\chi_i^{(2)})_{\text{eff}} = \chi^{(2)}_{NR} - NA^*_{fg}B_{fg}(\varrho_g - \varrho_f)/D\,,\\
&(\chi_j^{(2)})_{\text{eff}} = (\chi_k^{(2)})_{\text{eff}} = \chi^{(2)*}_{NR} - NA_{fg}B^*_{fg}(\varrho_g - \varrho_f)/D\,,\\
&D \equiv \hbar(\omega_i - \omega_{fg} + i\Gamma_{fg})\,.
\end{aligned}\right. \tag{39}$$

In the following we shall assume $\varrho = \varrho^{(0)}$ unless otherwise specified.

The physical implication of the expressions in eq. (39) is as follows. The e.m. wave $E(\omega_i)$ couples directly with the excitational wave $\psi(\omega_i)$. The coupling leads to a mixed mode known as polaritons [28]. In the present formalism, $k_i = \omega_i(\varepsilon_i)^{\frac{1}{2}}_{\text{eff}}/c$ actually describes the polariton dispersion (fig. 6) and attenuation. The expressions of $(\varepsilon_j)_{\text{eff}}$ and $(\varepsilon_k)_{\text{eff}}$ have a field-dependent term arising from two-photon transition; those of $\chi^{(2)}_{\text{eff}}$ show explicitly the contribution of the resonant excitation to the second-order nonlinear susceptibility. We then realize that in eq. (38) there are separate terms governing two-photon transitions and sum or difference frequency generation. The two-photon transition terms are proportional to E^3, while the second-order nonlinear interaction terms are proportional to E^2. We also notice that $(\chi_i^{(2)})_{\text{eff}}$ is in fact equivalent to $\chi^{(2)}(\omega_i = \omega_j + \omega_k)$ in eq. (14) with $\omega_i \sim \omega_{fg}$, as can be explicitly shown.

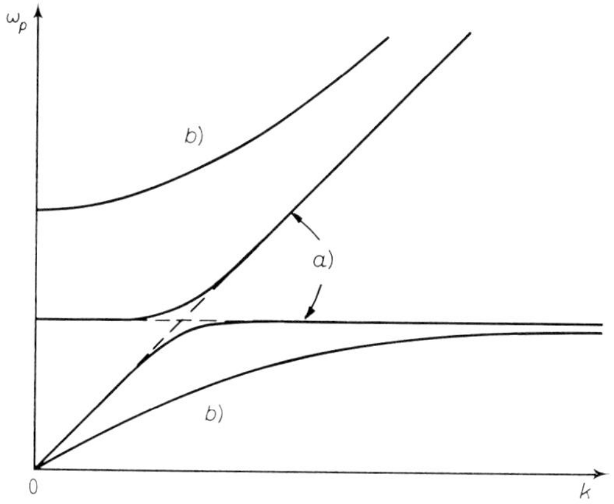

Fig. 6. – Polariton dispersion curves. *a*) Weak coupling and *b*) strong coupling between electromagnetic and material excitation waves.

A number of special cases are of interest:

a) $A_{fg} = B_{fg} = 0$. We have the ordinary case of second-order nonlinear effect in a transparent medium.

b) $A_{fg} = 0$. We have nonresonant second-order nonlinear effect under the influence of two-photon transition.

c) $A_{fg} = \chi_{\text{NR}}^{(2)} = 0$. We have a case of pure two-photon transition.

d) $B_{fg} = 0$. We have nonresonant second-order nonlinear effect with the field at $\omega_j + \omega_k$ propagating as a polariton wave.

e) $B_{fg} = \chi_{\text{NR}}^{(2)} = 0$. We have no nonlinear coupling between waves. The field at ω_i propagates as a polariton wave.

We can now use the formalism to discuss some specific problems which have recently attracted a great deal of attention. For simplicity, we shall assume that all beams are collinear and $-\partial^2/\partial z^2$ replaces $\nabla \times (\nabla \times)$ in eq. (38). This assumption can be easily removed. For the noncollinear case, we only consider phase mismatch along the z-direction. We shall also assume that the depletion of all pump fields is small and negligible in the first-order approximation.

A) Sum frequency generation and two-photon absorption in the polariton region.

We first consider the sum frequency generation. From eq. (38) with $E = \mathscr{E} \exp[ikz - i\omega t]$ and the approximation $|k_i \partial\mathscr{E}(\omega_i)/\partial z| \gg \partial^2 \mathscr{E}(\omega_i)/\partial z^2$,

$$\frac{\partial\mathscr{E}(\omega_i)}{\partial z} = \frac{2\pi\omega_i^2}{ic^2 k_i} (\chi_i^{(2)})_{\text{eff}} \mathscr{E}(\omega_j)\mathscr{E}(\omega_k) \exp[i\Delta k z] . \tag{40}$$

If $\mathscr{E}(\omega_i) = 0$ at $z = 0$, we find

$$\mathscr{E}(\omega_i, z) = \frac{2\pi\omega_i^2}{ic^2 k_i} (\chi_i^{(2)})_{\text{eff}}\, \mathscr{E}(\omega_j)\mathscr{E}(\omega_k) \frac{1}{i\Delta k} (\exp[i\Delta k z] - 1) , \tag{41}$$

where

$$k_i = k_i' + ik_i'' = [\omega_i^2(\varepsilon_i)_{\text{eff}}/c^2]^{\frac{1}{2}} ,$$

$$\Delta k = \Delta k' + i\Delta k'' = (k_j' + k_k' - k_i') + i(k_j'' + k_k'' - k_i'') .$$

The intensity of the sum frequency field at z is therefore given by

$$|\mathscr{E}(\omega_i, z) \exp[ik_i z]|^2 = \tag{42}$$
$$= \frac{4\pi^2\omega_i^2 |(\chi_i^{(2)})_{\text{eff}}|^2}{c^2|(\varepsilon_i)_{\text{eff}}|\,[(\Delta k')^2 + (\Delta k'')^2]} |1 - \exp[-i\Delta k z]|^2 |E(\omega_j)|^2 |E(\omega_k)|^2 .$$

Note that the phase-matching condition $\Delta k' = 0$ actually corresponds to resonant excitation of polariton wave. Sum frequency generation here is often described as the generation of a polariton wave which converts into light at the output crystal boundary. Experimentally, second-harmonic generation in the polariton region has been studied by HAUEISEN and MAHR [29].

We now consider the absorption of $E(\omega_j)$ or $E(\omega_k)$. There are two physical mechanisms responsible for the nonlinear attenuation of $E(\omega_j)$, one being two-photon absorption and the other being sum frequency generation. We can write $E(\omega_j) = \mathscr{E}(\omega_j) \exp[ik_j z - i\omega_j t]$ with $k_j^2 = \omega^2(\varepsilon_j)_{\text{eff}}/c^2$. From eqs. (38) and (41) with the approximation $|k_j \partial\mathscr{E}(\omega_j)/\partial z| \gg |\partial^2 \mathscr{E}(\omega_j)/\partial z^2|$, we find

$$\begin{cases} \dfrac{\partial\mathscr{E}(\omega_j)}{\partial z} = -K(\omega_j)|\mathscr{E}(\omega_k)|^2\mathscr{E}(\omega_j)[1 - \exp[-i\Delta k z]] \exp[-2k_k'' z]/i\Delta k , \\ K(\omega_j) = 4\pi\omega_i^2\omega_j^2 (\chi_i^{(2)})_{\text{eff}} (\chi_j^{(2)})_{\text{eff}}/c^4 k_i k_j . \end{cases} \tag{43}$$

For illustration, let us assume $\Delta k z \ll 1$ and $k_k'' z \ll 1$. Then, the solution of eq. (43) leads to

$$|E(\omega_j, z)|^2 = |E(\omega_j, 0)|^2 \exp[-2k_j'' z] \exp[-K(\omega_j)|E(\omega_k)|^2 z^2] , \tag{44}$$

where k_j'' is proportional to $|E(\omega_k)|^2$. Thus, the two exponential factors together can be considered as attenuation due to apparent (or effective) two-photon absorption. The factor $\exp[-2k_j''z]$ is due to true two-photon absorption, while the factor $\exp[-K|E(\omega_k)|^2z^2]$ is due to depletion by sum frequency generation. As a check, we notice that in case $A_{fg}=\chi^{(2)}_{\text{NR}}=0$, but $B_{fg}\neq 0$, we should still expect two-photon excitations to occur. Indeed, when $A_{fg}=\chi^{(2)}_{\text{NR}}=0$, we have $K=0$, but $k_j''\neq 0$. Bogget and Loudon [30] in their treatment of this problem have lumped the two parts of two-photon absorption together. They have neglected the growth of $E(\omega_i)$ along z. Two-photon absorption by polaritons has been studied experimentally by Fröhlich and his associates in CuCl [31].

B) Far–infra-red generation in the polariton region.

The above results and discussion (in *A*)) can apply equally well to the present case if we simply replace ω_k by $-\omega_k$ (and hence $E(\omega_k)$ by $E^*(\omega_k)$). The field generated here is the difference frequency field at $\omega_i=\omega_j-\omega_k$, which can be in the far infra-red. This happens, for example, when the final state $|f\rangle$ is a Zeeman-split excited magnetic state [26, 32, 33]. In this case the spin-flip transition from $\langle g|$ to $\langle f|$ can be induced both by one-photon magnetic-dipole transition and two-photon Raman transition. However, because of the weak magnetic-dipole oscillator strength (small $|A_{fg}|^2$), the direct coupling between the infra-red wave $E(\omega_i)$ and the excitational wave $\psi(\omega_i)$ is weak. Hence, the gap between the two polariton branches is small (fig. 6*b*)), and $(\varepsilon_i)_{\text{eff}}\simeq\varepsilon_i$. On the other hand, the Raman transition can be strong (large B_{fg}) and the resonant

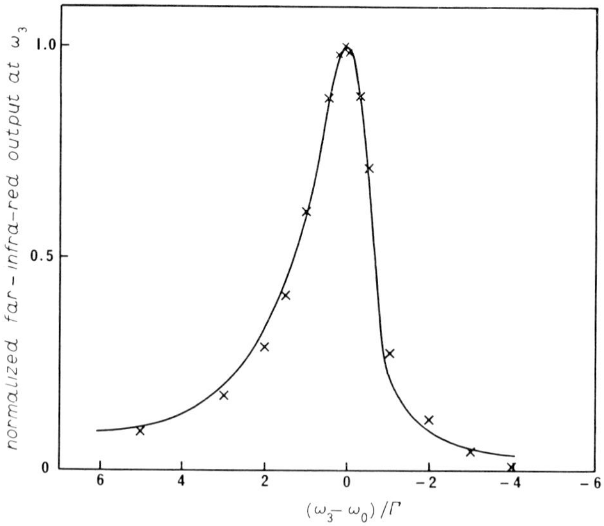

Fig. 7. – Theoretical results (crosses) of normalized far–infra-red output in optical mixing in comparison with the experimental results of Nguyen and Bridges [32]. (Taken from ref. [26].)

part of $(\chi_i^{(2)})_{\text{eff}}$ can dominate over the nonresonant part as $\omega \sim \omega_{fg}$. Then, from eq. (42), if Δk and k_i'' do not vary appreciably as ω_i is tuned over the resonant line, the intensity of the generated far–infra-red field is proportional to $|(\chi_i^{(2)})_{\text{eff}}|^2$, which has a resonant peak at $\omega_i \approx \omega_{fg}$. An example is shown in fig. 7 for the spin-flip transition in InSb. The theory [26] is in good agreement with experiment [32]. The resonant peak can be tuned by varying the Zeeman splitting. This then leads to a potential far–infra-red source which is coherent and tunable [32].

In many cases the polariton in the infra-red region is composed of photons and phonons. Then the material excitation corresponds to phonon excitation. The wave equation for an optical phonon is

$$\left[\frac{\partial^2}{\partial t^2} + 2\Gamma\frac{\partial}{\partial t} + \omega_{fg}^2\right] Q = \left(\frac{2\omega}{\hbar}\right)^{\frac{1}{2}} [A_{fg}E(\omega_i) + B_{fg}E(\omega_j)E^*(\omega_k)](\varrho_g - \varrho_f)\,, \tag{45}$$

where Q is the normal co-ordinate and the microscopic expressions for A_{fg} and B_{fg} in this case have been derived in the literature [26, 34]. If we are only interested in the region $\omega_i \sim \omega_{fg}$, then eq. (45) can be reduced to

$$\hbar\left[i\frac{\partial}{\partial t} - \omega_{fg} + i\Gamma\right]\left(\frac{2\omega}{\hbar}\right)^{\frac{1}{2}} Q = -[A_{fg}E(\omega_i) + B_{fg}E(\omega_j)E^*(\omega_k)](\varrho_g - \varrho_f)\,, \tag{46}$$

which is identical to eq. (35) for ψ, if we define $(2\omega/\hbar)^{\frac{1}{2}} Q \equiv \psi$. Therefore, eq. (42) can again be used to describe far–infra-red generation by difference frequency wave mixing in the phonon polariton region. DEMARTINI [35] has studied the problem both theoretically and experimentally in GaP. YANG *et al.* [36] have measured the dispersion of the far–infra-red output along the polariton curve.

C) Raman scattering by coherent excitation.

Recently, ROMESTAIN *et al.* [37] have found that when an n-type CdS sample is irradiated by microwaves at frequency ω_i close to the donor spin resonance, forward scattering of laser light at ω_j shows intense side bands at $\omega_j \pm \omega_i$ which are several orders of magnitude larger than spontaneous spin-flip Raman scattering. This phenomenon can be easily explained by mixing of the optical wave at ω_j with the polariton wave at ω_i. From eq. (38) with ω_i replaced by $\pm\,\omega_i$ and ω_j by ω_j we find, following the usual approximation,

$$|E(\omega_k = \omega_j \pm \omega_i, z)|^2 = \\ = \frac{4\pi^2\omega_k^2|(\chi_k^{(2)})_{\text{eff}}|^2}{c^2|(\varepsilon_k)_{\text{eff}}|\,|\Delta k_\pm|^2}|\exp[i\Delta k_\pm z] - 1|^2 \exp[-2k_k'' z]\,|E(\omega_j)|^2|E(\omega_i)|^2\,, \tag{47}$$

where $\Delta k_\pm = k_j \pm k_i - k_k$. With low laser power, $(\varepsilon_k)_{\text{eff}} = \varepsilon_k$. The generated field is most intense when $\Delta k_\pm' \simeq 0$, which occurs near the forward direction. The nonlinear susceptibility $(\chi_k^{(2)})_{\text{eff}}$ has a resonant peak at $\omega_i \simeq \omega_{fg}$ and can be calculated. With reasonable microwave power, the intensity of coherent scattering at $\omega_j \pm \omega_i$ as estimated from eq. (47) is indeed much larger than spontaneous scattering.

D) Stimulated Raman scattering.

Here we consider a special case where $A_{fg} = \chi^{(2)}_{\text{NR}} = 0$, but $B_{fg} \neq 0$. Then $\chi^{(2)}_{\text{eff}} = 0$ and we have pure two-photon transition. For two-photon Raman transition, we replace ω_k by $-\omega_k$ in eq. (38) and let $\omega_j > \omega_k$. With $E(\omega_j)$ being the pump field, we expect $E(\omega_k)$ to get amplified. The solution of eq. (38) for this case is simple. For negligible depletion of $|E(\omega_j)|^2$, we find immediately

$$(48)\qquad \begin{cases} E(\omega_i) = 0\,, \\ |E(\omega_k, z)|^2 = |E(\omega_k, 0)|^2 \exp[gz]\,, \\ g = -2k''_k = \dfrac{\omega_k}{c\sqrt{\varepsilon'_k}}\left[-\varepsilon''_k + \dfrac{4\pi N|B_{fg}|^2(\varrho_g - \varrho_f)\Gamma_{fg}}{(\omega_j - \omega_k - \omega_{fg})^2 + \Gamma^2_{fg}}|E(\omega_j)|^2\right]. \end{cases}$$

The result here applies not only to stimulated Raman scattering by electronic excitation, for example stimulated spin-flip Raman scattering, but also to stimulated Raman scattering by optical phonons, since the phonon wave equation has been shown to be identical in form to the wave equation for electronic excitation.

More generally, ψ can also represent other types of excitational waves, such as entropy waves, molecular orientational and librational waves, acoustic phonons, magnons, plasmons, etc. [34, 38]. There is a class of stimulated light scattering for each type of excitation. In general, the resonant frequency of ψ can have dispersion, *i.e.* ω_{fg} is a function of the wave vector k. In the equation for ψ we should then replace $\omega_{fg}(\boldsymbol{k})$ by $\omega_{fg}(i^{-1}\nabla)$. Because of the spatial derivatives in eq. (35) for ψ, we can no longer eliminate ψ to obtain eq. (37). We must now find the solution to the problem by solving eq. (35) together with eq. (36) [34, 38].

E) Stimulated polariton scattering.

What happens to stimulated Raman scattering when ψ is also directly excitable by infra-red radiation, *i.e.* $A_{fg} \neq 0$? We assume $E(\omega_j)$ to be the pump field with $|\omega_j| > |\omega_k|$ and replace ω_k by $-\omega_k$ in eq. (38). We also assume, in general, $\chi^{(2)}_{\text{NR}} \neq 0$. Equation (38) clearly shows that the Stokes field $E(\omega_k)$ is now generated simultaneously by pure stimulated Raman transition via $(\varepsilon_k)_{\text{eff}}$ and by parametric amplification process via $(\chi^{(2)}_k)_{\text{eff}}$. If $A_{fg} = B_{fg} = 0$ or $\chi^{(2)}_{\text{NR}}$ dominates, then we have ordinary pure parametric amplification, of which the solution from eq. (38) is well known [3]. Stimulated polariton scattering can therefore be considered as a combined stimulated Raman and parametric amplification process with the idler mode sitting on the polariton dispersion curve. We know that parametric amplification is inefficient when there is a large phase mismatch or strong attenuation at either the signal or the idler frequency.

On the other hand, stimulated Raman scattering is not affected by phase mismatch and by the nonresonant background absorption at the idler frequency ω_i. As a result, stimulated Raman scattering may dominate over parametric amplification in stimulated polariton scattering. This happens for stimulated polariton scattering connected with the spin-flip transition in InSb. There, A_{fg} is small because of the weak magnetic-dipole transition, and hence $|\chi_{\text{eff}}^{(2)}|$ is small even near resonance. The absorption coefficient at ω_i is large and, for collinear wave propagation, the phase mismatch $\Delta k' = k_j' - k_i' - k_k'$ is also large. Consequently, although strictly speaking generation of $E(\omega_k)$ and $E(\omega_i)$ is described by the parametric solution of the coupled equations in eq. (38) [26], it can also be physically described as a two-step process. First, $E(\omega_k)$ is generated by stimulated Raman scattering and then $E(\omega_i)$ is generated by optical mixing between $E(\omega_j)$ and $E(\omega_k)$. In such a process we expect to find both $E(\omega_i)$ and $E(\omega_k)$ in the output. Experimentally, however, only $E(\omega_k)$ has been observed in the stimulated spin-flip Raman scattering. But, then, only $E(\omega_i)$ propagating collinearly with $E(\omega_j)$ and $E(\omega_k)$ has been looked at, while $E(\omega_i)$ is expected to be large only in the phase-matched direction which is noncollinear.

In general, stimulated Raman scattering and parametric amplification are equally important in stimulated polariton scattering. The solution of eq. (38) is essentially the same as that for parametric amplification [39]. Assuming negligible depletion of pump power, we find

$$
(49)\quad \left\{
\begin{aligned}
&\{|E(\omega_k)|^2, |E(\omega_i)|^2\} \propto \exp[gz], \\
&g = 2\,\mathrm{Im}\,\varkappa_\pm, \\
&\varkappa_\pm = \tfrac{1}{2}(\gamma_k - \gamma_i) \pm [(\gamma_k + \gamma_i)^2 - 4A]^{\frac{1}{2}}, \\
&\gamma_k = (\omega_k/2c\sqrt{\varepsilon_k'})[i\varepsilon_k'' + 4\pi\chi_R^{(3)}|E(\omega_j)|^2], \\
&\gamma_i = -\Delta k' - i(\omega_i/2c\sqrt{(\varepsilon_i')_{\text{eff}}})(\varepsilon_i'')_{\text{eff}}, \\
&\chi_R^{(3)} = -N|B_{fg}|^2(\varrho_g - \varrho_f)/\hbar(\omega_i - \omega_{fg} - i\Gamma_{fg}), \\
&A = (4\pi^2\omega_k\omega_i/c^2\sqrt{\varepsilon_k'(\varepsilon_i')_{\text{eff}}})(\chi_i^{(2)})_{\text{eff}}(\chi_k^{(2)})_{\text{eff}}|E(\omega_j)|^2, \\
&|E(\omega_i)|^2/|E(\omega_k)|^2 = (\omega_i/\omega_k)^{\frac{1}{2}}[\varepsilon_k'/(\varepsilon_i')_{\text{eff}}]^{\frac{1}{4}}|A^{\frac{1}{2}}/(\varkappa + \gamma_3)|,
\end{aligned}
\right.
$$

where $\Delta k' = k_j' - k_k' - k_i'$ is the linear phase mismatch. For stimulated scattering, we only need to consider the solution corresponding to $g > 0$. We expect g to be a maximum at $\Delta k' \approx 0$. However, for a given ω_j and for a certain angle between $\boldsymbol{k}_j$ and $\boldsymbol{k}_k$, phase matching is satisfied only for a definite set of ω_k and ω_i. The output of stimulated polariton scattering can therefore be tuned over a range of frequencies by adjusting the angle between $\boldsymbol{k}_j$ and $\boldsymbol{k}_k$. Stimulated polariton scattering has been observed in many solids [40]. It has also been demonstrated as a possible means to generate coherent tunable far–infra-red radiation [41].

F) Wave-mixing spectroscopy.

Let us now assume the presence of two more fields at ω_j' and ω_k' in the medium with $\omega_j'+\omega_k'=\omega_j+\omega_k=\omega_i$. Then, eq. (35) for ψ should have an additional driving term on the right-hand side and it becomes

$$\hbar\left[i\frac{\partial}{\partial t}-\omega_{fg}+i\Gamma_{fg}\right]\psi(\omega_i)= \\ =-\left[A_{fg}E(\omega_i)+B_{fg}E(\omega_j)E(\omega_k)+B'_{fg}E(\omega_j')E(\omega_k')\right](\varrho_g-\varrho_f)\,. \tag{50}$$

Let $E(\omega_j)$, $E(\omega_j')$ and $E(\omega_k)$ be the pump fields, and we are interested in the generation of $E(\omega_i)$ and $E(\omega_k')$. The wave equations for $E(\omega_i)$ and $E(\omega_k')$ in the steady state can be obtained by substituting eq. (50) into eq. (36) (the equation for $E(\omega_k')$ is similar to that for $E(\omega_k)$):

$$\begin{cases}\left[\nabla\times(\nabla\times)-\dfrac{\omega_i^2}{c^2}(\varepsilon_i)_{\text{eff}}\right]E(\omega_i)= \\ \qquad =\dfrac{4\pi\omega_i^2}{c^2}\left[(\chi_i^{(2)})_{\text{eff}}E(\omega_j)\,E(\omega_k)\right]+(\chi_i^{(2)})'_{\text{eff}}E(\omega_j')E(\omega_k')\right]\,, \\ \left[\nabla\times(\nabla\times)-\dfrac{\omega_k^2}{c^2}(\varepsilon_{k'})_{\text{eff}}\right]E(\omega_k')= \\ \qquad =\dfrac{4\pi\omega_k'^2}{c^2}\left[(\chi_k^{(2)})'_{\text{eff}}E(\omega_i)E^*(\omega_j')+\chi^{(3)}E(\omega_j)E(\omega_k)E^*(\omega_j')\right]\,,\end{cases} \tag{51}$$

where

$$\chi^{(3)}=\chi^{(3)}_{\text{NR}}-NB_{fg}B'^*_{fg}(\varrho_g-\varrho_f)/D\,, \tag{52}$$

and $(\varepsilon_{k'})_{\text{eff}}$ and $(\chi^{(2)})'_{\text{eff}}$ are obtained from $(\varepsilon_k)_{\text{eff}}$ and $(\chi^{(2)})_{\text{eff}}$ by putting primes on appropriate quantities.

Consider first the simple case where $A_{fg}=0$ and $\chi^{(2)}_{\text{NR}}=0$. Then, all $\chi^{(2)}_{\text{eff}}$ vanish and $E(\omega_i)=0$. Only $E(\omega_k')$ is generated parametrically by a third-order four-wave mixing process. Assuming $\varepsilon_{\text{eff}}\approx\varepsilon$, we have at phase matching

$$|E(\omega_k',z)|^2\simeq\left(\frac{2\pi\omega_k'^2}{c^2k_k'}\right)^2|\chi^{(3)}|^2z^2|E(\omega_j)E^*(\omega_j')E(\omega_k)|^2\,. \tag{53}$$

The output is directly proportional to $|\chi^{(3)}|^2$. Measurements of $|E(\omega_k',z)|^2$ as a function of $\omega_i=\omega_j+\omega_k$ can therefore yield direct information about the dispersion of $|\chi^{(3)}(\omega_k'=\omega_j+\omega_k-\omega_j')|$. Around $\omega_i\sim\omega_{fg}$, if the dispersion of $\chi^{(3)}_{\text{NR}}$ is negligible, the maximum and minimum of $|\chi^{(3)}|$ occur at $(\omega_j+\omega_k)_\pm$ respectively with

$$(54)\qquad \begin{cases} (\omega_j+\omega_k)_\pm=\omega_{fg}+\left\{\dfrac{1}{2}\dfrac{a}{\chi_{\text{NR}}^{(3)}}\pm\left[\left(\dfrac{a}{\chi_{\text{NR}}^{(3)}}\right)^2+\Gamma^2\right]^{\frac{1}{2}}\right\}, \\ a=NB_{fg}B'^{*}_{fg}(\varrho_g-\varrho_f)/\hbar\,. \end{cases}$$

When $(a/\chi_{\text{NR}}^{(3)})^2\gg\Gamma^2$, we have $(\omega_j+\omega_k)_+=\omega_{fg}$. From the measured values of $(\omega_j+\omega_k)_\pm$ we can determine $a/\chi_{\text{NR}}^{(3)}$ and Γ, if ω_{fg} is known. Usually, a can also be deduced from the cross-section for two-photon transitions; then both the value and the sign of $\chi_{\text{NR}}^{(3)}$ can be deduced. This nonlinear spectroscopic technique has the advantage of accurately finding $\chi_{\text{NR}}^{(3)}$ through accurate frequency measurements.

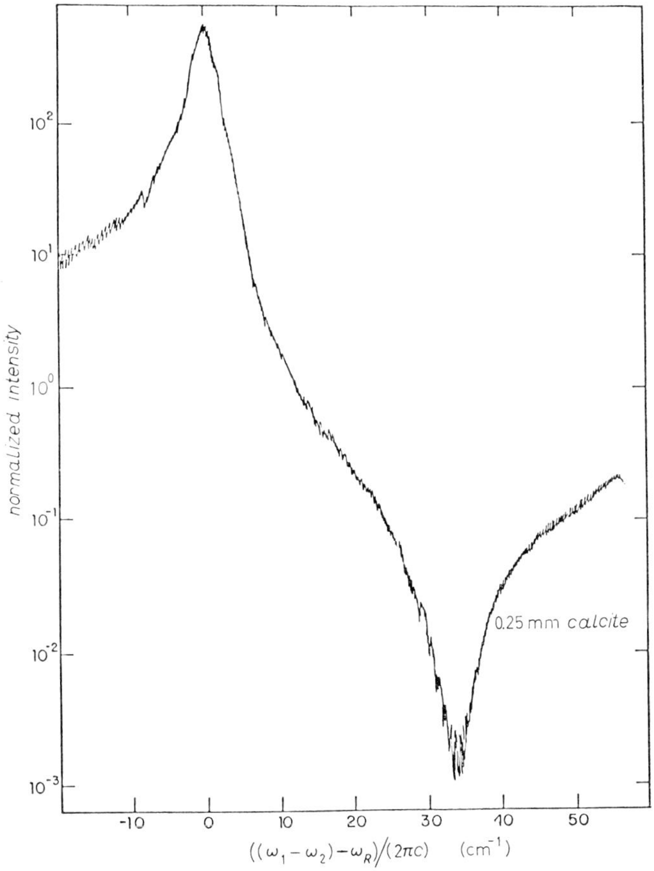

Fig. 8. – Generated anti-Stokes intensity in calcite using the four-wave–mixing spectroscopic technique. The polarizations were perpendicular to the optical axis of calcite (after M. D. Levenson: *IEEE Journ. Quantum Electron.*, QE-**10**, 110 (1974.))

The technique has recently been applied to a variety of materials, including gases [42], liquids and solids [43, 44]. In most cases Raman-active vibrational modes were studied. An example is shown in fig. 8. In these cases we should replace ω_k by $-\omega_k$ and ω'_j by $-\omega'_j$ in the above discussion, so that $\omega_j - \omega_k = \omega'_k - \omega'_j \sim \omega_{fg}$. For simple molecules or solids, when the resonance or other vibrational modes are far away, $\chi^{(3)}_{NR}$ is mainly of electronic origin.

It is seen in fig. 8 that the signal from four-wave mixing can be extremely large. The technique can therefore be used for detection of low-concentration substances from its characteristic resonances in $\chi^{(3)}$ [43, 45]. Régnier and Taran [42] have shown that, with input laser powers of about 1 MW, they can detect an H_2 concentration of about 10 ppm in an N_2 gas 1 atm pressure. The sensitivity can be greatly improved by letting ω_j and ω'_j approach resonance values.

The problem becomes somewhat more complicated when the one-photon excitation of ψ is also allowed ($A_{fg} \neq 0$) and the medium lacks inversion symmetry [43]. We must now also take into account $E(\omega_i)$ by solving the coupled equations in eq. (51). Physically, $E(\omega'_k)$ is generated not only directly by four-wave mixing through $\chi^{(3)}$, but also indirectly by first generating $E(\omega_i)$ from mixing of $E(\omega_j)$ and $E(\omega_k)$ and then generating $E(\omega'_k)$ from mixing of $E(\omega'_j)$ and $E(\omega_i)$. More generally, intermediate fields at other frequencies can be generated. They also contribute to indirect two-step generation of $E(\omega'_k)$ [43], but, here, we shall neglect their contribution. We can solve eq. (51) by iteration. First, we find

$$E(\omega_i) = \frac{2\pi\omega_i^2}{ic^2 k_i} (\chi_i^{(2)})_{\text{eff}} E(\omega_j) E(\omega_k) \frac{1}{i\Delta k} (1 - \exp[-i\Delta k_a z]) , \tag{55}$$

where $k_i^2 = \omega_i^2/c^2(\varepsilon_i)_{\text{eff}}$ and $\Delta k_a = k_j + k_k - k_i$, assuming $(\varepsilon_{k'})_{\text{eff}} = \varepsilon_{k'}$. Then, we have

$$\left\{\begin{aligned}
\partial\mathscr{E}(\omega'_k)/\partial z &= \frac{2\pi\omega_k'^2}{ic^2 k_{k'}} \left[\chi^{(3)} - \frac{2\pi\omega_i^2}{c^2 k_i \Delta k_a} (\chi_i^{(2)})_{\text{eff}} (\chi_k^{(2)})'_{\text{eff}} (1 - \exp[-i\Delta k_a z])\right] \cdot \\
&\qquad \cdot \mathscr{E}(\omega_j)\mathscr{E}(\omega_k)\mathscr{E}^*(\omega_i) \exp[i\Delta k_b z] , \\
\Delta k_b &= k_j + k_k - k_{j'} - k_{k'} , \\
|E(\omega'_k, z)|^2 &= \frac{4\pi^2\omega_k'^2}{c^2 \varepsilon_{k'}} \left| \frac{1}{i\Delta k_b} \left[\chi^{(3)} - \frac{2\pi\omega_i^2}{c^2 k_i \Delta k_a} (\chi_i^{(2)})_{\text{eff}} (\chi_k^{(2)})'_{\text{eff}}\right] (1 - \exp[-i\Delta k_b]) + \right. \\
&\quad \left. + \frac{2\pi\omega_i^2}{c^2 k_i \Delta k_a} (\chi_i^{(2)})_{\text{eff}} (\chi_k^{(2)})'_{\text{eff}} \frac{1}{\Delta k_c} (\exp[i\Delta k_c z] - 1) \exp[-i\Delta k_b z] \right|^2 \cdot \\
&\qquad \cdot |E(\omega_j) E(\omega_k) E^*(\omega'_j)|^2 , \\
\Delta k_c &= k_i - k_{j'} - k_{k'} .
\end{aligned}\right. \tag{56}$$

We notice that the two-step contribution from $(\chi_i^{(2)})_{\text{eff}}(\chi_k^{(2)})'_{\text{eff}}$ can become comparable to the direct four-wave mixing contribution from $\chi^{(3)}$ if $|\Delta k_a|$ is not too large. For example, if $\chi^{(2)}_{\text{eff}} \sim 10^{-7}$ e.s.u. and $\chi^{(3)} \sim 10^{-11}$ e.s.u., then the two contributions become comparable when $|k_i/\Delta k_a(\varepsilon_i)_{\text{eff}}| \sim 10^3$. It is possible to make either $\chi^{(3)}$ or $(\chi_i^{(2)})_{\text{eff}}(\chi_k^{(2)})'_{\text{eff}}$ vanish by choosing particular directions of beam propagation and polarization [43]. In general, however, the two terms will interfere with each other. By varying $\omega_j + \omega_k$, one often finds two special ranges of interest. First, when $\omega_j + \omega_k$ approaches ω_{fg}, phase mismatch and attenuation Δk_a may become so large that the two-step generation of $E(\omega_k')$ is negligible. The problem reduces to the simple case of four-wave mixing we discussed earlier. Second, if $\omega_j + \omega_k$ is sufficiently far away from ω_{fg} and Δk_a is sufficiently small, then $\chi^{(3)}$ is negligible and $E(\omega_k')$ is effectively generated by the two-step process. The expression of $|E(\omega_k', z)|^2$ in eq. (56) also shows that, if Δk_b for four-wave mixing is small but $\Delta k_c = k_i - k_{j'} - k_{k'}$ is large, then the output appears to be generated by an effective four-wave mixing process with $\chi_{\text{eff}}^{(3)} = \chi^{(3)} - 2\pi\omega_i^2(\chi_i^{(2)})_{\text{eff}}(\chi_k^{(2)})'_{\text{eff}}/c^2 k_i \Delta k_a$.

Experiments with $\omega_j + \omega_k$ near a dipole-allowed excitonic transition [46] or with $\omega_j - \omega_k$ near an infra-red–active phonon mode [43] have been reported. The results close to resonance can be approximately interpreted by considering only four-wave mixing via $\chi^{(3)}$.

G) Measurements of relaxation times of material excitation.

We now consider a transient case of four-wave mixing with $A_{fg} = \chi_{\text{NR}}^{(2)} = 0$. First, the medium is excited with a short pulse containing frequencies ω_j and ω_k such that $\omega_j + \omega_k = \omega_{fg}$. Both longitudinal and transverse excitations are excited according to eqs. (34) and (35), respectively. Given the laser pulse shape, $\Delta\varrho(t)$ and $\psi(\omega_i, t)$ can be calculated. At a later time t_D, a short probing pulse at ω_j' is propagated into the medium. Then, two output signals at ω_k' are expected. One from mixing of $E(\omega_j')$ with $\psi(\omega_i)$ is coherent and is peaked in the phase-matched direction for four-wave mixing. The other, due to anti-Stokes scattering from $\Delta\varrho(t)$ in the excited state, is incoherent and has a broad angular distribution. The two signals can therefore be disintuished by detection along different directions.

From the time-dependent wave equation for the coherent scattered field $E(\omega_k')$ we find

$$\left(\frac{\partial}{\partial z} + \frac{1}{V_{k'}}\frac{\partial}{\partial t}\right)\mathscr{E}(\omega_k', z, t) = i\left(\frac{2\pi\omega_k'^2}{c^2 k_{k'}}\right) B_{fg}'^{*}\psi(\omega_i, t + t_D) E^{*}(\omega_j', t)\exp[-ik_k' z + i\omega_k' t], \tag{57}$$

where $V_{k'}$ is the group velocity of $E(\omega_k')$, and hence, at phase matching, the coherent signal strength is

$$S^{\text{coh}}(t_D) = \text{const}\int \mathrm{d}t' \left|\int \mathrm{d}z\, \mathscr{E}^*(\omega_j', t')\,\psi(\omega_i, t + t_D)\right|^2. \tag{58}$$

The incoherent anti-Stokes signal, on the other hand, is given by

$$S^{\text{inc}}(t_D) = \text{const}\int \mathrm{d}t'\, \mathrm{d}z |\mathscr{E}(\omega_j', t')|^2 \Delta\varrho(t + t_D). \tag{59}$$

The variations of ψ and $\Delta\varrho$ as a function of time can be quite complicated, especially in the presence of the exciting pulse. However, we are often interested only in the relaxation times of the excitations. Then, if the exciting pulse width is smaller than, or comparable with, the relaxation times, $\Delta\varrho(t)$ and $\psi(t)$ should decay as $\exp[-t/T_1]$ and $\exp[-t/T_2]$, respectively, at sufficiently large t, where $T_2 = 1/\Gamma_{fg}$. With a short probing pulse, we should find $S^{\text{inc}}(t_D) \propto \propto \exp[-t_D/T_1]$ and $S^{\text{coh}}(t_D) \propto \exp[-t_D/T_2]$ for large t_D. Thus, we can directly measure T_1 and T_2 from the decays of $S^{\text{inc}}(t_D)$ and $S^{\text{coh}}(t_D)$.

DeMartini and Ducuing [47] first used such a method to measure T_1 of the vibrational excitation of gaseous H_2. More recently, Alfano and Shapiro [48] and Kaiser and his co-workers [49] have used picosecond laser pulses to measure T_1 and T_2 of molecular or lattice vibration in liquids and solids. In their cases the vibration is actually excited by transient stimulated Raman scattering with a laser pulse at ω_j. The underlying principle for the relaxation measurements is however the same. This provides the only method for direct measurements of vibrational relaxation times of condensed matter in the picosecond range. By detecting the incoherent anti-Stokes signals at various frequencies as a function of t_D, one can also study the decay routes of a particular excitation [50].

3. – Distinction between resonant stimulated Raman scattering and hot stimulated emission.

In the previous section we have assumed no intermediate resonance in the two-photon excitation, *i.e.* ω_j or ω_k is far away from the transition frequency ω_{ng} between the ground state $|g\rangle$ and an intermediate state $|n\rangle$. What happens when ω_j or ω_k approach ω_{ng}? Excitation of $|n\rangle$ will certainly make the problem even more complicated. We shall consider here only the special case of resonant stimulated Raman scattering (SRS). Generalization of the description to the other cases in sect. **2** is straightforward.

Let us first give a physical description of the problem. Figure 9 shows the Raman process in a three-level system. As ω_l approaches ω_{ng}, we should ex-

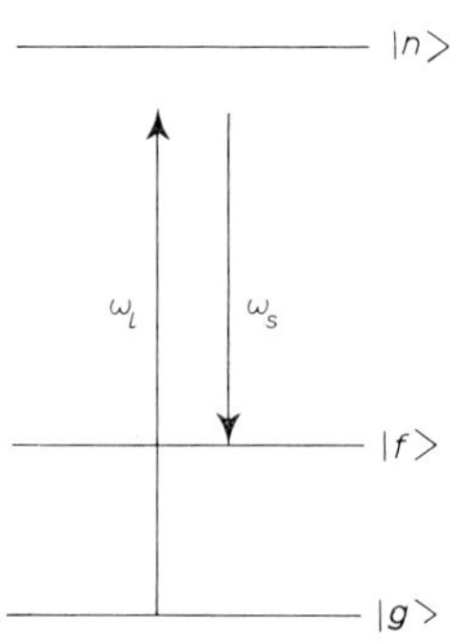

Fig. 9. – Sketch of Raman-Stokes process in a three-level system.

pect a resonant enhancement in the emission rate of $E(\omega_s)$, but then, if the transition from $|g\rangle$ to $|n\rangle$ becomes real, we should also expect hot emission of $E(\omega_s)$ in the form of three-level laser action. Now, the question arises as to whether the enhanced emission rate at ω_s is due to resonant enhancement in SRS or to hot stimulated emission (SE), or to both. Essentially, the same question has recently come up in the discussion of spontaneous resonant Raman scattering and of hot luminescence [51].

The answer to the above question is that both SRS and hot SE contribute to the enhanced rate of emission as ω_l approaches ω_{ng}. The two processes are, however, physically different. Resonant SRS is a direct two-photon process, while the hot SE is a two-step process. Hot SE arises from the excess population pumped into the excited intermediate state $|n\rangle$ by $E(\omega_l)$, but resonant SRS does not depend on the induced population change in the lower-order approximation. Hot SE is connected with the longitudinal excitation from $|g\rangle$ to $|n\rangle$, while resonant SRS is connected with the transverse excitation. Therefore, in the transient experiments, the two processes should show different dynamic behavior, as governed by the longitudinal and transverse relaxations, respectively. Since both processes contribute to emission at the frequency ω_s, they can interfere with each other in generating $E(\omega_s)$. In steady-state measurements it is then impossible to distinguish one from the other.

We shall now give a formal derivation to show explicitly the differences between resonant SRS and hot SE. It is most appropriate to use the density matrix formalism here, since, as it is a resonant phenomenon, we must properly treat relaxation of the excitations. Both resonant SRS and hot SE are third-order wave processes. We should, therefore, find the third-order polarization $\boldsymbol{P}^{(3)}(\omega_s) = N\ \mathrm{Tr}\,[\varrho^{(3)}(\omega_s)\boldsymbol{p}]$. We can use an iterative procedure to solve the equation of motion for the density matrix ϱ in eq. (11) with the Hamiltonian given in eq. (10), where $\boldsymbol{E} = \boldsymbol{E}(\omega_l) + \boldsymbol{E}(\omega_s)$ and $\omega_l - \omega_s = \omega_{fg}$. In each iterative process we only need to keep those density matrix elements which are near resonance. Thus, if we assume for simplicity $\varrho_g^{(0)} = 1$ and $\varrho_n^{(0)} = \varrho_f^{(0)} = 0$, the equations for the relevant density matrix elements of various order are

$$
(60)\quad \left\{
\begin{aligned}
&\hbar\left[i\frac{\partial}{\partial t} - \omega_{ng} + i\Gamma_{ng}\right]\langle n|\varrho^{(1)}(\omega_l)|g\rangle = -\langle n|\boldsymbol{p}\cdot\boldsymbol{E}(\omega_l)|g\rangle\,,\\
&\hbar\left[i\frac{\partial}{\partial t} - \omega_{fg} + i\Gamma_{fg}\right]\langle f|\varrho^{(2)}(\omega_l-\omega_s)|g\rangle = -\langle f|\boldsymbol{p}_s\cdot\boldsymbol{E}^*(\omega_s)|n\rangle\langle n|\varrho^{(1)}(\omega_l)|g\rangle\,,\\
&\left(\frac{\partial}{\partial t} + \frac{1}{T_n}\right)\langle n|\varrho^{(2)}(0)|n\rangle = \frac{1}{i\hbar}\big[-\langle n|\boldsymbol{p}_l\cdot\boldsymbol{E}(\omega_l)|g\rangle\langle g|\varrho^{(1)}(\omega_l)|n\rangle + \\
&\qquad\qquad + \langle n|\varrho^{(1)}(\omega_l)|g\rangle\langle g|\boldsymbol{p}_l\cdot\boldsymbol{E}^*(\omega_l)|n\rangle\big]\,,\\
&\hbar\left[i\frac{\partial}{\partial t} - \omega_{nf} + i\Gamma_{nf}\right]\langle n|\varrho^{(3)}(\omega_s)|f\rangle = -\langle n|\boldsymbol{p}_l\cdot\boldsymbol{E}(\omega_l)|g\rangle\langle g|\varrho^{(2)}(\omega_s-\omega_l)|f\rangle + \\
&\qquad\qquad + \langle n|\varrho^{(2)}(0)|n\rangle\langle n|\boldsymbol{p}_s\cdot\boldsymbol{E}(\omega_s)|f\rangle\,,
\end{aligned}
\right.
$$

where we have used $(\partial\langle n|\varrho^{(2)}|n\rangle/\partial t)_{\text{damping}} = -\langle n|\varrho^{(2)}|n\rangle/T_n$, which is a good approximation. The third-order nonlinear polarization is given by

$$P^{(3)}(\omega_s) \simeq \langle f|\boldsymbol{p}_s|n\rangle\langle n|\varrho^{(3)}(\omega_s)|f\rangle . \tag{61}$$

We realize, from the equation for $\langle n|\varrho^{(3)}(\omega_s)|f\rangle$ in eq. (60), that $P^{(3)}(\omega_l)$ has two parts of different origins, one from $\langle f|\varrho^{(2)}(\omega_l-\omega_s)|g\rangle$ and the other from $\langle n|\varrho^{(2)}(0)|n\rangle$. Note that $\langle f|p^{(2)}(\omega_l-\omega_s)|g\rangle$ is just the quantity we defined as $\psi(\omega_i=\omega_l-\omega)$ in sect. **2** and is responsible for SRS. On the other hand, $\langle n|\varrho^{(2)}|n\rangle$ is the excess population pumped by $E(\omega_l)$ into $|n\rangle$ and, therefore, is responsible for SE from $|n\rangle$ to $|f\rangle$.

We consider the solution of eq. (60) for a special case where $E(\omega_l)$ is switched on at $t=-\infty$ and then suddenly switched off at $t=t_0$. A Stokes beam of constant intensity is used to probe the gain at ω_s. With

$$E(\omega_l)=\mathscr{E}_l[1-u(t-t_0)]\exp[ik_lz-i\omega_lt] \quad\text{and}\quad E(\omega_s)=\mathscr{E}_s\exp[ik_sz-i\omega_st],$$

where $u(t-t_0)$ is a step function, we find from eq. (60)

$$\left\{\begin{aligned}
P^{(3)}(\omega_s) &= P^{(3)}_{\text{SRS}}(\omega_s)+P^{(3)}_{\text{SE}}(\omega_s),\\
P^{(3)}_{\text{SRS}}(\omega_s) &= \frac{N|\langle f|p_s|n\rangle\langle n|p_l\mathscr{E}_l|g\rangle|^2\mathscr{E}_s}{\hbar^3(\omega_l-\omega_s-\omega_{fg}-i\Gamma_{fg})(\omega_l-\omega_{ng}-i\Gamma_{ng})(\omega_s-\omega_{nf}+i\Gamma_{nf})}\cdot\\
&\quad\cdot\exp[ik_sz]\{[1-u(t-t_0)\exp[-i\omega_st]+\\
&\qquad\qquad+\exp[[-i\omega_{nf}-\Gamma_{nf}](t-t_0)-\omega_st_0]u(t-t_0)\},\\
P^{(3)}_{\text{SE}}(\omega_s) &= \frac{-2\Gamma_{ng}T_nN}{\hbar^3[(\omega_l-\omega_{ng})^2+\Gamma^2_{ng}]}|\langle f|p_s|n\rangle\langle n|p_l|g\rangle|^2|\mathscr{E}_l|^2|\mathscr{E}_s|\cdot\\
&\quad\cdot\left\{\frac{1-u(t-t_0)}{\omega_s-\omega_{nf}+i\Gamma_{nf}}\exp[-i\omega_st]+\frac{u(t-t_0)}{\omega_s-\omega_{nf}+i(\Gamma_{nf}-1/T_n)}\cdot\right.\\
&\quad\cdot\exp[(-i\omega_s-1/T_n)(t-t_0)-i\omega_st_0]+\\
&\quad+\left[\frac{1}{\omega_s-\omega_{nf}+i\Gamma_{nf}}-\frac{1}{\omega_s-\omega_{nf}+i(\Gamma_{nf}-1/T_n)}\right]\cdot\\
&\qquad\left.\cdot u(t-t_0)\exp[(-i\omega_{nf}-\Gamma_{nf})(t-t_0)-i\omega_st_0]\right\}.
\end{aligned}\right. \tag{62}$$

The gain coefficient at ω_s is then given by

$$g=(2\pi\omega_s/c\sqrt{\varepsilon'_s})\,\text{Im}\,[P^{(3)}(\omega_s)/E(\omega_s)]. \tag{63}$$

We notice in eq. (62) that, if $|\omega_l-\omega_{ng}|\simeq|\omega_s-\omega_{nf}|\gg\Gamma_{ng},\ \Gamma_{nf}$, then the contribution from SE is negligible and

$$P^{(3)}(\omega_s)\simeq P^{(3)}_{\text{SRS}}(\omega_s)=\left|\frac{\langle f|p_s|n\rangle\langle n|p_l|g\rangle}{\omega_l-\omega_{ng}}\right|^2\frac{|\mathscr{E}_l|^2\mathscr{E}_s}{\omega_l-\omega_s-\omega_{fg}-i\Gamma_{fg}}\cdot \tag{64}$$
$$\exp[ik_sz]\{[1-u(t-t_0)]\exp[-i\omega_st]+\exp[[-i\omega_{nf}-\Gamma_{nf}](t-t_0)]\,u(t-t_0)\}.$$

The results in eq. (62) also explicitly show the transient behavior of the two contributions for $t \gg t_0$; the gain due to both SRS and SE has a term which appears in the form of a damped oscillation with an oscillating frequency $|\omega_s - \omega_{nf}|$ and a decay rate Γ_{nf}. Physically, this is due to mixing of $E(\omega_s)$ with the transverse excitation at ω_{nf} induced by the sudden switch-off $E(\omega_l)$. The transverse excitation has a decay rate Γ_{nf}. The phenomenon is equivalent to the coherent Raman beat effect recently observed by SHOEMAKER and BREWER [52]. There is another contribution to the gain from stimulated emission. This one decays purely exponentially with a time constant T_n. Clearly, it reflects the decay of the excess population in $|n\rangle$. If $T_n \gg 1/\Gamma_{nf}$, then SE should show up with a much longer transient response than SRS.

The steady-state solution of eq. (60) is obtained by letting $u(t-t_0) = 0$ in eqs. (62). In general, the explicit expression for the steady-state gain g obtained from $P^{(3)}(\omega_s)$ is quite complicated. However, if the line widths of the states involved are dominated by lifetime broadening such that $\Gamma_{ij}\,(i \neq j) = (T_i^{-1} + T_j^{-1})/2$, where T_i and T_j are the lifetimes of $|i\rangle$ and $|j\rangle$, and if $T_g \gg T_n$, T_f such that $2\Gamma_{ng} T_n = 1$, then $P^{(3)}_{\text{SRS}}(\omega_s)$ and $P^{(3)}_{\text{SE}}(\omega_s)$ can be combined into a simple form. The corresponding gain becomes (under the assumption of negligible depletion of $|\mathscr{E}_l|^2$)

$$(65)\qquad g = (2\pi\omega_s/c\sqrt{\varepsilon_s'})\frac{N\Gamma_{fg}|\mathscr{E}_l|^2}{\hbar^3[(\omega_l-\omega_s-\omega_{fg})^2+\Gamma_{fg}^2]}\left|\frac{\langle f|p_s|n\rangle\langle n|p_l|g\rangle}{\omega_l-\omega_{ng}+i\Gamma_{ng}}\right|^2 .$$

This is the usual expression one would obtain for resonant stimulated Raman scattering from the golden-rule approach, which implicitly assumes lifetime broadening for the states and $T_g \gg T_n$, T_f. As we have seen from our derivation, eq. (65) actually includes contributions from both resonant SRS and hot HE.

It is interesting to compare this case with a case where an inverted population between $|n\rangle$ and $|f\rangle$ is established by some other means, *e.g.* discharge pumping. In a laser medium, the gain profile $g(\omega_s)$ should be proportional to $\Gamma_{nf}/[(\omega_s-\omega_{nf})^2+\Gamma_{nf}^2]$, but here, according to eq. (65), $g(\omega_s) \propto \Gamma_{fg}/[(\omega_l-\omega_s-\omega_{fg})^2+\Gamma_{fg}^2]$ is dominated by the Raman line shape even though the resonant excitation at ω_l has established an inverted population between $|n\rangle$ and $|f\rangle$. We should however emphasize that the results here may be mainly of academic interest, since they are derived for an ideal three-level system. A practical system often has many degenerate or nearly degenerate states. Cross-relaxation among degenerate or nearly degenerate states makes the problem more complicated. Consider a system with a set of intermediate states $|n\rangle$ which are close together and $E(\omega_l)$ at resonance with some of them but not all of them. In this case, we can still divide the emission into an SRS part due to transverse excitation $\langle f|\varrho^{(2)}(\omega_l-\omega_s)|g\rangle$ and an SE part due to excess populations $\langle n|\varrho^{(2)}(0)|n\rangle$ in the intermediate states $|n\rangle$. Even those intermediate states which are not resonantly excited by $E(\omega_l)$ are getting excess populations through relaxation of populations from the resonantly excited states. Emission from the unexcited intermediate states is, of course, not hot emission by definition, but its spectrum may overlap with that of hot emission, making the distinction between resonant SRS and hot SE more difficult. The problem of multiple intermediate states has been discussed in some detail in ref. [51]. Here we only want to point out that the usual expression for resonant stimulated Raman gain

$$g(\omega_s) = (2\pi\omega_s/c\sqrt{\varepsilon_s})\frac{N\Gamma_{fg}|\mathscr{E}_l|^2}{\hbar^3[(\omega_l-\omega_s-\omega_{fg})^2+\Gamma_{fg}^2]}\left|\sum_n\frac{\langle f|p_s|n\rangle\langle n|p_l|g\rangle}{\omega_l-\omega_{ng}+i\Gamma_{ng}}\right|^2, \tag{66}$$

which includes the contribution of hot SE, is strictly valid only if the cross-relaxation between the intermediate states is negligible and the states are dominated by lifetime broadening with $T_g \gg T_n$, T_f.

No experiment has yet been reported to distinguish resonant SRS and hot SE. As we have seen, their differences lie in the differences between longitudinal and transverse excitations. That longitudinal and transverse excitations are different is, of course, a well-established physical fact. For example, it is clearly demonstrated by the experiments discussed in sect. **2**, *G*). In the more general formalism, we should, of course, also include the higher-order effect [53].

REFERENCES

[1] L. D. LANDAU and E. M. LIFSCHITZ: *Electrodynamics in Continuons Media* (Reading, Mass., 1959), p. 252.
[2] Y. R. SHEN: *Phys. Rev.*, **133**, A 511 (1964).
[3] N. BLOEMBERGEN: *Nonlinear Optics* (New York, N. Y., 1965).
[4] N. BLOEMBERGEN and Y. R. SHEN: *Phys. Rev.*, **133**, A 37 (1964).
[5] See, for example, L. ALLEN and J. H. EBERLY: *Optical Resonance and Two-Level Atoms* (New York, N. Y., 1975) and references therein.
[6] P. N. BUTCHER: *Nonlinear Optical Phenomena* (Columbus, O., 1965), p. 43.
[7] J. ARMSTRONG, N. BLOEMBERGEN, J. DUCUING and P. S. PERSHAN: *Phys. Rev.*, **127**, 1918 (1962).
[8] D. BEDEAUX and N. BLOEMBERGEN: *Physica (Amsterdam)*, **69**, 67 (1973).
[9] F. N. H. ROBINSON: *Bell Syst. Tech. Journ.*, **46**, 913 (1967); *J. Phys. C*, **1**, 286 (1968).
[10] S. S. JHA and N. BLOEMBERGEN: *Phys. Rev.*, **171**, 891 (1968).
[11] C. FLYTZANIS and J. DUCUING: *Phys. Rev.*, **178**, 1218 (1969); J. DUCUING and C. FLYTZANIS: in *Optical Properties of Solids*, edited by F. ABELES (Amsterdam, 1971), p. 859.
[12] See, for example, J. C. PHILLIPS: *Covalent Bounding in Crystals, Molecules, and Polymers* (Chicago, Ill., 1969); *Bonds and Bands in Semiconductors* (New York, N. Y., 1973).
[13] B. F. LEVINE: *Phys. Rev. Lett.*, **22**, 787 (1969); *Phys. Rev. B*, **7**, 2600 (1973) and references therein.

[14] C. L. TANG and C. FLYTZANIS: *Phys. Rev. B*, **4**, 2520 (1971); C. L. TANG: *IEEE Journ. Quantum Electron.*, QE-**9**, 755 (1973); F. SCHOLL and C. L. TANG: *Phys. Rev. B*, **8**, 4607 (1973).
[15] See, for example, J. M. ZIMAN: *Principles of the Theory of Solids* (Cambridge, 1964).
[16] D. R. PENN: *Phys. Rev.*, **128**, 2093 (1962).
[17] J. C. PHILLIPS: *Phys. Rev.*, **166**, 832 (1968); **168**, 905 (1968); *Phys. Rev. Lett.*, **20**, 550 (1968).
[18] See, for example, C. A. COULSON: *Valence* (London, 1961).
[19] J. P. WALTER and M. L. COHEN: *Phys. Rev. Lett.*, **26**, 17 (1971).
[20] S. LOUIE and M. L. COHEN: private communications (unpublished).
[21] R. K. CHANG, J. DUCUING and N. BLOEMBERGEN: *Phys. Rev. Lett.*, **15**, 415 (1965); N. BLOEMBERGEN, R. K. CHANG and J. DUCUING: in *Physics of Quantum Electronics*, edited by P. L. KELLEY, B. LAX and P. E. TANNENWALD (New York, N. Y., 1966), p. 67.
[22] M. I. BELL: in *Electronic Density of States*, edited by L. H. BENETT, National Bureau of Standards (U.S.), Spec. Publ. No. 232 (Washington, D. C., 1971), p. 757; *Phys. Rev. B*, **6**, 516 (1972).
[23] C. Y. FONG and Y. R. SHEN: *Phys. Rev.* (to be published).
[24] M. L. COHEN and V. HEINE: in *Solid State Physics*, edited by H. ERENREICH, F. SEITZ and D. TURNBULL, Vol. **24** (New York, N. Y., 1970), p. 38; J. R. CHELIKOWSKY and M. L. COHEN: *Phys. Rev. Lett.*, **31**, 1582 (1973); **32**, 674 (1974).
[25] F. G. PARSONS and R. K. CHANG: *Opt. Comm.*, **3**, 173 (1971); H. LOTEM, G. KOREN and Y. YACOBY: *Phys. Rev. B*, **9**, 3532 (1974); D. BETHUNE, A. J. SCHMIDT and Y. R. SHEN: *Phys. Rev. B*, **11**, 3867 (1975).
[26] Y. R. SHEN: *Appl. Phys. Lett.*, **23**, 516 (1973); see also Y. R. SHEN: in *Raman Scattering in Solids*, edited by M. CARDONA (Berlin, 1975).
[27] See, for example, A. ABRAGAM: *Principles of Nuclear Magnetism* (Oxford, 1961).
[28] M. BORN and K. HUANG: *Dynamical Theory of Crystal Lattices*, Chap. II (Oxford, 1954); K. HUANG: *Proc. Roy. Soc.*, A **028**, 352 (1951); J. J. HOPFIELD: *Phys. Rev.*, **112**, 155 (1958).
[29] D. C. HAUEISEN and H. MAHR: *Phys. Rev. Lett.*, **26**, 838 (1971); *Phys. Lett.*, **36** A, 433 (1971); *Phys. Rev. B*, **8**, 2969 (1973).
[30] D. BOGGETT and R. LOUDON: *Phys. Rev. Lett.*, **28**, 1051 (1972).
[31] D. FRÖHLICH, B. STAGINNUS and E. SCHÖNHERR: *Phys. Rev. Lett.*, **19**, 1032 (1967); D. FRÖHLICH, E. MOHLER and P. WIESNER: *Phys. Rev. Lett.*, **26**, 554 (1971).
[32] V. T. NGUYEN and T. J. BRIDGES: *Phys. Rev. Lett.*, **29**, 359 (1972); T. J. BRIDGES and V. T. NGUYEN: *Appl. Phys. Lett.*, **23**, 107 (1973); N. BRIGNALL, R. A. WOOD, C. R. PIDGEON and B. S. WHERRET: *Opt. Comm.*, **12**, 17 (1974); V. T. NGUYEN and T. J. BRIDGES: *Appl. Phys. Lett.*, **26**, 452 (1975).
[33] T. L. BROWN and P. A. WOLFF: *Phys. Rev. Lett.*, **29**, 362 (1972).
[34] Y. R. SHEN and N. BLOEMBERGEN: *Phys. Rev.*, **137**, A 1786 (1965).
[35] F. DE MARTINI: *Phys. Lett.*, **30** A, 319, 547 (1969); *Phys. Rev. B*, **4**, 4556 (1971).
[36] K. H. YANG, J. R. MORRIS, P. L. RICHARDS and Y. R. SHEN: *Appl. Phys. Lett.*, **23**, 669 (1973).
[37] R. ROMESTAIN, S. GESCHWIND, G. E. DEVLIN and P. A. WOLFF: *Phys. Rev. Lett.*, **33**, 10 (1974).
[38] N. BLOEMBERGEN and Y. R. SHEN: *Phys. Rev.*, **141**, 298 (1966); Y. R. SHEN and N. BLOEMBERGEN: *Phys. Rev.*, **143**, 372 (1966).
[39] R. LOUDON: *Proc. Phys. Soc.*, **82**, 393 (1963); P. N. BUTCHER, R. LOUDON and T. P. MCLEAN: *Proc. Phys. Soc.*, **85**, 565 (1965); Y. R. SHEN: *Phys. Rev.*, **138**, A 1741 (1965); F. DE MARTINI: *Journ. Appl. Phys.*, **37**, 4503 (1966); B. A.

AKANEV, S. A. AKHMANOV and YU. G. KRONOPULO: *Sov. Phys. JETP*, **28**, 656 (1969); C. H. HENRY and C. G. B. GARRET: *Phys. Rev.*, **171**, 1059 (1968).
[40] S. K. KURTZ and J. A. GIORDMAINE: *Phys. Rev. Lett.*, **22**, 192 (1969); J. GELBWACHS, R. H. PANTELL, H. E. PUTHOFF and J. M. YARBOROUGH: *Appl. Phys. Lett.*, **14**, 258 (1969).
[41] J. M. YARBOROUGH, S. S. SUSSMAN, H. E. PUTHOFF, R. H. PANTELL and B. C. JOHNSON: *Appl. Phys. Lett.*, **15**, 102 (1969); M. A. PIESTRUP, R. N. FLEMING and R. H. PANTELL: *Appl. Phys. Lett.*, **26**, 418 (1975).
[42] F. DE MARTINI, G. P. GIULIANI and E. SANTAMATO: *Opt. Comm.*, **5**, 126 (1972); J. LUKASIK and J. DUCUING: *Phys. Rev. Lett.*, **23**, 1155 (1972); P. R. REGNIER and J. P. E. TARAN: *Appl. Phys. Lett.*, **23**, 240 (1973); F. MOYA, S. A. DRUET and J. P. E. TARAN: *Opt. Comm.*, **13**, 169 (1975).
[43] J. P. COFFINET and F. DE MARTINI: *Phys. Rev. Lett.*, **22**, 60 (1969); J. J. WYNNE: *Phys. Rev. Lett.*, **29**, 650 (1972); *Phys. Rev. B*, **6**, 534 (1972); E. YABLONOVITCH, C. FLYTZANIS and N. BLOEMBERGEN: *Phys. Rev. Lett.*, **29**, 865 (1972).
[44] M. D. LEVENSON, C. FLYTZANIS and N. BLOEMBERGEN: *Phys. Rev. B*, **6**, 3962 (1972); S. A. AKHMANOV, V. G. DMETRIEV, A. I. KOVRIGIN, N. I. KOROTEEV, V. G. TUNKIN and A. I. KHOLODYNYKH: *JETP Lett.*, **15**, 525 (1972); M. D. LEVENSON: *IEEE Journ. Quantum Electron.*, QE-**10**, 110 (1974); M. D. LEVENSON and N. BLOEMBERGEN: *Journ. Chem. Phys.*, **60**, 1323 (1964); S. A. AKHMANOV, N. I. KOROTEEV and A. I. KHOLODHYKH: *Journ. Raman Spectr.*, **2**, 239 (1974); I. ITZKAN and D. A. LEONARD: *Appl. Phys. Lett.*, **26**, 106 (1975).
[45] R. F. BEGLEY and D. A. LEONARD: *Appl. Phys. Lett.*, **25**, 387 (1974).
[46] S. D. KRAMER, F. G. PARSONS and N. BLOEMBERGEN: *Phys. Rev. B*, **9**, 1853 (1974).
[47] F. DE MARTINI and J. DUCUING: *Phys. Rev. Lett.*, **17**, 117 (1966).
[48] R. R. ALFANO and S. L. SHAPIRO: *Phys. Rev. Lett.*, **26**, 1247 (1971).
[49] D. VON DER LINDE, A. LAUBEREAU and W. KAISER: *Phys. Rev. Lett.*, **26**, 954 (1971); A. LAUBEREAU, D. VON DER LINDE and W. KAISER: *Phys. Rev. Lett.*, **27**, 802 (1971); **28**, 1162 (1972); *Opt. Comm.*, **1**, 173 (1973).
[50] R. R. ALFANO and S. L. SHAPIRO: *Phys. Rev. Lett.*, **29**, 1655 (1972); A. LAUBEREAU, G. KEHL and W. KAISER: *Opt. Comm.*, **9**, 182 (1973); A. LAUBEREAU, L. GREITER and W. KAISER: *Opt. Comm.*, **11**, 74 (1974).
[51] Y. R. SHEN: *Phys. Rev. B*, **9**, 622 (1973); M. V. KLEIN: *Phys. Rev. B*, **8**, 919 (1973); D. L. ROUSSEAU, G. D. PATTERSON and P. F. WILLIAMS: *Phys. Rev. Lett.*, **34**, 1306 (1975) and references therein.
[52] R. L. SHOEMAKER and R. G. BREWER: *Phys. Rev. Lett.*, **28**, 1430 (1972); R. G. BREWER and E. L. HAHN: *Phys. Rev. A*, **8**, 464 (1973).
[53] A. JAVAN: *Phys. Rev.*, **107**, 1579 (1957); K. SHIMODA and T. SHIMIZA: *Prog. Quant. Electr.*, **2**, 61 (1972).

Nonlinear Optics in a One-Dimensional Periodic Medium.

Y. R. SHEN

Department of Physics, University of California
Inorganic Materials Research Division, Lawrence Berkeley Laboratory - Berkeley, Cal. 94720

Recently, intense research effort has been devoted to the development of solids with a prescribed one-dimensional periodic structure, known as superlattice [1]. A medium with a superlattice can have very interesting optical and transport properties [1, 2]. Here we shall discuss some simple nonlinear optical effects in such a medium [3-6].

Wave propagation in a periodic structure is of course a well-known problem in solid-state physics. There, we deal with electron waves in a periodic lattice. The electron wave function takes the Bloch form $u(z)\exp[ikz - i\omega t]$ with $u(z) = u(z+d)$, where d is the period of the lattice [7]. If the Bragg condition for reflection is approximately satisfied, then propagation of the electron wave in the lattice becomes forbidden. This corresponds to the forbidden energy gap in the electron band structure. For electron-electron scattering in a periodic lattice, the crystal momentum defined as $G = 2\pi/d$ can participate in momentum matching. The process is then known as an umklapp process. All of these facts, true for electron waves, are also true for optical waves in a superlattice.

For nonlinear optics in a superlattice, the effect of the lattice periodicity is not so much on the strength of the wave interaction but rather on the phase-matching condition. With the help of the crystal momentum it is now possible to achieve the otherwise impossible collinear phase matching in a medium. For example, in second-harmonic generation we can have the phase-matching condition

$$|k(\omega)| \pm |k(\omega)| = \pm |k(2\omega)| + nG\,, \tag{1}$$

where n is an integer, and $+|k|$ and $-|k|$ refer to forward and backward wave propagation respectively. Equation (1) indicates that by properly adjusting the value of G we can now have phase-matched second-harmonic generation with not only fundamental and second-harmonic beams in the same direction, but also fundamental and second-harmonic beams in opposite directions, or two fundamental beams in opposite directions. In analogy to the electron case these processes with $n \neq 0$ can be called coherent optical umklapp processes.

We now give a more formal derivation of the effect. Assume a superlattice with a linear dielectric constant $\varepsilon(z)$ and a second-order nonlinear susceptibility $\chi^{(2)}(z)$ given by

$$(2)\qquad \begin{cases} \varepsilon(z) = \varepsilon(z+d) = \sum_n \varepsilon_n \exp[inGz]\,, \\ \chi^{(2)}(z) = \chi^{(2)}(z+d) = \sum_n \chi_n^{(2)} \exp[inGz]\,. \end{cases}$$

If $\chi^{(2)} = 0$, the fields at ω and 2ω should have the Bloch form

$$(3)\qquad E(\omega) = u_1(z)\exp[ik_1 z - i\omega t]\,, \qquad E(2\omega) = u_2(z)\exp[ik_2 z - i2\omega t]\,,$$

where

$$(4)\qquad u_j(z) = \sum_n A_{jn} \exp[inGz]\,.$$

The wave vector k_j and the coefficients A_{jn} (normalized to A_{j1}), obtained from the solution of the wave equation, are functions of ω, d and ε_n.

With $\chi^{(2)} \neq 0$ we can still let $E(\omega)$ and $E(2\omega)$ have the form of eq. (3), but the A_{jn}'s are now slowly varying functions of z as a result of nonlinear wave interaction. Second-harmonic generation is described by the wave equation

$$(5)\qquad \left[\frac{\partial^2}{\partial z^2} + \left(\frac{2\omega}{c}\right)^2 \varepsilon(2\omega, z)\right] E(2\omega) = -\frac{4\pi(2\omega)^2}{c^2}\chi^{(2)}(2\omega, z)E^2(\omega)\,.$$

Substituting the expressions of eqs. (2)-(4) into eq. (5) and neglecting the $\partial^2 A_{2n}/\partial z^2$ terms, we obtain

$$(6)\qquad \sum_{n_a}\left\{2i(k_2+n_aG)\frac{\partial A_{2n_a}}{\partial z} - (k_2+n_aG)^2 + \sum_{n_b}\varepsilon_{n_b}(2\omega)A_{2(n_a-n_b)}\right\}\cdot$$
$$\cdot\exp[i(k_2+n_aG)z] =$$
$$= -\frac{16\pi\omega^2}{c^2}\sum_{n_c,n_d,n_e}\chi_{n_c}^{(2)}A_{1n_d}A_{1n_e}\exp\left[i[(k_1+k_1') + (n_c+n_d+n_e)G]z\right]\,,$$

where k_1 and k_1' ($|k_1| = |k_1'|$) are the wave vectors of the two fundamental waves. We are interested in the second-harmonic generation near phase matching such that

$$\Delta k = k_1 + k_1' - k_2 - nG \sim 0\,.$$

From eq. (6), if we neglect the phase-unmatched terms and the depletion of the fundamental waves, then we find

$$(7)\qquad 2i(k_2+n_aG)\frac{\partial A_{2n_a}}{\partial z}-(k_2+n_aG)^2+\sum_{n_b}\varepsilon_{n_b}(2\omega)A_{2,(n_a-n_b)}= \\ =-\frac{16\pi\omega^2}{c^2}\sum_{n_d,n_e}\chi^{(2)}_{(n_a-n_d-n_e-n)}A_{1n_d}A_{1n_e}\exp[i\Delta kz]\,.$$

For different n_a, eq. (7) forms a set of coupled equations from which we can solve for $A_{2n_a}(z)$, subject to the appropriate boundary conditions.

The general solution of eq. (7) is difficult, but it does lead to a general conclusion. If $\varepsilon_0>\varepsilon_1>\varepsilon_2>\dots$ and $\chi^{(2)}_0>\chi^{(2)}_1>\chi^{(2)}_2>\dots$, and if $|k|$ is not very close to $G/2$ such that $A_{j0}>A_{j1}>A_{j2}>\dots$, then from eq. (7) the phase-matched second-harmonic generation decreases as n increases. In other words, n acts as a measure of the perturbation order.

If $\varepsilon(z)$ and $\chi^{(2)}(z)$ are simple sinusoidal functions of z, then the solution of eq. (7) is readily obtainable. This has been done by TANG and BEY [6], although there is some question as to whether the boundary conditions they used are correct. No such nonlinear optical experiment on artificial superlattice has yet been reported. There, however, exists in nature a number of substances which have built-in superlattice structure, for example, in crystals with periodic domain [9] or rotational twinning [9]. In those cases the difficulty of quantitative analysis usually lies in the proper description of $\varepsilon(z)$ and $\chi^{(2)}(z)$. In addition, the periodicity of the superlattice is often not easily tunable. It turns out that there is a case where none of these difficulties arises. This is the case of third-harmonic generation in cholesteric liquid crystals [4]. (Inversion symmetry forbids second-harmonic generation for waves propagating along z.) As we

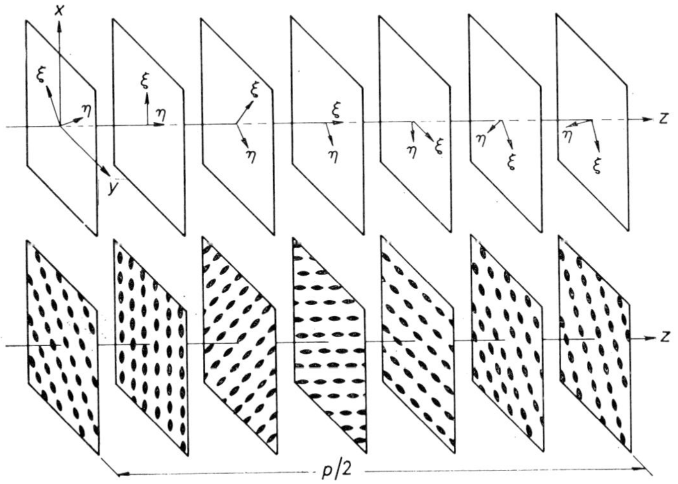

Fig. 1. – Sketch of molecular alignment in a cholesteric liquid crystal displacing the helical structure.

shall see, analytical solution of harmonic generation is available in this case and it provides an illuminating example for nonlinear optical effects in a superlattice.

In fig. 1 we show the average molecular arrangement in a cholesteric liquid crystal. In a layer perpendicular to the z-axis the molecules are aligned parallel to the layer, but as the layer advances along the z-axis the direction of molecular alignment gradually rotates. Consequently, the medium has an overall helical structure. It corresponds to a one-dimensional periodic lattice with the period equal to half of the helical pitch p. The most remarkable characteristic of cholesteric liquid crystals is that the helical pitch p can be easily adjustable from $\sim \pm 0.2$ μm to several hundred microns by almost any external perturbation, such as temperature, pressure and applied field.

Such a medium can be treated as a twisted birefringent material with a dielectric tensor

$$\boldsymbol{\varepsilon}(z) = \begin{pmatrix} \bar{\varepsilon}[1+\alpha\cos(4\pi z/p)] & \bar{\varepsilon}\alpha\sin(4\pi z/p) & 0 \\ \bar{\varepsilon}\alpha\sin(4\pi z/p) & \bar{\varepsilon}[1-\alpha\cos(4\pi z/p)] & 0 \\ 0 & 0 & \varepsilon_\eta \end{pmatrix}, \tag{8}$$

where $\bar{\varepsilon} = (\varepsilon_\xi + \varepsilon_\eta)/2$, $\alpha = (\varepsilon_\xi - \varepsilon_\eta)/2$. ε_ξ and ε_η are the principal dielectric constants in the directions parallel and perpendicular to the molecular alignment respectively. For this case, instead of using the Bloch-function formalism, it is actually much easier to solve the problem by a rotational transformation

$$\mathbf{R}(\theta = 2\pi z/p) = \begin{pmatrix} \cos\theta & \sin\theta & 0 \\ -\sin\theta & \cos\theta & 0 \\ 0 & 0 & 1 \end{pmatrix}, \tag{9}$$

which untwists the twisted helical structure. Then, in the rotating co-ordinate system the medium appears as a simple birefringent material with a dielectric tensor

$$\boldsymbol{\varepsilon}_T = \mathbf{R}:\boldsymbol{\varepsilon}(z):\mathbf{R}^{-1} = \begin{pmatrix} \varepsilon_\xi & 0 & 0 \\ 0 & \varepsilon_\eta & 0 \\ 0 & 0 & \varepsilon_\eta \end{pmatrix}. \tag{10}$$

After the rotational transformation the wave equation describing the third-harmonic generation is

$$\left[\frac{\partial^2}{\partial z^2} + \frac{4\pi}{p}\boldsymbol{\sigma}\frac{\partial}{\partial z} - \left(\frac{2\pi}{p}\right)^2 + \left(\frac{3\omega}{c}\right)^2 \boldsymbol{\varepsilon}_T(3\omega)\right]\cdot \boldsymbol{E}_T(3\omega) = -\frac{4\pi(3\omega)^2}{c^2}\boldsymbol{P}_T^{(3)}(3\omega)\,, \tag{11}$$

where

$$\boldsymbol{\sigma} = \begin{pmatrix} 0 & -1 & 0 \\ 1 & 0 & 0 \\ 0 & 0 & 0 \end{pmatrix}, \qquad \begin{aligned} &\boldsymbol{E}_T = R\boldsymbol{E}, \\ &\boldsymbol{P}_T^{(3)}(3\omega) = \chi_T^{(3)} : \boldsymbol{E}_T^3(\omega). \end{aligned}$$

$\chi_T^{(3)} = \mathbf{R} : \chi^{(3)}(z) : \mathbf{R}^{-1}\mathbf{R}^{-1}\mathbf{R}^{-1}$ is independent of z and has the form for a birefringent material such that we can write

$$(12) \qquad \begin{cases} P_{T\xi}^{(3)}(3\omega) = C_{11}E_\xi^3(\omega) + C_{12}E_\xi(\omega)E_\eta^2(\omega), \\ P_{T\eta}^{(3)}(3\omega) = C_{21}E_\eta(\omega)E_\xi^2(\omega) + C_{22}E_\eta^3(\omega). \end{cases}$$

Before solving eq. (11) we should find the form of $E_T(\omega)$ and $E_T(3\omega)$ in linear propagation. This can be obtained by solving eq. (11) with $\boldsymbol{P}^{(3)} = 0$. We find

$$(13) \qquad \begin{cases} \boldsymbol{E}_T(\omega) = [\mathscr{E}_+\hat{e}_+ \exp[i\varkappa_+ z] + \mathscr{E}_-\hat{e}_- \exp[-i\varkappa_- z]] \exp[-i\omega t], \\ \varkappa_\pm(\omega) = (\omega\bar{\varepsilon}^{\frac{1}{2}}/c)\, m_\pm, \\ m_\pm^2 \quad = (\lambda'^2 + 1) \pm (4\lambda'^2 + \alpha^2)^{\frac{1}{2}}, \\ \lambda' \quad = 2\pi c/\omega p\bar{\varepsilon}^{\frac{1}{2}}, \\ \hat{e}_\pm \quad = [1/(1 + |f_\pm|^2)^{\frac{1}{2}}](\hat{\xi} + if_\pm\hat{\eta}), \\ f_\pm \quad = 2m_\pm\lambda'/[m_\pm^2 + \lambda'^2 + (\alpha - 1)]. \end{cases}$$

In the laboratory co-ordinates the field becomes

$$(14) \qquad \boldsymbol{E}_\pm(\omega) = \mathbf{R}^{-1}\cdot\boldsymbol{E}_{T\pm}(\omega) =$$
$$= \tfrac{1}{2}\mathscr{E}_{\xi\pm}[(\hat{x} + i\hat{y})(1 + f_\pm) + (\hat{x} - i\hat{y})(1 - f_\pm)\exp[i4\pi z/p]] \exp[i(\varkappa_\pm - 2\pi/p)z - i\omega t].$$

Comparing with the Bloch form in eq. (3), we immediately recognize that

$$(15) \qquad k_\pm + nG = \varkappa_\pm - 2\pi/p,$$

where n is an integer, $G = 4\pi/|p|$, and $|k_\perp| \leqslant 2\pi/p$.

We can now solve eq. (11) by letting $\boldsymbol{E}_T(3\omega)$ have the form of eq. (13) with $\mathscr{E}_+$ and $\mathscr{E}_-$ being slowly varying functions of z. In the usual approximation the solution is straightforward. Under the near phase-matching condition

$$(16) \qquad \Delta\varkappa_{\pm lmn} = \varkappa_l(\omega) + \varkappa_m(\omega) + \varkappa_n(\omega) - \varkappa_\pm(3\omega) \sim 0$$

we find in the rotating co-ordinate system the output third-harmonic amplitude as

$$
(17)\quad \begin{cases}
\mathscr{E}_+(3\omega) = 4\pi\left(\dfrac{3\omega}{c}\right)^2\left(\dfrac{D\hat{e}_+^\dagger - B\hat{e}_-^\dagger}{AD-BC}\right)\cdot \boldsymbol{\chi}_T^{(3)}:\hat{e}_l\hat{e}_m\hat{e}_n\,\mathscr{E}_l(\omega)\,\mathscr{E}_m(\omega)\,\mathscr{E}_n(\omega)\cdot \\
\qquad\qquad \cdot\dfrac{1}{\Delta\varkappa_{+lmn}}\left[\exp\left[i\,\Delta\varkappa_{+lmn}z\right] + K_1\right], \\
\mathscr{E}_-(3\omega) = 4\pi\left(\dfrac{3\omega}{c}\right)^2\left(\dfrac{A\hat{e}_+^\dagger - C\hat{e}_-^\dagger}{AD-BC}\right)\cdot \boldsymbol{\chi}_T^{(3)}:\hat{e}_l\hat{e}_m\hat{e}_n\,\mathscr{E}_l(\omega)\,\mathscr{E}_m(\omega)\,\mathscr{E}_n(\omega)\cdot \\
\qquad\qquad \cdot\dfrac{1}{\Delta\varkappa_{-lmn}}\left[\exp\left[i\,\Delta\varkappa_{-lmn}z\right] + K_2\right],
\end{cases}
$$

where

$$
\begin{aligned}
A &= 2i\varkappa_+ + (4\pi/p)\,\hat{e}_+^\dagger(4\pi/p)\,\hat{e}_+^\dagger\cdot\boldsymbol{\sigma}\cdot\hat{e}_+\,, \\
B &= \hat{e}_+^\dagger\cdot[2i\varkappa_- + (4\pi/p)\,\boldsymbol{\sigma}]\cdot\hat{e}_-\,, \\
C &= \hat{e}_-^\dagger\cdot[2i\varkappa_+ + (4\pi/p)\,\boldsymbol{\sigma}]\cdot\hat{e}_+\,, \\
D &= 2i\varkappa_- + (4\pi/p)\,\hat{e}_-^\dagger\cdot\boldsymbol{\sigma}\cdot\hat{e}_-\,,
\end{aligned}
$$

and K_1 and K_2 are constants determined by the boundary conditions. The third-harmonic output intensity is given by the Poynting vector

$$
(18)\quad \begin{cases}
\boldsymbol{S}_\pm(3\omega) = \left(|\mathscr{E}_\pm|^2 c\bar{\varepsilon}^{\frac{1}{2}}/2\pi\right)\dfrac{\mathrm{Re}\,[q + |f|^2/q]_\pm}{1+|f_\pm|^2}\,\hat{z}\,, \\
q_\pm \qquad = m_\pm/\bar{\varepsilon}^{\frac{1}{2}} - \lambda' f_\pm\,.
\end{cases}
$$

Let us now discuss in more detail the phase-matching condition. Substitution of eq. (15) into eq. (16) gives for perfect phase matching

$$
(19)\qquad k_\pm(\omega) + k_\pm(\omega) + k_\pm(\omega) = k_\pm(3\omega) = nG\,.
$$

Note that $|k| \leqslant 2\pi/p$ is the wave vector of the optical Bloch wave in eq. (3). Equation (19) shows that allowing k to be either positive or negative for forward and backward propagation respectively, and having nG adjustable, we can in general find 15 different phase-matching conditions. If k is far from satisfying the Bragg condition such that $\lambda'^2 \gg \alpha^2$ and $(1-\lambda')^2 \gg \alpha^2/4\lambda'$, then we have $k_\pm(\omega) \simeq k_0(\omega) + n_\pm G$, where $k_0(\omega) = \omega\bar{\varepsilon}^{\frac{1}{2}}/c$ is the wave vector when the medium is in the isotropic phase. The phase-matching conditions become

$$
(20)\qquad \pm\,|k_0(\omega)| \pm |k_0(\omega)| \pm |k_0(\omega)| = \pm\,|k_0(3\omega)| + nG\,.
$$

This shows explicitly how an adjustable crystal momentum G can be used to compensate the phase mismatch between the fundamental and the third-harmonic waves. The value of G needed to satisfy eq. (19) or eq. (20) can be easily estimated (given n) knowing the linear dielectric tensor for the medium.

We can divide the phase-matching conditions in eq. (19) or eq. (20) into three groups: *a*) $E(\omega)$ and $E(3\omega)$ are propagating in the same direction. In this case, $|k_0(3\omega)-3k_0(\omega)|=G$ is small and the corresponding helical pitch p is long. *b*) $E(\omega)$ and $E(3\omega)$ are propagating in opposite directions. A very short pitch p is necessary to compensate the large phase mismatch $|k_0(3\omega)+3k_0(\omega)|$. *c*) Two fundamental waves are propagating in opposite directions. The pitch p for phase matching is also short since the mismatch $|k_0(3\omega)\pm k_0(\omega)|$ is large.

Phase-matched third-harmonic generation in cholesteric liquid crystals has been studied experimentally by SHELTON and SHEN [4]. Using mixtures of cholesteryl carbonate, cholesteryl nonanoate and cholesteryl chloride, they found good agreement between experimental results and theoretical predictions from eq. (19) or eq. (20). Temperature tuning of the helical pitch was used in the experiment to achieve phase matching. Some of the predicted and observed phase-matching conditions are given in table I. The phase-matching

TABLE I. – *Phase-matching conditions for third-harmonic generation in a mixture of cholesteric liquid crystals.*

Phase-matching condition		Predicted pitch for phase matching (m)	Predicted temperature for phase matching (°C)	Observed temperature for phase matching (°C)
Equation (85)	Equation (86)			
$3k_+(\omega)=$ $=k_+(3\omega)-G$	$3k_0(\omega)=$ $=k_0(3\omega)-G$	-17	49.4	49.3
		$+17$	54.2	54.1
$3k_-(\omega)=$ $=-k_-(3\omega)+G$	$3k_0(\omega)=$ $=-k_0(3\omega)+2G$	0.47	38.2	38.1
$2k_+(\omega)-k_-(3\omega)$	$2k_0(\omega)-k_0(\omega)=$ $=k_0(3\omega)-2G$	1.4	33.3	33.6
$-k_+(\omega)+2k_-(\omega)=$ $=k_-(3\omega)$	$-k_0(\omega)+2k_0(\omega)=$ $=k_0(3\omega)+2G$	0.7	31.1	31.2

curve of $3k_+(\omega)=k_+(3\omega)$ is shown in fig. 2 as an example. The cholesteric liquid-crystal mixture changes from left to right helicity as the temperature increases around 52 °C. The low-temperature and the high-temperature peaks in the figure correspond respectively to phase matching in left and right helical structure. Note that the observed peaks have a width of about ~ 0.2 °C. Without the accurate theoretical predictions, it would be quite difficult to find these peaks experimentally.

The above formalism is of course also valid for other wave-mixing problems. No other nonlinear optical effects in a superlattice have even been discussed in the literature. Presumably, using the Bloch wave functions, one can treat

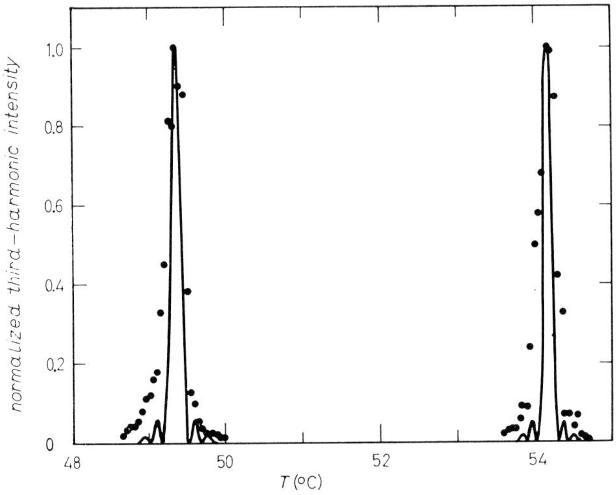

Fig. 2. – Normalized third-harmonic generation *vs.* temperature near phase matching in a mixture of 1.75 cholesteryl chloride and 1.00 cholesteryl myristate. The peak at lower temperature (corresponding to left helical structure) is generated by right circularly polarized fundamental waves and the one at higher temperature (corresponding to right helical structure) is generated by left circularly polarized fundamental waves. The solid line is the theoretical phase-matching curve and the dots are the experimental data points (ref. [4]).

these problems essentially in the same way as one would do with homogeneous media. As an example, let us consider stimulated Raman scattering in a superlattice. The wave equation for Stokes wave should have the form

$$\left[-\frac{\partial^2}{\partial z^2}+\frac{\omega_s^2}{c^2}\,\varepsilon_{s,\mathrm{eff}}(z)\right]E_s(z)=0\,, \tag{21}$$

where

$$\varepsilon_{s,\mathrm{eff}}(z)=\varepsilon_s(z)+\chi_R^{(3)}(z)|E_l(z)|^2=\varepsilon_{s,\mathrm{eff}}(z+d)\,,$$
$$|E_l(z)|^2=u_l^2(z)=u_l^2(z+d)\,,$$

assuming negligible depletion of $|E_l(z)|^2$. We can then express $E_s(z)$ also in the Bloch form

$$E_s(z)=u_s^2(z)\exp[ik_s z-i\omega_s t]\,.$$

The eigenvalue k_s can be solved from eq. (21) by expanding all the periodic functions into Fourier series and substituting them into eq. (21). For stimulated Raman scattering k_s is expected to have a negative imaginary part.

REFERENCES

[1] L. ESAKI and R. TSU: *IBM Journ. Res. Develop.*, **14**, 61 (1970); A. I. CHO: *Appl. Phys. Lett.*, **19**, 467 (1971); L. ESAKI: *Journn. Japan Soc. Appl. Phys. Suppl.*, **43**, 452 (1974).
[2] L. ESAKI and L. L. CHANG: *Phys. Rev. Lett.*, **33**, 495 (1974).
[3] N. BLOEMBERGEN and A. J. SIEVERS: *Appl. Phys. Lett.*, **17**, 483 (1970); R. TSU and L. ESAKI: *Appl. Phys. Lett.*, **19**, 246 (1971).
[4] J. W. SHELTON and Y. R. SHEN: *Phys. Rev. Lett.*, **25**, 23 (1970); **26**, 538 (1971); *Phys. Rev. A*, **5**, 1867 (1972).
[5] S. SONILEKH and A. YASIV: *Optics Comm.*, **6**, 301 (1972).
[6] C. L. TANG and P. P. BEY: *Journ. Quantum Electron.*, QE-**9**, 9 (1973).
[7] See, for example, C. KITTEL: *Introduction to Solid State Physics*, 4th ed. (New York, N. Y., 1971).
[8] I. FREUND: *Phys. Rev. Lett.*, **21**, 1404 (1968).
[9] C. F. DEWEY and L. O. HOCKER: *Appl. Phys. Lett.*, **26**, 442 (1975).

Nonlinear Optical Study of Pretransitional Behaviour in Liquid Crystalline Materials.

Y. R. Shen

Department of Physics, University of California
Inorganic Materials Research Division, Lawrence Berkeley Laboratory - Berkeley, Cal. 94720

Nonlinear optics can sometimes be used as a probe to study phase transition in a material. For example, second-harmonic generation, which is very sensitive to symmetry change, has been used to monitor the phase transition in crystalline NH_4Cl [1] and quartz [2]. Here we choose to discuss a case when the optical Kerr effect is used to probe the pretransitional behaviour of the isotropic $\rightarrow$ $\rightarrow$ nematic transition in a liquid crystal [3]. The same method has been applied to the study of pretransitional behaviour in plastic crystals [4].

A nematic substance is composed of long molecules. They are randomly oriented in the isotropic phase, but become more or less aligned along a common axis in the nematic phase. We can consider the substance as an analogue to a spin system with no spin polarity. Each molecule here plays the role of a spin. The liquid phase is then equivalent to the paramagnetic phase, and the nematic phase to the ferromagnetic phase. In a paramagnetic system, an applied magnetic field H can induce a spin alignment and hence a magnetization along the magnetic field. The induced magnetization M in the mean-field approximation obeys the well-known Curie's law

$$M = C_M H/(T - T_c)\,, \tag{1}$$

which exhibits a critically divergent behaviour as the temperature approaches a critical temperature T_c, where C_M is a constant. A close analogue exists in liquid crystalline materials. There, an applied optical field $E(\omega)$ can also induce a molecular alignment in the isotropic phase. However, since the field variation at optical frequencies is too fast for the moleculas to respond, the induced molecular alignment is proportional to $|E(\omega)|^2$ rather than to $E(\omega)$. We therefore have $|E|^2$ play the role of H, while the induced molecular alignment is manifested by the induced anisotropy δn in the optical refractive index. We then expect that δn should also obey the Curie's law

$$\delta n = C|E|^2/(T - T_c)\,, \tag{2}$$

where C is a constant and T_c is the critical temperature for a second-order isotropic → nematic transition in the mean-field approximation. (We should note that δn can also be induced by a d.c. electric field. However, the results are often more difficult to interpret because of the presence of ionic conduction in the medium which also affects the molecular alignment.)

Let us now give a more rigorous derivation of eq. (2). We use Landau's theory of second-order phase transition to describe the pre-transitional behavior of the isotropic → nematic transition [3, 5]. First, we define an order parameter Q by the relation

$$\delta\chi = \chi_{\parallel} - \chi_{\perp} = Q\,\Delta\chi\,, \tag{3}$$

where $\chi_{\parallel}$ and $\chi_{\perp}$ are the optical susceptibilities parallel and perpendicular to the applied linearly polarized field $\boldsymbol{E}(\omega)$ respectively, and $\Delta\chi$ is the anisotropy in χ when all molecules are perfectly aligned in one direction. Since the average susceptibility is $\chi_0 = (\chi_{\parallel} + 2\chi_{\perp})/3$, we have $\chi_{\parallel} = \chi_0 + 2Q\,\Delta\chi/3$ and $\chi_{\perp} = \chi_0 - Q\,\Delta\chi/3$. Then, following Landau's theory, we can write the free energy per unit volume in the isotropic phase ($Q \ll 1$) as

$$F = F_0 + \tfrac{1}{2}AQ^2 + BQ^3 + DQ^4 - \tfrac{1}{4}\chi_{\perp}|E(\omega)|^2\,, \tag{4}$$

where $A = a(T - T_0)$ and a, B and D are constant coefficients. At temperatures sufficiently far above the transition we expect Q to be so small that the BQ^3 and DQ^4 terms in eq. (4) become negligible. We can then readily obtain the Curie's law of eq. (2) from minimization of F with respect to Q. We find

$$\delta n = \frac{2\pi}{n}\,\delta\chi = \frac{\pi(\Delta\chi)^2}{3an}\,|E|^2/(T - T_c)\,. \tag{5}$$

It turns out that in nematic substances the BQ^3 and DQ^4 terms in F, which are responsible for the observed first-order isotropic → nematic transition in the medium, are negligibly small even when T is very close to the observed transition temperature T_k. In fact, in several substances, the observed T_k is only less than 1 °C above the fictitious critical temperature T_c.

We can also discuss the dynamics of the optical field-induced anisotropy. It is described by the simple equation [3-5]

$$\nu\,\delta Q/\delta t = -\,\delta F/\delta Q\,, \tag{6}$$

where ν is a viscosity coefficient. The solution of the equation is

$$(7) \qquad Q(t) = \int_{-\infty}^{t} \frac{\Delta\chi |E(\omega)|^2}{6\nu} \exp[-(t-t')/\tau]\,\mathrm{d}t' ,$$

where

$$(8) \qquad \tau = \nu/A = \nu/a(T-T_c)$$

is the relaxation time for the order parameter. Equation (7) shows that if $|E(\omega)|^2$ is a pulse shorter or comparable with τ, then at a sufficiently large t the order parameter Q, and hence the induced optical anisotropy $\delta n = 2\pi Q \Delta\chi/n$, will decrease exponentially with a time constant τ. From eq. (8) we note that, as T approaches T_c, the relaxation time increases rapidly as $(T-T_c)^{-1}$. This is again analogous to the well-known critical slowing-down behaviour of the induced magnetization in a paramagnetic crystal near the paramagnetic $\rightarrow$ $\rightarrow$ ferromagnetic transition.

It is easy to see from eq. (7) that, by measuring the transient response of $\delta n(t)$ induced by $|E(\omega)|^2$ of a given pulse shape, we can find the constants $\chi_Q^{(3)} = (\Delta\chi)^2/6a(T-T_c)$ and τ from which we can deduce $(\Delta\chi)^2/a$, ν/a and T_c. Note that $\chi_Q^{(3)}$ is a third-order nonlinear optical susceptibility which is connected to the optical Kerr constant K by $K = 2\omega\chi_Q^{(3)}/nc$ and to δn by $\delta n = 2\pi\chi_Q^{(3)}|E|^2/n$. Because of the critically divergent behaviour of $\chi_Q^{(3)}$ and τ, the field-induced δn and its relaxation time for an isotropic liquid crystalline material are much larger than those for an ordinary liquid, especially when the material approaches the isotropic $\rightarrow$ nematic transition.

The experimental arrangement for measuring $\delta n(t)$ is quite simple. An intense laser pulse is used to induce the optical anisotropy or Q, while a weak c.w. laser beam is used to probe the induced optical anisotropy. Since τ for liquid crystalline materials is of the order of a few ns or longer, the laser pulse needed for the experiment can be either a Q-switched or a weakly mode-locked one. Such experiments have been carried out for a number of nematic substances. Figure 1 shows, as an example, the experimental results and theoretical comparison of $\chi_Q^{(3)}$ and τ as functions of temperature for a nematic substance MBBA ($C_4H_9-(C_6H_4)-CH{:}N-(C_6H_4)-CH_3$). The curves clearly demonstrate the critically divergent behavior of $\chi_Q^{(3)}$ ($\beta = 2\chi_Q^{(3)}$ in fig. 1) and τ as the temperature approaches the isotropic $\rightarrow$ nematic transition. We note that $\chi_Q^{(3)}$ and τ shown in fig. 1 are much larger than those of ordinary liquids. At $T-T_c = 5$ °C, $\chi_Q^{(3)}(\text{MBBA}) = 2.7\cdot 10^{-10}$ e.s.u. is almost 100 times larger than that of the well-known Kerr liquid CS_2 at room temperature and $\tau(\text{MBBA}) \simeq 100$ ns is $5\cdot 10^4$ times longer than that of CS_2. The large values of $\chi^{(3)}$ and τ seem to be the characteristic properties of liquid crystalline materials as shown in table I for a number of nematic substances [6]. For MBBA τ varies from 40 to 800 ns as $T-T_c$ changes from 14.3 to 0.9 °C.

Isotropic liquid crystalline materials with their large optical Kerr constants and the temperature-dependent long relaxation time turn out to be ideal systems for investigation of self-focusing of light. Because of the large δn, a Q-switched laser pulse will self-focus readily in such a medium. By varying the temperature of an appropriate liquid crystalline material, it is now possible to change the relaxation time τ from a value much larger than the laser pulse width to a value shorter than the laser pulse width. This means that with the same Q-switched

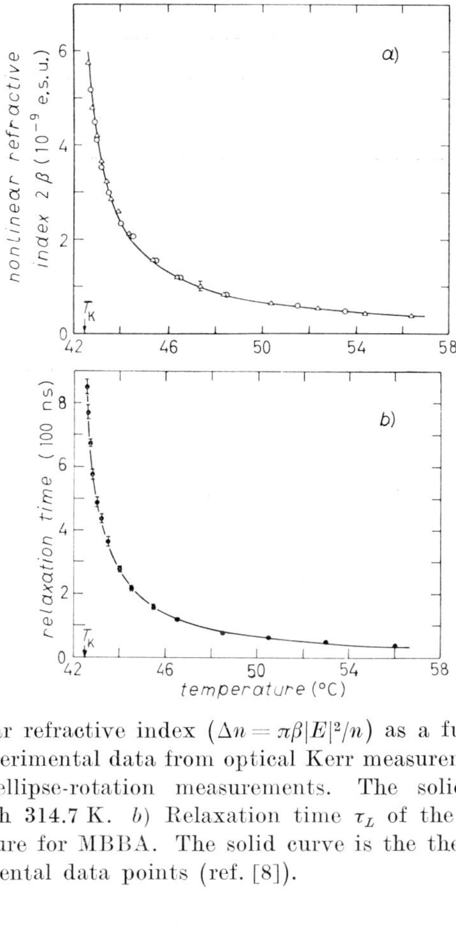

Fig. 1. – *a*) Nonlinear refractive index ($\Delta n = \pi\beta|E|^2/n$) as a function of temperature for MBBA. △ are experimental data from optical Kerr measurements and ○ are experimental data from ellipse-rotation measurements. The solid curve is given by $5.4\cdot 10^{-9}/(T - T_c)$ with 314.7 K. *b*) Relaxation time τ_L of the order parameter as a function of temperature for MBBA. The solid curve is the theoretical curve and the dots are the experimental data points (ref. [8]).

pulse we can now study not only the quasi–steady-state self-focusing phenomenon ($\tau \ll$ pulse width) but also the transient self-focusing phenomenon ($\tau \gg$ pulse width) and the intermediate cases between the two. In fact, study of self-focusing in MBBA has so far yielded the only quantitative experimental results on transient self-focusing in a liquid [7]. In ordinary liquids τ is in the picosecond range, and, therefore, picosecond laser pulses must be used to observe transient self-focusing. However, the present picosecond technology is not yet advanced enough for us to study transient self-focusing quantitatively.

TABLE I. – *Critical transition temperatures T_c, orientational relaxation times τ and nonlinear susceptibilities $\chi_Q^{(3)}$ for various nematic liquid crystalline compounds.*

Compound	R	T_c (°C)	$\tau(T-T_c)$ (10^{-8} s K)	$\chi_Q^{(3)}(T-T_c)$ (10^{-10} e.s.u. K)
C_4H_9—(C_6H_4)—CH:N—(C_6H_4)—R	CH_3	41.7	67	13.2
	C_2H_5	67.6	21	7.8
	CH_3	130.7	8.5	15
RO—(C_6H_4)—N_2O—(C_6H_4)—OR	C_2H_5	161.0	7.5	10.2
	C_5H_{11}	115.0	25	5.6
	C_6H_{13}	126.0	15	4.7
	C_7H_{15}	121.3	15	5.4

For a third-order nonlinear optical effect governed by a third-order nonlinear polarization $P^{(3)} = \chi^{(3)}|E(\omega)|^2 E(\omega')$ there is always a counterpart in light scattering. Thus, corresponding to the optical Kerr effect due to induced ordering, we have light scattering by order parameter fluctuation in an isotropic liquid crystalline material [3, 8]. Then, eqs. (4) and (6) with $|E(\omega)|^2 = 0$ can again be used to describe the light scattering phenomenon. Since $|E(\omega)|^2 = 0$, there is no induced ordering. However, as a result of thermal fluctuations, the mean square value $\langle Q^2 \rangle$ is not zero, but from the equipartition theorem applied to eq. (4) is equal to kT/A. Consequently, the scattering light intensity is

$$I \propto \left(\frac{\mathrm{d}\varepsilon}{\mathrm{d}Q}\right)^2 \langle Q^2 \rangle = \left(\frac{\mathrm{d}\varepsilon}{\mathrm{d}Q}\right)^2 \frac{kT}{a(T-T_c)}, \tag{9}$$

which shows the critically divergent behavior as $T \to T_c$. The dynamic equation for Q governs the spectrum of the scattered radiation. Equation (6) with $|E(\omega)|^2 = 0$ is in the form of a relaxation equation, and therefore it leads to a Lorentzian spectral line centered at the incoming laser frequency. The half-width of the Lorentzian line is given by

$$\Gamma = A/\nu' = a(T-T_c)/\nu . \tag{10}$$

We have, as expected, $\Gamma = 1/\tau$, which shows a critical narrowing behaviour as $T \to T_c$. The results in eqs. (9) and (10) have been experimentally verified by STINSON and LITSTER [7].

Light scattering by order parameter fluctuation appears as an intense narrow spectral line (half width $\leqslant 100$ MHz) centered at the laser frequency. However, as one looks more carefully at the scattering spectrum, one would find another much broader and much weaker component also centered at the laser frequency but with its wings extended out to tens of a cm^{-1} [9]. Such a broad component actually exists in all liquids and is believed to be due to light scattering by fluctuations of individual molecular orientations. This is known as Rayleigh-wing scattering [10]. Now, for liquid crystals there is apparently a dilemma, since the order parameter fluctuation is clearly also a fluctuation in the molecular orientation. How can the same fluctuation of molecular orientation give rise to both the narrow central component and the broad Rayleigh-wing component?

Qualitatively, the answer to the above question is simple. We remember that the Landau theory we have used is actually equivalent to the mean-field theory which takes into account only the average local-field correction. This can be seen as follows. Microscopically, the molecular orientation is induced by the local field

$$|E_{\text{loc}}|^2 = |E(\omega)|^2 + V\,, \tag{11}$$

where V is due to molecular interaction. In the mean-field approximation we have $V = \lambda Q$ with λ being a proportional constant. The average induced molecular orientation Q is proportional to $|E(\omega)|^2$, *i.e.* $Q = b(T)|E_{\text{loc}}|^2$. For $|T - T_c| \leqslant T_c$, by defining $b(T_c) = 1/\lambda$ and expressing $b(T) = b(T_c) + (T - T_c)(\mathrm{d}b/\mathrm{d}T)_{T_c}$ we can write

$$Q = \text{const}\,|E(\omega)|^2/(T - T_c)\,. \tag{12}$$

Equation (12) is of course the same as one would obtain from minimization of F in eq. (4).

The above derivation essentially follows the derivation of the magnetic susceptibility for a paramagnetic crystal near phase transition taking into account the mean local-field correction. This shows that the results derived from the free energy F in eq. (4) have not included the temporal and spatial dependence of the local field due to molecular interaction. Physically, the mean-field approximation describes the molecular orientational fluctuations under the influence of an average local field and predicts only a single Lorentzian spectral line with critical behaviour for the scattered radiation. In a more rigorous treatment we should take into account the local-field correction more rigorously. Then, the problem of N interacting molecules can be considered as a problem of N coupled oscillators (referring to the orientational motion) in the highly damped limit. It is well known that a system of N coupled harmonic oscillators has many resonant lines. In the highly damped limit the resonant frequencies of these lines are negligible compared to the line widths, but different lines have different line widths. Consequently, we expect to find a series of Lorentzian lines with different line widths centered at zero frequency. The narrowest line shows the critical behaviour arising from the average local-field correction, while the others are much broader and show no critical behaviour but more or less characteristics of orientational fluctuations of individual molecules. A more formal derivation of the above discussion is given in ref. [11].

REFERENCES

[1] I. FREUND: *Phys. Rev. Lett.*, **19**, 1288 (1967).
[2] J. P. BACHHEIMER and G. DOLINO: *Phys. Rev. B*, **11**, 3195 (1975).
[3] G. K. L. WONG and Y. R. SHEN: *Phys. Rev. Lett.*, **30**, 895 (1973); *Phys. Rev. A*, **10**, 1277 (1974).
[4] T. BISCHFBERGER and E. COURTENS: *Phys. Rev. Lett.*, **32**, 163 (1974).
[5] P. G. DE GENNES: *Phys. Lett.*, **30** A, 454 (1969); *Mol. Cryst. Liq. Cryst.*, **12**, 193 (1971).
[6] E. G. HANSON: to be published.
[7] T. W. STINSON and J. D. LITSTER: *Phys. Rev. Lett.*, **25**, 503 (1970).
[8] G. K. L. WONG and Y. R. SHEN: *Phys. Rev. Lett.*, **32**, 527 (1974).
[9] N. M. AMER, Y. S. LIN and Y. R. SHEN: *Solid State Comm.*, **16**, 115 (1975).
[10] See, for example, I. L. FABELLINSKI: *Molecular Scattering of Light*, Chap. II and VIII (New York, N. Y., 1968).
[11] C. FLYTZANIS and Y. R. SHEN: *Phys. Rev. Lett.*, **33**, 14 (1974).

Coherent Active Spectroscopy of Combinatorial (Raman) Scattering with Tunable Oscillators; Comparison with the Spontaneous-Scattering Technique.

S. A. Akhmanov

Department of Physics, Moscow State University - Moscow, USSR

1. – Introduction.

The method of active spectroscopy of combinatorial (Raman) scattering (ASCS) (*), successfully developed in the last 2 or 3 years, occupies an intermediate place with respect to spontaneous- and stimulated-combinatorial-scattering techniques. ASCS combines the wide spectroscopic potentialities of spontaneous spectroscopy and the high efficiency of scattering, strong excitation and phasing of molecular vibrations in a great volume of substance etc., that is the features inherent in stimulated Raman scattering.

Progress in ASCS is connected mainly with achievements in the development of tunable lasers.

The application of tunable lasers has made possible the new approach to the investigation of Raman scattering (RS). It has become possible to « prepare » in a proper way the scattering substance by means of tunable lasers, instead of the traditional investigation of thermal-equilibrium elementary excitations (internal molecular vibrations in gases and liquids, and phonons, polaritons, magnons, etc. in solids). The subject of ASCS is the investigation of light scattering off nonequilibrium excitations in the medium.

In a number of aspects ASCS possesses some advantages with respect to ordinary spectroscopy of spontaneous scattering, and thus provides the possibility of new physical and analytic applications.

(*) Other widely used terms for this method are a four-photon mixing spectroscopy and coherent anti-Stokes Raman scattering (CARS).

2. – The principles of ASCS.

The biharmonic laser pump is generally used for phasing and for strong excitation of Raman-active modes in ASCS (the frequencies of pump waves ω_1, ω_2 lie generally in the visible or infra-red ranges). The excitation of any vibrational transition with frequency Ω occurs when $\omega_1 - \omega_2 \simeq \Omega$. The beating of light waves excites elementary oscillators because of coupling between electron and nuclear motion in a molecule (in crystals excitation occurs due to electron-phonon interaction):

$$\frac{d^2Q}{dt^2} + 2\Gamma\frac{dQ}{dt} + \Omega^2 Q = \frac{1}{2M}\left(\frac{\partial \alpha_{mk}}{\partial Q}\right)_0 E_m^{(1)} E_k^{(2)*} \exp[-i(\omega_1 - \omega_2)t + i(\boldsymbol{k}_1 - \boldsymbol{k}_2)\boldsymbol{r}] \tag{1}$$

(see, for example, [1]). Here Q is the normal co-ordinate of a molecular vibration having the frequency Ω, $(\partial\alpha_{mk}/\partial Q)_0$ the derivative of the electronic polarizability with respect to the nuclear co-ordinate, taken in the equilibrium position, *i.e.* when $Q = 0$; Γ is the damping constant, determining the line width of the molecular oscillator; $\Delta\Omega = 2\Gamma$; M is the reduced mass of the molecule; $E^{(1)}$,

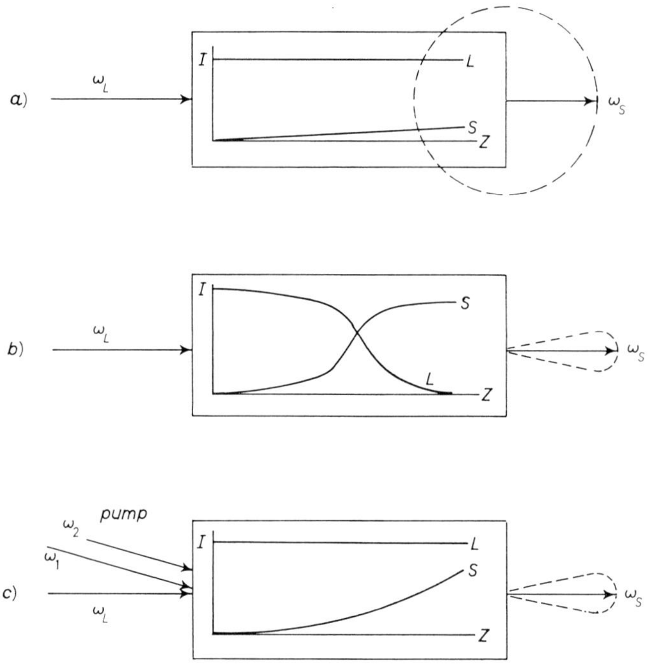

Fig. 1. – *a*) Spontaneous combinatorial scattering; *b*) stimulated combinatorial scattering; *c*) coherent active spectroscopy of RS, $\omega_1 - \omega_2 = \Omega$, $\boldsymbol{k}_S = \boldsymbol{k}_L - \boldsymbol{q}$, $\boldsymbol{q} = \boldsymbol{k}_1 - \boldsymbol{k}_2$.

$E^{(2)}$ are the amplitudes of pump waves with corresponding frequencies ω_1 and ω_2:

$$E_n^{(1,2)}(\boldsymbol{r}, t) = E_n^{(1,2)} \exp[-i(\omega_{1,2}t - \boldsymbol{k}_{1,2}\boldsymbol{r})] + \text{c.c.}$$

The so-called « probe » beam (frequency ω_L) is then used to detect coherently driven molecular vibrations. The Stokes or anti-Stokes probe intensity, obtained when $\omega_1 - \omega_2 \simeq \Omega$, prevails essentially over the intensity of the corresponding spontaneous scattering (see fig. 1). Nevertheless, a competition of lines and uncontrolled instabilities does not occur, because the SRS threshold is not exceeded. Besides, by tuning pump wave frequencies in ASCS it is possible to drive any normal mode of the molecule. So, using the ASCS method, one is able to investigate the full set of normal modes of the molecule (see fig. 2), in contrast with the stimulated-combinatorial-scattering method, where only the most intensive line of Raman scattering in the spontaneous spectrum is generally excited.

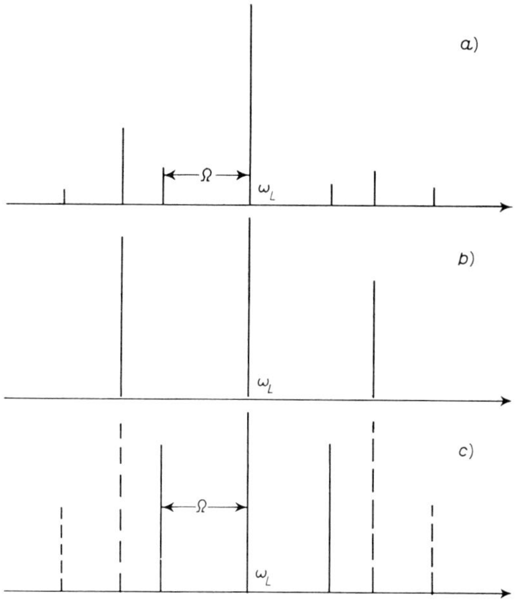

Fig. 2. – Spectra: *a*) spontaneous, *b*) stimulated, *c*) active. $\omega_1 - \omega_2 = \Omega$.

The coherent nature of the Raman-active–mode excitation gives rise to the possibility of effective separate probing of polarization and population difference (the behaviour of off-diagonal and diagonal elements of the density matrix) of the transition under investigation. In this connection, one is able to speak of two modifications of ASCS—the coherent and incoherent active spectroscopy of light scattering.

In coherent ASCS a Stokes or an anti-Stokes component of the probe beam is registered, when the probe wave is scattered off Raman-active vibrations being in phase all over the volume of a substance. The corresponding wave vectors of the pump waves ($\boldsymbol{k}_1$, $\boldsymbol{k}_2$), of the probe laser wave ($\boldsymbol{k}_L$) and of the scattered radiation ($\boldsymbol{k}_s$, $\boldsymbol{k}_a$) are connected to each other by the relations

$$\boldsymbol{k}_1 - \boldsymbol{k}_2 = \boldsymbol{k}_L - \boldsymbol{k}_s\,, \qquad \boldsymbol{k}_1 - \boldsymbol{k}_2 = \boldsymbol{k}_a - \boldsymbol{k}_L\,. \tag{2}$$

In incoherent ASCS an isotropic anti-Stokes component of the probe beam is registered. Its appearance is connected with real transitions from an upper level to a lower one, just as in the case of spontaneous scattering. Thus, coherent and incoherent ASCS give information complementary to one another.

Here only the coherent version of ASCS is considered.

3. – Active spectroscopy and tunable lasers.

ASCS has been first used by TERHUNE and MAKER [2] for studying the dispersion of Raman susceptibility in a number of organic liquids. In fact, they used coherent ASCS, and measurements have been carried out for a number of fixed frequencies ω_1, ω_2.

The work by GIORDMAINE and KAISER [3] devoted to light scattering off coherently driven molecular vibrations has also been of great importance for the development of the ASCS method. A series of works on coherent ASCS has been carried out in subsequent years (see, for example, [4-6]); the authors have used doublets with fixed frequencies to drive Raman-active transitions.

But it were frequency-tunable optical oscillators, used for driving molecular vibrations, that enabled a complete realization of all the advantages inherent in ASCS. The first smoothly tuned multifrequency oscillator for active spectro-

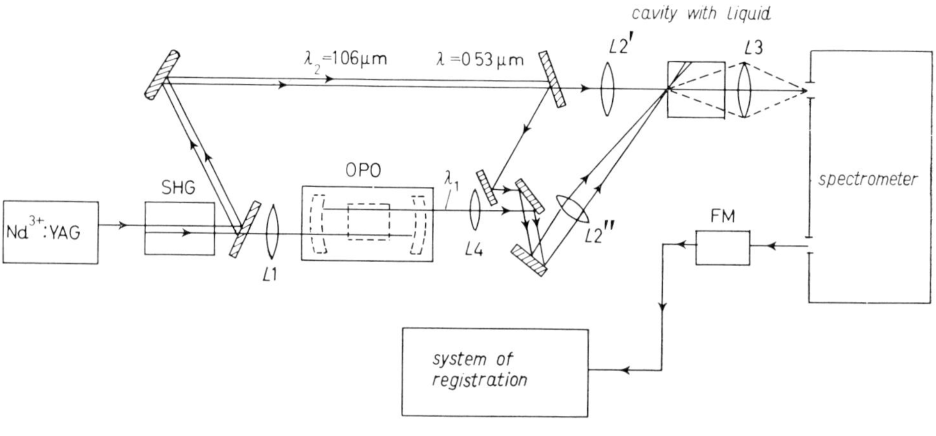

Fig. 3. – Experimental set-up for ASCS with the help of tunable OPO.

scopy with high pulse repetition rate was suggested in [7, 8] (see also [9]); it consisted of a repetitive YAG Nd^{3+} laser and of an optical parametric oscillator (OPO) excited by its second harmonic (see fig. 3). Later on, the multifrequency generators based on Nd^{3+}-glass laser second-harmonic radiation and on a dye laser [10] and also on two dye lasers, pumped by a molecular-nitrogen gas laser [11-13], have been created. Another scheme of biharmonic pumping now in use for ASCS includes the second harmonic of YAG Nd^{3+} laser radiation and the frequency-tunable dye laser radiation [14, 15] (fig. 4).

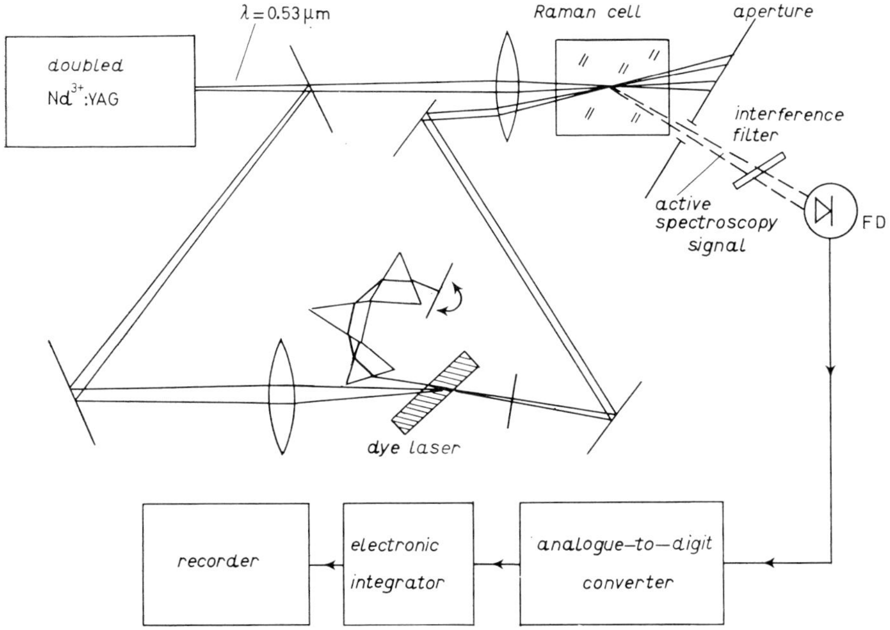

Fig. 4. – Experimental set-up for ASCS with the help of a tunable dye laser.

4. – Active spectroscopy in terms of nonlinear susceptibilities. Line shape of the active spectrum.

Phenomenologically the processes of Stokes or anti-Stokes Raman scattering of the probe beam in the presence of biharmonic pumping may be considered as four-photon interactions of the following types:

$$\omega_S = \omega_L - \omega_1 + \omega_2 \,, \qquad \omega_a = \omega_L + \omega_1 - \omega_2 \,. \tag{3}$$

These interactions in centrosymmetric medium are described in terms of cubic nonlinear susceptibilities $\chi^{(3)}$:

$$\begin{cases} \boldsymbol{E}^{(S)} \sim \chi^{(3)}(\omega_S = \omega_L - \omega_1 + \omega_2)\, \boldsymbol{E}\boldsymbol{E}^{(1)*}\boldsymbol{E}^{(2)} \,, \\ \boldsymbol{E}^{(a)} \sim \chi^{(3)}(\omega_a = \omega_L + \omega_1 - \omega_2)\, \boldsymbol{E}\boldsymbol{E}^{(1)}\boldsymbol{E}^{(2)*} \,. \end{cases} \tag{4}$$

In experiments the intensities

$$I^{(\mathrm{S,a})}_{\mathrm{ASCS}} \sim |\boldsymbol{E}^{(\mathrm{S,a})}|^2 \tag{5}$$

are registered, so the spectroscopic information, obtained in ASCS with the help of a tunable generator, consists in the determination of the modulus squared of the cubic nonlinear susceptibility $|\chi^{(3)}|^2$, while the spontaneous-Raman-scattering spectra correspond to the dispersion of the imaginary part of the cubic susceptibility:

$$I^{(\mathrm{S})}_{\mathrm{spont}} \sim \mathrm{Im}\,\chi^{(3)}(\omega_{\mathrm{S}} = \omega_L - \omega_L + \omega_S)\, I_L\,. \tag{6}$$

So the information, derived from « active » spectra, is more complete and rich in detail than the one derived from spontaneous-scattering spectra.

Different dynamic subsystems of the medium contribute to the nonlinear susceptibility $\chi^{(3)}$. We can select the contribution of internal molecular vibrations and derive

$$\chi^{(3)}_{ijkl} = \chi^{\mathrm{R}}_{ijkl} + \chi^{E}_{ijkl}\,. \tag{7}$$

Here

$$\chi^{\mathrm{R}}_{ijkl} = \frac{N}{2M} D \sum_n \overline{\left(\frac{\partial \alpha^{(n)}_{ij}}{\partial Q}\right)_0 \left(\frac{\partial \alpha^{(n)}_{kl}}{\partial Q}\right)^*_0} \{\Omega^2_n - (\omega_1 - \omega_2)^2 - 2i\Gamma_n(\omega_1 - \omega_2)\}^{-1} \tag{8}$$

is the Raman part of the cubic susceptibility (the index n numbers the normal molecular vibrations, the factor D is connected with the frequency degeneracy [2]); χ^{E}_{ijkl} depends on the electron subsystem contribution. The bar designates averaging over the orientations of the molecules.

In the case of molecules able to orient freely, it is easy to calculate the form of the nonlinear coupling coefficient for the four-frequency Raman process $\lambda_{\mathrm{R}} = \chi^{\mathrm{R}}_{ijkl} e^{(\mathrm{S})}_i e^{(L)}_j e^{(1)}_k e^{(2)}_l$ through the invariants of combinational scattering (CS) tensor:

$$\beta_{\mathrm{S}} = \frac{1}{3}\left[\sum_i \left(\frac{\partial \alpha_{ii}}{\partial Q}\right)_0 \left(\frac{\partial \alpha_{ii}}{\partial Q}\right)^*_0 + 2\sum_{i\neq k} \left(\frac{\partial \alpha_{ii}}{\partial Q}\right)_0 \left(\frac{\partial \alpha_{kk}}{\partial Q}\right)^*_0\right]$$

is the average polarizability, determining the scalar scattering;

$$\gamma^2 = \frac{1}{3}\left\{2\sum_i \left(\frac{\partial \alpha_{ii}}{\partial Q}\right)_0 \left(\frac{\partial \alpha_{ii}}{\partial Q}\right)^*_0 - \sum_{i\neq k}\left[\left(\frac{\partial \alpha_{ii}}{\partial Q}\right)_0 \left(\frac{\partial \alpha_{kk}}{\partial Q}\right)^*_0 + \left(\frac{\partial \alpha_{kk}}{\partial Q}\right)_0 \left(\frac{\partial \alpha_{ii}}{\partial Q}\right)^*_0\right]\right\}$$

is the anisotropy of the Raman-scattering tensor, connected with the quadrupole combinational scattering;

$$\beta_a = \sum_{i\neq k} \left(\frac{\partial \alpha_{ik}}{\partial Q}\right)_0 \left(\frac{\partial \alpha_{ik}}{\partial Q}\right)^*_0$$

is the value characterizing the antisymmetry of the RS tensor of the molecule and provides the magnetic-dipole scattering [16, 17].

Table I presents the magnitudes of the corresponding Raman coefficients of nonlinear coupling for different orientations of the polarization vectors $\mathbf{e}^{(\mathrm{S,a})}$, $\mathbf{e}^{(L)}$, $\mathbf{e}^{(1,2)}$. In studying crystals one should substitute the components of the crystalline tensor of RS into (8); no averaging over orientations is required in this case.

TABLE I. – *Coefficients of nonlinear wave coupling due to Raman effect:* $\lambda^{\mathrm{R}} = \sum_{ijkl} \tilde{\chi}^{\mathrm{R}}_{ijkl} e_i^{(\mathrm{S,a})} e_j^{(L)} e_k^{(1)} e_l^{(2)}$ *for different orientations of the polarization vectors* $\mathbf{e}^{(\mathrm{S,a})}$, $\mathbf{e}^{(L)}$, $\mathbf{e}^{(1,2)}$ *of light waves.* The index h shows that the corresponding polarization vector belongs to the scattering plane, which is assumed to be horizontal; the index v shows that the vector is normal to the scattering plane, *i.e.* its direction is vertical; α is the angle between probe and pump beams.

Scattered wave ($\omega_{\mathrm{S,a}}$)	Probe (ω_L)	Pump (ω_1)	Pump (ω_2)	λ^{R}
h	h	h	h	$\frac{1}{3}\beta_{\mathrm{S}} - \frac{1}{15}\gamma^2(1-3\cos^2\alpha)$
v	v	v	v	$\frac{1}{3}\beta_{\mathrm{S}} + \frac{1}{15}\gamma^2$
h	h	v	v	$\frac{1}{3}\beta_{\mathrm{S}} - \frac{1}{15}\gamma^2$
v	v	h	h	
v	h	h	v	$\left(\frac{1}{10}\gamma^2 - \frac{1}{6}\beta_{\mathrm{a}}\right)\cos\alpha$
h	v	v	h	
h	v	h	v	$\left(\frac{1}{10}\gamma^2 + \frac{1}{6}\beta_{\mathrm{a}}\right)\cos\alpha$
v	h	v	h	

To conclude this section we will give an expression for the line profile of the active spectrum. Using (7) and assuming the contribution of the electron subsystem χ^E to have a purely real value $\chi^E = \chi^{\mathrm{NR}}$ (in the absence of single- and multi-phonon absorption), one derives

$$\left|\frac{\chi^{(3)}(\Delta_1)}{\chi^{\mathrm{NR}}}\right|^2 = 1 + \frac{\alpha_1^2}{1+\Delta_1^2} - \frac{2\alpha_1\Delta_1}{1+\Delta_1^2}\,. \tag{9}$$

Here we assume for simplicity that the molecule has the only normal vibration;

$$\alpha_1 = \left[\left(\frac{ND}{2M}\sum_{ijkl}\left(\frac{\partial\alpha_{ij}}{\partial Q}\right)_0\left(\frac{\partial\alpha_{kl}}{\partial Q}\right)_0^* (2\Gamma\Omega)^{-1} e_i^{(\mathrm{S})} e_j^{(L)} e_k^{(\mathbf{1})} e_l^{(\mathbf{2})}\right)\Big/\left(\sum_{ijkl}\chi^{\mathrm{NR}}_{ijkl} e_i^{(\mathrm{S})} e_j^{(l)} e_k^{(1)} e_l^{(2)}\right)\right] \tag{10}$$

is the dimensionless quantity characterizing the relative contribution of resonant (Raman) and nonresonant (electron) parts into the net cubic susceptibility of the medium;

$$\Delta_1 = [(\omega_1 - \omega_2) - \Omega]/\Gamma$$

is frequency detuning from the centre of the Raman line. It turns out that the frequency difference between maximum and minimum of the active spectrum is connected with the quantity α_1 and the line width of spontaneous RS [9, 10]:

$$|\nu_{\max} - \nu_{\min}| = \Delta\Omega \left[1 + \left(\frac{\alpha_1}{2}\right)^2\right]^{\frac{1}{2}} . \tag{11}$$

The measurement of α_1 values for different lines of RS may replace the measurement of the corresponding spontaneous line intensities, because the presence of $\chi^{NR} \neq 0$ provides the common reference level for all the vibrational resonances.

When isotropic media are investigated, it is possible to measure different linear combinations of RS tensor invariants by changing the state of polarization of driving and probe waves. Thus, all three independent values of invariants are finally obtained.

In some cases the ASCS data indicate a considerable contribution of the antisymmetric part of the combinatorial scattering tensor in the intensities of the spontaneous lines [9].

5. – Intensity of active spectra.

We must emphasize that, in distinction with spontaneous RS, the indicatrix of the coherent scattering is highly anisotropic: the maximum of the Stokes (anti-Stokes) scattered probe ray lies in the phase-matched direction (see formulae (2)). In other directions, coherent scattering is negligibly small. As a result, the scattered-off coherent-molecular-vibration probe light represents itself a full spatial coherent and well-collimated beam.

Equations for the amplitudes $E_{S,a}$ of scattered Stokes or anti-Stokes waves have the following form [9] (*):

$$e_i^{(S,a)} \frac{dE_{S,a}}{dz} = i\frac{2\pi\omega_{S,a}}{cn_{S,a}} \sum \chi^{(3)}_{ijkl}(\omega_L, \mp \omega_1; \pm \omega_2) \cdot \cdot e_j^{(L)} e_k^{(1)} e_l^{(2)} E_L E_1^* E_2 \exp[-i\Delta k_{S,a} z] , \tag{11a}$$

where

$$\Delta k_S = k_{1z} - k_{2z} - (k_{Lz} - k_{Sz}) , \qquad \Delta k_a = k_{1z} - k_{2z} - (k_{az} - k_{Lz})$$

(*) Thus here and in the sequel we consider only the coherent part of the response of the medium; one can neglect the change in the populations of vibrational states involved in the case of relative weak optical fields and of condensed matter [18].

are the phase mismatches. Here we have not taken into account the term which is proportional to $|E_L|^2 E_{S,a}$ and corresponds to the effect of stimulated (or inverse) Raman scattering, because we believe that the threshold of SRS is not exceeded for each of the three input waves (frequencies ω_1, ω_2 and ω_L). If it is not so, then the active spectrum begins to distort and an extraction of interesting spectroscopic information about vibrational levels from the active spectrum becomes very difficult [19, 20].

The solution of eq. (11) is not the problem. It is simple to see that the intensities of Stokes and anti-Stokes scattered components are equal to each other in the case of exact phase matching: $\Delta k_{S,a} = 0$. If $\alpha_1 \gg 1$, then the intensity of a scattered beam is given by the formulae

$$I^{S,a}_{ASCS} \simeq \left(lN\frac{d\sigma}{do}\right)^2 2^4 c^4 \omega_{S,a}^{-6} (\hbar\Delta\Omega)^{-2} n^{-4} I_L I_1 I_2 \operatorname{sinc}^2(\Delta k_{S,a} l/2); \tag{12}$$

here $d\sigma/do$ is a spontaneous Stokes cross-section

$$\frac{d\sigma}{do} = \frac{1}{3}\frac{\omega_S^4}{c^4}\frac{\hbar}{2M\Omega}\beta_S;$$

$I_{1,2}$ are the intensities of the waves with respectively frequencies ω_L, ω_1, ω_2; $n(\simeq n_{S,L} \simeq n_{1,2})$ is the index of refraction; l is the length of the beam interaction; $\operatorname{sinc} x = (\sin x)/x$.

The enhancement of the total scattered power in ASCS as compared with spontaneous (Stokes)-scattering power is characterized by the following value:

$$\eta = P^{(S)}_{ASCS}/P^{(S)}_{spont} \simeq lN\frac{d\sigma}{do} 2^4 c^4 \omega_S^{-6} (\hbar\Delta\Omega)^{-2} n^{-4} I_1 I_2 (\delta o)^{-1}. \tag{13}$$

Here δo is the acceptance angle of the receiver. The coefficient η may be numerically large—depending upon the pump intensities; in our experiments $\eta \simeq 10^3 \div 10^5$.

6. – Spectral resolution of active spectroscopy.

If the excitation of Raman-active vibrations is due to the beating of two waves from narrow-band sources, then the spectral resolution of ASCS is determined by the ratio of Raman line width $\Delta\Omega$ to the line width of the tunable oscillator (*). The recording of the active spectrum occurs due to the tuning of the latter. If one uses in an experiment an extremely-narrow-band tunable

(*) The line width of a fixed-frequency source can be made very narrow: $\Delta\omega \sim \tau_p^{-1}$, where τ_p is the pulse duration.

dye laser, which presently has become commercially available (line width of about 10 MHz [21]), then a very high spectral resolution can be obtained, which is by no means obtainable in ordinary spontaneous Raman scattering.

However, in active spectroscopy, it is possible to use also broad-band tunable sources, without loss in spectral resolution compared with spontaneous RS. In such a case, one of the two pumping beams is provided by a broad-band tuned oscillator, so that its spectrum overlaps the spontaneous Raman line. It is evident that here we can simultaneously excite the coherent vibrations in the whole Raman line. If one wishes to resolve the line shape of $|\chi^{(3)}|^2$ in this case, one must normalize the active spectrum $S(\omega_{S,a})$ over the broad-band source spectrum $S(\omega_2)$ [22, 23]:

$$|\chi^{(3)}(\omega_{S,a} = \omega \mp (\omega_1 - \omega_2))|^2 \propto S(\omega_{S,a})/S(\omega_2) . \tag{14}$$

The recording of an active spectrum occurs due to the sweeping of the monochromator as in the spontaneous-scattering case. The spectral resolutionis determined by the performance of the monochromator. However, here very high values of signal-to-noise ratio can be obtained, due to the coherent excitation of the molecular vibration of interest.

7. – Active spectroscopy of resonant Raman scattering.

The intensities of Stokes and anti-Stokes scattered components are known to be enhanced when the exciting line is near to or lies inside of the absorption band of a substance. This phenomenon is called a resonant (spontaneous) Raman scattering [24]. Recently the number of laboratories interested in this kind of spectroscopy is growing. Active spectroscopy can be modified in such a way that it becomes possible to investigate also the resonant Raman spectra. For this purpose it is necessary to provide one-photon electronic absorption at one of the pump frequencies (for example, $\omega_1 \simeq \Omega_E$, where Ω_E is the central frequency of an electronic transition). It must be remembered that probe and scattered waves may be generally chosen far from the absorption band and thus there is the possibility of distinguishing between the resonant Raman scattering itself and background effects such as resonant fluorescence and Stokes luminescence.

If the medium under investigation is a homogeneous mixture (or solution) of two or more substances, there are two ways to provide absorption at the pump frequency ω_1:

1) absorption and scattering take place in the same molecule, *i.e.* in the same component of the mixture;

2) absorption of the pump wave occurs in a molecule of one component of the mixture and scattering of a probe wave occurs in a molecule of another

component; in the latter case the interference between the contributions of both components of the mixture into the scattered signal is due to the coherency of the scattering process.

It can be shown that any active spectrum of real resonant scattering of liquids and gases can be obtained in the transparent band of these substances by mixing to them the only properly chosen absorbing substance. Moreover, by choosing the proper concentration of the absorbing component and by frequency-tuning the pumping wave to the centre of the absorption band it becomes possible to obtain the strong increase of the relative contributions to $\chi^{(3)}$ from nuclear vibrations of the solvent. This is obtained by coherent subtraction of the nonresonant background of the active spectrum ($\chi^{\text{NR}} \neq 0$) due to the resonant susceptibility of the absorbing component (see fig. 5). This effect occurs also for any weak line of RS and for overtones.

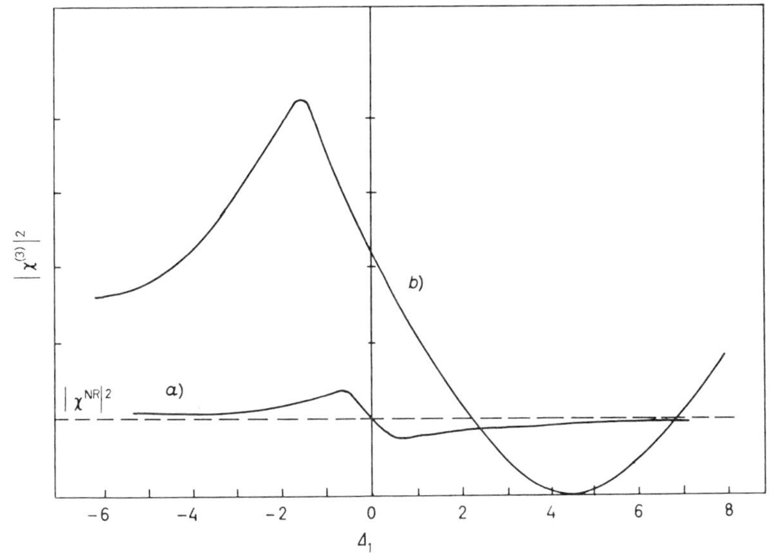

Fig. 5. – Transformation of the active spectra in the presence of the one-photon electronic resonance. $\Delta_1 = (\omega_1 - \omega_2 - \omega_{\text{R}})/\Gamma_{\text{R}}$. *a*) ω_1, ω_2 far from the electronic absorption band; *b*) ω_2 is tuned at the electronic resonance.

One can write down the following expression for the third-order susceptibility of a binary mixture:

$$\chi^{(3)}(\omega_{\text{S,a}} = \omega_L \mp \omega_1 \pm \omega_2) = \tag{15}$$

$$= N_1 \chi_1^{\text{NR}} + N_2 \chi_2^{\text{NR}} + \frac{N_1 \bar{\chi}^{\text{R}}}{\Omega_{\text{R}} - \omega_1 + \omega_2 + i\Gamma_{\text{R}}} + \frac{N_2 \bar{\chi}^E}{\Omega_E - \omega_1 + i\Gamma_E},$$

where N_1, N_2 are molecular number densities of the first (Raman active) and the second (absorbing) components;

(16) $$\bar{\chi}^R = \frac{c^4}{\hbar\omega_L\omega_S^3}\frac{d\sigma}{do};$$

and $\bar{\chi}^E$ is connected with the one-photon absorption cross-section σ_1:

(17) $$\bar{\chi}^E = \frac{4\sigma_1^2c^2\Gamma_E^2}{\hbar(\Omega_E^2-\omega_S^2)\,\omega_1\Omega_R}.$$

Let us introduce the following dimensionless values:

(18) $$\begin{cases} \alpha_1 = N_1\bar{\chi}^R/\Gamma_R(N_1\chi_1^{NR}+N_2\chi_2^{NR}), \\ \alpha_2 = N_2\bar{\chi}^E/\Gamma_E(N_1\chi_1^{NR}+N_2\chi_2^{NR}), \\ \Delta_2 = (-\Omega+\omega_1)/\Gamma_E, \end{cases}$$

then we can obtain the line shape of $|\chi^{(3)}|^2$ of the mixture as follows:

(19) $$\frac{|\chi^{(3)}(\omega_L, \mp\omega_1, \pm\omega_2)|^2}{(N_1\chi_1^{NR}+N_2\chi_2^{NR})^2} = 1 + \frac{\alpha_2^2}{1+\Delta_2^2} - \frac{2\alpha_2\Delta_2}{1+\Delta_2^2} + \frac{\alpha_1^2}{1+\Delta_1^2}\left(1+\frac{2\alpha_2/\alpha_1}{1+\Delta_2^2}\right) - \frac{2\alpha_1\Delta_1}{1+\Delta_1^2}\left(1-\frac{\alpha_2\Delta_2}{1+\Delta_2^2}\right).$$

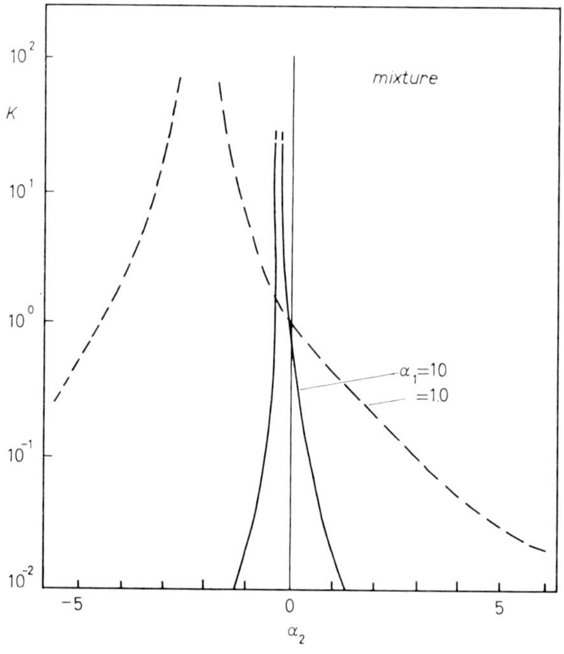

Fig. 6. – Contrast ratio $K = k/k(\alpha_2 = 0)$ of the dispersion curve of $|\chi^{(3)}|^2$ *vs.* the concentration α_2 of the absorber.

The contrast ratio of the dispersion curve $|\chi^{(3)}|^2$

$$k = \max|\chi^{(3)}(\omega'_{S,a})|^2/\min|\chi^{(3)}(\omega''_{S,a})|^2$$

substantially increases when

$$-\alpha_1\alpha_2 = 1 + (\Delta_2 - \alpha_2)^2 . \tag{20}$$

When eq. (20) is satisfied, $k \to \infty$ even in the case of an arbitrary small value of $\alpha_1 \sim (d\sigma/do)\cdot N_1$. In such a way the main disadvantage of ASCS can be

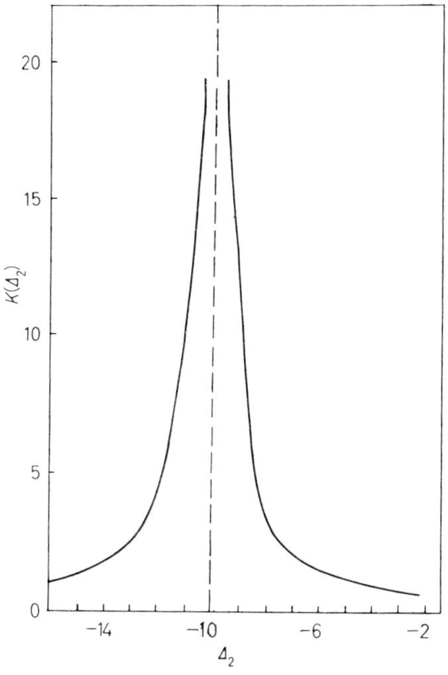

Fig. 7.

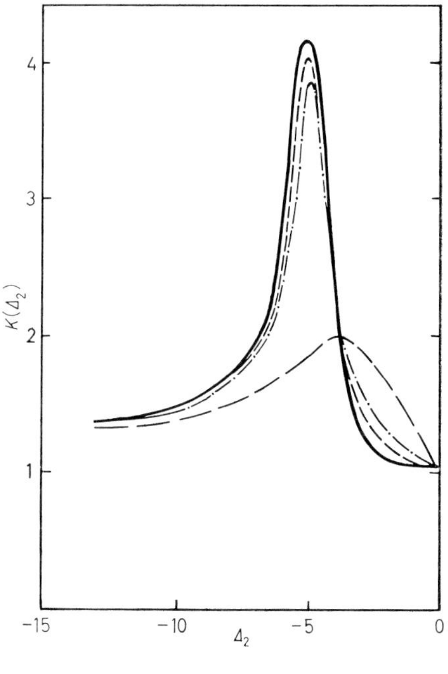

Fig. 8.

Fig. 7. – Contrast ratio K of the $|\chi^{(3)}(\Delta_1)|^2$ dispersion curve *vs.* the detuning Δ_2 from electronic resonance; optimal conditions. $\alpha_1 = 0.1$, $\alpha_2 = -10$.

Fig. 8. – Contrast ratio K of the $|\chi^{(3)}(\Delta_1)|^2$ dispersion curve *vs.* the detuning Δ_2 from electronic resonance; nonoptimal case. Influence of saturation: $\alpha_1 = 0.1$, $\alpha_2 = -5$. — — — $\gamma = 10$, — · · $\gamma = 1$, - - - $\gamma = 0.1$, ——— $\gamma = 0$; γ is the saturation factor.

overcome, which is connected with a poor selectivity of ASCS when one examines weak Raman lines, such as the second- and higher-order lines and overtones.

Some examples of calculated resonant active spectra and of frequency and concentration dependences of the contrast ratio k are represented in fig. 6-9.

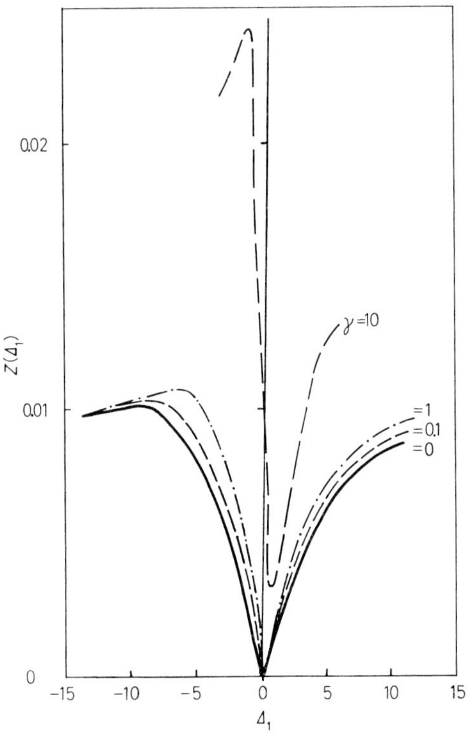

Fig. 9. – Active spectroscopy line shape in the presence of electronic resonance. Influence of saturation: $\alpha_1 = 0.1$, $\alpha_2 = 10$, $\Delta_2 = -10$; $Z(\Delta_1) = |\chi^{(3)}(\Delta_1)|^2/(N_1\chi_1^{NR} + N_2\chi_2^{NR})^2$.

8. – Active spectroscopy: experimental results.

To date more than ten papers have been published devoted to experimental investigation of vibrational and other excitations of molecules and crystals by means of the active-spectroscopy method (see, for instance, [2, 7, 9-15, 22]).

The main purposes of these works were both the examination of a correspondence between active and spontaneous spectra and the search for new spectroscopic and analytic applications of light scattering.

It has been pointed out earlier that active spectra possess more rich spectroscopic information than spontaneous ones. In particular, they contain data about both the real and imaginary parts of the third-order susceptibility. The interference of different contributions into the net cubic susceptibility provides the appearance in active spectra of more or less striking dips. The relative spacing of peaks and dips in an active spectrum enables one to determine the relative values and phases (*i.e.* signs) of Raman and electronic contributions. As an example, in fig. 10 are represented the active spectra of a calcite crystal [9, 22] (solid lines) as compared with spontaneous RS spectra (dashed lines).

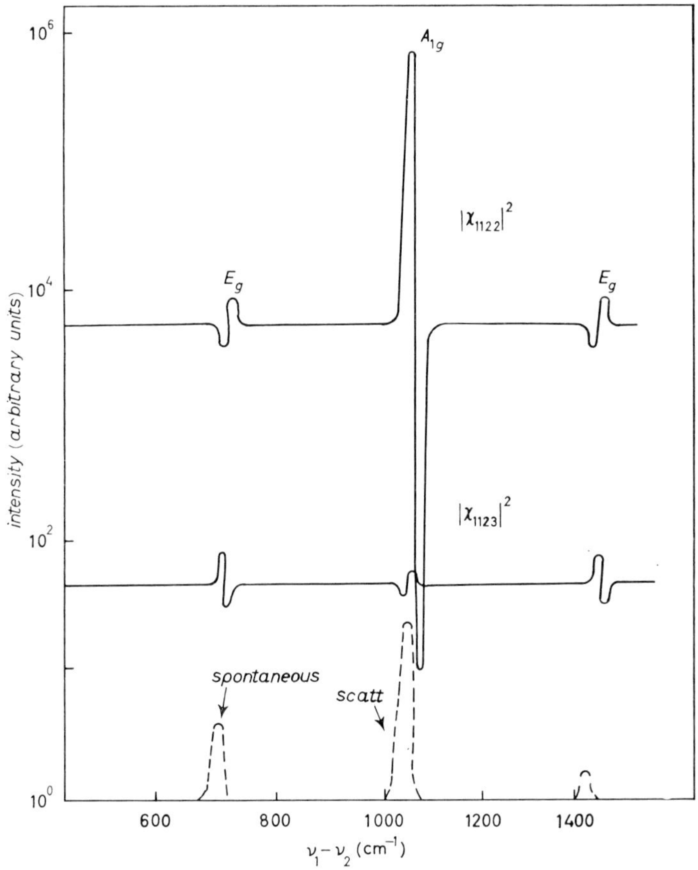

Fig. 10. – « Active » spectrum of a calcite crystal in comparison with spontaneous spectra.

Active spectra have been taken at different polarization directions of the interacting beams. One can conclude from these spectra that the contributions of the A_{1g} Raman line $(\chi^{\text{R}}_{1122}(A_{1g}))$ and of the electronic subsystem $(\chi^{\text{NR}}_{1122})$ of the crystal to the $\chi^{(3)}_{1122}$ component are of the same sign, and that the contributions of the E_g-lines $\chi^{\text{R}}(E_g)$ and of the electronic subsystem are of opposite signs. Since from vibrational symmetry one can attribute a positive sign to $\bar{\chi}^{\text{R}}_{1122}(A_{1g})$ and a negative one to $\bar{\chi}^{\text{R}}_{1122}(E_g)$, we can conclude that the sign of χ^{NR}_{1122} is positive [22].

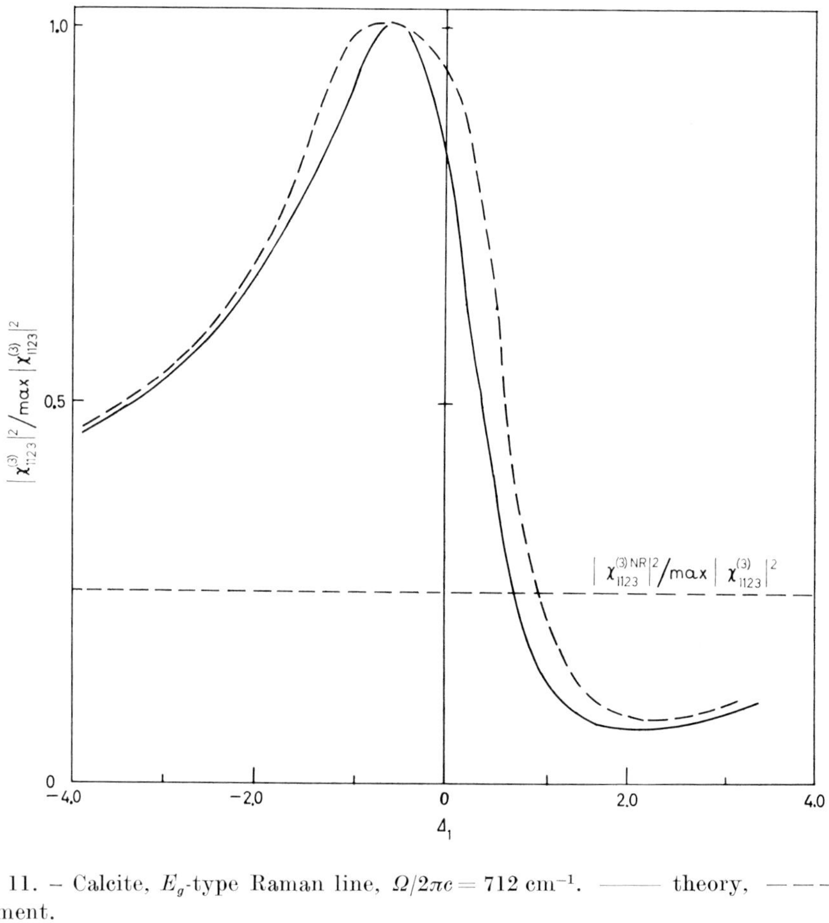

Fig. 11. – Calcite, E_g-type Raman line, $\Omega/2\pi c = 712\ \text{cm}^{-1}$. ——— theory, – – – experiment.

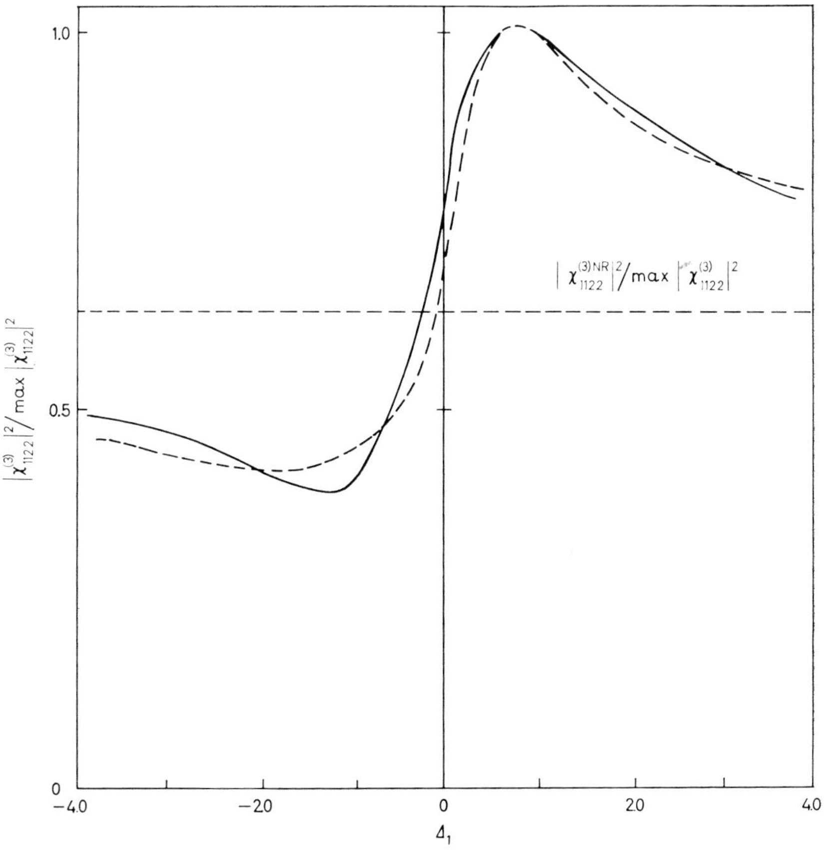

Fig. 12. – Calcite, E_g-type Raman line, $\Omega/2\pi c = 712\ \text{cm}^{-1}$, ——— theory, – – – experiment.

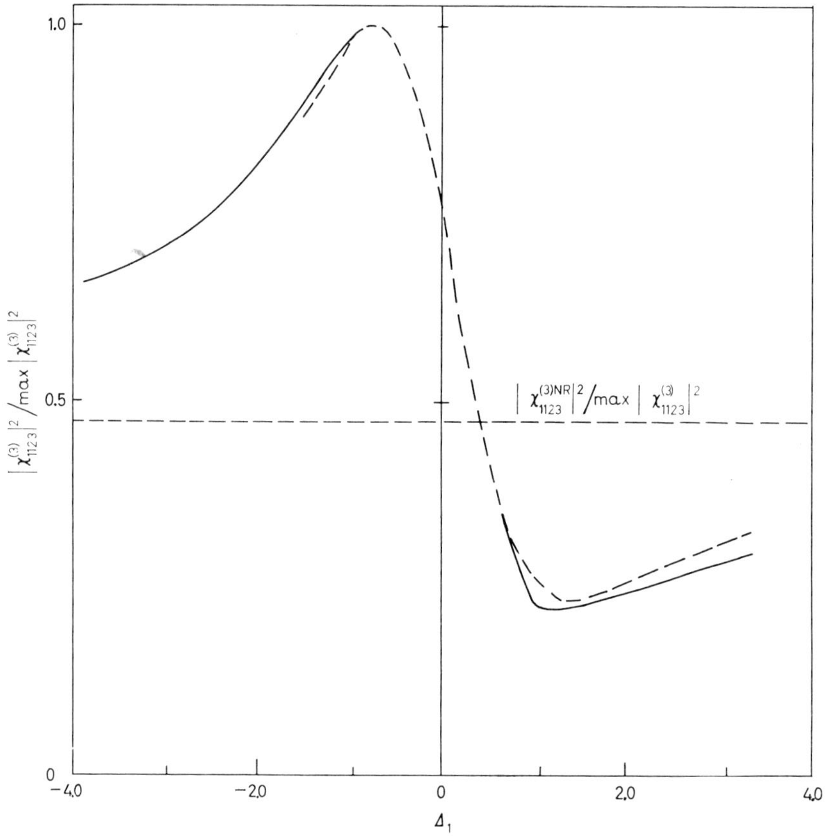

Fig. 13. – Calcite, E_g-type Raman line, $\Omega/2\pi c = 1436\ \text{cm}^{-1}$. ——— theory, – – – experiment.

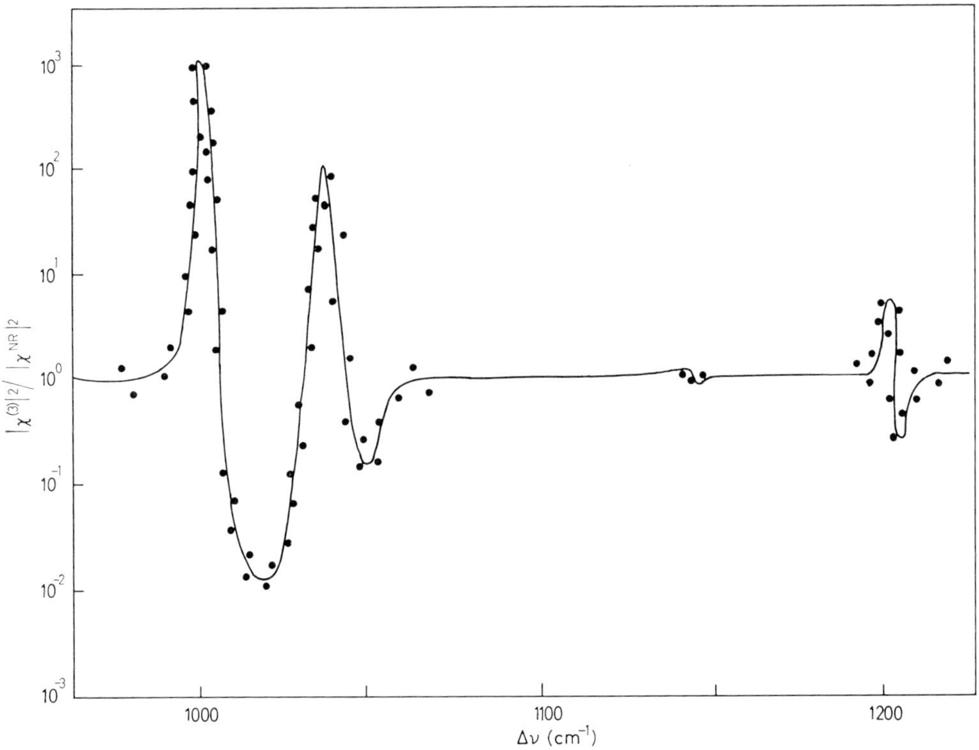

Fig. 14. – Pure toluene, active spectrum (experiment).

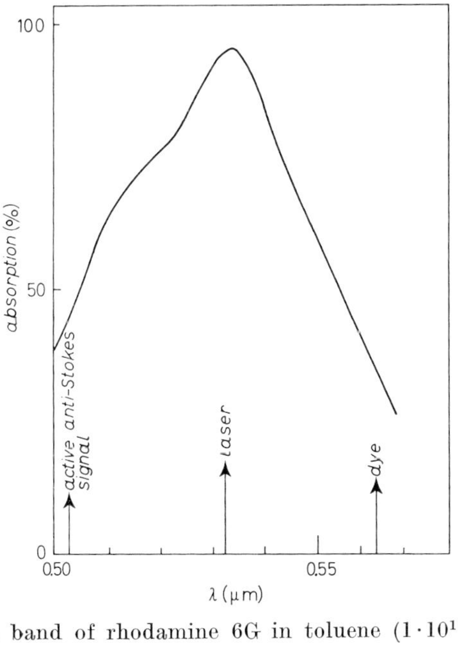

Fig. 15. – Absorption band of rhodamine 6G in toluene ($1\cdot10^{17}$ cm^{-3}).

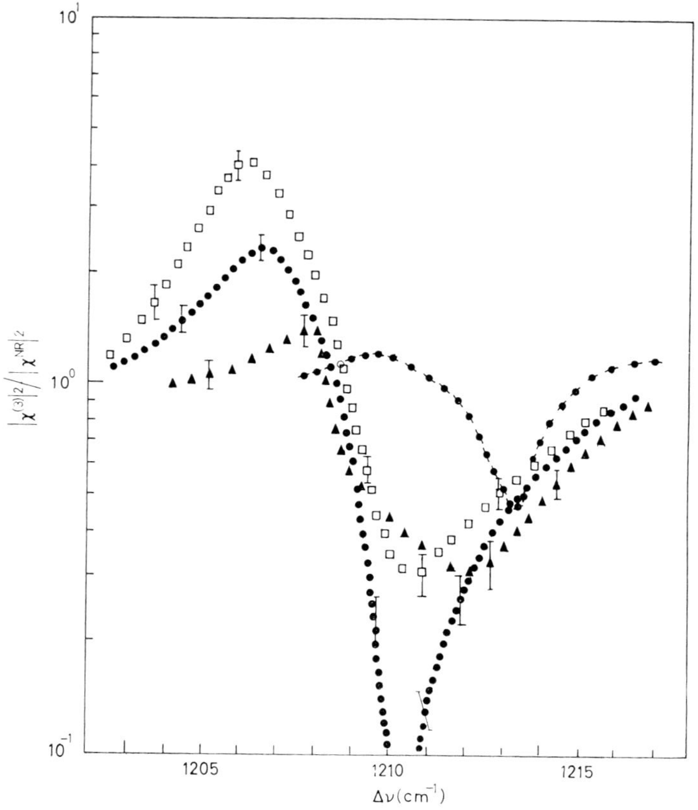

Fig. 16.

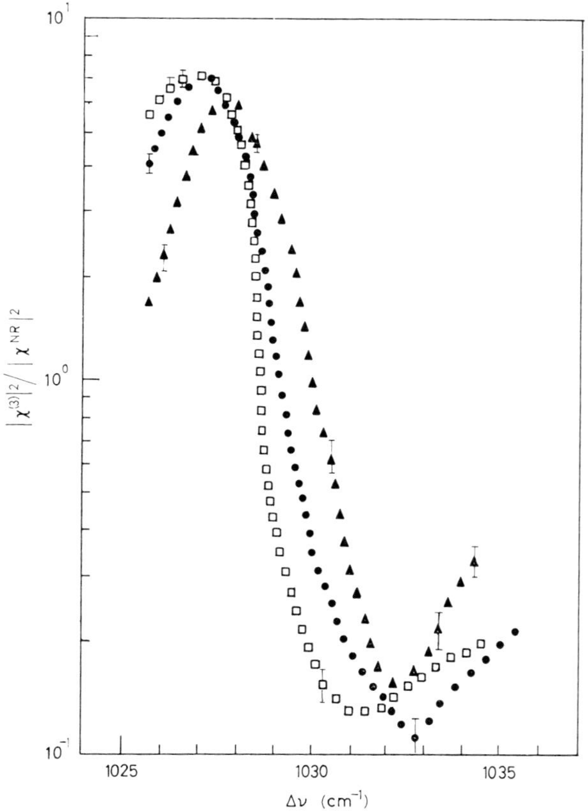

Fig. 17.

Fig. 16. – Influence of the absorption on the dispersion curve of $|\chi^{(3)}|^2$ of liquid toluene in the vicinity of the A_1 Raman line ($1206\ \mathrm{cm}^{-1}$) (experiment). $\Delta\nu = \nu_1 - \nu_2$. □ pure toluene, ● toluene + rhodamine 6G ($1\cdot10^{16}\ \mathrm{cm}^{-3}$), ▲ toluene + rhodamine 6G ($5\cdot10^{16}\ \mathrm{cm}^{-3}$), -•- ($1\cdot10^{17}\ \mathrm{cm}^{-3}$).

Fig. 17. – Influence of the absorption on the dispersion curve of $|\chi^{(3)}|^2$ of liquid toluene in the vicinity of the A_1 Raman line ($1027\ \mathrm{cm}^{-1}$) (experiment). □ pure toluene, ● toluene + rhodamine 6G ($1\cdot10^{16}\ \mathrm{cm}^{-3}$), ▲ toluene + rhodamine 6G) $1\cdot10^{17}\ \mathrm{cm}^{-3}$). $\Delta\nu = \nu_1 - \nu_2$.

From the dispersion curve of $|\chi^{(3)}_{1123}|^2$ in fig. 10 it is clear that off-diagonal elements of a combinatorial tensor of the totally symmetric vibration A_{1g} are practically equal to zero, and hence the off-diagonal elements observed in a number of earlier works [25] should be ascribed to imperfections in the experimental technique.

LEVENSON [12] has exploited an active-spectroscopy method to measure the values of χ^{NR}_{ijkl} for different transparent crystals by comparing it with $\bar{\chi}^{\text{R}}_{1111}(A_{1g})$ of a calcite crystal. In such a way he was able to determine the pure electronic contribution to the nonlinear index of refraction of some materials.

BLOEMBERGEN and LEVENSON, using the active-spectroscopy technique, have studied the influence of the two-photon absorption on the active spectra and have measured the values of the imaginary parts of the electronic contribution to the net cubic susceptibility [13]. Recently, experiments on active spectroscopy of resonant Raman scattering have been performed [15]. The results of these experiments are represented in fig. 16, 17. As a Raman-active medium liquid toluene was used and as absorbers, which have been added to Raman substance, some dyes (Rh6G, RhB and others) and molecular crystals of I_2 were used. It is clear from fig. 16 that by properly choosing the dye concentration one can enhance the contrast ratio of the $|\chi^{(3)}|^2$ dispersion curve.

REFERENCES

[1] Y. R. SHEN and N. BLOEMBERGEN: *Phys. Rev.*, **137**, A 1787 (1965).

[2] R. W. TERHUNE and P. D. MAKER: *Phys. Rev.*, **137**, A 801 (1965).

[3] J. A. GIORDMAINE and W. KAISER: *Phys. Rev.*, **144**, 676 (1966).

[4] F. DE MARTINI and J. P. COFFINET: *Phys. Rev. Lett.*, **22**, 60 (1969); F. DE MARTINI: *Phys. Rev. B*, **4**, 4556 (1971).

[5] G. CHARTIER and S. BIRAUD-LAVAL: *Phys. Rev. Lett.*, **21**, 1641 (1968); *Phys. Lett.*, **30** A, 177 (1969).

[6] J. J. WYNNE: *Phys. Rev.*, **178**, 1295 (1969); J. J. WYNNE and N. BLOEMBERGEN: *Phys. Rev.*, **188**, 1211 (1969).

[7] S. A. AKHMANOV, V. G. DMITRIEV, A. I. KOVRIGIN, N. I. KOROTEEV, V. G. TUNKIN and A. I. KHOLODNYKH: *JETP Lett.*, **15**, 600 (1972).

[8] S. A. AKHMANOV, V. G. DMITRIEV, A. I. KHOLODNYKH, A. I. KOVRIGIN, N. I. KOROTEEV and V. E. OGLUZDIN: report to the *VII International Quantum Electronics Conference* (Montreal, 1972), Digest of technical papers, A13.

[9] S. A. AKHMANOV and N. I. KOROTEEV: *Žurn Èksp. Teor. Fiz.*, **67**, 1306 (1974).

[10] M. D. LEVENSON, C. FLYTZANIS and N. BLOEMBERGEN: *Phys. Rev. B*, **6**, 3962 (1972).

[11] J. J. WYNNE: *Phys. Rev. Lett.*, **29**, 650 (1972); report to the *VII International Quantum Electronics Conference* (Montreal, 1972), Digest of technical papers, S9.

[12] M. D. LEVENSON: *IEEE Journ. Quant. Electr.*, QE-**10**, 110 (1974).

[13] M. D. LEVENSON and N. BLOEMBERGEN: *Journ. Chem. Phys.*, **60**, 1323 (1974); *Phys. Rev. B*, **10**, 4447 (1974).

[14] R. F. Begley, A. B. Harvey, R. L. Byer and B. S. Hudson: *Journ. Chem. Phys.*, **61**, 2466 (1974); *Appl. Phys. Lett.*, **25**, 387 (1974).

[15] N. I. Koroteev: report to the *IV Vavilov's Conference on Nonlinear Optics, Novosibirsk, 1975.*

[16] G. Placzek: *Marx Handbuch der Radiologie*, edited by E. Marx, 2nd ed., Vol. **6** (Leipzig, 1934).

[17] M. M. Sushchinsky: *Spectra of Combinatorial Scattering of Molecules and Crystals* (Moscow, 1969).

[18] S. A. Akhmanov, K. N. Drabovitch, A. P. Sukhorukov and A. S. Chirkin: *Žurn. Ėksp. Teor. Fiz.*, **59**, 485 (1970).

[19] N. I. Koroteev: *Opt. Spectrosc.*, **29**, 543 (1970).

[20] N. I. Koroteev and I. L. Shumay: *Quantum Electronics*, edited by N. G. Basov, Vol. **2** (1974), p. 2489.

[21] T. W. Hänsch: *Appl. Opt.*, **11**, 895 (1972).

[22] S. A. Akhmanov, N. I. Koroteev and A. I. Kholodnykh: *Journ. Raman Spectrosc.*, **2**, 239 (1974).

[23] N. I. Koroteev and A. I. Kholodnykh: *Izv. VUZ'ov, Radiophysica*, **17**, 814 (1974).

[24] P. P. Shorygin: *Usp. Fiz. Nauk*, **109**, 293 (1973).

[25] L. Couture: *Ann. de Phys.*, **2**, 5 (1947); S. Bhagavantam: *Proc. Ind. Acad. Sci.*, **11** A, 62 (1940).

Higher-Order Optical Nonlinearities.

S. A. AKHMANOV

Department of Physics, Moscow State University - Moscow, USSR

1. – Introductory remarks.

Since the Franken *et al.* [1] classical experiment, the optical nonlinearities of gases, liquids and solids have been of great interest both from the theoretical and experimental points of view. The interpretation of the Franken experiment, as well as those of all later experiments in this field, was based on the expansion of, generally speaking, an unknown nonlinear function (*)

$$\boldsymbol{P} = \boldsymbol{P}(\boldsymbol{E}) \tag{1}$$

(where $\boldsymbol{P}$ is the macroscopic polarization, and $\boldsymbol{E}$ the optical electric field) into the series

$$\boldsymbol{P} = \boldsymbol{P}_L + \boldsymbol{P}_{\text{NL}}, \qquad \boldsymbol{P}_L = \hat{\varkappa}\boldsymbol{E}, \tag{2}$$

$$\boldsymbol{P}_{\text{NL}} = \hat{\chi}^{(2)}\boldsymbol{E}\boldsymbol{E} + \hat{\chi}^{(3)}\boldsymbol{E}\boldsymbol{E}\boldsymbol{E} + \hat{\chi}^{(4)}\boldsymbol{E}\boldsymbol{E}\boldsymbol{E}\boldsymbol{E} + \hat{\chi}^{(5)}\boldsymbol{E}\boldsymbol{E}\boldsymbol{E}\boldsymbol{E}\boldsymbol{E} + \dots, \tag{3}$$

where $\hat{\varkappa} = \hat{\chi}^{(1)}, \hat{\chi}^{(2)}, \hat{\chi}^{(3)}, \hat{\chi}^{(4)}, \hat{\chi}^{(5)}$ are tensors of second, third, fourth, fifth, sixth and other successive orders. (Note that expansion (3) holds only for monochromatic and quasi-monochromatic light waves, for which the dispersion of the susceptibility tensors $\chi^{(n)}$ can be ignored.) From a theoretical point of view, expansion (3) corresponds to perturbation theory, which indicates that the terms in the polarization expansion (3) decrease as successive powers of the parameter E/E_a, where E_a is the characteristic « atomic » field. (For a hydrogen atom it is of the order $3\cdot10^8$ V/cm; for solids this characteristic field can be smaller.)

Expansion (3) offers the experimentalist a practically unlimited field of activity; measurements of different $\chi^{(n)}$ in gases, liquids and solids supply a great deal of new and important information.

(*) Generally speaking it should be considered as a nonlinear functional.

Now these data are of special interest in connection with the recent achievements of nonlinear susceptibility theory (see for example, the Flytzanis review [2]). Most theoretical and experimental papers in this field published in the course of the last 15 years, however, were devoted to the lower-order susceptibilities, namely $\chi^{(2)}$ and $\chi^{(3)}$. Higher-order nonlinearities were considered only in a few papers. Among them are:

1) The paper by LUKASIK and DUCUING [3], in which an attempt was made to measure the resonant susceptibility $\chi^{(5)}$ in a hydrogen molecule. They have observed the anharmonic effect in a coherently driven molecule H_2 with the help of coherent second-order Raman scattering.

In the Lukasik and Ducuing experiment, under strong excitation of the H_2 gas by two powerful light beams with the frequencies ω_1 and ω_2 ($\omega_1 - \omega_2 \simeq \simeq \Omega$ = resonant frequency of molecular vibrations) due to both electrical and mechanical anharmonicity, coherent modulation of the polarizability at frequency $2(\omega_1 - \omega_2)$ was created. The Raman scattering on this coherently driven molecular vibrations creates the light field at the anti-Stokes frequency

$$\omega_{\text{a-S}} = \omega_1 + 2(\omega_1 - \omega_2) = 3\omega_1 - 2\omega_2 \tag{4}$$

through the nonlinear source polarization

$$P^{(5)}_{\text{NL}}(3\omega_1 - 2\omega_2) = \chi^{(5)}(\omega_1, \omega_1, \omega_1, -\omega_2, -\omega_2)\, E^3(\omega_1)\, E^{*2}(\omega_2)\,. \tag{5}$$

Thus, the fifth-order nonlinearity $P^{(5)}_{\text{NL}}$ is the lower-order nonlinearity which should be measured to determine the coherently driven anharmonic processes in molecules with the help of Raman scattering (*).

The main problem which arises in experiments on higher-order nonlinearities is the problem of distinguishing between the « direct » process, which is determined by the susceptibility $\chi^{(5)}$ discussed here, and the « cascade » processes, which are connected with lower-order nonlinearities.

Thus, in an experiment performed by LUKASIC and DUCUING coherent radiation at frequency $\omega_{\text{a-S}} = \omega_1 + 2(\omega_1 - \omega_2)$ can also be produced through double coherent scattering, associated with the polarizability modulation at frequency $\omega_1 - \omega_2$. This process is determined by the lower-order nonlinearity $\chi^{(3)}$, $P^{(5)}_{\text{NL}} \sim |\chi^{(3)}|^2$. According to [3], for H_2 molecules

$$\chi^{(5)}(\omega_1, \omega_1, \omega_1, -\omega_2, -\omega_2) = N \cdot 0.5 \cdot 10^{-45} \text{ e.s.u.}\,,$$

where N is the molecular density.

(*) At present this can be done also with the help of powerful tunable infra-red sources, for example with powerful tunable optical parametric oscillators.

2) The Harris paper [4], according to which in low-pressure gases higher-order polarizations may be comparable with lower-order polarizations.

HARRIS claims [4] having observed the fifth-harmonic generation (5320 Å → → 1064 Å) in low-pressure xenon, the nonlinear process which is connected with the nonlinear polarization $P^{(5)}_{NL} = \chi^{(5)} E^5$. He also presented in [4] theoretical estimates of nonlinear polarizations $P^{(5)}_{NL}$, $P^{(7)}_{NL}$ and $P^{(15)}_{NL}$ in Li vapour.

The important point which HARRIS had mentioned is a large break-down threshold in low-pressure gases and vapours. Avalanche ionization here is negligible, and only multiphoton ionization would be taken into account. In Li vapour the limiting light intensity is of the order of $I_{thr} = 10^{15}$ W/cm², in xenon $I_{thr} = 2 \cdot 10^{12}$ W/cm² (in solids, due to avalanche ionization, $I_{thr} = 10^9$ W/cm² for nanosecond pulses and $I_{thr} = 10^{11}$ W/cm² for picosecond pulses).

3) Recently, in our laboratory, experiments on phase-matched fourth- and fifth-harmonic generation in LFM and $CaCO_3$ crystals were performed [5,6]. In these papers the corresponding nonlinear susceptibilities $\chi^{(4)}$ and $\chi^{(5)}$ were estimated. These results are discussed in detail in this lecture.

4) It should be mentioned that the problems connected with the « direct » and « cascade » processes were first formulated in connection with the measurements of the cubic susceptibility $\chi^{(3)}$ in noncentrosymmetric crystals.

YABLONOVITH, FLYTZANIS and BLOEMBERGEN [7] have shown that two-step, two-wave mixing makes a substantial contribution to the total three-wave mixing process in a GaAs crystal. ARAKELYAN *et al.* [8] have discussed this problem in connection with the experiments on four-photon parametric luminescence in $LiIO_3$ crystals, and HERRMANN [9] in connection with the third-harmonic generation in quartz.

2. – Higher-order nonlinearities in simple models of nonlinear polarization.

2·1. *The simple anharmonic-oscillator model.* – The first simple estimate of higher-order nonlinear polarizations was effected by BLOEMBERGEN [10]. Starting with the anharmonic-oscillation equation

$$\ddot{x} + \Gamma\dot{x} + \omega_0^2 x + v x^2 = \frac{e}{m} E \tag{6}$$

BLOEMBERGEN showed that in nonresonant situations

$$\frac{P^{(n+1)}_{NL}}{P^{(n)}_{NL}} = \frac{eE}{m\omega_0^4} v , \tag{7}$$

and if for $x \simeq a$ (a = characteristic atomic radius) the linear and nonlinear forces in (6) are equal, we obtain

$$\frac{P_{\text{NL}}^{(n+1)}}{P_{\text{NL}}^{(n)}} = \frac{E}{E_a}, \qquad \text{where } E_a = e/a^2 . \tag{8}$$

According to (8) $\chi^{(n+1)} = \chi^{(n)} E_a^{-1}$.

Actually, however (see also the data which are listed below),

$$\chi^{(n+1)} = \chi^{(n)} E_a^{-1} F(n) , \tag{9}$$

where $F(n)$ decreases, if n increases.

2'2. *Nonlinear oscillations of molecules.* – In classical terms, the nonlinear response of a diatomic molecule can be described by the anharmonic-oscillator equation with the Morse potential function

$$\ddot{x} + \Gamma\dot{x} + \omega_0^2[1 - \exp[-\beta x]]\beta^{-1}\exp[-\beta x] = \gamma E . \tag{10}$$

The cubic nonlinear susceptibility for this model was estimated by VEDA and SHIMODA [11]; the estimates of higher-order nonlinearities are straightforward.

The higher-order nonlinear susceptibilities of molecules are of special interest for polyatomic molecules. In this case, interactions of different modes should also be taken into account. It is interesting to note that under strong, near-resonant excitations, there is a competition between higher-order–harmonic generation and parametric excitation of molecular modes. Our estimates show that these processes are of great importance in the vicinity of the dissociation limit (9).

Thus, in a single molecule vibrational relaxation can be strongly affected by nonlinear and parametric interactions of molecular modes.

2'3. *Higher-order resonant effects; saturation.* – Resonant saturation is a nonlinear effect for which, generally speaking, all terms in expansion (3) should be simultaneously taken into account.

Thus for the index of refraction of the gas of two-level atoms, the well-known relation is derived (see for example [12])

$$n = 2\pi \frac{\varkappa'_{\text{res}}(\omega_0)(\omega - \omega_0)T_2}{1 + (\omega - \omega_0)T_2 + (p/\hbar^2)|E|^2 T_1 T_2}, \tag{11}$$

where T_1, T_2 = relaxation times, p = matrix dipole element, $\varkappa'_{\text{res}}(\omega_0)$ = real part of linear susceptibility. From (11) all the coefficients in the expansion of the nonlinear index $n_{\text{NL}} = n_2|E|^2 + n_4|E|^4 + n_6|E|^6 + \ldots$ can be determined.

2'4. *Miller's rule and higher-order optical susceptibilities.* – It is well known that in determining the quadratic nonlinear susceptitibilities of crystals, Miller's phenomenological rule is very useful.

As was shown in Miller's original paper [13], the quadratic nonlinear susceptibility $\chi^{(2)}_{ijk}(2\omega, \omega, \omega)$ can be expressed as

$$\chi^{(2)}_{ijk}(2\omega, \omega, \omega) = \varkappa_{ii}(2\omega)\varkappa_{jj}(\omega)\varkappa_{kk}(\omega)\,\Delta_{ijk}\,, \tag{12}$$

where $\varkappa_{ii}$ are the components of the linear susceptibility tensor and the « Miller delta » Δ_{ijk} is a third-rank tensor, which is almost constant over a broad class of noncentrosymmetric crystals.

A phenomenological rule for the susceptibility $\chi^{(3)}$ was proposed by WANG [14].

In table I are summarized the data on $\chi^{(4)}(4\omega)$ and $\chi^{(5)}(5\omega)$ obtained by means of the formal generalization of Miller's rule.

TABLE I. – *Fourth- and fifth-order nonlinear optical susceptibilities determined by the simple anharmonic-oscillator model.*

Crystal	$\chi^{(4)}$ (e.s.u.)	$\chi^{(5)}$ (e.s.u.)	Experimental value (e.s.u.)
LFM	$2\cdot 10^{-23}$	$2\cdot 10^{-30}$	$\chi^{(4)} \simeq 10^{-21}$
$CaCO_3$	—	10^{-29}	$\chi^{(5)} \simeq 10^{-27}$
$CdGeAs_2$	$2\cdot 10^{-18}$	$3\cdot 10^{-24}$	$\chi^{(4)} \simeq 10^{-18}$ (*)
Si	$2\cdot 10^{-18}$	$3\cdot 10^{-24}$	

(*) Estimated on the basis of the bond orbital model by BYER *et al.*

3. – Determination of nonlinear susceptibilities $\chi^{(4)}$ and $\chi^{(5)}$ in crystals by means of phase-matched fourth- and fifth-harmonic generation.

3'1. *Phase-matched higher-order–harmonic generation from a* Nd *laser.* – This part of my lecture is based on the results which were obtained in our laboratory during the last two years (*) and devoted to nonlinear susceptibilities $\chi^{(4)}$ and $\chi^{(5)}$ in crystals.

The rough estimates of the values of these nonlinearities with the help of the formulae of sect. **2**, for example with (8), show that in ordinary nonlinear crystals $\chi^{(4)}$ would be of the order of $(10^{-20}\div 10^{-22})$e.s.u. and $\chi^{(5)}=(10^{-27}\div 10^{-29})$e.s.u. It follows that only a small signal can be expected. This is the reason why we

(*) In collaboration with S. SALTIEL and V. TUNKIN.

choose the crystals LFM (lithium formate) and $CaCO_3$, which allow phase-matched fourth- and fifth-harmonic generation from a Nd glass laser.

As a result, using picosecond pulses with intensities near the avalanche break-down thresholds $I_1 \simeq (10^{10} \div 10^{11})$ W/cm^2, we perform reproducible measurements of the intensity, angular and spectral distributions of the fourth harmonic in LFM crystals ($\lambda_4 = 0.265$ μm) and of the fifth harmonic in $CaCO_3$ crystals ($\lambda_5 = 0.212$ μm).

In LFM crystals colour dispersion allows phase-matched fourth-harmonic generation:

$$4\gamma_0(\omega) = \gamma_e(4\omega)\,, \qquad K_4^e = 4K_1^0\,. \tag{13}$$

This phase-matched five-photon interaction occurs for an angle $\theta_m = 36°\,30'$.

In $CaCO_3$ crystals colour dispersion allows the phase-matched fifth-harmonic generation

$$5\gamma_0(\omega) = \gamma_e(5\omega)\,, \qquad K_5^e = 5K_1\,, \qquad \theta_m = 55°\,. \tag{14}$$

3·2. *Direct and cascade processes.* – As has already been mentioned above, the main problem which arises in experiments on higher-order nonlinearities is the interference between direct and cascade processes. For example, the fourth-harmonic generation can result either directly from the nonlinearity $\chi^{(4)}$, or from different cascade processes due to the nonlinearities $\chi^{(2)}$ and $\chi^{(3)}$. An important point is that, if the direct process is phase matched, cascade processes are also phase matched in the same direction (see [5-7]).

To explain this let us consider a cascade process

$$\omega + \omega + \omega \rightarrow 3\omega \rightarrow 3\omega + \omega \rightarrow 4\omega$$

in the fourth-harmonic generation. The intensity of this process is proportional to $[\chi^{(3)}(\omega+\omega+\omega)\,\chi^{(2)}(3\omega+\omega)]^2$. The third-harmonic field, which is created on the cubic nonlinearity $\chi^{(3)}$, can be presented as a sum of a driven and a free wave

$$E(3\omega) = A_d \exp[i(3\omega t - 3K_1 Z)] + A_f \exp[i(3\omega t - K_3 Z)]\,. \tag{15}$$

Generally $3K_1 \neq K_3$. This process is mismatched. It is evident, however, from (15), that if the driven third-harmonic wave beats against the incident wave via a quadratic nonlinearity, the nonlinear polarization with wave number $4k_1$ will be produced, *i.e.* with the same wave number as in direct phase-matched five-photon interaction. Since the cascade processes are phase matched in the direction (13), it is in this direction that the effective susceptibility is actually measured:

$$\chi^{(4)}_{\text{eff}}(4\omega) = \chi^{(4)}(4\omega) + b_1\chi^{(2)}(\omega+\omega)\,\chi^{(3)}(2\omega+\omega+\omega) + \qquad (16)$$

$$+ b_2\chi^{(3)}(\omega+\omega+\omega)\,\chi^{(2)}(3\omega+\omega) + b_3[\chi^{(2)}(\omega+\omega)]^2\,\chi^{(2)}(2\omega+2\omega)\,.$$

The constants b_1, b_2, b_3 are determined by the orientation of the crystal and by the refractive indices for the interacting waves.

In the fourth-harmonic generation, in addition to the direction $4K_1 = K_4$, there are also several directions of phase-matched fourth-harmonic generation corresponding to pure cascade processes including $2K_2 = K_4$, $K_1 + K_3 = K_4$, $2K_1 + K_2 = K_4$.

TABLE II. – *Phase-matched fourth-harmonic generation in* LFM *crystals.*

PM direction	$K_4 = 4K_1 \quad (\theta = 36°\ 30')$
direct process	$o_1 o_1 o_1 o_1 \to e_4$
cascades	$o_1 o_1 e_2 \to e_2 o_1 o_1 \to e_4$
	$o_1 o_1 o_1 \to o_3 \to o_3 o_1 \to e_4$
	$o_1 o_1 \to e_2 \to e_2 e_2 \to e_4$
$P_{\text{NL}}(4\omega) = [\chi^{(4)} + b_1\chi^{(2)}\chi^{(3)} + b_2\chi^{(3)}\chi^{(2)} + b_3(\chi^{(2)})^2\chi^{(2)}]E^4$	

Data on all these phase-matched interactions are listed in tables II and III. Corresponding data on fifth-harmonic generation in $CaCO_3$ crystals are listed in table IV.

TABLE III. – *Phase-matched fourth-harmonic generation in* LMF *crystals.*

1) PM direction	$K_4 = 2K_1 + K_2$	$(\theta = 22°\ 30')$
	$o_1 o_1 e_2 \to e_2 o_1 o_1 e_4$	
	$P_{\text{NL}}(4\omega) = b_1'\chi^{(2)}\chi^{(3)}E^4$	
2) PM direction	$K_4 = K_3 + K_1$	$(\theta = 45°\ 10')$
	$o_1 o_1 o_1 \to o_3 \to o_3 o_1 \to e_4$	
	$P_{\text{NL}}(4\omega) = b_2'\chi^{(3)}\chi^{(2)}E^4$	
3) PM direction	$K_4 = 2K_2$	$(\theta = 11°\ 10')$
	$o_1 e_1 \to o_2 \to o_2 o_2 \to e_4$	
	$P_{\text{NL}}(4\omega) = b_3'(\chi^{(2)})^2\chi^{(2)}$	

An important point which should especially be made is that measurement of relative powers of harmonics generated in different phase-matched directions makes it possible to express $\chi^{(4)}_{\text{eff}}$ or $\chi^{(5)}_{\text{eff}}$ in terms of the lower-order nonlinearities. If the absolute values and the signs of lower-order nonlinearities are known, it becomes possible to determine $\chi^{(4)}$ and $\chi^{(5)}$.

TABLE IV. – *Phase-matched fifth-harmonic generation in* $CaCO_3$ *crystals.*

1) PM direction	$K_5 = 5K_1$	$(\theta = 55°)$
direct process	$o_1 o_1 o_1 o_1 o_1 \to e_5$	
cascades	$o_1 o_1 o_1 \to e_3 \to e_3 o_1 o_1 — e_5$	
	$P_{NL}(5\omega) = [\chi^{(5)} + b|\chi^{(3)}|^2]E^5$	
2) PM direction	$K_5 = K_3 + K_1 + K_1$	$(\theta = 48°)$
only cascade processes	$o_1 o_1 o_1 \to o_3 \to o_3 o_1 o_1 \to e_5$	
	$P_{NL}(5\omega) = b'|\chi^{(3)}|^2 E^5$	

3'3. *Wave processes in higher-order optical-harmonic generation.* – For the plane monochromatic waves, generalization of the wave theory for higher-order harmonics generation is trivial. For the focussed beams one point should be mentioned: there is no « optimal focussing » for generating higher-order harmonics. Strong focussing, up to optical break-down, should be used to improve conversion efficiency.

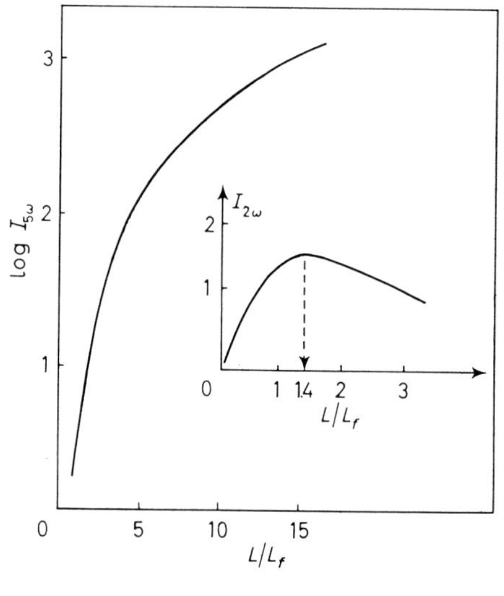

Fig. 1. – Fifth-harmonic generation in focused beams. L = crystal length, L_f = length of the focal spot. For this curve $L_f — \Delta K$ is optimal.

Figure 1 presents the fifth-harmonic intensity $I_{5\omega}$ *vs.* the focussing parameter $m = L/L_f$ (L = crystal length, L_f = length of the focal spot); in contrast with the well-known second-harmonic generation, $I_{5\omega}$ grows monotonically with L/L_f.

3'4. *Symmetry properties of the tensors $\chi^{(4)}$ and $\chi^{(5)}$; tables of nonzero components.* – Tensor equations for 4ω and 5ω nonlinear polarizations are constructed.

TABLE V. – *Nonlinear polarization* $\boldsymbol{P}_{\rm NL}^{(4)} = \chi^{(4)}\boldsymbol{EEEE}$.

$$\left.\begin{matrix} P_x^{(4)} \\ P_y^{(4)} \\ P_z^{(4)} \end{matrix}\right\} = \chi^{(4)}_{ijklmn} \left|\begin{matrix} E_x^4 \\ E_y^4 \\ E_z^4 \\ 4E_xE_y^3 \\ 4E_xE_z^3 \\ 4E_yE_x^3 \\ 4E_yE_z^3 \\ 4E_zE_x^3 \\ 4E_zE_y^3 \\ 6E_x^2E_y^2 \\ 6E_x^2E_z^2 \\ 6E_y^2E_z^2 \\ 12E_x^2E_yE_z \\ 12E_y^2E_xE_z \\ 12E_z^2E_xE_y \end{matrix}\right|$$

Table V presents the tensor equations for nonlinear polarization with frequency 4ω in the transparency region. Nonzero components of the $\chi^{(4)}$-tensor for the point group $mm2$ and the $\chi^{(5)}$-tensor for the point group $\bar{3}m$ are listed in tables VI and VII (see [17]).

TABLE VI. – *Nonzero components of fourth-order nonlinear susceptibility tensor for point group mm2.*

$x\,x\,x\,x\,z$	$y\,y\,y\,y\,z$	$z\,z\,z\,z\,z$
$x\,x\,z\,z\,z$	$y\,y\,z\,z\,z$	$z\,x\,x\,y\,y$

In our experiment:

$$P_z(4\omega) = \chi^{(4)}_{zyyyy}E_y^4(\omega)$$

TABLE VII. – *Nonzero components of fifth-order nonlinear susceptibility tensor for the point group* $\bar{3}m$.

$x\,x\,x\,x\,x\,x$	$y\,y\,y\,y\,y\,y$	$z\,z\,z\,z\,z\,z$
$x\,x\,x\,x\,y\,y$	$y\,y\,y\,y\,y\,z$	$z\,z\,z\,z\,x\,x$
$x\,x\,x\,x\,y\,z$	$y\,y\,y\,y\,z\,z$	$z\,z\,z\,y\,x\,x$
$x\,x\,y\,y\,y\,y$	$y\,y\,y\,z\,z\,z$	$z\,z\,x\,x\,x\,x$
$x\,x\,y\,y\,y\,z$	$y\,y\,z\,z\,z\,z$	$z\,z\,y\,y\,x\,x$

In our experiment:

$$P_z(5\omega) = \chi^{(5)}_{zyyyyy} E_y^5(\omega)$$

3'5. *Experiment.* – A block diagram of the experimental set-up is shown in fig. 2. Its principal elements are an Nd-glass picosecond laser, operating in a regime with zero transverse mode, and a sensitive recording system. Gating

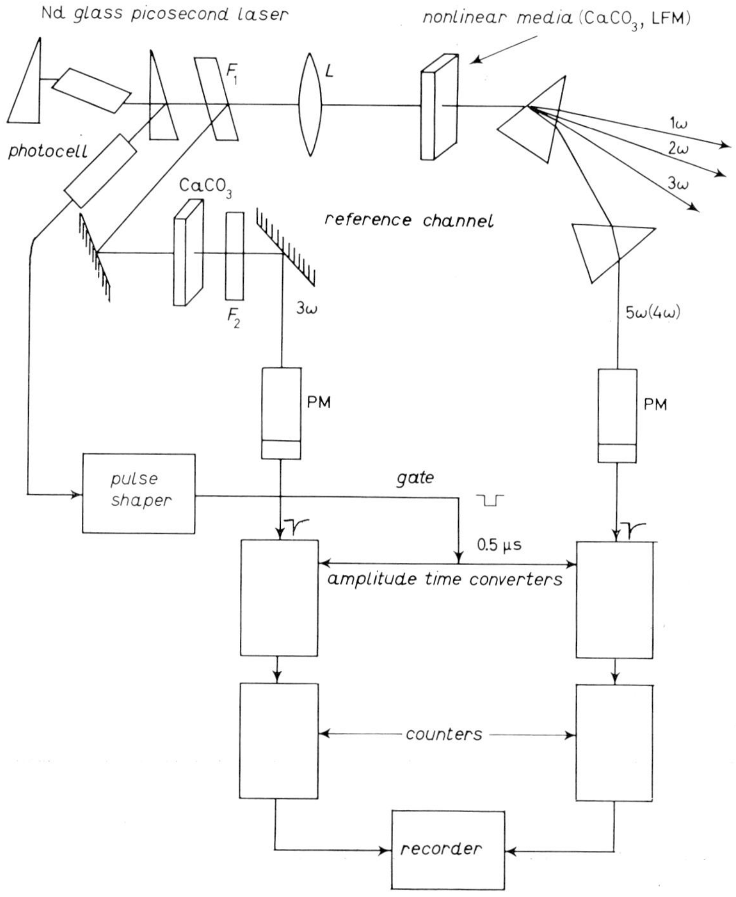

Fig. 2. – Experimental set-up for measurements of nonlinear susceptibilities $\chi^{(4)}$ and $\chi^{(5)}$.

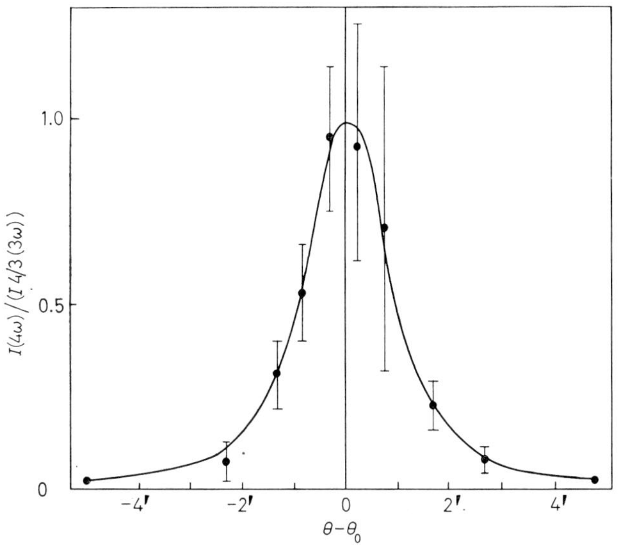

Fig. 3. – Angular dependence of the 4th-harmonic intensity in LFM ($4K_1^0 = K_4^e$). $\theta_0 = 36°\ 30'$, $2\,\Delta\theta_{exp} = 2'$, $2\,\Delta\theta_{theory} = 26''$.

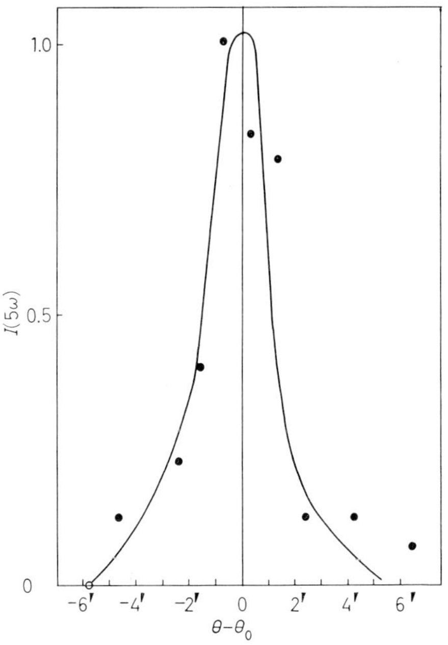

Fig. 4. – Angular dependence of the fifth-harmonic intensity in $CaCO_3$ crystals ($K_5^e = 5K_1^0$). $\theta_0 = 55°$, $2\,\Delta\theta_{exp} = 3'$, $2\,\Delta\theta_{theory} = 30''$.

of the amplitude converter with a gate duration 0.5 μs was used to cut off the photomultiplier noise. In all experiments reference channels are used.

Figure 3 shows the angular dependence of the intensity of the fourth harmonic in LFM near the phase-matching direction $4K_1^o = K_4^e$. The corresponding data for the fifth-harmonic generation in $CaCO_3$ crystals are presented in fig. 4.

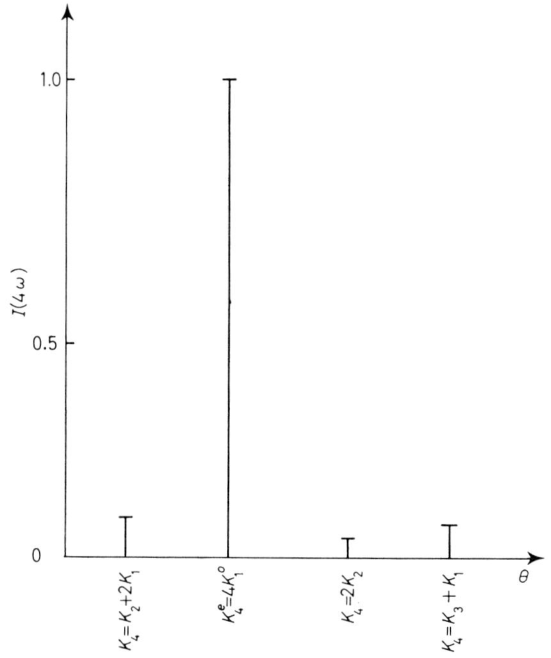

Fig. 5. – Relative intensities of the different phase-matched processes in fourth-harmonic generation in LFM crystals.

Figure 5 shows the relative intensities of the fourth-harmonic generation in the LFM crystal for different phase-matched interactions.

Experimental data are listed in tables VIII-IX. The uncertainities are connected with unknown signs of lower-order nonlinearities.

TABLE VIII. – *Effective fourth-order susceptibility* $\chi^{(4)}_{\text{eff}}$ *in* LFM *crystals.*

PM direction $K_4 = 4K_1$

$$\chi^{(4)}_{\text{eff}} = \chi^{(4)}_{zyyyy} - \frac{4\pi\chi^{(3)}\chi^{(2)}_{zyy}}{(n_2 - n_1)n_2} - \frac{6\pi\chi^{(3)}\chi^{(2)}_{zyy}}{(n_3 - n_1)n_3} - \frac{4\pi^2|\chi^{(2)}_{zyy}|^2\chi^{(2)}_{zyy}}{(n - n_2)^2 n_2^2},$$

where $\chi^{(3)} = 3\chi^{(3)}_{zzyy}\sin^2\theta + 3\chi^{(3)}_{yyxx}\cos^2\theta$.

Experimental results

$$\chi^{(4)}_{\text{eff}} = 7\cdot10^{-22}\ \text{e.s.u.}, \qquad \chi^{(4)}_{zyyyy} = (0.2\div2)\cdot10^{-21}\ \text{e.s.u.}$$

TABLE IX. – *Effective fifth-order subsceptibility* $\chi^{(5)}_{\text{eff}}$ *in* $CaCO_3$ *crystals.*

PM direction $K_5 = 5K_1$

$$\chi^{\text{eff}}_{(5)} = \chi^{(5)}_{zyyyyy} + \frac{6\pi C_{32}}{n_3^e(n_3^e - n_1^0)}(C_{18}\cos^2\theta + C_{16}\sin^2\theta) + \frac{6\pi C_{32}C_{11}}{n_3^0(n_3^0 - n_1^0)} .$$

Experimental results:

$$\chi^{(5)}_{\text{eff}} = (250 \pm 50)\, C_{11}C_{32} .$$

With the values $C_{11} = 1.8\cdot 10^{-14}$ e.s.u. and $C_{32} = 4\cdot 10^{-16}$ e.s.u.

$$\chi^{(5)}_{zyyyyy} = \begin{cases} 0.6\cdot 10^{-27}\ \text{e.s.u.}, & \text{if } C_{32}\chi^{(5)}_{zyyyyy} < 0 \\ 3.4\cdot 10^{-27}\ \text{e.s.u.}, & \text{if } C_{32}\chi^{(5)}_{zyyyyy} > 0 \end{cases}$$

4. – General discussion of the behaviour of higher-order nonlinearities in crystals. Concluding remarks.

Data on higher-order nonresonant susceptibilities now available are listed in table X. Although this information is hardly sufficient for a general discussion of the behaviour of higher-order nonlinearities, several conclusions could nevertheless be made.

TABLE X. – *Data on higher-order susceptibilities in crystals* (in e.s.u.).

Crystal	$\chi^{(2)}$	$\chi^{(3)}$	$\chi^{(4)}$	$\chi^{(5)}$
LFM	$2.8\cdot 10^{-9}$		10^{-21} (^{a})	
$CaCO_3$	—	$1.8\cdot 10^{-14}$	—	10^{-27} (^{a})
$CdGeAs_2$	$8.5\cdot 10^{-7}$		10^{-18} (^{b})	

(*a*) Our data (1974, 1975).
(*b*) Calculated by BYER *et al.* [15].

In table XI ratios between successive nonlinear susceptibilities for several crystals are presented. It is evident that for higher-order nonlinearities these ratios decrease.

TABLE XI. – *Behaviour of higher-order optical susceptivities* (all data are in e.s.u.).

Crystal	SiO_2	ADP	$LiNbO_3$	GaAs	$CdGeAs_2$
$\chi^{(3)}/\chi^{(2)}$	$0.9\cdot 10^{-5}$	$0.6\cdot 10^{-5}$	$1.5\cdot 10^{-5}$	$4\cdot 10^{-5}$	$3\cdot 10^{-4}$

$$\text{LFM}\ \frac{\chi^{(4)}_{zyyyy}}{\chi^{(2)}_{zyy}} \simeq 0.4\cdot 10^{-12}, \qquad CaCO_3\ \frac{\chi^{(5)}_{zyyyyy}}{\chi^{(3)}_{yyyy}} \simeq 10^{-13}, \qquad CdGeAs_2\ \frac{\chi^{(4)}_{xyyyz}}{\chi^{(2)}_{xyz}} \simeq 10^{-12}$$

In table XII normalized ultimate nonlinear polarizations of different orders in crystals are listed; it can be seen that in contrast with the rarified gases (cf. HARRIS [4]) in crystals, even at break-down, higher-order polarizations are much smaller than lower-order polarizations. The measurements of higher-order nonlinearities are a good check for the theories of nonlinear susceptibilities.

TABLE XII. – *Normalized ultimate nonlinear polarization of different orders in several crystals* (ultimate electric light field is equal to optical break-down field).

Crystal	E_{br}	$P^{(2)}/P^{(1)}$	$P^{(3)}/P^{(1)}$	$P^{(4)}/P^{(1)}$	$P^{(5)}/P^{(1)}$
$CaCO_3$	$2\cdot10^4$	—	$6\cdot10^{-5}$	—	$1.2\cdot10^{-9}$
LFM	$2\cdot10^4$	$6\cdot10^{-3}$	$0.4\cdot10^{-5}$ (^{a})	$0.1\cdot10^{-7}$	$0.2\cdot10^{-9}$ (^{a})
ADP	$2\cdot10^4$	$2.4\cdot10^{-3}$	$2.7\cdot10^{-5}$	$0.8\cdot10^{-7}$ (^{a})	$0.3\cdot10^{-9}$ (^{a})
$CdGeAs_2$	10^3	$8.5\cdot10^{-3}$	$3.4\cdot10^{-5}$	10^{-7} (^{a})	$0.2\cdot10^{-9}$ (^{a})
Ge	$2\cdot10^3$	—	$13\cdot10^{-5}$	—	$8\cdot10^{-9}$ (^{a})

(a) Polarization is calculated using a value of $\chi^{(n)}$ deduced from Miller's rule.

BYER *et al.* [15] have estimated the susceptibility $\chi^{(4)}$ for a $CdGeAs_2$ crystal, using the bond orbital model. Unfortunately, no experimental data on this value are yet available.

Theoretical estimations of the nonlinear susceptivities of $CaCO_3$ crystals were recently made by MEISSNER [16] in our laboratory. He used the classical model with Coulomb anharmonicity; the corresponding data are listed in table XIII.

TABLE XIII. – *Third- and fifth-order nonlinear susceptibilities for* $CaCO_3$ *crystals.*

Nonlinearity	Theory (^{a})	Experiment	References
$\chi^{(3)}_{yyyy}$	$+1.72\cdot10^{-14}$ e.s.u.	$1.8\cdot10^{-14}$	[18]
$\chi^{(3)}_{zzzz}$	$+1.0\cdot10^{-14}$ e.s.u.	$1.0\cdot10^{-14}$	[18]
$\chi^{(3)}_{zyyy}$	$-0.16\cdot10^{-14}$ e.s.u.	$0.4\cdot10^{-15}$	[19]
$\chi^{(5)}_{zyyyyy}$	$+0.36\cdot10^{-28}$ e.s.u.	$\sim10^{-27}$	our result

(a) Based on the classical model with Coulomb anharmonicity.

In conclusion I would like to mention that higher-order nonlinear susceptibilities can be measured not only in transmission, but also in reflection. It is very interesting to study higher-order nonlinearities in semiconductors with tunable laser (for example, with the powerful infra-red optical parametric oscillator) in reflection. These data should be of great importance for the theory of nonlinear susceptibilities, which is being successfully developed now.

NONLINEAR EFFECTS CONNECTED WITH $\hat{\chi}^{(4)}$ AND $\hat{\chi}^{(5)}$	
$\hat{\chi}^{(4)}$	1) Fourth-harmonic generation.
	2) Five-photon parametric luminescence.
	3) Second-order hyper-Raman scattering.
$\hat{\chi}^{(5)}$	1) Fifth-harmonic generation.
	2) Six-photon parametric luminescence.
	3) Anharmonic coherent Raman scattering.

FIFTH-HARMONIC GENERATION IN A LOW-PRESSURE XENON (ref. [4])

$$P_{\rm NL}^{(5)} = \chi_{\rm eff}^{(5)} E^5$$

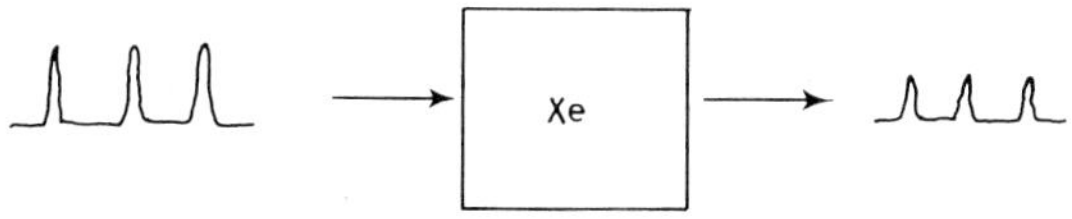

pulse train from picosecond laser $\lambda_1 = 5320$ Å, fifth harmonic $\lambda_5 = 1064$ Å

Theoretically estimated ultimate efficiency for phase-nonmatched fifth-harmonic generation in xenon:

$$\frac{I_{5\omega}}{I_{1\omega}} \simeq 0.05\% \,.$$

ANHARMONIC COHERENT RAMAN SCATTERING IN H_2 (ref. [3])

Anti-Stokes radiation from anharmonically driven molecules

$$E = E_1 + E_2 = E_1 \exp[i(\omega_1 t - k_1 r_1)] + E_2 \exp[i(\omega_2 t - k_2 r)] + \text{c.c.}$$

$\omega_1 - \omega_2 \simeq \Omega$, vibrational frequency of H_2

$$\xrightarrow{\omega_1},\ \xrightarrow{\omega_2}\ \boxed{H_2}\ \xrightarrow{\omega_a} \omega_a = \omega_1 + 2(\omega_1 - \omega_2) = 3\omega_1 - 2\omega_2$$

$$P_{\rm LN} = \chi^{(5)} E_1^3 (E_2^*)^2 \,.$$

GENERAL BEHAVIOUR OF THE HIGHER-ORDER POLARIZATION

In the simple anharmonic oscillator model

$$\frac{P_{\rm NL}^{(n+1)}}{P_{\rm NL}^{(n)}} \simeq \frac{E}{E_a} \qquad \text{and} \qquad \hat{\chi}^{(n+1)} = \chi^{(n)} \frac{1}{E_a}$$

really (see also listed data)

$$\chi^{(n+1)} = \frac{1}{E_a} \chi^{(n)} F(n) \,;$$

$F(n)$ is decreasing with incrasing n.

REFERENCES

[1] P. Franken, A. Hill, C. W. Peters and G. Weinreich: *Phys. Rev. Lett.*, **7**, 118 (1961).
[2] C. Flytzanis: paper in *Treatise on Quantum Electronics*, edited by H. Rabin and C. Tang (New York, N. Y., 1975).
[3] I. Lukasik and I. Ducuing: *Phys. Rev. Lett.*, **28**, 1155 (1972).
[4] S. E. Harris: *Phys. Rev. Lett.*, **31**, 341 (1974).
[5] S. Akhmanov, A. Dubovik, S. Saltiel, I. Tomov and V. Tunkin: *JETP Lett.*, **20**, 264 (1974).
[6] S. Akhmanov, L. Meissner, S. Saltiel and V. Tunkin: *Proceedings of the IV Vavilov Conference on Nonlinear Optics* (Novosibirsk, 1975); *JETP Lett.*, **22**, 143 (1975).
[7] E. Yablonovich, C. Flytzanis and N. Bloembergen: *Phys. Rev. Lett.*, **29**, 865 (1972).
[8] S. Arakelyan, V. Tunkin, A. Kholodnych and A. Tchirkin: *Sov. Journ. Techn. Phys.*, **64**, 1253 (1974).
[9] I. P. Herrmann: *Opt. Comm.*, **9**, 74 (1973).
[10] N. Bloembergen: *Nonlinear Optics* (New York, N. Y., 1965).
[11] Y. Veda and K. Shimoda: *Journ. Phys. Soc. Japan*, **28**, 196 (1970).
[12] A. Javan and P. Kelley: *IEEE Journ. Quant. Electr.*, QE-**2**, No. 9 (1966).
[13] R. Miller: *Appl. Phys. Lett.*, **5**, 17 (1964).
[14] C. Wang: *Phys. Rev.*, **2**, 2045 (1970).
[15] R. Begley, R. Byer, D. Chemla, R. Herbst, R. Feigelson and S. Ciraci: *Development of chalcopyryte crystals for nonlinear optical applications*, Techn. Report, Stanford University, Stanford, Cal. (December 1974).
[16] L. Meissner: *Žurn. Ėksp. Teor. Fiz.*, **69**, 2101 (1975).
[17] V. Zavelisko, V. Martinov, S. Saltiel and V. Tunkin: *Sov. Quantum Electr.*, **2**, 2541 (1975).
[18] M. Levenson and N. Bloembergen: *Phys. Rev. B*, **10**, 4447 (1974).
[19] P. D. Maker and R. Terhune: *Phys. Rev.*, **137**, A 801 (1965).

Statistical Effects in Resonant Nonlinear Optics.

S. A. AKHMANOV
Moscow State University - Moscow, USSR

1. – Introduction.

In spite of the coherence of the stimulated emission, even in the CW single-mode laser it is impossible to obtain fully coherent radiation.

Due to spontaneous emission there are deviations in the temporal [1] and spatial [2] coherence of CW single-mode laser from the ideal coherence. The situation with powerful multimode lasers is even more pronounced; in many cases their radiation can be treated as a powerful optical noise, both in space and in time. Thus, it is very important to understand the features of nonlinear optical effects, especially resonant nonlinear optical effects, when they are excited by the randomly modulated light fields.

There are several interesting problems in this field.

1) The question of the effectiveness of the noise pump in a nonlinear process is of great importance. There are two aspects here. On the one hand, the realization of conditions under which the effectiveness of the noise pump is equal to or exceeds that of the harmonic pump of the same power is of interest. This formulation is of importance for the nonlinear optics of the ultraviolet and X bands, where at present the construction of highly monochromatic sources meets with considerable difficulties. On the other hand, in many cases, nonlinear effects can be regarded as unwanted sources of instabilities of powerful radiation in media (see also [20]-[22]).

In this connection, it is of interest to find ways of modulating a high-intensity wave in such a manner as to facilitate its stabilization; as will be shown below, noise modulation proves to be one of the promising methods.

2) Nonlinear optical effects, and especially resonant effects, can be used to determine the statistics of light fields. It is a well-known topic; note that several new possibilities have recently been offered in this field. For example, the optical Stark effect in the three-level system can be used to study the more subtle statistical properties of powerful light fields [3].

3) Nonlinear interactions of broad-band light beams are strongly affected, of course, by the dispersion of linear and nonlinear susceptibilities inside the line widths of interacting beams. It follows that, in principle, the statistics at the output of nonlinear media contains some spectroscopic information. This can be regarded as an idea of the « noise nonlinear spectroscopy », which in a certain sense can be treated as an alternative to the nonlinear spectroscopy with the narrow-band tunable sources. Note that « linear noise spectroscopy » is well known. For example, DANIELMEYER and WEBER [4] have used the intensity correlation measurements for the direct determination of the group velocity of light.

As will be shown below, group velocity dispersion can be measured also by means of stimulated Raman scattering of « noise nonlinear spectroscopy », described also in [5], in connection with the active spectroscopy (or four-frequency-mixing spectroscopy) of Raman scattering.

For all these problems, both the temporal and the spatial statistics of a light field are of great importance [6]. The main part of my lecture is, however, devoted to the effects of temporal statistics—in a resonant nonlinear optics these effects dominate (*).

Propagation effects are markedly influenced by spatial statistics. Thus, spatial coherence is seen to affect optical harmonic generation in birefringent crystals as well as self-focusing and self-defocusing, stimulated scattering, etc.

It should also be mentioned that spatial coherence may play an important role in several schemes of nonlinear spectroscopy. For example, in Doppler-free two-photon absorption spectroscopy, the spatial coherence of opposing beams may influence spectral resolution—this problem was discussed in [7].

2. – Theoretical methods of statistical nonlinear optics.

In statistical nonlinear optics the set of nonlinear equations which describes the behaviour of the media (material equations) and the light fields should be solved with random external forces and random initial and boundary conditions. This greatly complicates the mathematical treatment of the problem. Two approaches may be used here. The first one consists in obtaining, if possible, an exact solution of the nonlinear equations and subsequently averaging this solution with respect to a known statistical ensemble. Unfortunately, this program can be realized only rarely. The closed-form solutions of nonlinear optics equations, especially equations which describe resonant nonlinear effects, are, as a rule, unknown.

(*) When dealing with spatial statistics, space-time analogy in the statistical nonlinear optics, which was established in [6], should be taken into account.

Another and, as will be shown below, very effective approach consists *in averaging equations*, instead of averaging the solutions. These averaged equations in many cases become much simpler than the primary corresponding equations for fluctuating quantities. The form of these averaged equations depends strongly on the statistics of the random fields involved.

Thus, in nonlinear optics, such special methods as the Fokker-Planck approximation [8-10] and the Dyson method [9-13] are effectively employed. They have been recently used in our laboratory to study such a complicated problem as stimulated Raman scattering excited by a broad-band pump under conditions of simultaneous manifestation of molecular relaxation and of medium dispersiveness, as well as the behaviour of a two-level system driven by a random field. Several results of these investigations are presented below.

3. – The theory of SRS in the field of a noise pump.

In the given field approximation (we also neglect population movements), SRS is described by two equations for the off-diagonal element of the density matrix Q and the complex amplitude A_S of the Stokes wave:

$$e_S \frac{\partial A_S}{\partial z} + \frac{1}{u_S}\frac{\partial A_S}{\partial t} + \delta_S A_S = \sigma_1 A_p(\Theta) Q^* , \tag{1}$$

$$\frac{\partial Q}{\partial t} + \frac{Q}{T_2} = \sigma_2 A_p(\Theta) A_S^* + N(\Theta, z) . \tag{2}$$

Here u_S is the group velocity, δ_S is the damping constant of the Stokes wave, σ_1 and σ_2 are coupling constants, $\Theta = t - z/u_p$ is the running time connected with the pump, and T_2 is the transverse relaxation time. In eq. (1), $e_S = +1$, if the Stokes wave and the pumping wave are codirectional, and $e_S = -1$, if they propagate in opposite directions. In eq. (2) $N(\Theta, z)$ is the stochastic force describing the intrinsic noise of the medium.

The nature of SRS in the field of a noise pump in a medium of characteristic dimension l is determined by the relation between the correlation time τ_{cor} of the pump, on the one hand, and the transverse relaxation time T_2 and the characteristic group lag time T_3 on the other. For co-moving waves

$$T_3 = l\left(\frac{1}{u_p} - \frac{1}{u_S}\right),$$

while for opposing waves

$$T_3 = l\left(\frac{1}{u_p} + \frac{1}{u_S}\right) \approx \frac{2l}{c} .$$

Comparing the indicated characteristic times, we are able to distinguish four characteristic stimulated-scattering regimes, which are all of practical interest:

1) $\tau_{cor} > T_2, T_3$; the quasi-static regime.

2) $T_2 < \tau_{cor} < T_3$; SRS in a dispersive medium with broad Raman lines. Such a situation is realized, for example, in experimental investigations of SRS in certain liquids.

3) $T_2 > \tau_{cor} > T_3$; nonstationary SRS in a nondispersive medium. A fairly typical case of this scattering is forward scattering in gases. In condensed media this regime is usually realized when the scattering is observed in focused beams ($L \simeq L_f$ is the focal length of the lens).

4) $\tau_{cor} < T_2, T_3$; a lag appears in the molecular vibrations and, at the same time, the medium becomes dispersive. This is the most important regime in the investigation of SRS in large volumes.

Let us proceed to consider the above-enumerated cases.

3·1. *The quasi-static* SRS *regime in the field of a noise pump. Stochastic instability.* – Under conditions in which the pump can be regarded as a slow (in the time scales T_2 and T_3) function, we have from (1) and (2), for the instantaneous intensity of the Stokes wave, the expression

$$I_S(z) = I_{S0} \exp[gI_p z], \qquad g = 2T_2\sigma_1\sigma_2. \tag{3}$$

For a Gaussian pump $W(I_p) = \bar{I}_p^{-1} \exp[I_p/\bar{I}_p]$, and the mean intensity of the Stokes wave is given by

$$\bar{I}_S = \int_0^\infty I_S W(I_p) \, dI_p = \frac{I_{S0}}{1 - g\bar{I}_p z}. \tag{4}$$

It is evident from (4) that in the quasi-static regime the Raman amplification in the field of a harmonic pump is of the same mean power. Moreover, it follows from (4) that $\bar{I}_S \to \infty$ as $g\bar{I}_p z \to 1$. This implies that the prescribed field approximation becomes inapplicable at $g\bar{I}_p z = 1$; the Gaussian-pump excursions lead to the divergence of the moments of the Stokes intensity. To eliminate the divergences we must take into account the counter-reaction of the Stokes wave on the pump. Then, instead of (3), we have

$$\begin{cases} I_S(z) = I_{S0} \dfrac{\exp[A\xi]}{1 + \exp[A(\xi - 1)]}, \\ A = \ln(I_{p0}/I_{S0}), \quad x = gI_{p0}z, \quad \xi = x/A, \end{cases} \tag{4a}$$

and for the Gaussian pump we obtain

$$\bar{I}_S(z) = \int_0^\infty I_S(z)\, W(I_p)\, \mathrm{d}I_p = \frac{I_{S0}}{x_1} \exp\left[A_1 - \frac{1}{\xi_1}\right] \left\{ \frac{\alpha^{1-1/x_1} - 1}{1 - 1/x_1} + \frac{\pi}{\sin \pi (1 - 1/x_1)} \right\}, \tag{5}$$

$$A_1 = \ln (\bar{I}_{p0}/I_{S0}), \qquad x_1 = g \bar{I}_{p0} z, \qquad \xi_1 = \frac{x_1}{A_1}, \qquad \alpha = \frac{I_{S0}}{\bar{I}_{p0}}. \tag{6}$$

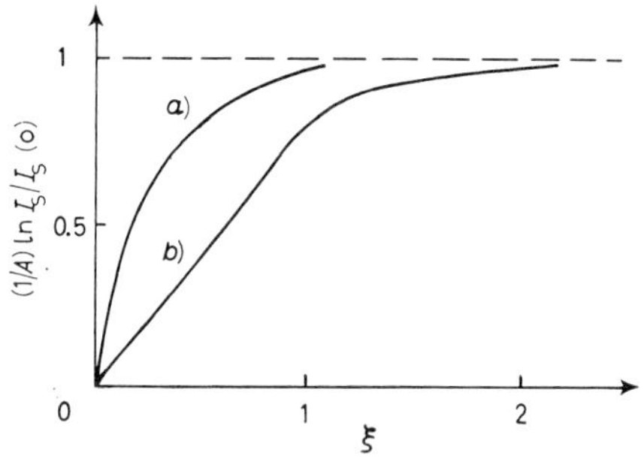

Fig. 1. – The averaged Stokes intensity as a function of the normalized amplification factor ζ: *a*) noise pump, *b*) harmonic pump.

In fig. 1 we present graphs characterizing (on a logarithmic scale) the growth of the mean intensity of the Stokes component for a noise pump and a harmonic pump of the same power. It can be seen that the growth for the noise pump is considerably more rapid. This is due to the high sensitivity of the exponentially growing process to the excursions of the Gaussian pump. For the observation of the mean quantities described by the formula (5) the observation time T_{obs} must be much longer than τ_{cor} (*i.e.* we must have $T_{obs} \gg \tau_{cor}$). Using (3) we can also easily compute the Stokes wave and phonon spectral widths.

3'2. SRS *in a dispersive medium with broad Raman lines.* – Since the derivative $\partial Q/\partial t$ in eq. (2) can be neglected, as we are primarily interested in the growth rate of the Stokes wave, we set $N(\Theta, z) = 0$ and $A_S(0, t) = A_{S0}(t)$. Then from (1), for the co-moving waves, we have

$$A_S(t, z) = A_{S0}\left(t - \frac{z}{u_S}\right) \exp\left[-\delta_S z + \frac{g}{2}\int_0^z I_p\left(t - \frac{z}{u_S} - \mu z'\right) \mathrm{d}z'\right], \tag{7}$$

where $I_p = |A_p|^2$ and $\mu = 1/u_p - 1/u_S$ is the detuning of the group velocities.

Representing I_p in the form

$$I_p = \bar{I}_p + \tilde{I}_p, \qquad \langle \tilde{I}_p \rangle = 0, \tag{8}$$

we can easily verify that the amplification of the Stokes component in the field of the noise pump exceeds the amplification in the field of a harmonic pump of intensity $\bar{I}_p$ by the factor $F(t, z)$, where

(9) $$F(t,z) = \frac{1}{2} g \int_0^z \tilde{I}_p \left(t - \frac{z}{u_S} - \mu z'\right) dz' .$$

By virtue of (9), the fluctuations in the pump average out when $\tau_{cor} \ll T_3 = \mu z$; therefore, as the pump spectrum broadens (under conditions in which $\tau_{cor} > T_2$), the increment tends to the static increment, which is determined by the mean pump intensity $\Gamma_0 = g\bar{I}_p$.

The computation of the increment for $\tau_{cor} < T_3$ can be performed under the assumption that the fluctuations in the pump in this case are δ-correlated:

(10) $$\langle \tilde{I}_p(t) \tilde{I}_p(t+\tau) \rangle = \frac{\bar{I}_p^2}{\pi c \Delta\nu_p} \delta(\tau) ,$$

where $\Delta\nu_p$ is the spectral width of the pump (in cm^{-1}), while the function F is a standard random process (the standardization, when $\tau_p < T_3$, occurs owing to the integration, so that F is a Gaussian process independent of the pump distribution). Using the above-indicated circumstance, we can determine the mean intensity and the correlation function (and, cosequently, the spectrum) of the Stokes wave.

If $A(t) = A_0 \exp[i\Omega t]$ when $z = 0$, then

(11*a*) $$\bar{I}_S(Z) = |A_0|^2 \exp[(\Gamma_n - 2\delta) z] ,$$

where the increment in the field of the noise pump

(11*b*) $$\Gamma_n = \Gamma_0(1 + \Gamma_0 L_{coh}/2\pi) \equiv \Gamma_0 + 2\Gamma' , \qquad \Gamma_0 = g\bar{I}_p ,$$

$L_{coh} = (c\,|\mu|\,\Delta\nu_p)^{-1}$ being the coherence length. It follows from (11*b*) that the increment in the noise pump field exceeds the static value Γ_0, which is determined by the mean intensity of the pump; the excess is determined by the value of the amplification over the coherence length. Since the coherence lengths in the forward and backward directions are different, the latter circumstances lead to an asymmetry in the scattering indicatrix. Moreover, $L_{coh} \to 0$, $\Gamma_n \to \Gamma_0$ as $\Delta\nu_p \to \infty$ independent of the scattering direction. This is the fundamental difference between the noise pump and the regular pulsed pump, where the shortening of the pulse duration (the broadening of the spectrum) leads to a sharp increase in the asymmetry of the scattering indicatrix (see [3]). This is one of the examples of situations in which the properties of SRS can be used to draw conclusions about the statistics of the envelope of the pump.

For the spectral width of the Stokes component we obtain

$$\Delta\nu_S(z) = \frac{\exp[\Gamma' z] - 1}{\exp[\Gamma' z] - 1 - \Gamma' z} \frac{\Gamma'}{\pi c|\mu|} . \tag{12}$$

It follows from (12) that the spectrum of the Stokes wave narrows down with increasing z: $\Delta\nu_S/\Delta\nu_p \to \Gamma_0 L_{coh}/\pi$ as $z \to \infty$. Since $Q \approx T_2 \sigma_2 A_p A_S^*$, the spectrum of the phonon wave is broader than the spectrum of the scattered light.

3'3. *A noise pump in a nondispersive medium with slowly relaxing molecular vibrations.* – If the group velocities of the pump and of the Stokes wave coincide (*i.e.* if $u_p = u_S$), then $T_3 = 0$, the corresponding «dispersion» band $\Delta\nu_d = (\pi c T_3)^{-1} \to \infty$, and the solution to eqs. (1)-(2) can be obtained in the form [2, 7]

$$A_S(\Theta, z) = \sigma_1 A_p(\Theta) \int_0^\Theta dt \int_0^z dz' \exp\left[-\frac{t}{T_2}\right] I_0 \left[\frac{2z'}{T_2} \int_{\theta - t}^{\Theta} \Gamma_0(y)\, dy\right]^{\frac{1}{2}} N(\Theta - t, z - z') , \tag{13}$$

where $\Gamma_0(\Theta) = g I_p(\Theta)$ and I_0 is the modified Bessel function. Introducing the natural assumption that

$$\langle N^*(t', z') N(t, z)\rangle = G\delta(t - t')\delta(z - z') ,$$

we obtain from (13)

$$\bar{I}_S(t, z) = G\sigma_1 \bar{I}_p(\Theta) \int_0^\Theta \exp\left[-\frac{t}{T_2}\right] dt \int_0^z I_0^2 \left[\frac{2z'}{T_2} \int_{\Theta - t}^{\Theta} \Gamma_0(y)\, dy\right]^{\frac{1}{2}} dz' . \tag{13a}$$

The increment in the field of the noise pump is evidently determined by the argument of the Bessel function in (13*a*). By writing the intensity of the pump in the form (8), we can verify that the influence of the fluctuations in the pump is described by the integral

$$Y(t, \Theta) = \frac{1}{t\bar{I}_p} \int_0^t \tilde{I}_p(\Theta' - \Theta)\, d\Theta' . \tag{14}$$

The upper limit of the interval of integration in (14) is equal to $g\bar{I}_p z T_2$. In the case $\tau_{cor} < T_2$ under consideration, as a result of the averaging due to the integration, the quantity Y is small and

$$\bar{I}_S = \bar{I}_{S0} \exp[g\bar{I}_p z] = I_{S0} \exp[\Gamma_0 z] .$$

Thus, a broad-band noise pump turns out to be as effective as a harmonic pump of intensity equal to $\bar{I}_p$.

The spectrum of the Stokes wave has, according to (13), the same width as the spectrum of the pump. In fact, for $\tau_{cor} < T_2$, the quantity $A_S(t)$ can be represented in the form $A_S(t) = A_p(t)\,\Phi(t)$, where $\Phi(t)$ is a slowly (in comparison with $A_p(t)$) varying function. Therefore, to a high degree of accuracy,

$$\Delta\nu_S \approx \Delta\nu_p\,. \tag{15}$$

Moreover, the spectral width of the phonon wave is considerably narrower. Since

$$\frac{dQ}{dt} + \frac{Q}{T_2} = \sigma_2[\bar{I}_p + \tilde{I}_p]\,\Phi(t)\,,$$

then $\Delta\nu_Q = \Delta\nu_0(g\bar{I}_p z)^{-\frac{1}{2}}$, $\Delta\nu_0 = (\pi c T_2)^{-1}$.

The corresponding spectra are presented in fig. 2.

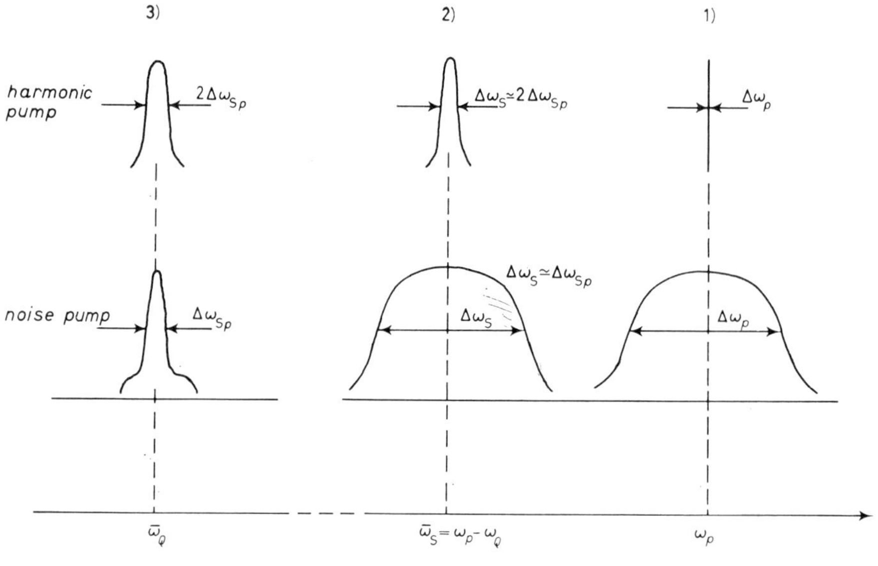

Fig. 2. – Spectra of the pump, Stokes and molecular oscillations for the SRS with a regular pump and the noise pump (dispersion is negligible).

3·4. *The noise pump under conditions of simultaneous manifestation of molecular relaxation and of medium dispersiveness. Noncoherent scattering.* – It is not possible in the case under consideration to solve the dynamical equations (1) and (2) exactly. Therefore, the data presented below are based on a different stochastic approach in which we seek at once the equations for the mean amplitudes or intensities, the correlation functions, etc.

One of the variants of the stochastic approach to the SRS equations (the so-called Fökker-Planck approximation) is based on the fact that, for $L > L_{coh}$, the fluctuations in the prescribed noise pump can be assumed to be δ-correlated. For $L > L_{coh}$ the correlation time of the pump is clearly much shorter than the correlation times of the Stokes wave and of the phonons; there occurs a smoothing-out of the waves being amplified as they move relative to the pump wave. The phase correlations between the interacting waves are largely lost: the scattering becomes noncoherent.

Another variant of the stochastic approach to the analysis of the system (1)-(2) is connected with the use of the Dyson-equation technique. Here the equations for the mean quantities can be derived for an arbitrary correlation of the pump; as a result, the passage to both the δ-correlation pump and the harmonic-pump limits proves to be possible.

A) *The characteristics of noncoherent scattering* (*the Fökker-Planck approximation*). We shall assume, in accordance with the foregoing, that A_p is a δ-correlated Gaussian noise:

$$\left\{\begin{aligned} &\langle A_p(\Theta)\rangle 0\,, \qquad \langle A_p(\Theta_1)\,A_p(\Theta_2)\rangle = 0\,,\\ &\langle A_p(\Theta_1)\,A_p^*(\Theta_2)\rangle = \frac{S(\Theta)}{c}\,\delta(\Theta_1-\Theta_2)\,,\\ &S(\Theta) = \bar{I}_p(\Theta)/\Delta\nu_p\,. \end{aligned}\right. \tag{16}$$

Assuming that the A_S-Q correlation times will be $\tau_{S,Q} \gg \tau_p$, and using (1)-(2), we can separate out those small corrections to the amplitudes that correlate with the amplitude of the pump:

$$\tilde{A}_S = \sigma_1 Q^* \int_0^\infty A_p(\Theta + \mu z')\,dz'\,, \qquad \tilde{Q} = \sigma_2 A_S^* \int_0^\infty A_p(\Theta - \tau')\,d\tau'\,. \tag{17}$$

Using (17), we can express the mixed moments of the amplitudes A_S, Q and A_p in terms of only A and Q. In consequence, we can derive the equations for the mean intensities of the Stokes wave $\bar{I}_S = \langle A_S A_S^*\rangle$ and the molecular vibrations $\overline{W} = \langle QQ^*\rangle$:

$$\frac{d\bar{I}_S}{dz} + \frac{1}{u_S}\frac{\partial \bar{I}_S}{\partial t} + \left[2\delta_S - \frac{gS(\Theta)}{2T_2c}\right]\bar{I}_S = \frac{g\omega_S S(\Theta)}{\omega_Q T_2 \mu'}\,W\,, \tag{18}$$

$$\frac{\partial W}{\partial t} + \left[\frac{2}{T_2} - \frac{gS(\Theta)}{2T_2\mu'}\right]\overline{W} = \frac{\omega_Q g S(\Theta)}{4\omega_S T_2 c}\,\bar{I}_S\,, \tag{19}$$

where $\omega_Q = \omega_p - \omega_S$ and $\mu' = \mu c$ is the relative dispersion of the group velocities.

In the stationary case it follows from (18) and (19) that

$$\bar{I}_S(z) = I_{S0}\exp[\Gamma z]\,, \qquad \overline{W}(z) = \frac{\omega_Q g S}{8\omega_S(1 - S/S_{cr})}\bar{I}_S(z)\,, \tag{20}$$

$$\Gamma = g\bar{I}_p \frac{\Delta\nu_0}{\Delta\nu_p}\frac{\pi/2}{1 - S/S_{cr}}\,, \qquad S_{cr} = \frac{4\mu'}{g}\,. \tag{21}$$

Thus, the increment in the noise pump field is a nonlinear function of the intensity of the group. The key parameter in this case turns out to be the critical value of the spectral density of the pump defined by the formula (21). For $S \ll S_{cr}$, $\Gamma \simeq \Gamma_0 \Delta\nu_0/\Delta\nu_p \ll 1$; the increment increases sharply when $S \simeq S_{cr}$ (see fig. 3). The jump in the increment at $S = S_{cr}$ is connected with a corresponding decrease in the damping of the optical phonons.

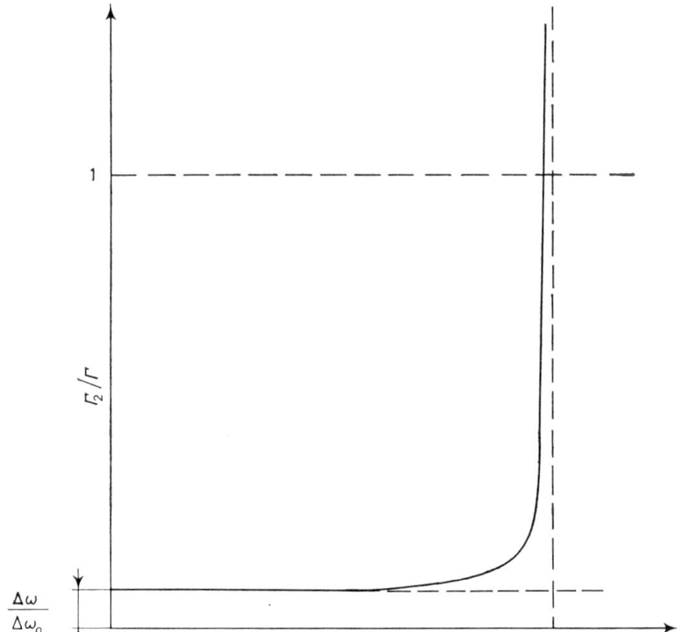

Fig. 3. – Increment of SRS in a noise pump *vs.* pump spectral intensity. $S_{cr} = 4c|1/u_p - 1/u_S|g^{-1}$.

According to (19), in the field of the noise pump the effective transverse relaxation time increases:

$$T_{2\,\mathrm{eff}} = \frac{T_2}{1 - S/S_{cr}}\,. \tag{22}$$

For the spectral width of the optical phonons we obtain, in accordance with (22), the expression $\Delta\nu_Q = \Delta\nu_0(1 - S/S_{cr})$, *i.e.* the quantity $\Delta\nu_Q \ll \Delta\nu_0$ for $S \to S_{cr}$.

The spectral width of the Stokes component

$$\left\{\begin{aligned} \Delta\nu_S &= \frac{(\exp[x]-1)x}{\exp[x]-1-x}\frac{1}{\pi z\mu'}\,, \\ x &= 2\pi\Delta\nu_0 z\mu'(S/S_{cr})^2((1-S)/S)^{-1}\,. \end{aligned}\right. \tag{23}$$

It follows from (23) that as $S \to S_{cr}$ the spectrum of the Stokes component rapidly broadens. The theory developed in this section is valid so long as $\Delta\nu_S < \Delta\nu_p$, *i.e.* so long as $(1 - S/S_{cr})^{-1} < \Delta\nu_p/2\Delta\nu_0$. In the case when (here the noncoherent scattering becomes coherent) stimulated Raman scattering cannot be considered in the framework of the Fökker-Planck approximation, this can, however, be done by using the Dyson-equation technique.

B) *Coherent and noncoherent* SRS (*the Dyson-equation method*). We shall now assume that the pump wave is a Gaussian stochastic process, but with an arbitrary correlation function $K(\tau) = \langle A_p^*(t)A_p(t+\tau)\rangle$ and spectrum

$$G(\omega) = \frac{1}{2\pi}\int_{-\infty}^{+\infty} K(\tau)\exp[i\omega\tau]\,d\tau\,.$$

In (24) we have retained only terms of definite parity with respect to A_p; it follows from the structure of eqs. (1) and (2) for $N(\Theta, z) = 0$ and the boundary condition $A_S(t, z=0) = A_{S0}$ that A_S is an even, while Q is an odd function of A_p:

$$\left\{\begin{aligned} &A_S = \sum_{n=0}^{\infty} A_{S,2n}\,, \qquad Q = \sum_{n=0}^{\infty} Q_{2n+1}\,, \\ &A_m, Q_m \sim \langle A_p(t_1, z_1)A_p(t_2, t_2) \dots A_p(t_m, z_m)\rangle\,. \end{aligned}\right. \tag{24}$$

The approximate representation of the amplitudes in the form of a few leading terms of the series (24) is, in the case of SRS, ineffective: even the steady-state solution $A_S(z) = A_{S0}\exp[gI_p z]$, in which usually $gI_p z \simeq 10 \div 25$, cannot be well approximated in this way. Another method of estimating the amplitudes consists in separating out from the series (24) certain infinite subsequences that are exactly summable. Such an approach is analogous to the method employed in, for example, the theory of multiple scattering, and is connected with the derivation of the so-called Dyson equations for the mean amplitudes or the Bethe-Salpeter equations for the correlation functions. The Dyson-equation method can, in principle, be developed for application to linear equations of the types (1) and (2), as well as to nonlinear equations that take saturation into account. The representation in the form of a finite series in

A_p is then not used to determine directly, for example, $\bar{A}_S$ or $\bar{I}_S = \langle A_S A_S^* \rangle$, but to compute approximately the coefficients of those equations which these mean quantities satisfy. In the first approximation—it is sometimes called the Bouret approximation—only quantities that are of second order in A_p, *i.e.* certain linear functionals of the correlation function $K(\tau)$, enter into the coefficients of the equations for the mean quantities.

The equation for the mean amplitude of the Stokes wave in a medium without losses has, in this approximation, the form

$$\frac{\partial \bar{A}_S}{\partial z} = \bar{A}_S \frac{g}{2} \int_{-\infty}^{+\infty} \frac{G(\omega)\, d\omega}{1 + \omega^2 T_2^2}, \tag{25}$$

and describes both the coherent ($\tau_{cor} > T_2, T_3$) and the noncoherent ($\tau_{cor} < T_2, T_3$) SRS regimes if

$$\frac{\Gamma_0}{2(1 + \Delta\nu_p/\Delta\nu_0)} \, \frac{1 - \exp[-2\pi\, \Delta\nu_p \mu' z]}{2\pi\, \Delta\nu_p \mu'} \ll 1 .$$

The fulfilment of the last condition is necessary if A_p is a Gaussian stochastic process.

If the complex amplitude of the pump contains only one stochastic parameter—a diffusing phase—, *i.e.* if

$$A_p(t) \sim \exp\left[i\int_{-\infty}^{t} \xi(t')dt'\right], \qquad \langle \xi(t)\xi(t+\tau)\rangle = D\delta(\tau),$$

then it can be shown that eq. (25) is exact.

The estimation of the increment Γ for the mean intensity $\bar{I}_p(z)$ is more complicated and shows that Γ satisfies the following transcendental equation:

$$\Gamma = g\left(\frac{1}{T_2} + \frac{\Gamma}{2\mu}\right) \int_0^{\infty} \exp\left[-\left(\frac{1}{T_2} + \frac{\Gamma}{2\mu}\right)\tau\right] K(\tau)\, d\tau .$$

For a Lorentz-pump spectrum this equation goes over into a quadratic equation and determines two values for the increment ($d = \Delta\nu_D/\Delta\nu_p$):

$$\frac{\Gamma_{1,2}}{\Gamma_0} = \frac{1}{2}\left[1 - (1+d)\frac{S_{cr}}{S}\right] \pm \left\{\frac{1}{4}\left[1 - (1+d)\frac{S_{cr}}{S}\right]^2 + \frac{S_{cr}}{S} d\right\}^{\frac{1}{2}}, \tag{26}$$

$$\begin{cases} \Gamma_1 \approx \Gamma_0 - \Gamma_{cr}, \\ \Gamma_{cr} = g S_{cr} \Delta\nu_p = 4\mu' \Delta\nu_p . \end{cases} \tag{27}$$

Allowance for saturation leads to the nonlinear equation for $\bar{A}_S$

$$\frac{\partial \bar{A}_S}{\partial z} = \bar{A}_S \frac{g}{2} \int_{-\infty}^{+\infty} \frac{G(\omega)}{1+\omega^2 T_2^2} \exp\left[-\frac{\omega_p}{\omega_S}\frac{g}{1+\omega^2 T_2^2}\int_0^z |\bar{A}_S|^2 dz'\right] d\omega ,$$

from which it follows, in particular, the possibility of a complete transfer of the energy of the broad pump line to the narrow SRS line in a highly dispersive medium.

3'5. *The experiment.* – The experimental verification of the theory presented here was carried out by our group [13], as well as two groups from the Lebedev Physical Institute [14, 15].

It should be mentioned that the case $\tau_{cor} > T_3$, but $\tau_{cor} < T_2$, has been investigated by many experimentalists. This situation was carefully investigated for the first time in [16]. In [13-15] more attention is given to the case when both dispersion of the media and molecular relaxation must be taken into account. These experiments are of great importance, because in this case the theory is based on several assumptions, which are not well grounded.

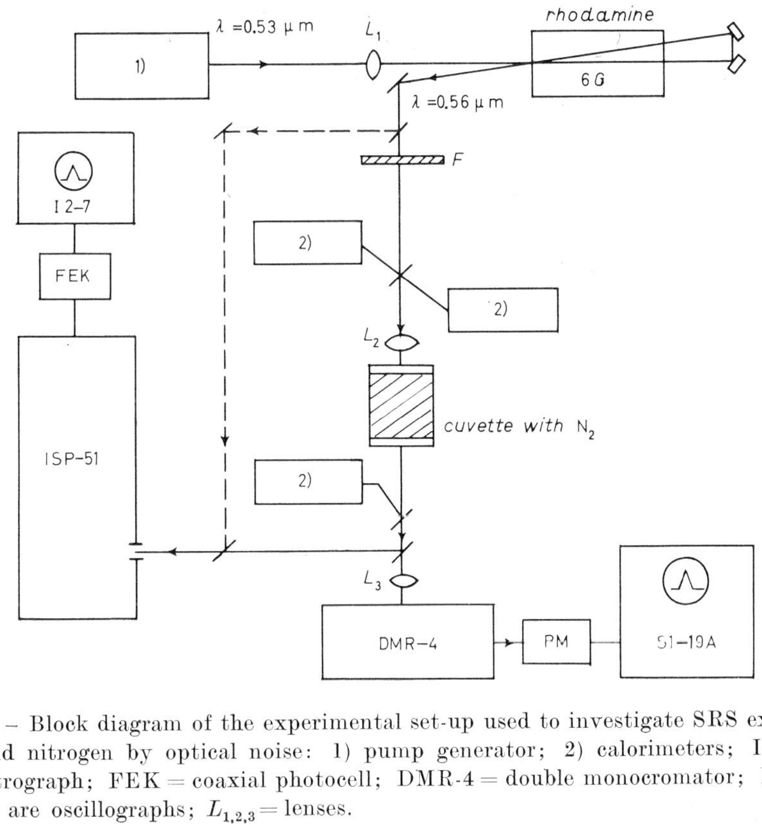

Fig. 4. – Block diagram of the experimental set-up used to investigate SRS excitation in liquid nitrogen by optical noise: 1) pump generator; 2) calorimeters; ISP-51 = = spectrograph; FEK = coaxial photocell; DMR-4 = double monocromator; I2-7 and S1-19A are oscillographs; $L_{1,2,3}$ = lenses.

In our experiments the broad-band pump source was the radiation of a rhodamine 6*G* dye laser operating in the two-pass superradiance regime. A block diagram of the set-up is shown in fig. 4. The line width $\Delta\nu_p$ of the dye radiation was then equal to approximately 200 cm^{-1}. Liquid nitrogen was chosen as the dispersive Raman-active medium. Here, the Raman amplification is relatively high; for $\lambda_p = 0.56$ μm, vibrational frequency $\nu_Q = 2326$ cm^{-1}, $g = 2.15 \cdot 10^{-2}$ cm/MW. The line width of the spontaneous scattering is equal to $\Delta\nu_D = 0.067$ cm^{-1}; the relative dispersion of the group velocities $\mu' = c\mu = 6 \cdot 10^{-3}$.

Therefore for the optical noise used in our experiments $L_{coh} = 0.67$ cm in the forward direction and $L_{coh} = 2 \cdot 10^{-3}$ cm in the backward direction. Consequently for the critical-intensity values $I_{cr} = S_{cr}\Delta\nu_p$ we have

$$I_{cr} = 280 \text{ MW/cm}^2 \qquad \text{(for forward scattering)},$$

$$I_{cr} = 93 \text{ GW/cm}^2 \qquad \text{(for backward scattering)}.$$

The experimental data are in very good agreement with the theory. This corresponds to critical spectral density, forward-backward etc. Figure 5 shows

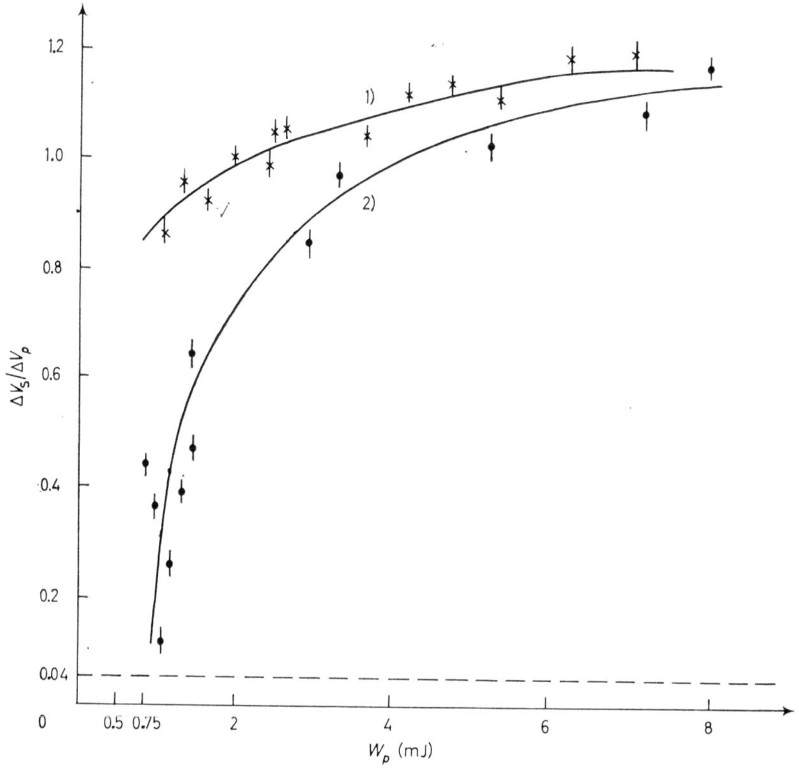

Fig. 5. – The relative spectral line width of the first Stokes component excited by the noise pump as a function of the pump energy. Curve 1) corresponds to the case when $L \approx L_{coh}$, curve 2) to the case when $L \gg L_{coh}$. The dashed line indicates the theoretical value of the minimum band width of the Stokes wave for $L \gg L_{coh}$.

the relative spectral line width of the first Stokes component as a function of the pump energy. For $L > L_{coh}$ the transition between incoherent ($S \ll S_{cr}$) and coherent ($S \gg S_{cr}$) scattering is distinctly visible.

The formulae obtained are directly applicable to stimulated Brillonin scattering; the experimental verification of this fact was given in the POPOVICHEV *et al.* paper [15].

4. – The two-level system driven by a random field.

In this section I would like to present some theoretical results, connected with the behaviour of the two-level systems under strong random excitation. In contrast to the material of sect. **3**, the equation for a population difference is also taken into account. At present there is full understanding about the process of interaction between the two-level system and the monochromatic-light field. Such phenomena, as, for example, the saturation of the absorption, nutations, etc., are well known. To a great extent also the propagation of a regular light pulse in a system of two-level atoms [17] has been studied.

In this section these results are generalized for a random pump. We start with the well-known equations (see [23])

$$T_1 \frac{\partial n}{\partial \Theta} + n = QA^* + Q^* A + n_0 , \tag{28}$$

$$T_2 \frac{\partial Q}{\partial \Theta} + Q = -\frac{1}{2} An , \tag{29}$$

$$\frac{\partial A}{\partial z} + \frac{\gamma}{2} A = Q , \tag{30}$$

for normalized amplitude A, polarization Q and the population difference n. Here γ is the nonresonant losses, $T_{1,2}$ the relaxation times and n_0 the equilibrium value of n.

4·1. *Randomly excited nutations.* – It is well known that short time oscillations of population difference n can be obtained in a strong electromagnetic field.

Let us consider the case when the light field is randomly modulated. The useful model to study the influence of statistics of the exciting light field on the nutations in the two-level system is the frequency-modulated light, which is abruptly switched on:

$$A_0(t) = \begin{cases} a_0 \exp[i\varphi(t)] , & t \geqslant 0 , \\ 0 , & t < 0 , \end{cases} \tag{31}$$

$$\varphi(t) = \int_0^t \xi(t')\,\mathrm{d}t' , \qquad \langle \xi(t)\xi(t') \rangle = 2D\delta(t - t') , \tag{31a}$$

$\overline{\varphi^2} = 2Dt$. Thus, this is a random process with a Lorentzian spectrum, of spectral width $\Delta\omega_0 = D$. For this model the Fokker-Planck approximation can be effectively used (ξ is a δ-correlated process). For the problem under consideration eq. (3) can be omitted. So for the average population difference $\bar{n}$ we have

$$\ddot{\bar{n}} + (T_1^{-1} + T_2^{-1} + \Delta\omega_0)\,\dot{\bar{n}} + T_1^{-1}\left(T_2^{-1} + \Delta\omega_0 + \frac{I_0}{T_2}\right)\bar{n} = \tag{32}$$
$$= T_1^{-1}(T_2^{-1} + \Delta\omega_0)n_0\,, \qquad \dot{\bar{n}}(0) = n_0\,, \qquad \bar{n}(0) = 0\,.$$

It follows from (32) that the saturation intensity (which corresponds to $n(t\to\infty) = n_0/2$) depends strongly on the spectral width

$$I_{\text{sat}} = 1 + T_2\Delta\omega_0 \tag{33}$$

(note that intensities are normalized to saturation intensity in the monochromatic field).

In the randomly modulated field the threshold of nutations also depends on the line width:

$$I_{\text{nut}} \simeq \frac{T_1}{4T_2}(1 + T_2\Delta\omega_0)^2\,. \tag{34}$$

Usually $T_1 > T_2$ and $I_{\text{sat}} \ll I_{\text{nut}}$.

From eqs. (28) and (29) for $\bar{n}^2$, $\bar{n}^3$ and higher-order moments can also be obtained. Thus, the statistics of the oscillations of a two-level system in a random light field can be fully understood.

4·2. *Random waves in a resonant two-level medium.* – In this subsection propagation of noise light pulse through the resonant two-level system is considered with the assumption that the light pulse is relatively long ($t_p \gg T_2$). It is assumed also that the peak intensity value does not exceed the nutation threshold I_{nut} (see (34)).

For the average $y = \int_0^z \bar{n}\,\mathrm{d}z$ the following nonlinear differential equation may be obtained:

$$T_1\dot{y} + y = F_0(t)\int_{-\infty}^{+\infty}\left[\exp\left[-\frac{y}{1+\omega^2T_2^2}\right] - 1\right]S_0(\omega)\,\mathrm{d}\omega + n_0 z\,. \tag{35}$$

The solution for (35) being known, the average output light intensity is determined by

$$\bar{I} = F_0(t)\int_{-\infty}^{+\infty}\exp\left[-\frac{y}{1+\omega^2T_2^2}\right]S_0(\omega)\,\mathrm{d}\omega\,. \tag{36}$$

In the quasi–stady-state limit ($t_p \gg T_{1,2}$) these equations can be transformed into one transcendental equation for

$$\bar{I} = F_0(t) \int_{-\infty}^{+\infty} \exp\left[-\frac{\bar{I} - \bar{I}_0 + n_0 z}{1 + \omega^2 T_2^2}\right] S_0(\omega)\, d\omega \,. \tag{37}$$

The integrand in (37) may be identified as the optical noise spectrum at a given distance z:

$$S(\omega, z) = S_0(\omega) \exp\left[-\frac{\bar{I} - \bar{I}_0 + n_0 z}{1 + \omega^2 T_2^2}\right]. \tag{38}$$

The well-known result

$$I = I_0 \exp\left[I_0 - I + n_0 z\right] \tag{39}$$

follows from (37) in the particular case of monochromatic light.

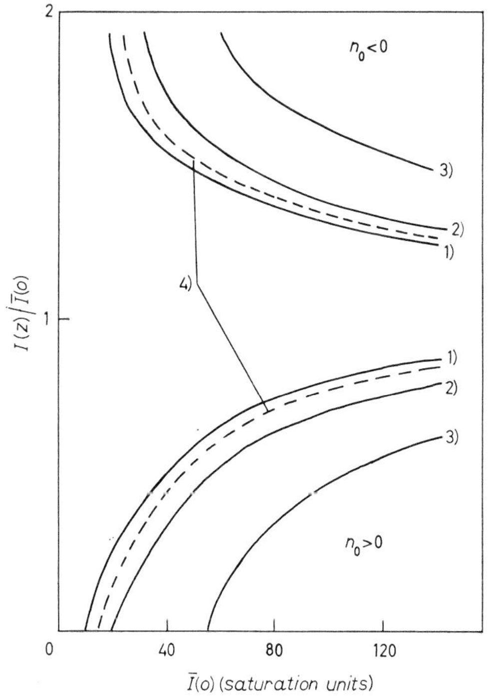

Fig. 6. – Influence of the spectrum band width upon the light amplification ($n_0 < 0$) or absorption ($n_0 > 0$) in the two-level system. $z = 20$ (absorption or amplification units): 1) $\Delta\omega_0 T_2 = 0$, 2) $\Delta\omega_0 T_2 = 2$, 3) $\Delta\omega_0 T_2 = 5$, 4) $\Delta\omega_0 T_2 = 10$.

The curves represented in fig. 6, which were calculated from eq. (37), describe the saturation effect for optical noise. It is interesting to note that amplification ($n_0 < 0$) or absorption ($n_0 > 0$) reach the extremal value at some *finite* band width $\Delta\omega_0$ depending on I_0 and z.

5. – Self-actions of randomly modulated waves.

Self-actions (self-modulation, self-steepening, self-focusing and self-defocusing) are important nonlinear optical effects. Let us start with the problem of self-modulation of a randomly modulated light field in a medium with a cubic nonlinearity.

In the first approximation of the dispersion theory, for the complex amplitude of the propagating wave, we have the following equation:

$$\left(\frac{\partial}{\partial z} + \frac{1}{u}\frac{\partial}{\partial t}\right)A = -i\gamma|A|^2 A\,, \qquad \gamma = k\varepsilon_2/4\varepsilon_0\,. \tag{40}$$

The solution of this equation is

$$A(t, z) = A_0\left(t - \frac{z}{u}\right)\exp\left[-i\gamma\left|A_0\left(t - \frac{z}{u}\right)\right|^2 z\right]. \tag{41}$$

It is instructive to note that, if the light field can be treated as a stationary Gaussian process, a closed-form expression exists for the field correlation function at an arbitrary point in the nonlinear medium:

$$R(\tau, z) = \frac{\langle A(t_1, z)A^*(t_2, z)\rangle}{I_0} = \frac{R_0(\tau)}{\{1 + (\gamma T_0 z)^2[1 - |R_0(\tau)|^2]\}^2}\,, \tag{42}$$

where I_0 is the mean intensity (see also [24]).

Correspondingly, the power spectrum broadens during the propagation:

$$\Delta\omega(z) = \Delta\omega(0)[1 + 4(\gamma I_0 z)^2]^{\frac{1}{2}} \tag{43}$$

(it is assumed that $S(\Omega, 0) = \exp[-\Omega^2/\Delta\omega(0)^2]$).

This symmetrical broadening differs from the broadening which is well known for the regular pulses; it should be mentioned that formulae (42) and (43) must be used to interpret the experimental data, obtained with the multimode lasers. Many experiments of this kind have been reported, but the interpretation was often based on the formulae obtained with regular signals. Frequency broadening of a Gaussian optical noise in cubic media is a good example of a situation where the result of averaging the solution of nonlinear optics equations is clear and simple. If, however, second derivatives in equations are taken into account (it is necessary, if self-focusing or self-steepening is under consideration), only the methods based on the averaging of the equations can be used.

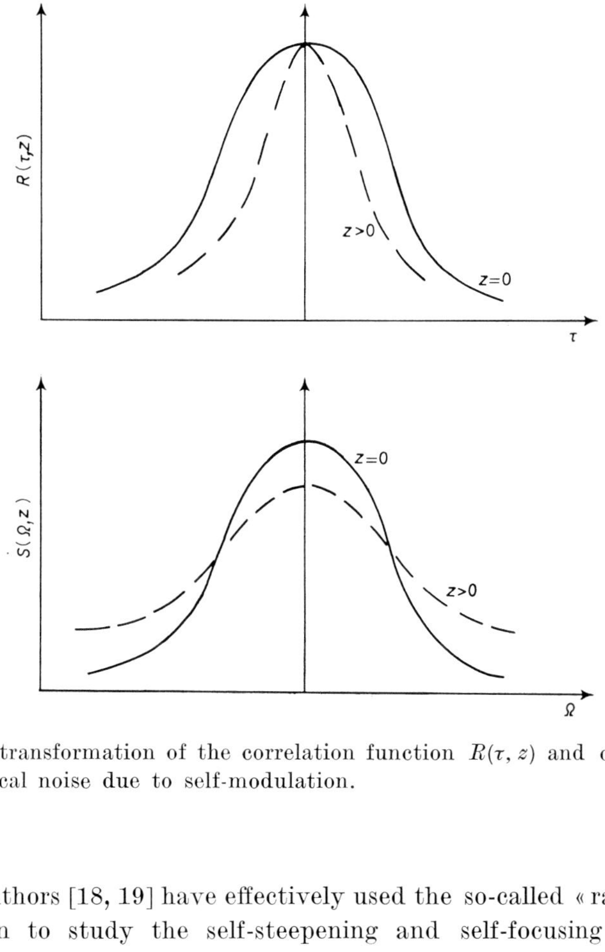

Fig. 7. – The transformation of the correlation function $R(\tau, z)$ and of the spectrum $S(\Omega, z)$ of optical noise due to self-modulation.

Several authors [18, 19] have effectively used the so-called « random phase » approximation to study the self-steepening and self-focusing of Gaussian optical noise.

Assuming that, in spite of self-actions, light statistics still remains Gaussian, LYAKHOV [18] has studied the propagation of the plane-wave Gaussian noise in cubic media in the second-order approximation of the diffraction theory. Starting with the equation

$$\frac{\partial A}{\partial z} = ig\frac{\partial^2 A}{\partial \Theta^2} - i\gamma |A|^2 A\,, \qquad \text{where } g = \frac{1}{2}\frac{\partial^2 k}{\partial \omega^2}\,,$$

they have shown that the intensity fluctuations of light can grow in the media according to the law

$$d = \frac{[\bar{I}^2 - (\bar{I})^2]^{\frac{1}{2}}}{\bar{I}} = 1 - \frac{g\gamma I_0 z^2}{\tau_{\text{cor}}^2},$$

which can be regarded as a random analogue of the pulse self-steepening.

Self-focusing of the Gausian noise was treated in the « random phase » approximation by PASMANIK [19].

6. – Concluding remarks.

The facts discussed in this lecture show that at present a fair degree of understanding has been reached of the main features of nonlinear optical effects in random light fields.

It can hardly be doubted that the methods described here may be effectively used in the theory of resonant harmonic generation and parametric interactions, in the Doppler-free two-phonon absorption (where the spatial coherence of the interacting waves should be taken into account) resonant self-actions, etc.

REFERENCES

[1] Any textbook or monograph in quantum optics, for example R. KLAUDER and S. SUDARSHAN: *Fundamentals of Quantum Optics*.
[2] S. ARAKELYAN, S. AKHMANOV, V. TUNKIN and A. CHIRKIN: *JETP Lett.*, **19**, 571 (1974).
[3] A. BONCH-BRUEVICH, V. KHODOVOI and V. KHROMOV: *JETP Lett.*, **11**, 431 (1970).
[4] H. G. DANIELMEYER and H. WEBER: *Phys. Rev. A*, **3**, 1708 (1971).
[5] S. A. AKHMANOV: this volume, p. 239.
[6] S. A. AKHMANOV and A. S. CHIRKIN: *Statistical Effects in Nonlinear Optics* (Moscow, 1971).
[7] S. ARAKELYAN, S. AKHMANOV, A. CHIRKIN and V. PAKHALOV: *Ultimate spatial coherence of gas laser radiation and its use in physical experiments*, paper III.8 of the *Second Conference on Gas Lasers, Novosibirsk, June 16-18, 1975*.
[8] S. AKHMANOV, YU. D'YAKOV and A. CHIRIKIN: *JETP Lett.*, **13**, 514 (1971).
[9] YU. D'YAKOV: *Kratikie soobshcheniya po fizike* (FTAN), No. 7, 49 (1971).
[10] S. AKHMANOV: *Izv. Vysch. Uchebn. Zaved. Radiofiz.* (*Radiophysics*), **17**, 540 (1974).
[11] YU. D'YAKOV: *Kratkie soobshcheniya po fizike* (FTAN), No. 4, 23 (1973).
[12] S. A. AKHMANOV and YU. D'YAKOV: *JETP Lett.*, **18**, 305 (1973).
[13] S. A. AKHMANOV, YU. D'YAKOV and L. PAVLOV: *Žurn. Ėksp. Teor. Fiz.*, **66**, 520 (1974).
[14] A. GRASYUK, T. ZUBAREV and N. SUYAZIV: *JETP Lett.*, **16**, 166 (1972).

[15] V. Popovichev, V. Ragulski and F. Faizulov: *JETP Lett.*, **19**, 350 (1974).
[16] V. Bocharov, A. Grasyuk, T. Zubarev and V. Mulikov: *Sov. Phys. JETP*, **29**, 235 (1969).
[17] P. Krikov and V. Letokhov: *Sov. Phys. Usp.*, **99**, 169 (1969).
[18] S. A. Akhmanov and G. L'yakhov: *Statistical effects in nonlinear optics*, paper presented at the *Conference on Lasers in Information Processes, Kiev, October, 1974.*
[19] G. Pasmanik: *Sov. Phys. JETP*, **66**, 490 (1974).
[20] J. Ducuing and N. Bloembergen: *Phys. Rev.*, **133**, A 1493 (1964).
[21] J. Ducuing and J. Armstrong: *C — r International Quantum Electronics Conference* (Paris and New York, N. Y., 1964).
[22] S. A. Akhmanov and R. Khokhlov: *Problems of Nonlinear Optics* (Moscow, 1964).
[23] Yu. D'yakov and N. Iskanderov: paper presented at the *XII All Union Conference on Nonlinear Optics* (Tashkent, 1974).
[24] V. Babenko, B. Zeldovich, V. Malyshev and A. Sychev: *Kvantovaya Electronika*, **14**, 19 (1973).

Optical Nonlinearities in Conjugated One-Dimensional Systems.

J. DUCUING
Laboratoire d'Optique Quantique, Ecole Polytechnique
Route de Saclay - 91120 Palaiseau, France

1. – Introduction.

The dominant contribution to the polarizability of an electronic system comes quite generally from the most weakly bound electrons which are more sensitive to the perturbation produced by the electromagnetic field. Thus the outer or valence electrons determine overwhelmingly the optical polarizability. These electrons are also the ones with the most extended orbitals. This connection between polarizability and orbital size is well known. It is illustrated by the expression for the polarizability of a hydrogenic system with Bohr radius a

$$\alpha = \frac{9}{2} a^3 . \tag{1}$$

The influence of orbital delocalization will be felt even more strongly on higher-order polarizability. For the same hydrogenic system one obtains [1] for the cubic hyperpolarizability γ

$$\gamma = \frac{3555}{16} \frac{a^7}{e^6} . \tag{2}$$

We show now that this strong enhancement of hyperpolarizabilities is a very general property. The tensorial polarizabilities are defined through the relation

$$\underline{p} = \underline{\underline{\alpha}}\,\underline{E} + \underline{\underline{\underline{\beta}}}\,\underline{E}\,\underline{E} + \underline{\underline{\underline{\underline{\gamma}}}}\,\underline{E}\,\underline{E}\,\underline{E} + \dots .$$

We restrict our considerations to centro-symmetric systems for which $\underline{\underline{\underline{\beta}}}$ vanishes. The longitudinal components of α and γ are given by

$$\alpha_{xx} = \frac{2e^2}{\hbar} \sum_{i \neq 0} \frac{x_{0i} x_{i0}}{\omega_{i0}} , \tag{3}$$

$$\gamma_{xxxx} = \frac{4e^4}{\hbar^3} \left(\sum_{i,j,k \neq 0} \frac{x_{0i} x_{ij} x_{jk} x_{k0}}{\omega_{i0} \omega_{j0} \omega_{k0}} - \sum_{i,j \neq 0} \frac{x_{0i} x_{i0} x_{0j} x_{j0}}{\omega_{i0}^2 \omega_{j0}} \right) . \tag{4}$$

As usual, in these sums 0 refers to the ground state, and i, j, k to intermediate states with excitation energy $\hbar\omega_{i0}$.

Following UNSOLD [2] we assume that the excitation energies can be replaced by an average energy $\hbar\Omega$. This in turn is obtained from the Thomas-Kuhn sum rule as

$$\Omega = \frac{N\hbar}{2m\langle x^2\rangle}, \tag{5}$$

where N is the number of electrons in the system. In the following brackets indicate the expectation value in the ground state. Here we have assumed $\langle x\rangle = 0$.

Then

$$\alpha \simeq \frac{4e^2 m}{\hbar^2}\frac{\langle x^2\rangle}{N}, \tag{6}$$

$$\gamma \approx \frac{2^5 e^4 m^3}{\hbar^6}\frac{(\langle x^4\rangle - 2\langle x^2\rangle^2)\langle x^2\rangle^3}{N^3}. \tag{7}$$

If L is a characteristic dimension of the system

$$\alpha \approx \frac{L^4}{N}, \tag{8}$$

$$\gamma \approx \frac{L^{10}}{N^3}. \tag{9}$$

A good example of strong delocalization is provided by the π electrons in organic conjugated systems. Whereas the σ-orbitals form covalent bonds, overlapping with just one neighbouring orbital, the π-orbitals overlap with two, one on each side. Whereas the σ electrons can be assigned in pairs to individual bonds which can, more or less, be treated as independent units, this is no longer true of the π electrons which are delocalized over all the conjugated carbon atoms. These delocalized electrons are responsible for the peculiar features of conjugated molecules [3]: high reactivity, large diamagnetic susceptibility and most of all their unusual optical properties such as enhancement of the polarizability and strong visible absorption.

Nonlinear optical properties of conjugated systems have been extensively studied [4-14]. However the majority of investigations have dealt with planar molecules in which the π electrons are spread over a two-dimensional network. In that case $L \propto N^{\frac{1}{2}}$ and the average energy $\hbar\Omega$ does not depend strongly on the size of the molecule, in contrast with the one-dimensional case (this behaviour is confirmed for instance by the study of the spectra of aromatic compounds). As a result one expects γ to vary moderately with molecular size. From (9) $\gamma \propto N^2$. In the case of chainlike, one-dimensional systems $L \propto N$

and (9) predicts $\gamma \propto N^7$. Although the results obtained by the Unsold approximation should be used cautiously, this dimensional argument suggests a strong enhancement of optical nonlinearities with size in conjugated one-dimensional systems, a fact first noted in ref. [12, 13] and supported by the early molecular-orbital calculations of SCHWEIG on two polyene compounds [4]. In the following we describe some theoretical and experimental investigations of this problem.

2. – Optical nonlinearities of the free-electron model.

The description of the π electrons of conjugated molecules as a free-electron gas delocalized over a branched network has been extensively used by a number of authors [15]. Thus KUHN [16] has met with considerable success in interpreting and predicting the main features in the spectra of organic dyes. Among these are some with chainlike structures, such as the symmetrical cyanine dyes, which can be represented by a one-dimensional model in which the π electrons behave as free electrons in a potential well. Figure 1 shows a comparison [17] between the theoretical predictions and the experimental results for six compounds of this family.

Let us consider such a model with $2N$ electrons in a one-dimensional box of length $2L$ ($0 \leqslant x \leqslant 2L$). Neglecting electron interaction, one has one-electron eigenstates with wave function and energy

$$\psi_q = \frac{1}{\sqrt{L}} \sin q \frac{\pi}{2} \frac{x}{L}, \tag{10}$$

$$w_q = \frac{\hbar^2 \pi^2}{8mL^2} q^2 . \tag{11}$$

The first N levels are each filled by two electrons with opposite spins.

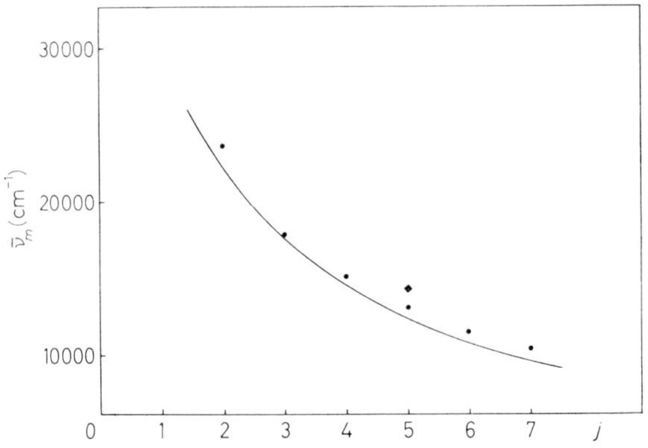

Fig. 1. – The wave number of maximum absorption plotted *vs.* the number j of double bonds in a cyanine series [17].

Our interest in this simple model stems from the fact that it lends itself to an exact calculation of α and γ [13]. Assume that an electric field is applied to the system. In the electric-dipole approximation it induces a perturbation represented by the Hamiltonian $H^{(1)} = -eEx$. If we are able to calculate the perturbed one-electron energies w_q to fourth order in the field, we can obtain the perturbed total energy W and determine α and γ through the expansion

$$W(E) = 2\sum_{q=1}^{N} w_q(E) = W(0) - \tfrac{1}{2}\alpha E^2 - \tfrac{1}{4}\gamma E^4 - \dots .$$

Analytical expressions for the perturbed one-electron energies can be obtained in a straightforward way. Call $H^{(0)}$ the unperturbed Hamiltonian, $\psi^{(n)}$ and $w^{(n)}$ the perturbed one-electron wave function and energy in n-th order. Then

$$(H^{(0)} - w^{(0)})\,\psi^{(n)} = (w^{(1)} - H^{(1)})\,\psi^{(n-1)} + \dots + w^{(n)}\,\psi^{(0)} .$$

If one chooses phases in such a way that $\langle\psi^{(n)}|\psi^{(0)}\rangle = \delta_{n0}$, then

$$w^{(2)} = \langle\psi^{(0)}|H^{(1)}|\psi^{(1)}\rangle ,$$

$$w^{(4)} = \langle\psi^{(2)}|H^{(1)}|\psi^{(1)}\rangle - w^{(2)}\langle\psi^{(1)}|\psi^{(1)}\rangle - w^{(1)}\langle\psi^{(2)}|\psi^{(1)}\rangle ;$$

$\psi^{(1)}$ and $\psi^{(2)}$ are obtained by solving the inhomogeneous equations

$$(H^{(0)} - w^{(0)})\,\psi^{(1)} = (w^{(1)} - H^{(1)})\,\psi^{(0)} ,$$

$$(H^{(0)} - w^{(0)})\,\psi^{(2)} = (w^{(1)} - H^{(1)})\,\psi^{(1)} + w^{(2)}\,\psi^{(0)} .$$

These equations are particularly simple in our case where

$$H^{(0)} = -\frac{\hbar^2}{2m}\frac{\mathrm{d}^2}{\mathrm{d}x^2} , \qquad H^{(1)} = -eEx .$$

This procedure yields analytical expressions for α and γ:

$$\alpha = \frac{4L^4}{a_0}\sum_{n=1}^{N}\left(\frac{-2}{3\pi^2 n^2} + \frac{10}{\pi^4 n^4}\right) , \tag{12}$$

$$\gamma = \frac{128L^{10}}{a_0^3 e^2}\sum_{n=1}^{N}\left(\frac{-2}{9\pi^6 n^6} + \frac{140}{3\pi^8 n^8} - \frac{440}{\pi^{10} n^{10}}\right) . \tag{13}$$

When L is kept constant and N increases to infinity α and γ tend toward zero as shown by the following relations:

$$\lim_{N\to\infty} \alpha = \frac{4L^4}{a_0}\left(\frac{-2}{3}B_1 + \frac{10}{3}B_2\right) = 0 ,$$

$$\lim_{N\to\infty} \gamma = \frac{128L^{10}}{a_0^3 e^2}\left(\frac{-4}{9\times 45}B_3 + \frac{140}{3\times 315}B_4 - \frac{2^9\times 440}{10!}B_5\right) = 0 ,$$

where the B_m are the Bernoulli numbers. This is a very natural result as the energy of the first excited state increases without limit (filled Fermi sea). An asymptotic expansion gives for large N

$$\alpha = \frac{8L^4}{3a_0\pi^2 N}, \tag{14}$$

$$\gamma = \frac{256L^{10}}{45a_0^3 e^2 \pi^6 N^5}. \tag{15}$$

For systems in a homologous series $L \simeq Nd$, where d is the spacing between conjugated atoms. Thus $\alpha \propto L^3$ and $\gamma \propto L^5$. This last result differs slightly from that given by eq. (9). This quantitative discrepancy is due to the fact that the Unsold approximation gives an imperfect account of the interferences occurring in eq. (4) between terms corresponding to different intermediate states. The suggestion of a rapid increase of γ with chain length is, however, confirmed.

As mentioned earlier, the spectrum of the cyanine dyes is well predicted by such a model. This is considered to be due to a resonance [16] occurring between two symmetrical structures of the cyanine ion (fig. 2). As a result of this res-

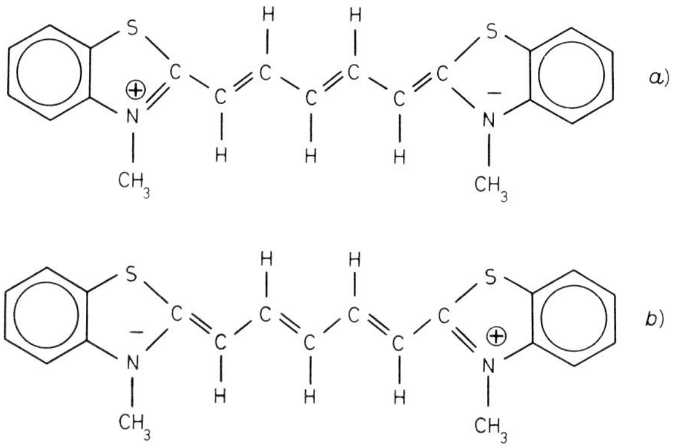

Fig. 2. – The two resonant forms of a cyanine ion.

onance there is no bond length alternation and the potential seen by the π electrons shows little modulation. This situation is however rather exceptional and in most conjugated compounds bond alternation will appear, which leads to a variation of the π electron potential along the chain. To take this into account we can modify our model by constraining the electrons to move within the box in a sinusoidal potential of amplitude V_0 and of period equal to twice the average distance between carbon atoms d

$$V(x) = V_0 \cos \pi \frac{x}{d}. \tag{16}$$

This is the method used by KUHN [16] to describe the spectrum of polyenes. With $V_0 = 2.4$ eV excellent agreement is obtained with experiment (fig. 3). In strong contrast with the free-electron case (cyanine dyes fig. 1) the wavelength of the main absorption peak tends toward a finite limit when the chain length increases. A Brillouin gap occurs [18] due to the periodic potential: the lowest excitation energy of the system tends to V_0. With this periodic

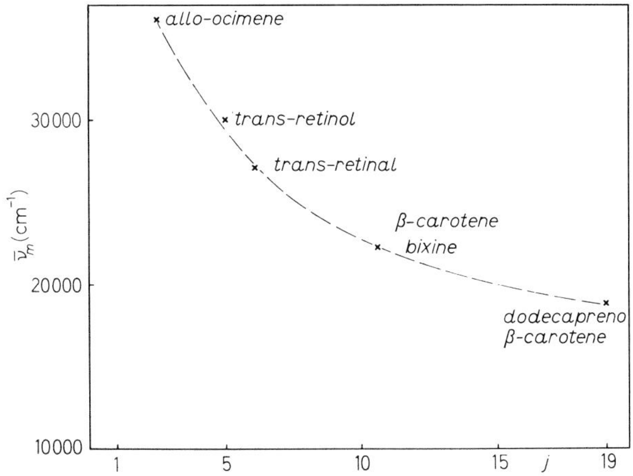

Fig. 3. – The wave number of maximum absorption plotted *vs.* the number j of double bonds for polyenes.

potential the analytical solution of our problem is rather involved. One can resort to a numerical procedure [13] in which the complete Hamiltonian is diagonalized in a finite set of unperturbed ψ_q. The net induced electric-dipole moment μ is then obtained, from which α and γ are calculated using the relation $\mu = \alpha E + \gamma E^3$.

From the values of γ obtained from these simple models we can calculate the hyperpolarizabilities for the cyanine dyes and the polyenes. One considers that the conjugated skeleton forms a zig-zag chain with bond angles of 120°. One then assumes that an electric field E along the straight line joining the centres of the bonds is represented in our model by a constant field $(\sqrt{3}/2)\,E$. This neglect of the components normal to the bonds is in agreement with our treatment of the π electron gas as purely one-dimensional [15]. With this assumption the longitudinal component $\gamma_{xxxx} = (9/16)\gamma$. The numerical results obtained for polyenes and those deduced from eq. (13) for a homologous series of cyanine dyes (a member of which is shown in fig. 2) are represented in fig. 4.

It is seen that for long chains γ increases less rapidly than for cyanines. This is consistent with the idea that the periodic potential reduces the electron delocalization, hence the polarizability. This reduction should become very strong when V_0 is not negligible compared to the Fermi energy of the homologous free-electron system $w_F = (\hbar^2\pi^2/8m)(N^2/L^2) \simeq \hbar^2\pi^2/8m\,d^2$. In our case with $d \simeq 1.4$ Å $w_F \simeq 5$ eV. For short chains the situation is not as simple. Indeed for $N < 10$, γ is smaller for the free-electron case. Following RUSTAGI [9] one can argue that this is due to a competition between anharmonicity and delocalization. In fact for the free-electron gas, near the Fermi level, the matrix elements and energy levels approximate those of a harmonic oscillator,

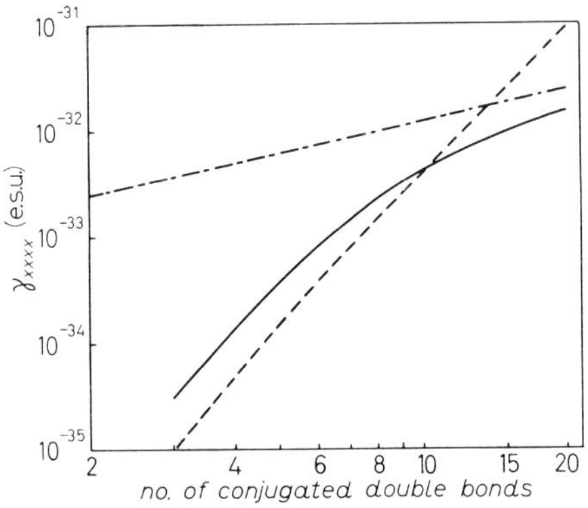

Fig. 4. – Nonlinear polarizability γ_{xxxx} for symmetric cyanines of the homologous series shown in fig. 2 (— — —) and polyenes (———) [13]. The asymptotic variation of polyenes has also been indicated (see sect. **4**).

i.e. of a perfectly linear system for which the terms in eq. (4) exactly cancel. In this range we find the states which dominate the summations in eq. (4) and this results in a substantial reduction of γ. In the polyenes, on the contrary, the periodic potential strongly perturbs the matrix elements and the distribution of energy levels in the same range making the system more anharmonic, and thus reducing the cancellations. This discussion strongly suggests that for a given length there is an optimum value of the delocalization (or V_0) for which γ is a maximum. For long chains this value of V_0 is intermediate between that of polyenes and cyanines, and decreases, tending toward the free-electron case as $N \to \infty$. At this optimum delocalization γ should increase faster than L^5 (*).

Numerical calculations give no indication about the asymptotic behaviour of α and γ for polyenes. To get some insight into this problem one can seek an approximate expression for the matrix elements of x appearing in eq. (3) and (4). The periodic potential V (eq. (16)) has matrix elements between states q and q' given by

$$\langle q|V|q'\rangle = \frac{V_0}{2}\{\delta(q'-q+2N) + \delta(q-q'+2N) - \delta(q+q'-2N)\} .$$

(*) This is confirmed by recent tight-binding calculations [33].

If we restrict our basis set to the first $2N$ states (N of which are occupied) we see that V couples state q only to state $2N-q$. In this approximation we obtain for the one-electron energies

$$w_q = w_q^0 - \frac{V_0}{2} f(r) , \tag{17}$$

$$w_{2N-q} = w_{2N-q}^0 + \frac{V_0}{2} f(r) , \tag{18}$$

where

$$r = N - q \geqslant 0 ,$$

$$f(r) = \left(\left(\frac{r}{Na}\right)^2 + 1\right)^{\frac{1}{2}} - \frac{r}{Na} ,$$

and the parameter

$$a = V_0 \frac{2md^2}{\hbar^2 \pi^2} = \frac{V_0}{4w_F}$$

expresses the magnitude of the periodic potential in terms of the Fermi energy $w_F = \hbar^2\pi^2/8md^2$ of the free-electron gas. Expressions (17) and (18) show that for $N \to \infty$ filled and empty levels are separated by a finite gap $\simeq V_0$. The corresponding wave functions are

$$\psi_q = \frac{\psi_q^0 + f(r)\,\psi_{2N-q}^0}{\sqrt{1+f^2(r)}} , \tag{19}$$

$$\psi_{2N-q} = \frac{\psi_q^0 - f(r)\,\psi_{2N-q}^0}{\sqrt{1+f^2(r)}} . \tag{20}$$

Using the matrix elements between unperturbed states

$$\langle q^0|x|q'^0\rangle = \frac{-16L}{\pi^2} \frac{qq'}{(q^2-q'^2)^2} ,$$

one can then calculate the matrix elements of x between perturbed states, substitute in eq. (3) and (4) and study the limiting behaviour for large N. This procedure applied to α_{xx} gives the following results.

For $Na < 1$, the dominant terms reduce to

$$\alpha \simeq \frac{2^8}{\pi^6} \frac{L^4}{a_0} \frac{1}{N} (1 - 2Na) . \tag{21}$$

In the limit $a \to 0$ one finds (within 1%) the expression (14) obtained for the free-electron case.

For $Na > 1$

$$\alpha \simeq \frac{8}{3\pi^2} \frac{e^2 d^2}{a V_0} N . \tag{22}$$

The situation is rather different from that in the free-electron case. The polarizability increases linearly with length. For a macroscopic medium consisting of such infinite chains this result leads to a well-defined susceptibility. This does not occur for the free-electron case $a = 0$ where it is well known that the susceptibility exhibits a singularity in the low-frequency limit. One can interpret eq. (22) by saying that in the presence of a periodic potential the conjugated chain behaves as a succession of N/m unconjugated free-electron units of length md. Then one finds $m = (\pi/\sqrt{6})(1/a)$. In this picture the electrons are delocalized over a distance $D \sim ad = (4w_F/V_0)\, d$ and the polarizability is enhanced by a factor $\sim a^{-2}$ from its value in the bound-electron case.

Note that the condition $Na > 1$ can be given a very simple interpretation. For large N and near the Fermi energy, the level spacing is $\Delta E \simeq 2w_F/N$. For $V_0 \ll \Delta E$ the perturbing potential introduces a weak perturbation and the behaviour is essentially that of a free-electron gas. When $V_0 > \Delta E$, the situation is reversed and the electrons tend to become localized around the minima of the periodic potential.

3. – Experimental studies of molecules with conjugated chains.

The study of third-harmonic generation yields a measurement of the hyperpolarizability. This method has been used by HERMANN [14, 17] to obtain an experimental confirmation of the previous considerations on the optical nonlinearities of conjugated chains. Third-harmonic generation by an infra-red laser was studied in the liquid phase. The source, obtained through stimulated Raman scattering of Nd^{3+} glass in H_2 gas, gave a linearly polarized 6 MW, $3 \cdot 10^{-8}$ s pulse at $\lambda = 1.89\ \mu m$ ($5290\ cm^{-1}$). The measurements yielded the value of the longitudinal component $\chi^{(3)}_{xxxx}$ of the third-order susceptibility. A description of the experimental arrangement and procedure can be found in ref. [12, 17]. The orientation-averaged polarizability $\langle\gamma\rangle$ is related to the susceptibility through

$$\chi^{(3)}_{xxxx} = \mathcal{N} \mathcal{L}^4 \langle\gamma\rangle ,$$

where $\mathcal{N}$ is the number of molecules per unit volume and $\mathcal{L}$ a local-field correction factor [19, 20], for which we use the Lorentz expression $\mathcal{L} = (n^2 + 2)/3$, n being the refractive index of the liquid.

The very peculiar behaviour of conjugated compounds is best appreciated when compared to that of the saturated ones. The values of $\langle\gamma\rangle$ for liquid compounds of the alkane and cycloalkane families are shown in fig. 5. A fairly linear variation of $\langle\gamma\rangle$ with the number of carbon atoms is observed. This is consistent with our picture of the σ electrons as being assigned to individual bonds which behave as independant polarizable units. As the polarizabilities of these bonds are not very anisotropic, one expects little dependence of $\langle\gamma\rangle$ on molecular geometry. This is observed experimentally: there is little difference between the alkane and cycloalkane for a given number of carbon atoms.

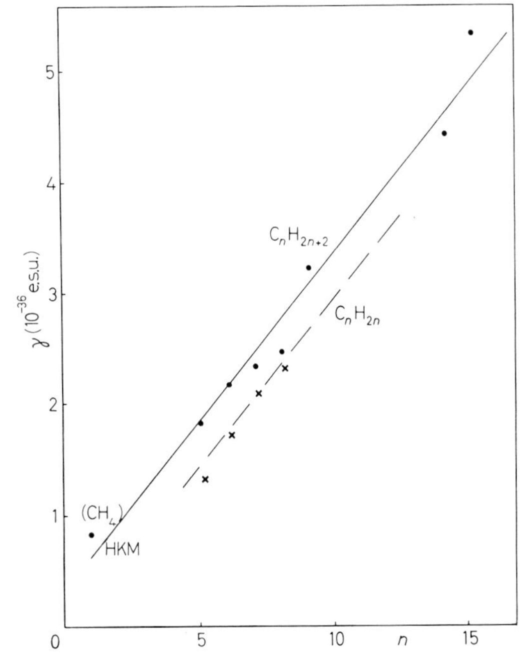

Fig. 5. – The orientation-averaged hyperpolarizability of alkanes (•) and cycloalkanes (+) *vs.* the number of carbon atoms [14, 17]. The value of $\langle\gamma\rangle$ for CH_4 obtained by HAUCHECORNE, KER HERVÉ and MAYER (*J. Physique*, **32**, 47 (1971)) is shown for comparison.

Linear conjugated chains show a strikingly different behaviour. Six different compounds with a number of double bonds j extending from 3 to 19 were studied [14, 17]. Alloocimene ($j = 3$) is a liquid at room temperature, retinol ($j = 5$) and retinal ($j = 6$) melt around 60 °C. These were studied in the liquid phase. The substances corresponding to larger chains which are solids at STP were dissolved in benzene or dimethylsulfoxide. The relative value $\langle\gamma\rangle/\langle\gamma\rangle_{solvent}$ was measured, the values of $\langle\gamma\rangle_{solvent}$ being deduced from a comparison with quartz [10, 14].

The measurements yield the value of the orientation-averaged hyperpolarizability $\langle\gamma\rangle$. In order to compare these results to the predictions of the preceding section, we must first subtract the contribution of the σ electrons.

This is assumed to be given, to a good approximation, by the value of $\langle\gamma\rangle$ for the alkane with the same number of carbon atoms. Further, molecular-orbital calculations such as those of SCHWEIG [4] indicate that, even for $j=3$, γ_{xxxx} far dominates the other components of the γ-tensor. Thus one can take $\gamma_{xxxx}=5\langle\gamma_\pi\rangle$. These experimental values are compared to the theoretical predictions in fig. 6. It should be noticed that there is no adjustable parameter

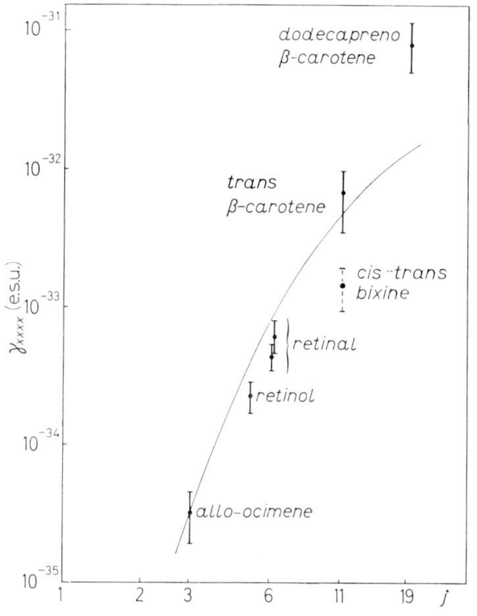

Fig. 6. – The longitudinal component γ_{xxxx} of polyenes *vs.* the number of double bonds [14, 17]. The curve represents the theoretical predictions of the nearly free-electron model [13] discussed in sect. **2**.

in the calculation of the previous sections, in the sense that the value of the only parameter of the model V_0 has been determined from the absorption spectra of the compounds (see fig. 3). In view of the extreme simplicity of the model one should not attach as much importance to the quantitative agreement as to the correct prediction of the variation of γ_{xxxx} with chain length. Two points can be noted in connection with these results. One is the influence of molecular geometry: bixine was obtained as a mixture of cis and trans stereo-isomers. In the cis form, departure from a linear geometry appreciably reduces the induced dipole moment and accounts at least partially for the smaller γ. The second is the influence of resonances. The main peak in the absorption spectrum of dodecapreno β-carotene ($j=19$) lies at 18100 cm^{-1}, only 2300 cm^{-1} above the third-harmonic energy. Measurements repeated with sources of lower frequency indicated a substantial drop in the value of γ for this compound (about 4 times at 2.47 μm). In contrast, β-carotene, for which the absorption peak lies at 22 000 cm^{-1}, showed no appreciable variation between 1.9 and 2.5 μm.

These experimental results show that the enhancement of γ due to electron delocalization can be extremely large: the hyperpolarizability of β-carotene is 3000 times that of the corresponding alkane. How are these large nonlinearities reflected on a macroscopic scale? Molten retinol and retinal as well as β-carotene glass do not exhibit an anomalously large refractive index. In these substances the contribution of the π electrons to α remains of the same magnitude as that of the σ electrons and the linear susceptibility $\chi^{(1)}$ is not very different for polyenes and alkanes. Thus superficial, empirical considerations based on the polarizability of the medium would predict a low $\chi^{(3)}$. This conclusion is however incorrect as the situation is very different for α and γ. In the latter case, as we have seen, the π electron contribution completely dominates the hyperpolarizability. It is sufficiently large to more than compensate for the decrease in density due to the size and geometry of the molecules. This results in large electronic $\chi^{(3)}$ for these substances. In table I the values

TABLE I. – *The longitudinal component* $\chi^{(3)}_{xxxx}$ *of the third-order susceptibility tensor and the ratio* $\chi^{(3)}_{xxxx}/(\chi^{(1)}_{xx})^4$ *for conjugated and nonconjugated compounds.* Linear optical data have been included for reference.

	Transparency range	n (infra-red)	$\chi^{(3)}_{xxxx}$ (e.s.u.)	$\chi^{(3)}_{xxxx}/(\chi^{(1)}_{xx})^4$ in 10^{-10} e.s.u.
LiF	$\lambda > 0.12$ μm	1.38	$7 \cdot 10^{-15}$	2.6
quartz	$\lambda > 0.12$ μm	1.55	$3 \cdot 10^{-14}$	1.9
molten retinol	$\lambda > 0.47$ μm	1.58	$5 \cdot 10^{-13}$	25
molten retinal	$\lambda > 0.53$ μm	1.61	$1.1 \cdot 10^{-12}$	42
β-carotene glass	$\lambda > 0.70$ μm	1.50	$1.6 \cdot 10^{-12}$	160
GaAs ([a])	$\lambda > 1$ μm	3.3	$4.8 \cdot 10^{-11}$	1.2
silicium ([a])	$\lambda > 1.2$ μm	3.4	$2.4 \cdot 10^{-11}$	0.5
germanium ([a])	$\lambda > 1.8$ μm	4	$4 \cdot 10^{-10}$	2

(*a*) J. J. WYNNE: *Phys. Rev.*, **178**, 1295 (1969).

deduced from third-harmonic generation of the 1.89 μm source are compared to that of other substances. It should be noted that they strongly deviate from the generalized Miller's rule proposed by WYNNE and BOYD [21] which predicts a universal value of the ratio $\chi^{(3)}_{xxxx}/(\chi^{(1)}_{xx})^4$. This is due, among other things, to the previously mentioned fact that different polarization mechanisms are responsible for the magnitudes of $\chi^{(1)}$ and $\chi^{(3)}$.

These large values of $\chi^{(3)}$ imply a strong optical Kerr effect. It has indeed been demonstrated experimentally [12] that the optical Kerr coefficient n_2

TABLE II. – *Optical Kerr effect in β-carotene glass and* CS_2.

	β-carotene glass	CS_2
nonlinear index (e.s.u.)	$8\cdot10^{-12}$	$1.3\cdot10^{-11}$
mechanism	electronic	molecular orientation
response time	$? < 10^{-14}$ s	$2\cdot10^{-12}$ s

is of the same magnitude in β-carotene glass and in CS_2 (see table II). An important difference exists however between the mechanisms responsible for the effect in these two substances. In CS_2 the variation of the refractive index results from the orientation of the molecules in the optical field. Naturally such an orientation does not take place in the glass where the variation of the refractive index can be attributed essentially to the distortion of the π electron distribution [12]. As this is a much faster mechanism than molecular reorientation, optical devices such as shutters [22] or modulators, using conjugated substances, will have a negligible time constant. In this respect molten retinal with a transparency range which extends into the green part of the spectrum might be of special interest.

Even more spectacular values of $\chi^{(3)}$ should be expected in ordered media like crystals. Several factors should contribute to enhance the nonlinearity in this case. The orientation of molecules along a preferred direction, a better packing which increases the density and an enhancement of the local-field factor due to the larger refractive index are probably the most important. On can also expect, in the crystal, a reduction of the partial cis isomerization which occurs in the glass. Indeed preliminary measurements performed in our laboratory by HERMANN on β-carotene crystals [23, 24] indicate a very large value of $\chi^{(3)}_{xxxx}\simeq 2\cdot10^{-10}$, *i.e.* of the same magnitude as in germanium.

4. – Optical nonlinearities in one-dimensional crystals.

We now extend the considerations of sect. **2** and consider the case of infinite conjugated chains. Such chains with extended electron delocalization are encountered in one-dimensional solids whose conducting properties have received much attention recently. Examples of these are the TCNQ salts and transition metal complexes [25] and most recently conjugated polymers such as the polydiacetylenes [26]. The simplest model for such a system is the one-dimensional Mathieu crystal [27]. Again electrons are assumed to move in a periodic pseudopotential

$$V(x) = V_0 \cos \pi \frac{x}{d},$$

which couples free-electron states with wave vectors k and $\pi/d - k$. Restricting ourselves, as in sect. **2**, to a two-band model [28], we find for these valence and conduction bands the energies

$$E_v(k) = E_0(k) - \frac{V_0}{2} f(k) ,$$

$$E_c(k) = E_0(k - \pi/d) + \frac{V_0}{2} f(k) ,$$

and wave functions in Bloch form

$$\psi_{c,v}(k; x) = \exp[ikx]\, u_{c,v}(k; x) ,$$

$$u_v(k; x) = \frac{1}{\sqrt{(1+f^2)2d}} \left(1 - f \exp\left[-i\frac{\pi x}{d}\right]\right) ,$$

$$u_c(k; x) = \frac{1}{\sqrt{(1+f^2)2d}} \left(f + \exp\left[-i\frac{\pi x}{d}\right]\right) ,$$

where $E_0(k) = (\hbar^2/2m)\, k^2$,

$$f(k) = \frac{a}{(1 - 2kd/\pi) + \sqrt{(1-2kd/\pi)^2 + a^2}} ,$$

and the delocalization parameter $a = V_0/(\hbar^2\pi^2/2md^2) = V_0/4w_{\text{F}}$, as defined in sect. **2**. The interband transition energy $E_{cv} = E_c(k) - E_v(k)$ has its maximum value $\simeq \hbar^2\pi^2/2md^2$ at $k = 0$ and decreases to the finite value V_0 at the edge of the Brillouin zone $k = \pi/2d$. At this edge the joint density of states per unit length

$$\varrho(E_{cv}) = \frac{1}{2\pi}\left(\frac{\mathrm{d}E_{cv}}{\mathrm{d}k}\right)^{-1} \propto \frac{1}{\sqrt{E_{cv}^2 - V_0^2}} \tag{23}$$

exhibits a singularity, characteristic of one-dimensional systems.

In the case of bands, the straightforward application of formulae (3) and (4) leads to difficulties connected with the application of perturbation theory to a continuum of states. One can circumvent this by using the approach of GENKIN and MEDNIS [29]. The following expressions are obtained for the susceptibilities per unit length $\varkappa^{(1)}$ and $\varkappa^{(3)}$:

$$\varkappa^{(1)} = \frac{2e^2}{\pi} \int\limits_{-\pi/2d}^{+\pi/2d} \mathrm{d}k \frac{\Omega_{vc}\Omega_{cv}}{E_{cv}} , \tag{24}$$

$$\varkappa^{(3)} = \frac{4e^4}{\pi} \int\limits_{-\pi/2d}^{+\pi/2d} \mathrm{d}k \left(\frac{1}{E_{cv}} \left|\frac{\mathrm{d}}{\mathrm{d}k}\left(\frac{\Omega_{cv}}{E_{cv}}\right)\right|^2 - \frac{\Omega_{cv}^4}{E_{cv}^3}\right) \tag{25}$$

where

$$\Omega_{rs} = \int\limits_0^{2d} u_r^* \frac{\partial}{\partial k} u_s \, dx \,,$$

and the result $\Omega_{cc} = \Omega_{vv} = 0$ has been taken into account.

For $a < 1$ this gives for the linear susceptibility

$$\varkappa^{(1)} = \frac{4}{3\pi^2} \frac{e^2 d}{a V_0} \,, \tag{26}$$

and for a chain segment of length $2Nd$

$$\alpha = \frac{8}{3\pi^2} \frac{e^2 d^2}{a V_0} N \,,$$

which is identical to expression (22) of sect. **2**.

The two terms under the integral sign in eq. (25) correspond respectively to the intra- and interband contributions to $\varkappa^{(3)}$. They are of opposite sign, as already noted by VAN VECHTEN *et al.* [30]. Again for $a < 1$ a straightforward integration gives

$$\varkappa^{(3)} = \frac{2^{14}}{45\pi^4} \frac{e^4 d^3}{w_F^3} \left(\frac{w_F}{V_0}\right)^6 , \tag{27}$$

or equivalently

$$\varkappa^{(3)} = \frac{2^5}{45} \pi^2 \left(\frac{a_0}{d}\right)^3 \frac{e^{10}}{V_0^6} \,. \tag{28}$$

The *positive* sign of $\varkappa^{(3)}$ is due to the stronger intraband contribution. This contrasts with the result obtained for a localized two-level system.

The salient feature in these results is the strong enhancement of $\varkappa^{(3)}$ with respect to a system of localized electrons. This enhancement goes as $(w_F/V_0)^6$ and can reach an extremely large value for systems having a small bond alternance. This dependence of $\varkappa^{(3)}$ on V_0^{-6} can be traced directly to the variation of the integrand in eq. (25), which is strongly concentrated in a range $\simeq a(\pi/2d)$ around the limits of the Brillouin zone. In this range the matrix element $\Omega_{cv} = \big(1/(1+f^2)\big)(df/dk)$ has a value $(d/\pi)(4w_F/V_0)$ which is enhanced by delocalization. VAN VECHTEN *et al.* [30] have studied the third-order susceptibility $\chi^{(3)}$ of a three-dimensional semiconductor. They use a three-part model band structure [31] and find that one of the contributions coming from the Penn « threshold » E_G [32] is proportional to E_G^{-6}. A similar result is obtained in a direct study [33] of the three-dimensional Penn's model. This is not surprising as Penn's model postulates expressions for the energy and wave functions which are identical to those of the present one-dimensional model. Two points should be noted, however. First, whereas in Penn's model E_G has the meaning of an average interband energy, V_0 is the lowest interband energy and coincides with the optical gap. In germanium $E_G \simeq 4.5$ eV whereas the optical gap E_0 lies around 0.6 eV. This is of practical importance. Second, the representation of the E_2 structure by a Penn threshold is only approximate and it is not obvious that the features which in Penn's model give rise to the E_G^{-6}-dependence are all reproduced in three-dimensional semiconductors.

Equation (27) implies a sixth-power dependence of $\varkappa^{(3)}$ on the delocalization length $L_d \simeq d/a$. This differs from the result obtained by considering the infinite system as a succession of independent free-electron units of length L_d, which is $\varkappa^{(3)} \propto L_d^4$. This is related to the comments made in sect. **2** about the balance between delocalization and anharmonicity. It should also be noted from eq. (28) that $\varkappa^{(3)} \propto d^{-3}$. The nonlinearity will be increased in chains with smaller spacing of atoms. This, all other factors being given, favours the carbon chains ($d \simeq 1.4$ Å) in comparison with the TCNQ salts or the transition metal complexes ($d \simeq 3.5$ Å). Finally eqs. (27) and (28) yield the limiting behaviour of finite chains when the number of double bonds j tends to infinity. For polyenes, taking into account the geometry of the chain, we obtain

$$\gamma_{xxxx}(\text{e.s.u.}) = 1.2 \cdot 10^{-33} j \,, \tag{29}$$

which gives the asymptote shown in fig. 4.

The solid-state polymerization of diacetylenes R—C≡C—C≡C—R leads to a highly conjugated structure exhibiting infinite linear chains [26*a*, 26*b*]. Upon thermal annealing or γ irradiation each monomer molecule joins with two neighbouring molecules (fig. 7). The electron redistribution leads to an infinite conjugated chain similar to that of polyenes. Polymer single crystals of large size and good optical quality can be obtained. The nonlinear optical properties of such a compound TCDU where $R = (CH_2)_6OCONHC_6H_5$ have been investigated in our laboratory [34]. Again third-harmonic generation from an infra-red laser was used to determine the third-order susceptibility $\chi^{(3)}$. The colourless monomer single crystals exhibit a normal nonlinearity, only slightly higher than predicted by Miller's rule with $\chi^{(3)}$-components $\simeq 2 \cdot 10^{-13}$ e.s.u. In strong contrast the yellow polymer crystals obtained upon γ-ray irradiation showed a strong highly anisotropic nonlinearity with the longitudinal component along the chain $\chi^{(3)}_{xxxx} = 3.7 \cdot 10^{-11}$ e.s.u. dominating all other components by more than two orders of magnitude. With a gap energy $V_0 = 2.15$ eV, a distance $d = 1.35$ Å and a cross-sectional area per chain $\sigma = 116$ Å^2, eq. (28) predicts

$$\chi^{(3)}_{xxxx} = \frac{\varkappa^{(3)}}{\sigma} = 1.4 \cdot 10^{-11} \text{ e.s.u.}$$

This gives a good order-of-magnitude agreement. Obviously one should not expect more in view of the simplicity of the model which neglects the supermodulation of bond lengths which takes place in polydiacetylenes. Whereas

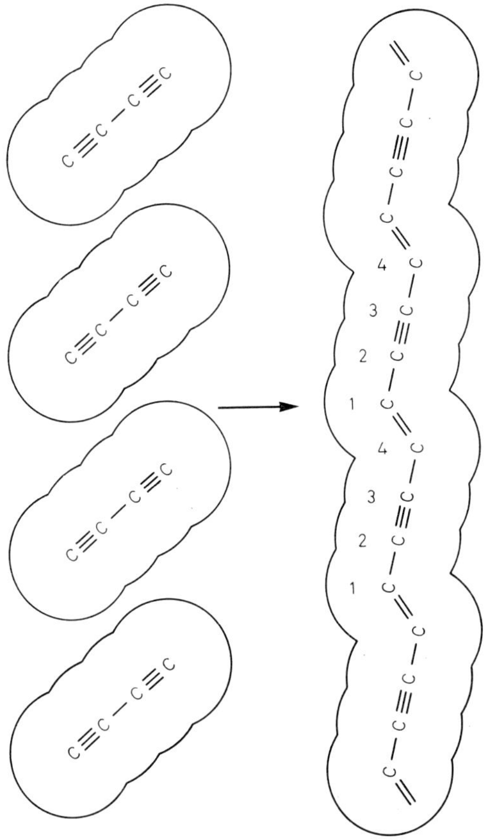

Fig. 7. – Solid-state polymerization of diacetylenes. Four monomer molecules and the corresponding segment of the all-transpolymer chain are shown (after R. H. BAUGHMAN and K. C. YEE).

in polyenes a simple alternance occurs, the situation is more complicated here. Owing to the presence of an additional π bond between atoms 2 and 3 the corresponding interatomic distance is shortened with respect to that between atoms 4 and 1 (fig. 7). Additional Fourier components should be introduced in the pseudopotential to take this into account. A treatment of this problem by the Hückel method can be found in ref. [33]. Also the local-field correction has been omitted: although difficult to estimate it should not be very large for highly delocalized electrons. Resonance effects could also somewhat enhance the experimental value.

The high value of $\chi^{(3)}$, comparable to those observed in semiconductors, obtained in this polydiacetylene is especially remarkable in view of the small relative number of π electrons: roughly 1 out of 40 valence electrons takes part in the conjugated chain. They produce a comparatively small infra-red anisotropy with a refractive index $n = 1.81$ for light polarized along the chains, compared to $n_{\perp} = 1.65$ for light polarized normal to the chains. However they completely dominate the nonlinear properties as could be expected from eqs. (26) and (27) which predict a much stronger dependence on delocalization for $\varkappa^{(3)}$ than for $\varkappa^{(1)}$. The small density of π electrons is due to the bulky side groups $R = (CH_2)_4OCONHC_6H_5$. Proper choice of these groups could reduce the chain cross-section, this with a lowering of the optical gap could lead to very large values of $\chi^{(3)}$. From a practical point of view the moderate values of the refractive index are of importance. It should be noted that for nonlinear interactions the important quantity is not $\chi^{(3)}$ but the ratio $\chi^{(3)}/n^2$. This factor of merit is practically the same in TCDU and Ge. However the transmission ranges are very different for the two compounds. Whereas Ge has an optical gap of 0.6 eV, TCDU presents a very steep absorption edge (characteristic of one-dimensional compounds) at 2.2 eV. A last favourable point can be mentioned. In the picosecond regime, the damage threshold measured at 1.06 μm is in the range of 50 GW/cm², about 20 times the corresponding value for Ge.

Note added in proofs.

Since these lecture notes have been written, other polymers of same family have been studied [34]. Poly-PTS exhibit even higher nonlinearities ($\chi^{(3)}_{xxxx} = 1.6 \cdot 10^{-10}$ e.s.u. and $\chi^{(3)}/n^2 = 4.5 \cdot 10^{-11}$).

5. – Prospects.

Further systematic studies of semiconducting one-dimensional compounds should uncover very large nonlinearities. The large variety of systems which are available or can be designed offers attractive prospects. The considerations developed here could be extended to the case when resonances take place. Large multiphoton cross-sections are expected. The sharp edges and peaks observed in the linear properties of one-dimensional compounds [37] should also appear in the dispersion of $\chi^{(3)}$. In order to underline the essential physical ideas we have limited ourselves to simple centro-symmetric systems. It would be interesting to study $\chi^{(2)}$ in one-dimensional noncentro-symmetric chains such as asymmetrically substituted carotenoides along the lines followed by LEVINE [35] and OUDAR and CHEMLA [36] for two-dimensional systems. This raises the problem of the influence of asymmetry on electron delocalization. One-dimensional systems should also exhibit a strong enhancement of the higher-order nonlinear susceptibilities discussed by AKHMANOV in this volume.

A high damage threshold combined with a large nonlinearity and a low refractive index may allow the study of unusual situations. Consider for instance a crystal with $\chi^{(3)} = 5 \cdot 10^{-10}$, a linear susceptibility $\chi^{(1)} = 0.2$ ($n = 1.87$) and a damage threshold of 50 GW/cm^2. At this maximum intensity level the ratio of the nonlinear to the linear polarization is 0.25. This value can be compared to the much smaller ones given by AKHMANOV for currently available compounds [38] and raises the question of the validity of the perturbation expansion of the polarization. Wave propagation is also expected to take up new aspects: for instance the interaction length for third-harmonic generation becomes of the order of the harmonic wavelength and can be shorter than the coherence and absorption lengths even in strongly absorbing media.

* * *

The author takes pleasure in acknowledging several fruitful discussions with Drs. R. H. BAUGHMAN, C. FLYTZANIS, J.-P. HERMANN and K. C. RUSTAGI.

REFERENCES

[1] G. L. SEWELL: *Proc. Camb. Philos. Soc.*, **45**, 678 (1949).
[2] A. UNSOLD: *Ann. der Phys.*, **43**, 388 (1927).
[3] L. SALEM: *The Molecular Orbital Theory of Conjugated Systems* (New York, N. Y., 1966).
[4] A. SCHWEIG: *Chem. Phys. Lett.*, **1**, 195 (1967).
[5] L. D. DERKACHEVA, A. I. KRYMOVA and N. P. SOPINA: *Pis'ma Žurn. Ėksp. Teor. Fiz.*, **11**, 319 (1970).
[6] J. JERPHAGNON: *IEEE Journ. Quantum Electron.*, QE-**7**, 42 (1971).
[7] J. R. GOTT: *J. Phys. B*, **4**, 116 (1971).
[8] M. BASS, D. BUA, R. R. MONCHAMP and R. MOZZI: *Appl. Phys. Lett.*, **15**, 393 (1969).
[9] P. D. SOUTHGATE and D. S. HALL: *Journ. Appl. Phys.*, **43**, 2765 (1972).
[10] J. G. BERGMAN, G. R. CRANE, B. F. LEVINE and C. G. BETHEA: *Appl. Phys. Lett.*, **20**, 21 (1972); B. F. LEVINE and C. G. BETHEA: *Appl. Phys. Lett.*, **24**, 445 (1974); *Journ. Chem. Phys.*, **60**, 3856 (1974).
[11] J. L. OUDAR and D. S. CHEMLA: *Opt. Comm.*, **13**, 164 (1975).
[12] J. P. HERMANN, D. RICARD and J. DUCUING: *Appl. Phys. Lett.*, **23**, 178 (1973).
[13] K. C. RUSTAGI and J. DUCUING: *Opt. Comm.*, **10**, 258 (1974).
[14] J. P. HERMANN and J. DUCUING: *Journ. Appl. Phys.*, **45**, 5100 (1974).
[15] For a complete bibliography see ref. [3] and J. R. PLATT *et al.*: *Free Electron Theory of Conjugated Molecules* (New York, N. Y., 1964).
[16] H. KUHN: *Fortsch. Chem. Org. Nat.*, **16**, 169 (1958); **17**, 404 (1959).
[17] J. P. HERMANN: Thèse, Université de Paris-Sud, Orsay (1974).
[18] J. R. PLATT: in *Handbuch der Physik*, edited by S. FLÜGGE, Vol. **37**/2 (Berlin, 1959).
[19] J. A. ARMSTRONG, N. BLOEMBERGEN, J. DUCUING and P. S. PERSHAN: *Phys. Rev.*, **127**, 1918 (1962).
[20] D. BEDEAUX and N. BLOEMBERGEN: *Physica*, **69**, 57 (1973).
[21] J. J. WYNNE and G. D. BOYD: *Appl. Phys. Lett.*, **12**, 191 (1968).

[22] M. A. DUGUAY and J. W. HANSEN: *Appl. Phys. Lett.*, **15**, 192 (1969).
[23] J.-P. HERMANN: private communication.
[24] The author is indebted to Prof. MADJID of Pennsylvania State University for kindly providing the β-carotene crystals.
[25] I. F. SHCHEGOLEV: *Phys. Stat. Sol.* (*a*), **12**, 9 (1972).
[26] a) G. WEGNER: *Zeits. Naturforsch.*, **24** *b*, 924 (1969); *Makromol. Chem.*, **134**, 219 (1970); *b*) R. H. BAUGHMAN: *Journ. Appl. Phys.*, **43**, 4362 (1972).
[27] J. C. SLATER: *Phys. Rev.*, **81**, 385 (1951).
[28] L. PINCHERLE: *Electronic Energy Bonds in Solids*, edited by MACDONALD (London, 1971).
[29] V. N. GENKIN and P. M. MEDNIS: *Sov. Phys. JETP*, **27**, 609 (1968).
[30] J. A. VAN VECHTEN, M. CARDONA, D. E. ASPNES and R. M. MARTIN: *Proceedings of the Tenth International Conference on Semiconductors* (Cambridge, Mass., 1970), p. 82.
[31] M. CARDONA, K. L. SHAKLEE and F. H. POLLAK: *Phys. Rev.*, **154**, 696 (1967).
[32] D. R. PENN: *Phys. Rev.*, **128**, 2093 (1962).
[33] D. AGRAWAL and C. FLYTZANIS: to be published.
[34] C. SAUTERET, J. P. HERMANN, R. FREY, F. PRADÈRE, J. DUCUING, R. H. BAUGHMAN and R. R. CHANCE: *Phys. Rev. Lett.*, **36**, 956 (1976).
[35] B. F. LEVINE: *Journ. Chem. Phys.*, **63**, 115 (1975); *Chem. Phys. Lett.*, to be published.
[36] J. L. OUDAR and H. LE PERSON: *Opt. Comm.*, **15**, 258 (1975).
[37] D. BLOOR, D. J. ANDO, F. H. PRESTON and G. C. STEVENS: *Chem. Phys. Lett.*, **24**, 407 (1974).
[38] S. AKHMANOV: this volume, p. 239.

Nonlinear Optics with Polaritons.

R. LOUDON

Physics Department, Essex University - Colchester CO4 3SQ, England

1. – Introduction.

Polaritons are coupled modes of the electromagnetic field and a crystal excitation [1]. They occur for all crystal excitations which have transverse electric-dipole coupling to the electromagnetic field. Figure 1 shows a plot in dashed lines of the frequency ω as a function of the wave vector Q for an electromagnetic wave and a crystal excitation. For the electromagnetic wave

$$\omega = cQ/\varkappa_\infty^{\frac{1}{2}}\,, \tag{1}$$

where c is the velocity of light and $\varkappa_\infty$ is the constant relative permittivity of the crystal in the absence of the excitation, which has itself been drawn with constant frequency ω_T.

Interaction between the electromagnetic field and the excitation dipole moment results in new dispersion curves shown by the continuous lines in fig. 1. The crystal excitation considered is usually either an exciton, where typical co-ordinates of the cross-over of the dashed lines are $\omega_T \approx 6\cdot 10^{14}$ Hz and $Q = \omega_T \varkappa_\infty^{\frac{1}{2}}/c \approx 2\cdot 10^6\ \text{m}^{-1}$, or a phonon, where typically $\omega_T \approx 1.5\cdot 10^{13}$ Hz and $Q \approx 5\cdot 10^4\ \text{m}^{-1}$. The wave vectors in both cases are very small compared to zone boundary values ($\approx 10^{10}\ \text{m}^{-1}$), and any dependence of the bare exciton or phonon properties on the wave vector can generally be ignored.

The polaritons represented by the continuous lines in fig. 1 can be called exciton-polaritons or phonon-polaritons depending on the parent excitation. The dispersion curve has two branches separated by a gap extending from the bare excitation frequency ω_T to a higher frequency ω_L which coincides with the frequency of the longitudinal crystal excitation. This excitation does not mix with the transverse photons but its frequency is raised from the value ω_T it would have for an excitation of zero dipole moment.

Ordinary linear optical processes are fully described by the frequency-dependent linear susceptibility $\chi^{(1)}$ or the relative permittivity

$$\varkappa = 1 + \chi^{(1)}\,. \tag{2}$$

These functions show resonant effects for frequencies ω close to the frequency ω_T of an electric-dipole excitation. They do not explicitly depend on the wave vector, and there is generally no need to consider the coupled-mode properties

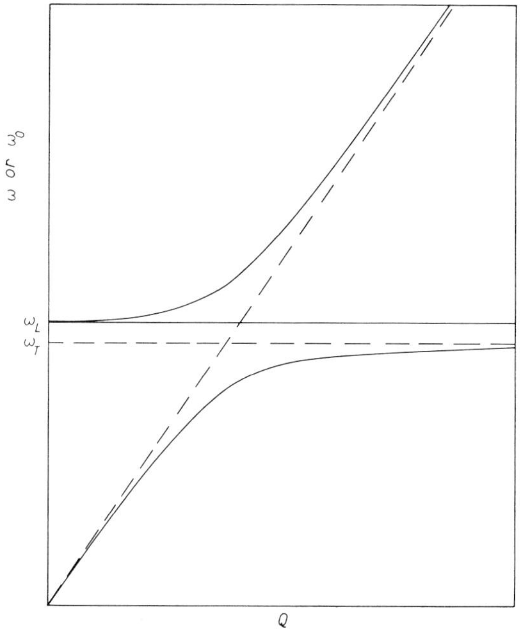

Fig. 1. – Dashed lines: dispersion relations of the uncoupled photon and crystal excitation. Continuous lines: the two branches of the polariton dispersion relation and the constant-frequency branch of the longitudinal excitation.

of the excitations in the vicinity of ω_T. Indeed, it can be misleading to introduce the idea of polaritons in this connection since $\varkappa$ and $\chi^{(1)}$ contain no direct reference to the dispersion curves of fig. 1.

Nonlinear optical processes are described by higher-order components of the susceptibility, $\chi^{(2)}$, $\chi^{(3)}$, etc., which also show resonant effects associated with the electric-dipole frequencies ω_T of the crystal. However, in contrast to *linear* processes, the resonances in *nonlinear* processes depend upon both the wave vector Q and the frequency ω, and they are centred on the polariton dispersion curve of fig. 1. Nonlinear optical processes thus provide a means of studying the properties of polaritons, and the concept of polaritons makes a useful contribution to the interpretation of experiments.

The importance of polaritons in nonlinear processes stems from the form of the response of a crystal to a driving polarization, and sect. **2** is devoted to a study of the properties of polaritons by means of linear-response theory.

Section **3** is concerned with a detailed study of a particular nonlinear process, that of two-photon absorption by exciton-polaritons. Two-photon absorption by an electric-dipole excitation can occur only in crystals which lack a centre of inversion symmetry—noncentrosymmetric crystals—since parity requirements otherwise restrict two-photon absorption to even excitations and electric-dipole moments to odd excitations. The process is a combination of two effects: 1) sum frequency generation of the photon part of the polariton at a rate determined by $\chi^{(2)}$, and 2) two-photon absorption by the exciton part of the polariton at a rate determined by $\chi^{(3)}$.

Section **4** is concerned with nonlinear processes involving phonon-polaritons. The main effect is that of stimulated Raman scattering, a combination of difference frequency generation of the photon part of the polariton and stimulated Raman scattering by the phonon part. The theory is very similar to that of the two-photon absorption process. The remainder of the section is concerned with a two-stage process which has been studied in several experiments.

2. – Polaritons.

2·1. *Basic theory.* – Consider the dipole resonance associated with an exciton or phonon in a cubic insulating crystal of isotropic optical properties. The crystal excitation is characterized by a displacement amplitude $\boldsymbol{W}$ which satisfies the equation of motion

$$\ddot{\boldsymbol{W}} + \Gamma\dot{\boldsymbol{W}} + \omega_T^2 \boldsymbol{W} = Z\boldsymbol{E}, \tag{3}$$

where ω_T is the resonance frequency, Γ is the damping, and Z is an effective charge (dimensions = charge/mass$^{\frac{1}{2}}$). For an optical vibration in a diatomic lattice, $\boldsymbol{W}$ is the relative displacement of the two types of atom multiplied by the square root of their reduced mass; for an exciton it is a similar quantity defined with respect to the electron and hole.

The field $\boldsymbol{E}$ is that which necessarily accompanies the excitation; conversely, an externally applied field excites an excitation amplitude $\boldsymbol{W}$ given by (3). If ω is the frequency of the motion,

$$(\omega_T^2 - \omega^2 - i\omega\Gamma)\,\boldsymbol{W} = Z\boldsymbol{E}. \tag{4}$$

The dielectric polarization is

$$\boldsymbol{P} = Z\boldsymbol{W}/\Omega + \varepsilon_0(\varkappa_\infty - 1)\,\boldsymbol{E}, \tag{5}$$

where Ω is the unit cell volume. The first term is the contribution of the resonance considered and the second term is the polarization in the absence of the dipole excitation.

The relative permittivity $\varkappa$ is defined by

$$\varkappa \varepsilon_0 \boldsymbol{E} = \varepsilon_0 \boldsymbol{E} + \boldsymbol{P}, \tag{6}$$

and use of (4) and (5) leads to

$$\varkappa = \varkappa_\infty + \frac{\omega_P^2}{\omega_T^2 - \omega^2 - i\omega\Gamma}, \tag{7}$$

where

$$\omega_P^2 = Z^2/\varepsilon_0 \Omega \tag{8}$$

is the square of the plasma frequency associated with the excitation. The relative permittivity at zero frequency is thus

$$\varkappa_0 = \varkappa_\infty + \omega_P^2/\omega_T^2 . \tag{9}$$

The absorption of transverse electromagnetic waves in linear optics is entirely determined by Im $\varkappa$ and the absorption maximum occurs at the frequency ω_T. The frequency ω of the single electromagnetic wave is the sole variable in an experiment on an isotropic medium.

The electric field $\boldsymbol{E}$ is also related to the polarization $\boldsymbol{P}$ by Maxwell's equations, and the general form is

$$\nabla \times \nabla \times \boldsymbol{E} - (\omega/c)^2 \boldsymbol{E} = (\omega^2/\varepsilon_0 c^2)\, \boldsymbol{P}, \tag{10}$$

or equivalently

$$(\omega/c)^2 \boldsymbol{P} = -\varepsilon_0 \{\nabla^2 + (\omega/c)^2\} \boldsymbol{E} + \varepsilon_0 \nabla(\nabla \cdot \boldsymbol{E}) . \tag{11}$$

The wave equation (10) can also be cast into the forms

$$\nabla \times \nabla \times \boldsymbol{E} - \varkappa_\infty (\omega/c)^2 \boldsymbol{E} = (\omega^2/\varepsilon_0 c^2)(Z\boldsymbol{W}/\Omega), \tag{12}$$

$$\nabla \times \nabla \times \boldsymbol{E} - \varkappa (\omega/c)^2 \boldsymbol{E} = 0 \tag{13}$$

with the help of (5) and (6). It follows from (13) that the wave vector q of free waves in the medium must satisfy

$$q^2 = \varkappa \omega^2/c^2 \tag{14}$$

for transverse polarization.

2'2. *Linear response.* – Consider the effect on the crystal of an applied polarization

$$\boldsymbol{P}^{\text{ext}} \exp[-i\omega t + i\boldsymbol{Q}\cdot\boldsymbol{r}], \tag{15}$$

where the frequency and wave vector are independent real quantities. In all cases, real polarizations and electric fields are obtained from their complex representations by addition of the complex conjugates.

The applied polarization induces an electric field $\overline{\boldsymbol{E}}$ in the crystal and an associated induced polarization $\overline{\boldsymbol{P}}$ related to $\overline{\boldsymbol{E}}$ by (6). Thus (11) becomes

$$(\omega/c)^2(\overline{\boldsymbol{P}} + \boldsymbol{P}^{\text{ext}}) = \varepsilon_0\{Q^2 - (\omega/c)^2\}\overline{\boldsymbol{E}} - \varepsilon_0\boldsymbol{Q}(\boldsymbol{Q}\cdot\overline{\boldsymbol{E}}). \tag{16}$$

Take a Cartesian component and use (6) to remove $\overline{\boldsymbol{P}}$:

$$(\omega/c)^2 P_j^{\text{ext}} = \varepsilon_0\{Q^2 - \varkappa(\omega/c)^2\}\overline{E}_j - \varepsilon_0 Q_j(\boldsymbol{Q}\cdot\overline{\boldsymbol{E}}). \tag{17}$$

This is a system of equations for the three components of $\overline{\boldsymbol{E}}$, which can be solved without difficulty. It is convenient to express the solutions in terms of the linear response functions defined as

$$S_{ij}(\boldsymbol{Q}\omega) = \frac{\overline{E}_i}{VP_j^{\text{ext}}} = -\frac{1}{\varepsilon_0 V}\frac{Q_iQ_j - \varkappa(\omega/c)^2\delta_{ij}}{\varkappa\{Q^2 - \varkappa(\omega/c)^2\}}, \tag{18}$$

where V is the sample volume.

The response functions defined in this way determine the power spectra of the electric-field fluctuations in the crystal. The connection is made via the fluctuation-dissipation theorem [2] which we here state without proof in the form

$$\langle E_i E_j^* \rangle_\omega = (\hbar/\pi)\{n(\omega) + \tfrac{1}{2}\} \operatorname{Im} S_{ij}(\boldsymbol{Q}\omega), \tag{19}$$

where $n(\omega)$ is the Bose-Einstein thermal factor

$$n(\omega) = \{\exp[\hbar\omega/k_B T] - 1\}^{-1}. \tag{20}$$

The quantity on the left of (19), the power spectrum, is the contribution of frequency ω and wave vector $\boldsymbol{Q}$ to the time-dependent correlation function of the i and j field components

$$\langle E_i(0) E_j^*(t) \rangle = \int d\omega \exp[i\omega t] \langle E_i E_j^* \rangle_\omega, \tag{21}$$

where the angle brackets denote an ensemble average over the probability distribution for the field components.

The denominator of the linear response function (18) can be factorized by introducing three orthogonal unit vectors $\boldsymbol{\lambda}$, $\boldsymbol{\mu}$ and $\boldsymbol{\nu}$, of which $\boldsymbol{\lambda}$ is parallel to $\boldsymbol{Q}$. Then if we use the property

$$\lambda_i \lambda_j + \mu_i \mu_j + \nu_i \nu_j = \delta_{ij}, \tag{22}$$

(18) can be written

$$S_{ij}(\boldsymbol{Q}\omega) = \frac{1}{\varepsilon_0 V}\left\{-\frac{\lambda_i\lambda_j}{\varkappa} + \frac{\mu_i\mu_j + \nu_i\nu_j}{(cQ/\omega)^2 - \varkappa}\right\}. \tag{23}$$

This result is not restricted to the case where $\varkappa$ has a single resonance as in (7), but applies generally for cubic crystals. It is shown in later sections that the strengths of nonlinear effects involving polaritons are proportional to the imaginary parts of the expressions which occur on the right of (23).

Consider the form of the response function in the case of a single resonance frequency ω_T where (7) is appropriate. The first term in (23) contributes for field components polarized parallel to $\boldsymbol{Q}$ and it is thus the longitudinal term. With the use of (7) we obtain

$$-\operatorname{Im}\frac{1}{\varkappa} = -\frac{1}{\varkappa_\infty}\operatorname{Im}\frac{\omega_T^2 - \omega^2 - i\omega\Gamma}{\omega_L^2 - \omega^2 - i\omega\Gamma}, \tag{24}$$

where the longitudinal frequency ω_L is given by

$$\omega_L^2 = \omega_T^2 + \omega_P^2/\varkappa_\infty = \varkappa_0\,\omega_T^2/\varkappa_\infty\,. \tag{25}$$

Provided that the separation between ω_L and ω_T is much larger than Γ, the longitudinal part of the response function is approximately

$$-\operatorname{Im}\frac{1}{\varkappa} \approx \frac{\omega\Gamma\omega_P^2/\varkappa_\infty^2}{(\omega_L^2 - \omega^2)^2 + \omega^2\Gamma^2} \approx \frac{\Gamma\omega_P^2/4\omega_L\varkappa_\infty^2}{(\omega_L - \omega)^2 + (\frac{1}{2}\Gamma)^2}\,. \tag{26}$$

The longitudinal contribution thus has a peak centred at frequency ω_L and with a frequency line width equal to Γ. It does not depend on the wave vector Q.

The second term in (23) contributes for field components polarized perpendicular to $\boldsymbol{Q}$ and it is thus the transverse term. Its magnitude depends on both Q and ω, and its form is illustrated in fig. 2. The function was first analysed by Henry and Garrett [3] and later by Barker and Loudon [4]. For small Γ it can be approximated by

$$\operatorname{Im}\frac{1}{(cQ/\omega)^2 - \varkappa} \approx \frac{\omega_0(\omega_T^2 - \omega_0^2)^2}{4\{\varkappa_\infty(\omega_T^2 - \omega_0^2)^2 + \omega_T^2\omega_P^2\}}\,\frac{\Gamma(\omega_0)}{(\omega_0 - \omega)^2 + [\frac{1}{2}\Gamma(\omega_0)]^2}, \tag{27}$$

where ω_0 is defined in terms of the wave vector Q by

(28) $$\left(\frac{cQ}{\omega_0}\right)^2 = \varkappa_\infty + \frac{\omega_P^2}{\omega_T^2 - \omega_0^2},$$

and the line width parameter in the second factor is

(29) $$\Gamma(\omega_0) = \frac{\omega_0^2 \omega_P^2 \Gamma}{\varkappa_\infty(\omega_T^2 - \omega_0^2)^2 + \omega_T^2 \omega_P^2}.$$

The transverse contribution thus has peaks at wave vectors Q and frequencies ω_0 which are related by (28), and this is taken to define the dispersion relation

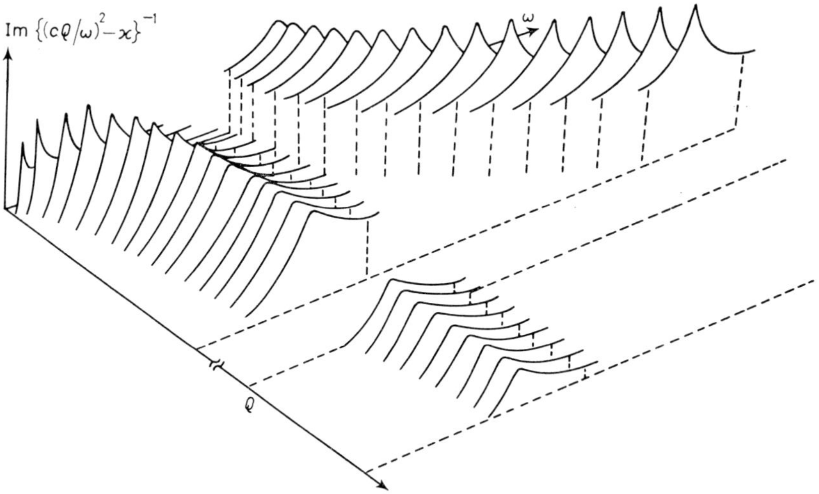

Fig. 2. – Imaginary part of the transverse polariton response as a function of Q and ω. Note that the vertical scale is logarithmic (after BARKER and LOUDON [4]).

of the transverse polaritons. Figure 1 shows the dispersion relation defined in this way, together with the longitudinal peak frequency ω_L from (26). Note that the function on the right of (28) is the same as the relative permittivity $\varkappa$ given by (7) but with the damping Γ removed; the damping plays no role in determining the peaks of the transverse response function and (28) is sometimes misleadingly called the « undamped » polariton dispersion relation. It is not to be confused with the dispersion relation (14) for propagation of free waves.

The line width of the transverse response at constant Q is given by (29). Note that the maximum line widths are

(30) $$\Gamma(\omega_T) = \Gamma(\omega_L) = \Gamma,$$

and the line width decreases as ω_0 moves up the upper branch away from ω_L or down the lower branch away from ω_T. However, the basic damping parameter Γ varies with frequency in real crystals, and the polariton line width can vary with frequency in a complicated way (see, for example, [5]).

A similar calculation of the linear response can be carried out for crystal structures which are not optically isotropic. The response functions for a biaxial crystal have the forms [4]

$$(31)\qquad S_{xx}(\boldsymbol{Q}\omega) = -\{Q^2Q_x^2 - [\varkappa_y(Q_x^2+Q_y^2)+\varkappa_z(Q_z^2+Q_x^2)](\omega/c)^2+\varkappa_y\varkappa_z(\omega/c)^4\}/\varepsilon_0 VD\,,$$

$$(32)\qquad S_{xy}(\boldsymbol{Q}\omega) = S_{yx}(\boldsymbol{Q}\omega) = -Q_xQ_y\{Q^2-\varkappa_z(\omega/c)^2\}/\varepsilon_0 VD\,,$$

where

$$(33)\qquad D = Q^2(\varkappa_xQ_x^2+\varkappa_yQ_y^2+\varkappa_zQ_z^2) - \\ -[\varkappa_x(\varkappa_y+\varkappa_z)Q_x^2+\varkappa_y(\varkappa_z+\varkappa_x)Q_y^2+\varkappa_z(\varkappa_x+\varkappa_y)Q_z^2](\omega/c)^2+\varkappa_x\varkappa_y\varkappa_z(\omega/c)^4\,.$$

These expressions reduce to (18) in the isotropic case ($\varkappa_x=\varkappa_y=\varkappa_z=\varkappa$). The imaginary parts of the response functions in the general case have peaks at

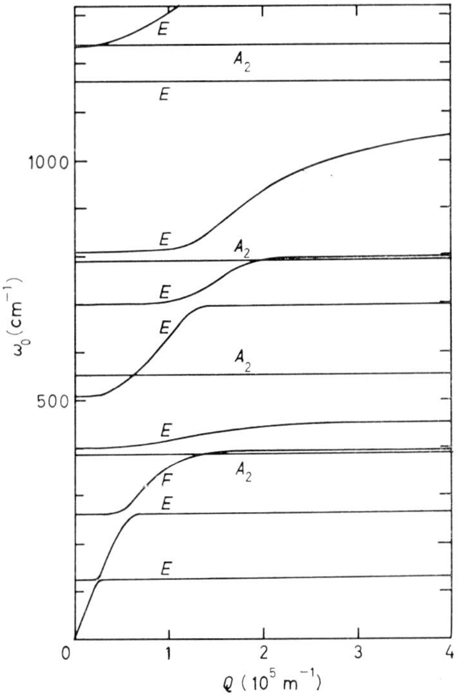

Fig. 3. – Polariton dispersion relations in quartz for $\boldsymbol{Q}$ parallel to the c-axis. The branches labelled E are transverse phonon-polaritons, those labelled A_2 are longitudinal phonons (after LOUDON [6]).

frequencies and wave vectors which satisfy

$$D = 0\,, \tag{34}$$

where all damping factors are removed from the relative permittivities which occur in D. Thus (34) is a generalization of (28). The polariton dispersion curves obtained from (34) depend in general on the direction of $\boldsymbol{Q}$ and can be complicated in low-symmetry crystals with many resonance frequencies. Figure 3 shows the phonon-polariton dispersion relations in uniaxial quartz for $\boldsymbol{Q}$ parallel to the c-axis.

3. – Two-photon absorption by exciton polaritons.

3·1. *Equations of nonlinear optics.* – Suppose that two light beams with fields $\boldsymbol{E}_1$ and $\boldsymbol{E}_2$, having frequencies ω_1 and ω_2, are introduced into a crystal. We consider the attenuation of the incident beams caused by the generation of exciton-polaritons of frequency ω given by

$$\omega = \omega_1 + \omega_2\,. \tag{35}$$

The two-photon absorption process for exciton-polaritons can be treated [7, 8] by means of a theory based on the use of an energy density function [9-11]

$$U = -\,\varepsilon_0\{dE_2^* E_1^* W + bE_2^* E_1^* E\} + \text{c.c.} \tag{36}$$

to describe the coupling of the incident fields with the exciton-polariton variables. The nonlinear susceptibility parameters d and b are tensor quantities, but we first consider a scalar theory.

The nonlinear polarizations at the frequencies ω_1, ω_2 and ω are now given respectively by

$$P_1^{\text{NL}} = -\,\partial U/\partial E_1^* = \varepsilon_0(dE_2^* W + bE_2^* E)\,, \tag{37}$$

$$P_2^{\text{NL}} = -\,\partial U/\partial E_2^* = \varepsilon_0(dE_1^* W + bE_1^* E)\,, \tag{38}$$

$$P^{\text{NL}} = -\,\partial U/\partial E^* = \varepsilon_0 b^* E_2 E_1\,, \tag{39}$$

and there is also a nonlinear force on the displacement W:

$$F^{\text{NL}} = -\,\Omega\,\partial U/\partial W^* = \varepsilon_0\,\Omega d^* E_2 E_1\,. \tag{40}$$

Let the crystal be transparent at frequencies ω_1 and ω_2 so that the corresponding relative permittivities $\varkappa_1$ and $\varkappa_2$ are real, and the wave vectors $\boldsymbol{q}_1$ and $\boldsymbol{q}_2$ of the incident beams are also real. The wave equations for the incident fields are

$$\nabla\times\nabla\times \boldsymbol{E}_1 - \varkappa_1(\omega_1/c)^2\boldsymbol{E}_1 = (\omega_1^2/\varepsilon_0 c^2)\boldsymbol{P}_1^{\text{NL}}, \tag{41}$$

$$\nabla\times\nabla\times \boldsymbol{E}_2 - \varkappa_2(\omega_2/c)^2\boldsymbol{E}_2 = (\omega_2^2/\varepsilon_0 c^2)\boldsymbol{P}_2^{\text{NL}}. \tag{42}$$

The corresponding equations of motion for $\boldsymbol{W}$ and $\boldsymbol{E}$ are given by modified forms of (4) and (12):

$$(\omega_T^2 - \omega^2 - i\omega\Gamma)\boldsymbol{W} = Z\boldsymbol{E} + \boldsymbol{F}^{\text{NL}}, \tag{43}$$

$$\nabla\times\nabla\times \boldsymbol{E} - \varkappa_\infty(\omega/c)^2\boldsymbol{E} = (\omega^2/\varepsilon_0 c^2)[Z\boldsymbol{W}/\Omega + \boldsymbol{P}^{\text{NL}}]. \tag{44}$$

Thus if $\boldsymbol{W}$ is eliminated, the equation for the field $\boldsymbol{E}$ is

$$\nabla\times\nabla\times \boldsymbol{E} - \varkappa(\omega/c)^2\boldsymbol{E} = \frac{\omega^2}{\varepsilon_0 c^2}\left\{\frac{Z\boldsymbol{F}^{\text{NL}}/\Omega}{\omega_T^2 - \omega^2 - i\omega\Gamma} + \boldsymbol{P}^{\text{NL}}\right\}. \tag{45}$$

The displacement $\boldsymbol{W}$ can also be removed from the nonlinear polarizations (37) and (38) with the help of (43). Then if we use also (40), P_2^{NL} can be expressed in the form

$$P_2^{\text{NL}} = \varepsilon_0\,\chi^{(2)}E_1^* E + \varepsilon_0\,\chi^{(3)}|E_1|^2 E_2, \tag{46}$$

and P_1^{NL} is given by a similar expression with 1 and 2 interchanged, where

$$\chi^{(2)} = b + \frac{dZ}{\omega_T^2 - \omega^2 - i\omega\Gamma}, \tag{47}$$

$$\chi^{(3)} = \frac{\varepsilon_0\Omega|d|^2}{\omega_T^2 - \omega^2 - i\omega\Gamma}. \tag{48}$$

These quantities are just the second- and third-order nonlinear susceptibilities defined in accordance with the standard formulation of nonlinear optics [11, 12].

The second-order susceptibility is associated with sum frequency generation in which the photon (E-field) part of the polariton is enhanced by absorption of one ω_1 photon and one ω_2 photon. The third-order susceptibility is associated with two-photon absorption by the exciton (W displacement) part of the polariton. These processes occur simultaneously in the interaction of the incident beams with the polariton. The nonlinear susceptibilities obtained from (36) by the method outlined above have the advantage of showing explicitly the resonant effects occurring at $\omega = \omega_T$.

3·2. *Symmetry considerations.* – The nonlinear coefficients in tensor form can be written d^{ijk} and b^{ijk}, where i, j and k refer respectively to the Cartesian components of $\boldsymbol{E}_2$, $\boldsymbol{E}_1$ and $\boldsymbol{W}$ or $\boldsymbol{E}$. The energy density given by (36) is then summed over the Cartesian components. It is not difficult to show that the tensorial nature of the coefficients is correctly taken into account if the quantities d and b which occur in the expressions for the two-photon absorption coefficient given later are calculated from

$$d = \sum_{i,j,k} e_2^i \, e_1^j \, e^k \, d^{ijk} , \tag{49}$$

with a similar equation for b, where $\boldsymbol{e}_1$, $\boldsymbol{e}_2$ and $\boldsymbol{e}$ are the unit polarization vectors of the two incident beams and the polariton.

The electric-field vectors have odd parity in any centrosymmetric crystal structure. In this case b in (36) must vanish by symmetry and d can be nonzero only for excitons which have an even-parity displacement amplitude $\boldsymbol{W}$. Such excitons have a zero dipole moment, that is $Z = 0$. Thus $\chi^{(2)}$ given by (47) is zero in crystals which have inversion symmetry. The third-order susceptibility $\chi^{(3)}$ given by (48) governs the two-photon absorption by even-parity excitons, but these excitons do not have the associated polariton excitations of interest here. The longitudinal and transverse exciton frequencies are degenerate in this case and the two-photon absorption resulting from $\chi^{(3)}$ alone is always centred on the bare-exciton frequency ω_T. Crystals such as the group-IV semiconductors (Si, Ge) and the alkali halides (NaCl, KBr) fall in this category.

The coefficient b does not in general vanish for crystal structures which do not have inversion symmetry, and such crystals also have excitons for which d and Z are simultaneously nonzero. It is thus possible to observe two-photon generation of polaritons. The most complete experimental results are available for CuCl, a cubic material of zincblende $\bar{4}3m$ symmetry, where [12]

$$b^{xyz} = b^{yzx} = b^{zxy} = b^{zyx} = b^{yxz} = b^{xzy} , \tag{50}$$

and the remaining coefficients are zero. The symmetry restrictions on d^{ijk} are identical.

3'3. *Two-photon absorption coefficient.* – The three wave equations (41), (42) and (45) describe the coupled time development of the fields $\boldsymbol{E}_1$, $\boldsymbol{E}_2$ and $\boldsymbol{E}$. They can be reduced to two coupled equations for two of the fields in an approximation which corresponds to an experimental arrangement in which beam 1 is obtained from a laser and beam 2 from a conventional source. In this case the two-photon absorption depletes the laser beam by an insignificant fraction of its initial intensity and the experiment records the absorption coefficient K for the weaker beam.

Suppose then that beam 1 has a constant amplitude E_1^0 and take the spatial dependence of E_2 in the form

$$E_2^0 \exp\left[- i\omega_2 t + i\boldsymbol{q}_2 \cdot \boldsymbol{r} - \boldsymbol{\alpha} \cdot \boldsymbol{r}\right] , \tag{51}$$

where $\boldsymbol{\alpha}$ is a parameter to be determined which describes the nonlinear loss associated with the two-photon absorption. Then, with the assumption of transverse polarization for beam 2, (42) gives

$$\{(\boldsymbol{q}_2+i\boldsymbol{\alpha})^2-\varkappa_2(\omega_2/c)^2\}\,E_2^0=(\omega_2/c)^2\,\{\chi^{(2)}\,E_1^{0*}\,E+\chi^{(3)}|E_1^0|^2\,E_2^0\}\,, \tag{52}$$

where (46) has been used. We recall that $\boldsymbol{q}_2$ and $\varkappa_2$ have been assumed real so that there is no linear loss at frequency ω_2.

The spatial dependence of the polariton field $\boldsymbol{E}$ is now determined by (45). The nonlinear terms on the right of the equation drive the field with a wave vector whose real part is

$$\boldsymbol{Q}=\boldsymbol{q}_1+\boldsymbol{q}_2\,. \tag{53}$$

There is also a nonlinear decay term in the spatial dependence of $\boldsymbol{E}$, but it can be neglected in comparison with the large linear decay associated with the imaginary part of $\varkappa$. Thus (45) gives

$$Q^2E-\varkappa(\omega/c)^2E=\frac{\omega^2}{c^2}\left\{b^*+\frac{d^*Z}{\omega_T^2-\omega^2-i\omega\Gamma}\right\}E_2^0E_1^0 \tag{54}$$

for the case of a transverse polariton.

Elimination of E_2^0 and E from (52) and (54) leads to an equation whose imaginary part provides the two-photon absorption coefficient of beam 2 in the form

$$K=2\boldsymbol{q}_2\cdot\boldsymbol{\alpha}/q_2= \tag{55}$$

$$=\frac{\omega_2^2|E_1^0|^2}{c^2q_2}\,\mathrm{Im}\left\{\frac{\left[b+\dfrac{dZ}{\omega_T^2-\omega^2-i\omega\Gamma}\right]\left[b^*+\dfrac{d^*Z}{\omega_T^2-\omega^2-i\omega\Gamma}\right]}{(cQ/\omega)^2-\varkappa}+\frac{\varepsilon_0\Omega|d|^2}{\omega_T^2-\omega^2-i\omega\Gamma}\right\},$$

where (47) and (48) have been used. The expression in the large bracket is unfortunately somewhat complicated, but it is characteristic of nonlinear processes which involve polaritons and it is necessary to consider its structure in a little detail.

Only the final term in the bracket contributes for a centrosymmetric crystal, giving an absorption line centred on $\omega=\omega_T$ as mentioned earlier. Both terms contribute for a noncentrosymmetric crystal, and K appears at first sight to have two kinds of peak whose positions are given by the zero-damping limits of the two denominators:

$$\begin{cases}\text{i) } (cQ/\omega)^2=\varkappa & (\text{for } \Gamma=0),\\ \text{ii) } \omega=\omega_T\,. & \end{cases} \tag{56}$$

Examination of (55), using the expression (7) for $\varkappa$, shows however that the two terms in the large bracket make equal and opposite contributions in the vicinity of condition ii) and there is *no* corresponding peak in the absorption coefficient K.

Thus only the peaks given by condition i) in (56) remain, and the terms in (55) can be combined over a common denominator to give

$$(57)\qquad K = (\omega_2^2|E_1^0|^2/c^2 q_2)\cdot \cdot\mathrm{Im}\left\{\frac{|b|^2 + (Z/\omega_P^2)(\varkappa-\varkappa_\infty)(db^* + d^*b) + (Z^2|d|^2/\omega_P^4)[(cQ/\omega)^2 - \varkappa_\infty](\varkappa-\varkappa_\infty)}{(cQ/\omega)^2 - \varkappa}\right\}.$$

Since the pole at $\omega = \omega_T$ resulting from the $\varkappa$ in the numerator makes no contribution, it is permissible to ignore the damping Γ therein, and then, if we replace $(cQ/\omega)^2$ by $\varkappa$ in the numerator, the expression reduces to

$$(58)\qquad K = \frac{\omega_2 I_1}{2\varepsilon_0 c^2(\varkappa_1\varkappa_2)^{\frac{1}{2}}}\left|b + \frac{dZ}{\omega_T^2-\omega^2}\right|^2 \mathrm{Im}\left\{\frac{1}{(cQ/\omega)^2-\varkappa}\right\},$$

where

$$(59)\qquad I_1 = 2\varepsilon_0 c\varkappa_1^{\frac{1}{2}}|E_1^0|^2$$

is the cycle-averaged intensity of the stronger beam.

The final factor in (58) is just the linear response function for the transverse polaritons considered in subsect. **2**·2. Thus if we use (27) and set $\omega = \omega_0$ in

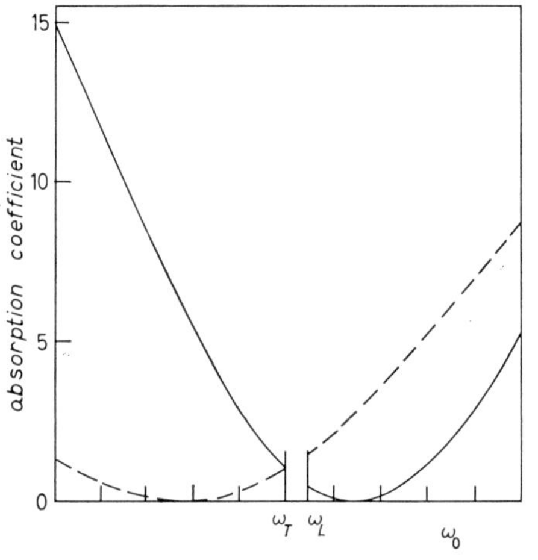

Fig. 4. – Integrated two-photon absorption coefficient as a function of the polariton frequency ω_0 at the peak of the absorption line. A horizontal scale division is $\omega_P^2/\varkappa_\infty\omega_T$, and the magnitude of $\varkappa_\infty\omega_T^2/\omega_P^2$ is 300. The integrated absorption coefficient is normalized to its value at $\omega_0 = \omega_T$. The two curves refer to the values $b\omega_T^2/\mathrm{d}Z = 105$ (———) and -70 (– – –) (after BOGGETT and LOUDON [7]).

the square modulus of (58)

$$K = \frac{\omega_2 \omega_0 Z^2 I_1}{8\varepsilon_0 c^2 (\varkappa_1 \varkappa_2)^{\frac{1}{2}}} \frac{|d + (b/Z)(\omega_T^2 - \omega_0^2)|^2}{\varkappa_\infty(\omega_T^2 - \omega_0^2)^2 + \omega_T^2 \omega_P^2} \frac{\Gamma(\omega_0)}{(\omega_0 - \omega)^2 + [\frac{1}{2}\Gamma(\omega_0)]^2} . \tag{60}$$

The variation of peak absorption frequency ω_0 with Q is given by the polariton dispersion relation (28) and is illustrated in fig. 1. The numerator of the second factor of (60) vanishes for some frequency ω_0, and fig. 4 shows two examples of the variation of the integral of K over frequency as a function of the peak frequency ω_0.

The corresponding result for the longitudinal exciton is easily found. The double curl in (45) is zero for longitudinal polarization and the results are obtained from the transverse case by simple removal of Q. From (58)

$$K = -\frac{\omega_2 I_1}{2\varepsilon_0 c^2 (\varkappa_1 \varkappa_2)^{\frac{1}{2}}} \left| b + \frac{dZ}{\omega_T^2 - \omega^2} \right|^2 \operatorname{Im} \frac{1}{\varkappa} = \tag{61}$$

$$= \frac{\omega_2 \omega_L Z^2 I_1}{8\varepsilon_0 c^2 (\varkappa_1 \varkappa_2)^{\frac{1}{2}}} \frac{|d - (b/Z)(\omega_P^2/\varkappa_\infty)|^2}{\omega_L^2 \omega_P^2} \frac{\Gamma}{(\omega_L - \omega)^2 + (\frac{1}{2}\Gamma)^2} ,$$

where (25) and (26) have been used in the second step.

FRÖHLICH and his co-workers [13-15] have made detailed measurements of two-photon absorption in CuCl, and fig. 5 shows their results for the positions of the peaks in the absorption coefficient. The solid curve is the upper branch of the dispersion relation (28) and the horizontal line is at the longitudinal ex-

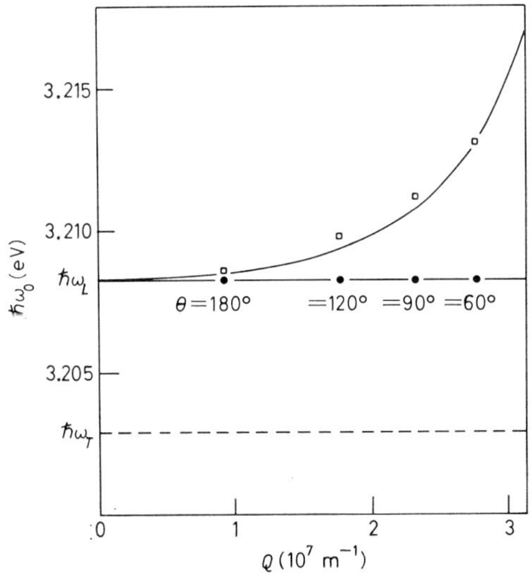

Fig. 5. – Experimental results for the peak frequencies of two-photon absorption in CuCl, θ being the angle between $\boldsymbol{q}_1$ and $\boldsymbol{q}_2$ (FRÖHLICH *et al.* [14]).

citon frequency ω_L. Note that, for $\varkappa_1 = \varkappa_2$, the frequency and wave vector matching relations (35) and (53) allow ω_0 and Q to sweep over the region to the *left* of the oblique dashed line in fig. 1 as the angle θ between $\boldsymbol{q}_1$ and $\boldsymbol{q}_2$ is varied.

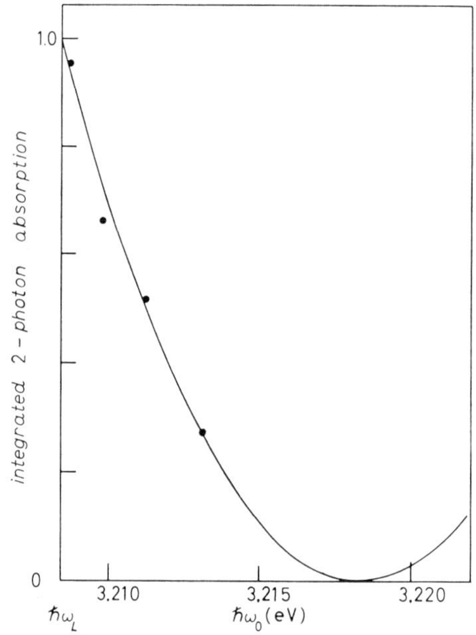

Fig. 6. – Experimental results for the variation of integrated two-photon absorption coefficient with polariton-frequency ω_0 in CuCl (FRÖHLICH *et al.* [15]).

Thus only the *upper* branch of the polariton dispersion curve is accessible. Figure 6 is the measured variation of the integrated absorption coefficient with peak frequency ω_0, showing behaviour in agreement with the continuous theoretical curve of fig. 4.

BOGGETT and LOUDON [8] have given a generalization of the above calculation to crystals of general symmetry which contain arbitrary numbers of exciton dipole resonances. The method of calculation, which is somewhat different from that employed here, leads directly to two-photon absorption coefficients expressed in terms of the polariton response functions of the kind given in (31) and (32).

3·4. *Sum frequency generation.* – The same nonlinear process considered above in terms of the two-photon absorption coefficient of one of the incident beams can also be studied, theoretically and experimentally, in terms of the generation of the beam at the sum frequency. We consider only the transverse case where, with the same assumptions as made in the preceding subsection,

(45) gives

(62) $$-\nabla^2 E - \varkappa(\omega/c)^2 E = (\omega^2/\varepsilon_0 c^2) R^{\mathrm{NL}} \exp[i\boldsymbol{Q}\cdot\boldsymbol{r} - \boldsymbol{\alpha}\cdot\boldsymbol{r}],$$

where

(63) $$R^{\mathrm{NL}} = \varepsilon_0 \left\{ b^* + \frac{d^* Z}{\omega_T^2 - \omega^2 - i\omega\Gamma} \right\} E_2^0 E_1^0$$

is the amplitude of a nonlinear polarization which drives the electric field E of the wave of frequency ω.

The solution of equations of the general form of (62) is treated in detail in chapter 4 of ref. [11]. We assume the simple case of perpendicular incidence of beams 1 and 2 on the surface of a crystal in the xy-plane of a co-ordinate system, so that $\boldsymbol{q}_1$ and $\boldsymbol{q}_2$ are parallel to z. The solution contains two contributions, the complementary function and the particular integral, corresponding to free and driven waves of frequency ω. Matching of fields at the boundary, including a reflected component in the usual way, gives in the interior of the crystal for $z > 0$

(64) $$E = \left\{ \frac{(\mathfrak{q} - Q)\exp[iqz]}{(Q^2 - q^2)(q - \mathfrak{q})} + \frac{\exp[iQz - \alpha z]}{(Q + i\alpha)^2 - q^2} \right\} \frac{\omega^2}{\varepsilon_0 c^2} R^{\mathrm{NL}},$$

where

$$\mathfrak{q} = \omega/c$$

is the free-space wave vector at frequency ω, and q given by (14) can be written

(65) $$q = q' + iq'' = \varkappa^{\frac{1}{2}} \omega/c.$$

The treatment is restricted to frequencies ω in the region of strong linear absorption where q'' is much larger than α. Then, except for very close to the crystal surface at $z = 0$, the free-wave first term of (64) can be neglected, and also α can be neglected in the denominator of the driven-wave term. These are, in a little more detail, essentially the same approximations as made in the derivation of the two-photon absorption coefficient. They show that the presence of the surface produces very little effect in the interior of the crystal. The resulting intensity of the sum frequency beam at co-ordinate z obtained from the approximate form of (64) is

(66) $$I = (2\varepsilon_0 c^2 Q/\omega)|E|^2 = \frac{Q\omega^3 I_1 I_2}{2\varepsilon_0 c^4 (\varkappa_1\varkappa_2)^{\frac{1}{2}}} \frac{|b^* + d^* Z/(\omega_T^2 - \omega^2 - i\omega\Gamma)|^2}{|Q^2 - (q' + iq'')^2|^2},$$

where I_1 and I_2 are the intensities of the incident beams at co-ordinate z.

The denominator of (66) produces a resonance at $Q = q'$ provided $q'' \ll q'$. The Γ in the numerator can be neglected close to this resonance, and approximations similar to those made previously reduce (66) to

$$I = \frac{\omega^3 I_1 I_2}{8\varepsilon_0 c^4 (\varkappa_1 \varkappa_2)^{\frac{1}{2}} q'} \frac{|b + dZ(\omega_T^2 - \omega^2)^{-1}|^2}{(Q - q')^2 + q''^2}. \tag{67}$$

This form of the intensity shows clearly the dependence on the phase mismatch.

HAUEISEN and MAHR [16, 17] have studied the sum frequency beam in CuCl. They used in fact only a single incident beam so that the « sum frequency » was the second harmonic. They successfully fitted their data to an expression similar to (67) and determined a polariton dispersion curve in agreement with that obtained by measurement of the two-photon absorption coefficient of an incident beam [14].

4. – Nonlinear optics with phonon-polaritons.

4'1. *Stimulated Raman effect.* – Table I summarizes the various nonlinear experiments which have been performed on polaritons. The references are in

TABLE I. – *Summary of nonlinear experiments performed on polaritons.* The columns show respectively 1) name of the experiment, 2) representation of the light beams involved, 3) relation satisfied by the frequencies, 4) crystal used, 5) frequency of beam on which observations were made and 6) references.

1)	2)	3)	4)	5)	6)
Two-photon absorption sum generation (exciton-polaritons)	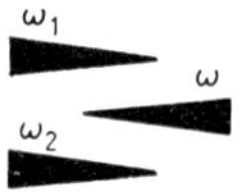	$\omega_1 + \omega_2 = \omega$	CuCl	ω	[16, 17]
			CuCl	ω_2	[13-15]
Stimulated Raman effect difference generation (phonon-polaritons)	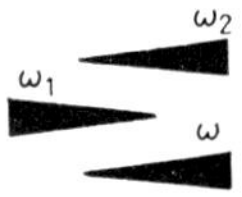	$\omega_1 - \omega_2 = \omega$	SiO_2	ω	[18]
			$LiNbO_3$	ω_2	[19, 20]
	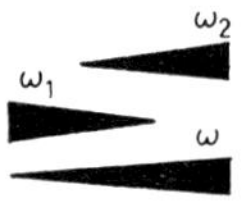		GaP	ω_2	[21]
	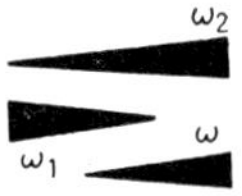		SiO_2	ω	[22]
			GaP	ω	[23]
			$LiNbO_3$	ω	[24]
			SiO_2	ω_2	[25, 26]
Two-stage process (phonon-polaritons)	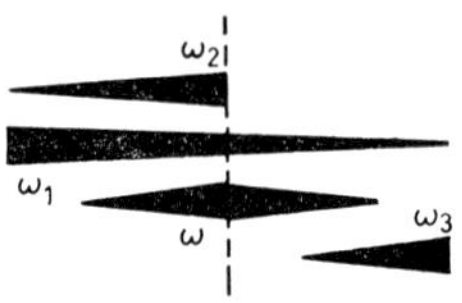		GaP	ω_3	[27, 28]
			GaAs	ω_3	[29]
			$LiNbO_3$	ω_3	[30]

no way intended to be complete or even to show the first observation of a given effect, but rather indicate a selection of representative measurements. The wedges show which beams are present initially at the left-hand side and, moving with time towards the right, which new frequencies occur and whether a beam is amplified or attenuated by the nonlinear process.

The first section of the table shows the two-photon absorption by polaritons, already discussed in detail. We consider here the second section of the table, which represents three different but related kinds of experiment on phonon-polaritons. In all three cases there are two light beams whose frequencies ω_1 and ω_2 are related to a phonon-polariton frequency ω by

$$\omega_1 - \omega_2 = \omega\,. \tag{68}$$

Nonlinear coupling between the waves results in attenuation of beam ω_1 accompanied by amplification of beams ω_2 and ω. The coupling is a combination of difference frequency generation of the photon part and stimulated Raman scattering by the phonon part of the polariton. The combined effect is again observable only in noncentrosymmetric crystals.

The three experiments differ in the initial conditions. In the first case only a single laser beam of frequency ω_1 is incident on the crystal. Spontaneous processes provide in principle a continuous range of pairs of generated beams whose frequencies ω_2 and ω satisfy (68). In practice, the geometrical details of an experiment generally favour particular directions for the beam wave vectors, and maximum power transfer in the stimulated Raman effect is then limited to selected frequencies ω_2 and ω by the polariton dispersion relation. For example in $LiNbO_3$, the frequencies can be varied by adjustment of the orientation of an optical resonator containing the crystal sample relative to the direction of the incident beam [20].

The second and third cases have two beams incident on the crystal. The first experiment of this kind [21] had a second incident beam at a frequency ω in the phonon-polariton region of GaP, and the generation of the light beam of frequency ω_2 was observed. However, most experimental work has used two incident beams ω_1 and ω_2, whose difference is equal to a polariton frequency ω. This kind of two-beam method was pioneered by Biraud-Laval and Chartier [22] and De Martini [23]. It has the advantages of no threshold requirements for production of the polariton beam and of careful control of the frequency and wave vector of the generated polariton by selection of the parameters of the two incident beams.

There has been a large amount of theoretical work done on these processes. The essential ideas required to take account of the photon and phonon parts of the polariton were given by Faust and Henry in the analysis of their experiments on GaP [21]. Various approaches to a more detailed theory are possible; the method we outline here is closely related to the theory given in sect. **3** for two-photon absorption and is originally due to HENRY and GARRETT [3]. The theory of spontaneous scattering of light by polaritons [1, 4] is also very closely related.

The theory can be based on an energy density function similar to (36)

$$U = -\varepsilon_0(d' E_2 E_1^* W + b' E_2 E_1^* E\} + \text{c.c.}, \tag{69}$$

where W and E are now the displacement amplitude and electric field of a phonon-polariton, and the change from E_2^* to E_2 reflects the change of process from one entailing decay of beam 2 to one entailing growth of beam 2. The assumption of crystal transparency at frequencies ω_1 and ω_2 is retained, and the real wave vector at which the polariton is driven is now

$$\boldsymbol{Q} = \boldsymbol{q}_1 - \boldsymbol{q}_2 \,. \tag{70}$$

Only trivial changes of complex conjugations and minus signs are required to convert the equations of sect. **3** to those which describe the stimulated Raman effect, and we consider only the results of the calculations.

For experiments in which there is only a single incident beam of frequency ω_1, the most important scattered beams occur for the values of ω and ω_2 which maximize the growth. If the nonlinear attenuation of the incident beam is neglected, the gain coefficient for scattering by transverse polaritons is given by an expression similar to (55), and the approximations leading to (60) remain valid. Maximum gain occurs for ω_0 and Q which satisfy (28) and thus lie on the polariton dispersion curve. The maximum gain coefficient for beam 2 is

$$G_{\text{max}} = \{\omega_2 \Omega I_1/2c^2 \omega_0(\varkappa_1\varkappa_2)^{\frac{1}{2}} \Gamma\} |d' + (b'/Z)(\omega_T^2 - \omega_0^2)|^2 , \tag{71}$$

where (8) and (29) have been used.

Figure 7 shows a plot of the theoretical gain for GaP as a function of ω_0 [6], where the ratio $d'Z/b'\omega_T^2$ is given the value -0.53 [21]. The figure is similar to fig. 4 for the integrated two-photon absorption coefficient. Note that the frequency and wave vector matching relations (68) and (70) allow only the *lower*-branch polaritons to participate in Raman scattering for a cubic crystal like GaP where the polariton dispersion relation has the form shown in fig. 1.

The theory of the two-beam experiment is also very similar to that of sect. **3**. In this case $\boldsymbol{q}_1$, $\boldsymbol{q}_2$, ω_1 and ω_2 are fixed by the experimental conditions and it is the gain for given $\boldsymbol{Q}$ and ω which is required. The gain expression is similar to (60) if the attenuation of beam 1 is again neglected. The greater flexibility of the two-beam method allows the gain to be studied at $\boldsymbol{Q}$ and ω which do not necessarily correspond to the $\boldsymbol{Q}$ and ω of maximum gain on the polariton dispersion curve.

An alternative approximation to neglecting the attenuation of beam 1 is the assumption of constancy of the product $E_2^* E_1$. Since beam 1 is attenuated while beam 2 grows, the latter approximation corresponds more closely to the

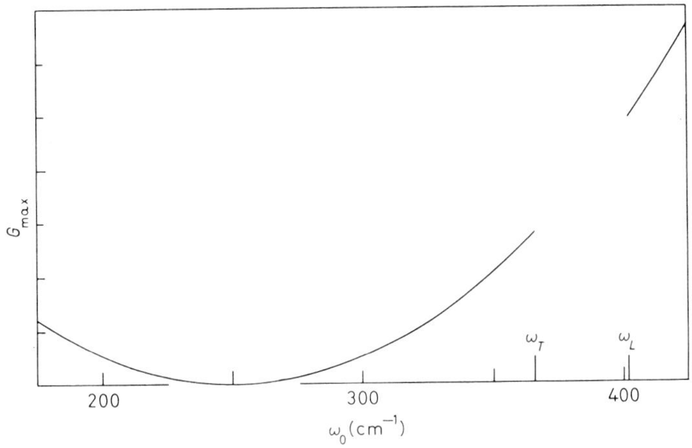

Fig. 7. – Maximum gain of beam 2 for GaP as a function of the polariton frequency ω_0. For a given incident frequency ω_1, the phase-matching relations can be satisfied experimentally for only part of the frequency range lying below ω_T.

experimental situation. The theory of the growth of the polariton beam in this case [26, 31] is similar to that given in subsect. 3˙4, leading to (66). There are two contributions to the field E as in (64), a free wave which is significant only close to the entrance surface and a driven wave which provides the major contribution to the polariton beam. The driven wave again has largest intensity when $\boldsymbol{Q}$ and ω lie on the polariton dispersion curve.

As is shown in table I, some experiments detect the stimulated Raman effect by observation of the component of the polariton of frequency ω which leaves the crystal as far–infra-red radiation, while others observe the effect in the growth of the beam of frequency ω_2. It was realized some time ago [32] that the generation of polaritons by stimulated Raman scattering could provide a tunable source of far–infra-red radiation. Although other types of source have proved to be more versatile, the possibility has been pursued experimentally, and for example YANG *et al.* [24] have been able to tune the polariton emission frequency in $LiNbO_3$ from 20 to 190 cm^{-1}.

4˙2. *Two-stage processes.* – The final entry in table I shows a more complicated two-stage process. In the first stage, two incident beams of frequencies ω_1 and ω_2 generate a polariton at the difference frequency ω, as in the previous entry in the table. In the second stage, the polariton couples to beam 1 to generate another beam at the sum frequency

$$\omega_3 = \omega_1 + \omega = 2\omega_1 - \omega_2 . \tag{72}$$

Several kinds of measurement have been made with this experimental arrangement.

The first example is the work of COFFINET and DE MARTINI [27] on GaP, where the second stage was used essentially as a means of examining the polariton beam generated in the first stage. They made series of observations, for each of which q_1, q_2, ω_1 and ω_2 were kept fixed but the angle between $\boldsymbol{q}_1$ and $\boldsymbol{q}_2$ was varied. It is possible in this way to examine the polariton dispersion on a scan of constant ω but variable Q. The Q-dependence of the phonon-polariton response is similar to that of (67) and the experiment determines q' and q''. The technique provides a novel method for measuring the linear refractive index and absorption coefficient at frequency ω.

A similar experimental set-up has been used by LAUBEREAU *et al.* [28] to study the polariton lifetime in GaP. In their experiment, the first stage was triggered by simultaneous picosecond light pulses at frequencies ω_1 and ω_2, while the second stage was initiated by a further, delayed pulse of ω_1 light. The decay of the intensity of the ω_3 pulse with increasing delay between the ω_1 pulses provides a direct determination of the polariton lifetime ($5.5 \cdot 10^{-12}$ s).

The process illustrated in table I is only one of several possible schemes whereby two incident beams of frequencies ω_1 and ω_2 can mix to give a beam of frequency ω_3 as in (72). For example, there exists another two-stage process in which second-harmonic generation at frequency $2\omega_1$ is followed by difference mixing with the beam of frequency ω_2. Again, the beams can couple to form ω_3 by a direct process controlled by the purely electronic susceptibility $\chi^{(3)}$, in which there are no real intermediate states. These processes all contribute simultaneously to the generation of the beam of frequency ω_3, but the various contributions can in principle be separated experimentally by their different variation with ω_1 and ω_2 and with polarization and wave vector directions.

Experiments of this kind have been carried out on GaAs [29] and on $LiNbO_3$ [30]. The conditions of the latter experiment particularly gave rise to enhancement of the intensity of beam ω_3 when the two incident beams were such that ω and $\boldsymbol{Q}$ given by (68) and (70) coincided with the many-branch polariton dispersion curve of $LiNbO_3$. We do not write down any equations for these experiments, but the general principles are the same as for the simpler processes considered earlier. The contribution of the two-stage polariton process of table I to the nonlinear susceptibility $\chi^{(3)}$ is indeed given by an expression similar to the large bracket of (55). The experiments thus provide a further method for studying the linear and nonlinear characteristics of the polaritons.

All the above discussion refers to phonon-polaritons. A two-stage process can also be used to study exciton-polaritons. The principles of the experiment remain the same except that the conditions are now arranged so that $2\omega_1$ lies close to the exciton-polariton frequency ω, and the frequency ω_3 is obtained by difference mixing of the second harmonic with the ω_2 beam, as described above. KRAMER *et al.* [33] have performed an experiment of this kind on CuCl. They observe resonant effects when $2\omega_1$ is close to the longitudinal exciton frequency ω_L, and the experiment could also in principle be used to investigate the transverse polariton branches.

5. – Conclusion.

The most striking characteristic of nonlinear optical processes which involve polaritons is their resonant behaviour at wave vectors $\boldsymbol{Q}$ and frequencies ω_0 which lie on the polariton dispersion curve, given by (28) for a cubic crystal with a single dipole resonance or more generally by (34). No similar effect occurs in linear optical processes, where the behaviour is governed by the linear susceptibility or relative permittivity; these functions depend only on frequency (apart from some rare exceptions), and they produce resonant effects at the bare-dipole excitation frequencies.

The polariton dispersion curve, defined as the frequency *vs.* wave vector relation for maximum electromagnetic response, is given in general by a theoretical relation (34) involving components of the relative permittivity in which all damping constants are to be set equal to zero. Incorrect curves are obtained by any other procedure, for example taking the real parts of the relative permittivities with finite damping constants retained.

The summary in table I of the kinds of experiment which have been performed lists a selection of references. There are other experimental references, not listed, particularly for the central section of the table. However, it is fair to say that the sum total of knowledge of polariton properties obtained by nonlinear spectroscopy is not very high. For example, the two-photon absorption experiments on exciton-polaritons in CuCl [13-15] are very elegant and they provide new information on the exciton properties of the crystal. However, to the author's knowledge, CuCl remains the only crystal to be successfully studied in this way. As a second elegant example, the measurements of phonon-polariton lifetime in GaP by the two-stage process [28] were restricted to a single polariton frequency, and the author does not know of any other applications of what is potentially a very powerful technique. Both these examples concern more direct ways of obtaining data whose acquisition by other methods requires careful and sometimes uncertain interpretation of results obscured by other effects.

It is to be hoped that the techniques for these difficult experiments will improve to a point where they can be applied more widely as tools for studying crystals. It may then be worth considering even more difficult polariton experiments, for example nonlinear optical processes with surface polaritons.

REFERENCES

[1] D. L. Mills and E. Burstein: *Rep. Prog. Phys.*, **37**, 817 (1974).
[2] L. D. Landau and E. M. Lifshitz: *Statistical Physics*, Chap. 12 (Oxford, 1969).
[3] C. H. Henry and C. G. B. Garrett: *Phys. Rev.*, **171**, 1058 (1968).
[4] A. S. Barker jr. and R. Loudon: *Rev. Mod. Phys.*, **44**, 18 (1972).
[5] S. Ushioda and J. D. McMullen: *Solid State Comm.*, **11**, 229 (1972).
[6] R. Loudon: *Light Scattering Spectra of Solids* (New York, N. Y., 1969), p. 25.
[7] D. Boggett and R. Loudon: *Phys. Rev. Lett.*, **28**, 1051 (1972).
[8] D. Boggett and R. Loudon: *J. Phys. C*, **6**, 1763 (1973).
[9] D. A. Kleinman: *Phys. Rev.*, **126**, 1977 (1962).
[10] P. S. Pershan: *Phys. Rev.*, **130**, 919 (1963).
[11] N. Bloembergen: *Nonlinear Optics* (New York, N. Y., 1965).
[12] P. N. Butcher: *Nonlinear Optical Phenomena* (Columbus, O., 1965).
[13] D. Fröhlich, B. Staginnus and E. Schönherr: *Phys. Rev. Lett.*, **19**, 1032 (1967).
[14] D. Fröhlich, E. Mohler and P. Wiesner: *Phys. Rev. Lett.*, **26**, 554 (1971).
[15] D. Fröhlich, Ch. Uihlein and E. Mohler: *Phys. Stat. Sol.*, **55**, 175 (1973).
[16] D. C. Haueisen and H. Mahr: *Phys. Rev. Lett.*, **26**, 838 (1971).
[17] D. C. Haueisen and H. Mahr: *Phys. Lett.*, **36** A, 433 (1971).
[18] I. M. Aref′ev, S. V. Krivokhizha, Yu. I. Kyzylasov, V. S. Starunov and I. L. Fabelinskii: *JETP Lett.*, **8**, 84 (1968).
[19] S. K. Kurtz and J. A. Giordmaine: *Phys. Rev. Lett.*, **22**, 192 (1969).
[20] J. Gelbwachs, R. H. Pantell, H. E. Puthoff and J. M. Yarborough: *Appl. Phys. Lett.*, **14**, 258 (1969).
[21] W. L. Faust and C. H. Henry: *Phys. Rev. Lett.*, **17**, 1265 (1966).
[22] S. Biraud-Laval and G. Chartier: *Phys. Lett.*, **30** A, 177 (1969).
[23] F. De Martini: *Phys. Lett.*, **30** A, 319, 547 (1969).
[24] K. H. Yang, J. R. Morris, P. L. Richards and Y. R. Shen: *Appl. Phys. Lett.*, **23**, 669 (1973).
[25] S. Biraud-Laval, R. Reinisch, N. Paraire and R. Laval: *Phys. Rev. B*, **4**, 1797 (1976).
[26] R. Reinisch, S. Biraud-Laval and N. Paraire: *J. Physique*, **37**, 227 (1976).
[27] J. P. Coffinet and F. De Martini: *Phys. Rev. Lett.*, **22**, 60 (1969).
[28] A. Laubereau, D. von der Linde and W. Kaiser: *Opt. Comm.*, **7**, 173 (1973).
[29] E. Yablonovitch, C. Flytzanis and N. Bloembergen: *Phys. Rev. Lett.*, **29**, 865 (1972).
[30] J. J. Wynne: *Phys. Rev. Lett.*, **29**, 650 (1972).
[31] R. Reinisch, N. Paraire and S. Biraud-Laval: *Compt. Rend.*, **275**, 829 (1972).
[32] R. Loudon: *Proc. Phys. Soc.*, **82**, 393 (1963).
[33] S. D. Kramer, F. G. Parsons and N. Bloembergen: *Phys. Rev. B*, **9**, 1853 (1974).

Nonlinear Spectroscopy of Bulk and Surface Polaritons (*).

F. De Martini

Istituto di Fisica dell'Università - 00185 Roma, Italia

1. – Introduction.

In recent years the advent of mode-locked and narrow-band frequency-tunable lasers has made possible a new kind of investigation in solid-state spectroscopy: the coherent excitation of elementary excitations of various kinds by resonant interaction with laser fields via a linear or nonlinear mechanism. Obviously, the fundamental laws of nonlinear optics, as they are presented in ref. [1], apply to the new situation, but they should be complemented by a new class of theories. Processes like resonant fluorescence, transient coherent propagation phenomena like self-induced transparency, local coherence effects like photon-echo and optical nutation, and some aspects of the propagation of polariton fields in solids often require new theoretical approaches leading to problems that are sometimes unsolvable with present techniques of mathematical physics. However, in crystalline solids the coherence relaxation time T_2 of most of the elementary excitations is generally so short with respect to the amount of energy that can be stored in a laser pulse without damage to the medium, that transient coherence effect can hardly be detected. Therefore, we limit ourselves in these lectures to considering some simple aspects of the theory of the polariton field generation and propagation, starting from conventional classical electrodynamics and quantum mechanics. In view of recent nonlinear experiments involving bulk polaritons, some of which are being presented at the present school, we devote our attention to only few effects, namely to nonlinear mixing processes such as sum frequency generation (SFG), difference frequency generation (DFG), and to coherent-scattering processes like those involved in the experiments of 4-photon coherent spectroscopy. We shall first outline a linear transient propagation theory of the fields in SFG and DFG and obtain the expression for parametric gain valid in either case.

(*) Work supported by Gruppo di Ricerca Elettronica Quantistica e Plasmi of the C.N.R. and by NATO 1155 joint program between Università di Roma and University of California at Berkeley.

We conclude with a brief presentation of the most relevant experiments that have been carried out recently in the field, including some of the author's own results. We shall present, at the end of the paper, the first formulated theory of the nonlinear excitation of the surface polaritons and the results of the first experiment on 4-photon excitation of surface polaritons in GaP.

2. – Transient propagation theory of the bulk polariton sum and difference frequency generation.

The parametric sum and difference frequency generation processes involving the scattering of a coherent polariton wave may be described by starting either from the classical coupled-wave theory, largely used in nonlinear optics [1], or from a quantum transition rate formulation [2]. The theory of the gain affecting the « signal » (optical) wave in a parametric process, in which the frequency of one of the interacting fields is near a lattice resonance, has been considered by SHEN [3], BUTCHER and McLEAN [4] and by HENRY and GARRETT [5], according to the classical approach. The application of the classical method, which is usually concerned with e.m. fields rather than with energy flow, becomes difficult when a complete description of the dynamics of a three-wave interaction is required. In order to overcome serious mathematical problems, some approximations are usually introduced in the theory, for example the lack of the depletion of the pump and the *a priori* hypothesis of a single exponential gain solution for the waves in the case of the difference frequency generation. As we shall see below, these approximations are indeed not appropriate to a transient analysis involving polariton particles rather than photons.

As long as we limit ourselves to considering a negligible change of population of the exciton state, the interacting e.m. and polariton fields may be expressed in terms of boson creation and annihilation operators, which satisfy the commutation relations

$$[b_{k_i}, b^\dagger_{k_j}] = \delta_{k_i k_j} \qquad (i, j = 1, 2, 3). \tag{1}$$

If m_{k_1}, m_{k_2}, m_{k_3} are the mode occupation numbers for the three fields, the matrix elements of these operators in the Fock space are

$$\begin{cases} \langle m_{k_i} - 1 | b_{k_j} | m_{k_j} \rangle = m_{k_i}^{\frac{1}{2}}\, \delta_{k_i k_j}\,, \\ \langle m_{k_i} + 1 | b^\dagger_{k_j} | m_{k_j} \rangle = (m_{k_i} + 1)^{\frac{1}{2}}\, \delta_{k_i k_j}\,. \end{cases} \tag{2}$$

In the above expressions the symbols k_i represent the momenta of the fields $\boldsymbol{E}_i$.

The fields, assumed to be monochromatic plane waves, may be written in the following form [6]:

$$(3)\qquad \begin{cases} \boldsymbol{E}_1(\omega_1) = \left(\dfrac{h\omega_1^3}{k_1^2 c^2 V}\right)^{\frac{1}{2}} \boldsymbol{p}_1\left(b_{k_1} \exp[i\boldsymbol{k}_1\cdot\boldsymbol{r}] + b_{k_1}^\dagger \exp[-i\boldsymbol{k}_1\cdot\boldsymbol{r}]\right), \\ \boldsymbol{E}_2(\omega_2) = \left(\dfrac{h\omega_2^3}{k_2^2 c^2 V}\right)^{\frac{1}{2}} \boldsymbol{p}_2\left(b_{k_2} \exp[i\boldsymbol{k}_2\cdot\boldsymbol{r}] + b_{k_2}^\dagger \exp[-i\boldsymbol{k}_2\cdot\boldsymbol{r}]\right), \\ \boldsymbol{E}_q(\omega_3) = \left(\dfrac{h\omega_3 v_p v_g}{c^2 V}\right)^{\frac{1}{2}} \boldsymbol{p}\left(b_{k_3} \exp[i\boldsymbol{k}_3\cdot\boldsymbol{r}] + b_{k_3}^\dagger \exp[-i\boldsymbol{k}_3\cdot\boldsymbol{r}]\right), \\ \boldsymbol{Q}(\omega_3) = (4\pi N\mu\omega_0^2)^{-\frac{1}{2}}\cdot(c^2/v_p v_g - \varepsilon_\infty)^{\frac{1}{2}} \boldsymbol{E}_q(\omega_3)\,. \end{cases}$$

In the above expressions V is a volume of the solid with linear dimensions small compared to the inverse of the absorption coefficient $\alpha(\omega_3)$; N is the volume density of the primitive cells of the crystal; v_p and v_g are, respectively, the phase and the group (energy) velocity of the undamped polariton wave; μ is the reduced mass of the ionic or electronic oscillator relative to one primitive cell of the crystal and associated with the TO lattice mode at frequency $\omega_{\text{TO}} \equiv \omega_0$ or, correspondingly, with the TO excitonic mode. ε_∞ is the high-frequency dielectric constant and the expressions of $v_p(\omega_3)$ and $v_g(\omega_3)$ may be obtained by the classical Huang dispersion theory [7] or by its quantum-mechanical counterpart [8]. If we allow for a phase mismatch of the waves, $\Delta\boldsymbol{k} = \boldsymbol{k}_1 \pm \boldsymbol{k}_2 - \boldsymbol{k}_3'$ $(\boldsymbol{k}_3 = \boldsymbol{k}_3' + i\boldsymbol{k}_3'')$, an interaction Hamiltonian density for the coupled-field system may be expressed in the following form (*):

$$(4)\qquad \mathscr{H} = A\boldsymbol{p}_1\boldsymbol{p}_2\boldsymbol{p} \vdots \underline{d}\left(b_1 b_2^\dagger b_3^\dagger \exp[i\Delta\boldsymbol{k}\cdot\boldsymbol{r}] + b_1^\dagger b_2 b_3 \exp[-i\Delta\boldsymbol{k}\cdot\boldsymbol{r}]\right)$$

for DFG and

$$(4')\qquad \mathscr{H} = A p_1 p_2 p_3 \vdots \underline{d}\left(b_1 b_2 b_3^\dagger \exp[i\Delta\boldsymbol{k}\cdot\boldsymbol{r}] + b_1^\dagger b_2^\dagger b_3 \exp[-i\Delta\boldsymbol{k}\cdot\boldsymbol{r}]\right)$$

for SFG. A is a constant to be determined on the basis of the coefficients of the fields of (3) and d is the nonlinear susceptibility tensor that accounts for the coupling. A significant expression for the nonzero elements of $\underline{d}$ near the TO resonance has been given by FAUST and HENRY [9] in the form $\underline{d} = \underline{d}_E(1 + C\omega_0^2 D^{-1})$, where $C = e^* N\underline{d}_Q/\mu\omega_0^2 \underline{d}_E$ is a tensor parameter that characterizes the nonlinear response of the crystal, e^* is the effective charge of the oscillators and $D = \omega_0^2(k_3) - \omega^2 - i\omega\Gamma$, Γ being the damping parameter of the oscillator.

(*) The upper sign appearing in the expression of $\Delta\boldsymbol{k}$ corresponds to SFG. Hereafter in the paper we shall adopt this convention.

The interaction Hamiltonian H_1 is found by integrating $\mathscr{H}$ over the volume V. If we consider one initial state $|m_{k_1}, m_{k_2}, m_{k_3}\rangle$ for the fields, then the transition probability, over a unit frequency range per unit time corresponding to the scattering of a pump photon, is given by the following expression [10]:

$$(5)\qquad w = \frac{4\pi^2}{\hbar^2}\left(|\langle m_{k_1}-1, m_{k_2}\mp 1, m_{k_3}+1|H_1|m_{k_1}, m_{k_2}, m_{k_3}\rangle|^2 - \right.$$
$$\left. - |\langle m_{k_1}+1, m_{k_2}\pm 1, m_{k_3}-1|H_1|m_{k_1}, m_{k_2}, m_{k_3}\rangle|^2\right)\frac{\frac{1}{2}\alpha(\omega_3)}{|\Delta\boldsymbol{k}|^2+[\frac{1}{2}\alpha(\omega_3)]^2}\,\delta(\Delta\omega)\,,$$

where $\Delta\omega = \omega_1 \pm \omega_2 - \omega_3$, and $\alpha(\omega_3) = 2|\boldsymbol{k}_3''(\omega_3)|$ is the absorption coefficient. The Lorentzian factor depending on $\Delta\boldsymbol{k}$ in eq. (5) accounts for the damping of the final polariton momentum state that is characterized by the absorption coefficient α. The damping affecting the final-energy state of the interaction is characterized by a frequency-dependent damping parameter $\tilde{\Gamma}(\omega_3)$ that is related to $\alpha(\omega_3)$ through the polariton group velocity:

$$(6)\qquad \tilde{\Gamma}(\omega_3) = v_g(\omega_3)\,\alpha(\omega_3)\,.$$

If we make use of eq. (2), w becomes

$$(7a)\qquad w = \frac{B}{V}\,\frac{(\frac{1}{2}\alpha)^2}{|\Delta\boldsymbol{k}|^2+(\frac{1}{2}\alpha)^2}\,\delta(\Delta\omega)\times[\,]\,,$$

where

$$[\,] \equiv [m_{k_1}m_{k_2} - m_{k_3}(m_{k_1}+m_{k_2}+1)] \qquad \text{(SFG)}\,,$$

$$[\,] \equiv [m_{k_1}(m_{k_2}+m_{k_3}+1) - m_{k_2}m_{k_3}] \qquad \text{(DFG)}$$

and

$$(7b)\qquad B = \frac{32\pi^5\hbar\omega_1^3\omega_2^3\omega_3^3 v_p v_g |d|^2}{|\boldsymbol{k}_1|^2|\boldsymbol{k}_2|^2c^6}\,.$$

In view of the discussion of our experiment dealing with a cubic crystal (GaP) belonging to the $\bar{4}3m$ class, we have omitted in (7b) the tensor notation for the nonlinear coupling coefficient d. Hereafter we shall consider d, d_E, d_Q as scalar quantities, because, for the given symmetry and crystal class, these tensors are determined by only one parameter [11].

By integrating eq. (7a) over the distribution of the final-energy states, we obtain the transition probability per unit time

$$(8)\qquad W = \frac{B}{\pi^2 v_g V}\,\frac{(\frac{1}{2}\tilde{\Gamma})^2}{(\Delta\omega)^2+(\frac{1}{2}\tilde{\Gamma})^2}\,\frac{(\frac{1}{2}\alpha)^2}{|\Delta\boldsymbol{k}|^2+(\frac{1}{2}\alpha)^2}\times[\,]\,.$$

Equation (8) is valid when $\Delta\omega$ and $|\Delta\boldsymbol{k}|$ are not too much larger than the corresponding damping parameters $\tilde{\Gamma}$ and α.

As will be shown, the polariton density is much smaller than the photon densities at frequencies ω_1 and ω_2. Hence for $m_{k_3} \gg 1$ we can write a simplified expression of W in the form

$$W = \frac{g m_{k_1} m_{k_2}}{V}, \tag{9a}$$

where

$$g = \frac{B}{\pi^2 v_g} \frac{(\frac{1}{2}\alpha)^2}{|\Delta\boldsymbol{k}|^2 + (\frac{1}{2}\alpha)^2} \frac{(\frac{1}{2}\tilde{\Gamma})^2}{(\Delta\omega)^2 + (\frac{1}{2}\tilde{\Gamma})^2}. \tag{9b}$$

We consider in this section a collinear interaction of the fields along the direction $\boldsymbol{z}$ normal to the entrance face of the crystal. The fields are written as functions of the corresponding z–co-ordinate, which is positive in the crystal and zero at the surface.

In this section we also consider a phase-matched interaction. The equation describing the evolution of the densities of the pump photons $\varrho_1(z, t) \equiv m_{k_1}/V$ and $\varrho_2(z, t) \equiv m_{k_2}/V$ may be written, owing to (9), in the following form:

$$\frac{\partial \varrho_2}{\partial t} + v_2 \frac{\partial \varrho_2}{\partial z} = \pm \left(\frac{\partial \varrho_1}{\partial t} + v_1 \frac{\partial \varrho_1}{\partial z} \right) = \mp g \varrho_1 \varrho_2, \tag{10}$$

where v_1 and v_2 are the group velocities of the two waves. For negligible optical dispersion of the medium at the frequencies ω_1 and ω_2 we have $v_1 = c/n_1$ and $v_2 = c/n_2$, where n_1 and n_2 are the refractive indices. Our present theory is greatly simplified if we set $v_1 = v_2$ in eq. (10).

We can now write the fields as functions of the propagation variables:

$$w_1 = t + z/v_1, \qquad \overline{w}_1 = t - z/v_1. \tag{11}$$

The corresponding inverse relations are $t = \frac{1}{2}(w_1 + \overline{w}_1)$ and $z = (w_1 - \overline{w}_1)\frac{1}{2}v_1$. A first integral of (10) is given by the following expression:

$$\varrho_1(w_1, \overline{w}_1) \mp \varrho_2(w_1, \overline{w}_1) = f(\overline{w}_1). \tag{12}$$

The difference (sum) of the pump photon wave packets is a forward-propagation arbitrary function. We impose the following boundary condition on the solution. At $z = 0$, $w_1 = \overline{w}_1$, the ratio of the two fields is a prescribed function of time, $\varrho_1(0, t)/\varrho_2(0, t) = \varepsilon(t)$. We rewrite that condition in the notation of the propagating frame, eq. (11), and we make use of eq. (12) in order to write the driving term $\varrho_1\varrho_2$ appearing in eq. (10) as a function of one of the fields. The solution of (10) is now straightforward. By restoring the initial variables but still keeping the useful notation $\overline{w}_1 = t - z/v_1$, the product of the fields is found to be

$$(13)\qquad \begin{cases} \varrho_1\varrho_2 = [\tfrac{1}{2}f(\overline{w}_1)]^2 \operatorname{cosech}^2\{G_1 f(\overline{w}_1)(z-\bar{z}(\overline{w}_1))\} & \text{(SFG)}\,, \\ \varrho_1\varrho_2 = [\tfrac{1}{2}f(\overline{w}_1)]^2 \operatorname{sech}^2\{G_1 f(\overline{w}_1)(z-\bar{z}(\overline{w}_1))\} & \text{(DFG)}\,. \end{cases}$$

In the above expressions we have introduced the parameter $G_1 \equiv g/4v_1$ and the function $\bar{z}(\overline{w}_1) \equiv \mp [\ln \varepsilon(\overline{w}_1)]/[2G_1 f(\overline{w}_1)]$ that accounts for the initial conditions. Note that $\bar{z}$ is always negative for SFG. A simple equation relates the two fields:

$$(14)\qquad \varrho_2(z, t) = \varrho_1(z, t) \exp[\mp 2G_1 f(\overline{w}_1)(z-\bar{z}(\overline{w}_1))]\,,$$

which, in the case of DFG, accounts for the stimulated-scattering exponential growth.

The continuity equation for the polariton density $\varrho_q(z, t) = m_{k_3}/V$ is given in the following form:

$$(15)\qquad \frac{\partial \varrho_q}{\partial t} + v_g \frac{\partial \varrho_q}{\partial z} = g(\omega_3)\varrho_1\varrho_2 - \tilde{\Gamma}'(\omega_3)\varrho_q\,,$$

where $\tilde{\Gamma}'(\omega_3)$ is a phenomenological « damping » parameter that accounts for the relaxation of the polaritons to the ground state. Making use of the notation of magnetic resonance [12], we can associate $\tilde{\Gamma}'(\omega_3)$ with the longitudinal relaxation time T_1, and write $\tilde{\Gamma}'(\omega_0) = 1/T_1$. We notice that, in general, $\tilde{\Gamma}'(\omega_0) \equiv \Gamma'$ is smaller than the parameter $\tilde{\Gamma}(\omega_0) \equiv \Gamma$ that appears in the expression for $g(\omega_0)$ given by eq. (9*b*). In effect, $\tilde{\Gamma}(\omega_0)$ is the line width of the polariton resonance and is determined, competitively, by the inhomogeneous broadening process and by the effect of the coherence relaxation. When this last effect is overwhelming, the line width of the TO resonance $\Gamma \equiv \tilde{\Gamma}(\omega_0)$ is approximately equal to the inverse of the transverse relaxation time T_2. Obviously the present considerations on the effects of the different homogeneous broadening processes are physically relevant only when we are dealing with quantum systems that are not too strongly coupled by collisions, *e.g.* excited molecules in a gas. In that case $T_2 \ll T_1$. For optical phonons in solids at normal temperature we can assume $T_1 \simeq T_2$.

In the present section we shall consider a driving pulse $f(\overline{w}_1)$ that is a regular analytic function of $\overline{w}_1$. With no loss of generality and for simplicity's sake, we shall mainly deal with a symmetric pulse characterized by a time duration $\Delta\tau = (k')^{-1}$, that is approximately equal to the inverse of the time derivative of $f(\overline{w}_1)$, *e.g.* $f(\overline{w}_1) = \operatorname{sech}(k'\overline{w}_1)$.

In order to solve eq. (15), we again apply a transformation of variables from the set $\{z, t\}$ to the propagation set $\{w, \overline{w}\}$, where $w = t + z/v_g$ and $w = t - z/v_g$. The functions $f(\overline{w}_1)$ and $\bar{z}(\overline{w}_1)$ appearing in eq. (13) are now written in terms of the new variables according to the equation $\overline{w}_1 = \tilde{\alpha}w + \tilde{\delta}\overline{w}$, with $\tilde{\alpha} \equiv (1-\tilde{\beta})/2$, $\tilde{\delta} \equiv (1+\tilde{\beta})/2$ and $\tilde{\beta} \equiv v_g/v_1 \ll 1$.

In the new frame, eq. (15) is a standard linear equation whose general solution may be expressed in the following integral form:

$$(16)\qquad \varrho_q(w, \overline{w}) = \exp[-\tilde{\Gamma}' w/2]\left[C_2(\overline{w}) + (g/2)\int \varrho_1 \varrho_2 \exp[\tilde{\Gamma}' w/2]\,dw\right],$$

where $C_2(\overline{w})$ is an arbitrary function of $\overline{w}$, to be determined by the boundary conditions.

The integral appearing in (16) may be calculated in successive integrations by parts, taking at each step the exponential function as the differential factor of the integration. For $\tilde{\Gamma}' \gg k'$ and $g\tilde{\beta} \ll \tilde{\Gamma}'$ we obtain in this way a rapidly converging series with terms proportional to increasing powers of $k'/\tilde{\Gamma}'$. As the minimum line width of the polariton resonance in solids is of the order of $(5\div 10)\ \text{cm}^{-1}$, taking into account in the theory only the first term of the expansion is legitimate, if we consider $f(\overline{w}_1)$ pulses with time duration $\Delta\tau \approx (k')^{-1}$ larger than $(10^{-11}\div 10^{-12})$ s, namely pulses generated by the common mode-locked lasers. We determine $C_2(\overline{w})$ for a pulse $f(\overline{w}_1)$ of finite length by imposing the condition $\varrho_q(z, t) = 0$ at $t = -\infty$. The free solution of eq. (15) disappears and the general solution is given by the following expressions:

$$(17a)\qquad \begin{cases} \varrho_q(z, t) = \dfrac{g}{4\tilde{\Gamma}'}\{[f(\overline{w}_1)]^2 \operatorname{cosech}^2[G_1 f(\overline{w}_1)(z - \overline{z}(\overline{w}_1))]\}, \\[2ex] \varrho_q(z, t) = \dfrac{g}{4\tilde{\Gamma}'}\{[f(\overline{w}_1)]^2 \operatorname{sech}^2[G_1 f(\overline{w}_1)(z - \overline{z}(\overline{w}_1))]\}, \end{cases}$$

valid respectively for SFG and DFG, again having kept the useful notation $\overline{w}_1 = t - z/v_1$.

The time evolution of the solutions (17a) corresponding to the conditions $f(z, t) \equiv \operatorname{sech}[k(z - v_1 t)]$ and $\overline{z}(\overline{w}_1) \equiv \text{const}/[G_1 f(\overline{w}_1)]$ is shown in fig. 1 and 2. Figure 1 shows that the polariton density spatial distribution in SFG is always decreasing quasi-exponentially in space and time and does not reproduce the shape of the exciting optical pulse. In the case of DFG, if $\overline{z} > 0$, the solution is represented by a pulse having a shape that is rapidly changing in time but that keeps quasi-stationary in space the position of its absolute maximum. For a regular symmetric pulse $f(\overline{w}_1)$ having its maximum at $\overline{w}_1 = 0$ and for $\overline{z}(\overline{w}_1) f(\overline{w}_1) G_1 \equiv \eta(\overline{w}_1) = \text{const}$, the polariton density function reaches its absolute maximum at the co-ordinate $z = \overline{z}(0) = \eta/(G_1 f(0))$ and at the time $t = \overline{z}(0)/v_1$. For $t > \overline{z}(0)/v_1$, the polariton pulse spreads out in space at a rate proportional to the coupling parameter G_1. That same quantity determines the width of the zone of maximum $\varrho_q(z, t)$, if the length of the pulse $f(\overline{w}_1)$ is larger than $1/G_1 f(0)$, and it is proportional to the maximum value of $\varrho_q(z, t)$, which

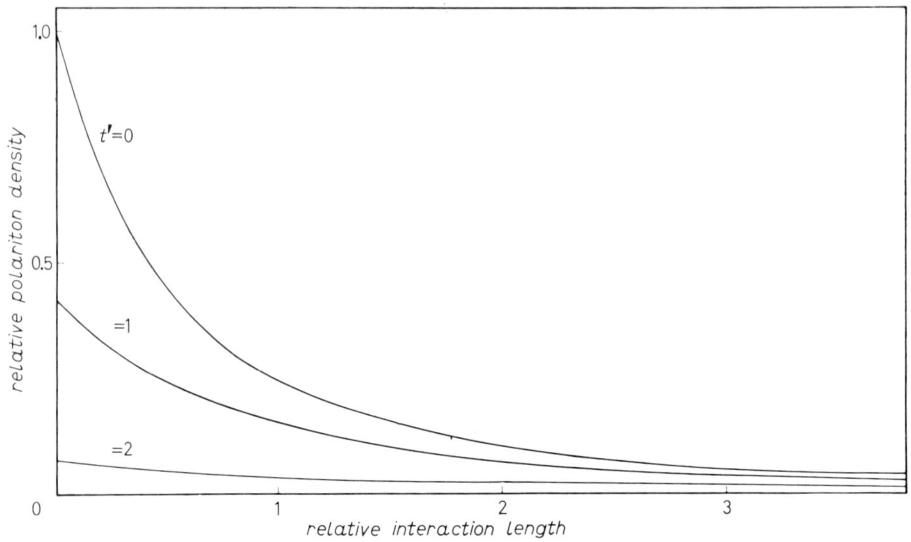

Fig. 1. – Time evolution of the polariton density pulse ϱ_q created by sum frequency generation and corresponding to the driving pulse $\varrho_1(z,t)=\varrho_2(z,t)=\frac{1}{2}\,\text{sech}\,k(z-v_1t)$. The above curves are drawn as functions of the dimensionless quantities $z'=kz$, $t'=kv_1t$ and correspond to $G_1/k=1$.

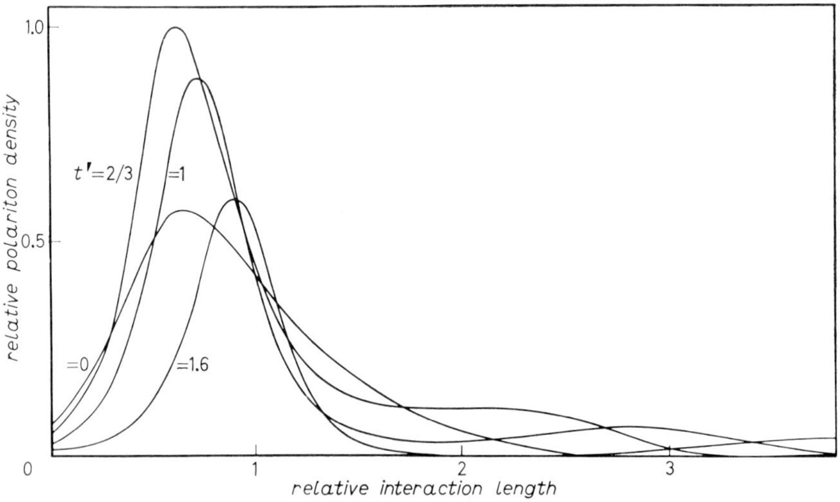

Fig. 2. – Time evolution of the polariton density pulse ϱ_q created by difference frequency generation and corresponding to the driving pulse $f(z,t)=\text{sech}\,k(z-v_1t)$ and to the condition $\bar{z}(\overline{w}_1)=\eta/(G_1f(\overline{w}_1))$, η being constant. The above curves are drawn as functions of the dimensionless quantities $z'=kz$, $t'=kv_1t$ and correspond to the following values of the parameters: $\eta=2$ and $G_1/k=3$. We note that the maximum value of ϱ_q is reached at the time $t'=\eta k/G_1$ and at the co-ordinate $z'=\bar{z}'(0)=t'$.

is equal to $g/(4\tilde{\Gamma}')$. Obviously more complex situations arise when η is a variable function of $\bar{w}_1$.

Owing to the above considerations, we can conclude that the effect of IR generation on the excitation of polaritons is a somewhat critical process, in particular when short pulses (*e.g.* generated by mode-locked lasers) are made to interact in a highly nonlinear medium. Referring to the conditions corresponding to fig. 2, we notice that the maximum value of $\varrho_q(z, t)$ will never be reached during the interaction if the co-ordinate $\bar{z}(0)$, which depends on the amplitude of the optical fields, lies outside the nonlinear crystal. Furthermore, when the DF pulse emitted from a crystal of length L is detected in the forward direction, the maximum efficiency of the DFG process corresponds to the condition $\bar{z}(0) \simeq L$, namely the exciting fields before the interaction must be related as follows: $\varrho_1^0(\bar{w}_1) \simeq \varrho_2^0(\bar{w}_1) \exp[G_1 f(0) L]$. Similarly, when the DF wave is detected in the backward direction owing to a process of nonlinear reflection from the entrance boundary located at $z = 0$ (see ref. [1]), the maximum efficiency corresponds to $\bar{z}(0) = 0$ and $\varrho_1^0(\bar{w}_1) = \varrho_2^0(\bar{w}_1)$. The propagation theory we have just formulated is greatly simplified if we consider in the equations $\varepsilon(\bar{w}_1) = \text{const}$ and $f(\bar{w}_1) \equiv \varrho^0 = \text{const}$, *i.e.* if we are dealing with the steady-state interaction of the fields. In that case eq. (17*a*) may be rewritten in the following form:

$$(17b)\qquad \begin{cases} \varrho_q(z) = \dfrac{(\varrho^0)^2 g}{4\xi\tilde{\Gamma}'} \operatorname{cosech}^2\{\varrho_0 G_2(z-\bar{z})\} & \text{(DFG)}, \\[2ex] \varrho_q(z) = \dfrac{(\varrho^0)^2 g}{4\xi\tilde{\Gamma}'} \operatorname{sech}^2\{\varrho^0 G_2(z-\bar{z})\} & \text{(SFG)}, \end{cases}$$

where $\xi \equiv n_1/n_2$, $\varrho^0 = \varrho_1^0 \pm \varrho_2^0 \xi$, $\varrho_1^0 \equiv [\varrho_1(z)]_{z=0}$, $\varrho_2^0 \equiv [\varrho_2(z)]_{z=0}$ are the fields at the crystal boundary. $G_2 = g_2/(4v_2)$ and $\bar{z} = (2G_2\varrho^0)^{-1} \ln y$, where $y \equiv \varrho_1^0/(\xi\varrho_2^0)$. Again, $\bar{z}$ is always negative and different from zero for SFG, but can be of either sign for DFG. Furthermore, for SFG, the maximum value of $\varrho_q(z)$ is reached at $z = 0$: $\varrho_q(0) = \varrho_1^0 \varrho_2^0 g/\tilde{\Gamma}'$, further decreasing quasi-exponentially with $z > 0$ as shown in fig. 1. For DFG we have again $\varrho_q(0) = \varrho_1^0 \varrho_2^0 g/\tilde{\Gamma}'$, but the maximum value of $\varrho_q(z)$ is larger if $\bar{z} > 0$, and is given by $(\varrho_q)_{\max} \equiv \varrho_q(\bar{z}) = \varrho_q(0)(y+1)^2/(4y)$. We verify that, for $y = 1$, $\bar{z} = 0$ and $(\varrho_q)_{\max} = \varrho_q(0)$.

The general formulation in terms of polariton particles we have adopted has led us to build a transient propagation theory that applies entirely to the pure Raman and 2-photon absorption interaction [13, 14]: $\omega_0 = \omega_1 \pm \omega_2$, even when the lattice or electronic oscillator is not e.m. active. In that case $\varrho_q(z, t)$ has to be interpreted as the space-time distribution of the density of the crystal excitation in the medium. In the absence of spatial dispersion (*e.g.* Raman scattering from 0-phonon and 2-photon absorption from impurity levels), the phase-matching condition $\Delta\boldsymbol{k} = 0$ is always satisfied for e.m. inactive transitions and we can set $\alpha(\omega) = 0$ owing to (6). In that case the quantity $\varrho_q(\omega) = (\varrho^0)^2 g/4\xi\tilde{\Gamma}'$ appearing in (17*b*) may be conveniently expressed as a function of the frequency mismatch $\Delta\omega = \omega_1 \pm \omega_2 - \omega$ in the following form:

$$\bar{\varrho}_q(\omega) = \frac{\pi\hbar\omega_1\omega_2\omega_0(\varrho^0)^2}{2n_1^2n_2^2}\,\frac{Sd_Q'^2}{(\Delta\omega)^2+\Gamma^2}\,\frac{\Gamma}{\Gamma'}\,. \tag{18}$$

We have introduced in eq. (18) the strength of the oscillator $S \equiv \omega_p^2/\omega_0^2 \equiv \beta\varepsilon_\infty$ and $d_Q' \equiv d_Q/e^*$, where $\omega_p = (4\pi N e^{*2}/\mu)^{\frac{1}{2}}$ is the plasma frequency. We have shown above that, for the Raman effect, $\varrho_q(0)$ coincides with the maximum value $(\varrho_q)_{\max}$ of the exciton density created in the medium if $\bar{z} \geqslant 0$, while for 2-photon absorption $(\varrho_q)_{\max} = \varrho_q(0)$ and $(\varrho_q)_{\max} = \bar{\varrho}_q(\omega)[4y/(y-1)^2] < \bar{\varrho}_q(\omega)$.

The most complete expression for $\bar{\varrho}_q(\omega)$ valid for e.m. active excitations, in the presence of spatial dispersion, is the following [2]:

$$\bar{\varrho}_q(\omega) = \bar{\varrho}_q(\omega_0)\,\frac{[D' + C\omega_0^2]^2[D'^2+\beta\omega_0^4]}{C^2\beta\omega_0^5\omega^3}\,\frac{[\frac{1}{2}\alpha(\omega)]^2}{|\Delta\boldsymbol{k}|^2+[\frac{1}{2}\alpha(\omega)]^2}\,\frac{[\frac{1}{2}\tilde{\Gamma}(\omega)]^2}{(\Delta\omega)^2+[\frac{1}{2}\tilde{\Gamma}(\omega)]^2}\,, \tag{19a}$$

where

$$\bar{\varrho}_q(\omega_0) = \frac{\pi\hbar\omega_1\omega_2\omega_0(\varrho^0)^2}{2n_1^2n_2^2}\,\frac{Sd_Q'^2}{\Gamma\Gamma'}\,, \tag{19b}$$

after having neglected damping in the expression of D and set $D \simeq D' \equiv \omega_0^2 - \omega^2$. In eq. (19$a$) spatial dispersion is taken into account by the explicit expression of the group velocity $v_g = \partial\omega/\partial k$ that relates $\tilde{\Gamma}(\omega)$ to $\alpha(\omega)$ according to eq. (6). If we are dealing with an exciton state with small k near the valence band edge of a cubic semiconductor, assuming a diagonal tensor for ε_∞ and lack of degeneracy in $\omega_0(\boldsymbol{k})$, the group velocity is approximately given by [15]

$$v_g = \frac{1}{\partial k/\partial\omega} \approx \frac{c}{\sqrt{\varepsilon_\infty}}\left[1+\frac{\beta}{4}\left(\frac{\omega_0(\boldsymbol{k})}{\omega_0(\boldsymbol{k})-\omega}\right)^2\right]^{-1} \quad \text{for } |\omega_0-\omega| > \beta\omega\,. \tag{20}$$

3. – Electromagnetic fields in the medium. Nonlinear gain.

The formulation in terms of polariton particles adopted in section **2** does not permit a complete description of the generated e.m. field in the medium. The above analysis must be completed by additional conditions that lead to reflected and transmitted electromagnetic waves at the boundaries of the crystal at the fundamental frequencies ω_1, ω_2 as well as at the frequency of the wave ω_3. In this section we shall deal with classical waves rather than photons, in order to take into account the phase of the fields and to consider in a more transparent way the kinematics of the process. We limit ourselves here to considering a steady-state regime for the field, namely the value of the interaction path L in the medium is supposed to be much smaller that the width of the driving pulse $f(\overline{w}_1)$ and of the value of the coupling parameter $[G_1 f(0)]^{-1}$. Furthermore, we make the simplifying hypothesis that the optical fields at frequencies ω_1 and ω_2 are not affected by the nonlinear « gain » or « absorption » processes. The propagation equation of the polariton field at the frequency ω_3 may be written as follows:

$$(21)\qquad \begin{cases} \left[\nabla^2 + \left(\frac{\omega_3}{c}\right)^2 \varepsilon_\infty\right] \boldsymbol{E}_q(\omega_3) = -4\pi N e^* \left(\frac{\omega_3}{c}\right)^2 \boldsymbol{Q}(\omega_3) - 4\pi \left(\frac{\omega_3}{c}\right)^2 \boldsymbol{P}^{\rm NL}(\omega_3)\,, \\ \mu D(\omega_3)\boldsymbol{Q}(\omega_3) = e^* \boldsymbol{E}_q(\omega_3) + \boldsymbol{F}^{\rm NL}(\omega_3)\,. \end{cases}$$

$\boldsymbol{F}^{\rm NL}(\omega_3)$ and $\boldsymbol{P}^{\rm NL}(\omega_3)$ are, respectively, the nonlinear force and polarization, that can be defined in terms of the nonlinear energy density $U^{\rm NL}$ in the following way:

$$(21b)\qquad \begin{cases} \boldsymbol{F}^{\rm NL}(\omega_3) = -\dfrac{\partial U^{\rm NL}}{\partial[N\boldsymbol{Q}^*(\omega_3)]}\,, \\ \boldsymbol{P}^{\rm NL}(\omega_3) = -\dfrac{\partial U^{\rm NL}}{\partial \boldsymbol{E}_q^*(\omega_3)}\,. \end{cases}$$

According to HENRY and GARRETT [5], $U^{\rm NL}$ may be expressed for DFG in the following form:

$$(22)\qquad U^{\rm NL} = -\,[\underline{d}_E : \boldsymbol{E}_1(\omega_1)\,\boldsymbol{E}_2^*(\omega_2)\,\boldsymbol{E}_q^*(\omega_3) + N\underline{d}_Q : \boldsymbol{E}_1(\omega_1)\,\boldsymbol{E}_2^*(\omega_2)\,\boldsymbol{Q}^*(\omega_3) + \text{c.c.}]\,.$$

Generalization of (22) to SFG is obvious. We verify that eq. (22) is consistent with the definition of the Hamiltonian density given in eq. (4). The tensor nonlinear coupling coefficients appearing in eq. (22) may be taken, to the first approximation, as real and frequency independent over a large range of frequencies centred on ω_0. The complete solution of eq. (21) is a linear superposition of the solutions of the homogeneous Maxwell equation, written in terms of the electric fields, and of the inhomogeneous one. We consider, in the present section, the parametric excitation of a polariton wave in a nonlinear cubic crystal bounded by two plane faces that are not necessarily parallel. We assume that the linear medium in which the crystal is immersed has refractive index $n \simeq 1$ and that the optical fields $\boldsymbol{E}_1(\omega_1)$ and $\boldsymbol{E}_2(\omega_2)$ are undamped infinite plane waves interacting through the entire thickness of the crystal. They give rise, in the crystal, to a nonlinear polarization $\boldsymbol{P}^{\rm NL}(\omega_3) = \boldsymbol{p}P^{\rm NL}(\omega_3)$. We limit ourselves to considering the case in which the (real) wave vectors $\boldsymbol{k}_1$ and $\boldsymbol{k}_2$ of the optical fields in the medium belong to the same incidence plane, which is orthogonal to the crystal boundaries, with $\boldsymbol{p}$ orthogonal to that plane.

A particular solution $\boldsymbol{E}_3(\omega_3)$ of eq. (21) may be given in the following form:

$$(23)\qquad \boldsymbol{E}_s(\omega_3) = \frac{[d'_E + d'_Q(\omega_p^2/D)]|E_1 E_2|}{(k_s c/\omega_3)^2 - (\varepsilon_\infty + \omega_p^2/D)} \left(\boldsymbol{p} - \frac{\boldsymbol{k}_s(\boldsymbol{k}_s\cdot\boldsymbol{p})}{k_T^2}\right) \exp[i(\boldsymbol{k}_s\cdot\boldsymbol{r} - \omega_3 t)] + \text{c.c.}$$

We have introduced in eq. (23) the notation for the moduli of the vectors: $E_i \equiv |\boldsymbol{E}_i|$, $k_i \equiv |\boldsymbol{k}_i|$, etc. In our particular case the term $k_s(k_s \cdot p)/k_T^2$ is zero throughout the crystal. The wave vector of the driving polarization wave is

$$\boldsymbol{k}_s = \boldsymbol{k}_1 \pm \boldsymbol{k}_2 \equiv (n_s \omega_3/c)(\boldsymbol{k}_s/k_s)\,, \qquad n_s^2 \equiv \varepsilon_s\,, \qquad k_T^2 \equiv \left(\frac{\omega_3}{c}\right)^2 (\varepsilon_\infty + \omega_p^2/D)\,,$$

and the nonlinear coefficients are rewritten in the form $d'_E = 4\pi d_E$, $d'_Q = d_q/e^*$. The solution of the homogeneous Maxwell equation for the ω_3-field near the entrance boundary is composed of a transmitted wave $\boldsymbol{E}_T(\omega_3)$, travelling in the nonlinear medium with wave vector $\boldsymbol{k}_T = \boldsymbol{k}'_T + i\boldsymbol{k}''_T$, and a nonlinearly reflected wave $\boldsymbol{E}_R(\omega_3)$, travelling in the backward direction, in the linear medium. If we call φ'_s, φ'_T, φ'_R the angles made by the $\boldsymbol{k}_s$, $\boldsymbol{k}_3 \equiv \boldsymbol{k}'_T$, $\boldsymbol{k}_R \equiv n(\omega_3/c)\,\boldsymbol{k}_R/k_R$ wave vectors with the $\boldsymbol{z}$-axis as defined above, the continuity of the transverse components of the momenta of the waves at the boundary leads to the generalized Snell equations $n_s \sin\varphi'_s = n_T \sin\varphi'_T = n \sin\varphi'_R$, where $n_T = n'_T(\omega_3) + i n''_T(\omega_3)$ is the complex refractive index of the nonlinear medium at the frequency of the wave created in the medium. We can show that, in a zone of large dispersion of the linear response of the medium, the $\boldsymbol{k}'_T$ wave is an inhomogeneous wave. In order to show that, it is convenient to express $\cos\varphi'_T$ in the form $\cos\varphi'_T = p\exp[i\psi]$. Expressions for p and ψ in terms of the relative (scalar) phase mismatch $\Delta = (n_s - n'_T(\omega_3))/n'_T(\omega_3)$ and of the relative absorption parameter $\delta = n''_T(\omega_3)/n'_T(\omega_3)$ of the nonlinear medium are immediately obtained from the generalized Snell equation given above:

$$\text{(24)} \qquad \begin{cases} p^2\cos 2\psi = 1 - (1+\Delta)^2 \dfrac{1-\delta^2}{(1+\delta^2)^2}\sin^2\varphi'_s\,, \\[2ex] p^2\sin 2\psi = \dfrac{2\delta(1+\Delta)^2}{(1+\delta^2)^2}\sin^2\varphi'_s\,. \end{cases}$$

Furthermore, the spatial argument of the $\boldsymbol{k}_T$-wave, $\boldsymbol{k}_T\cdot\boldsymbol{r}$, may be written in the following form:

$$\text{(25)} \qquad \boldsymbol{k}_T\cdot\boldsymbol{r} = k_T(x\sin\varphi'_T + z\cos\varphi'_T) = \\ = k'_T\{x(1+\Delta)\sin\varphi'_s + zp[(\cos\psi - \delta\sin\psi) + i(\sin\psi + \delta\cos\psi)]\}\,,$$

x and z being the spatial co-ordinates corresponding, respectively, to the $\boldsymbol{x}$-axis, belonging to the incidence plane and to the boundary plane, and to the $\boldsymbol{z}$-axis.

We verify that the surfaces of constant amplitude of the $\boldsymbol{k}_T$-wave are given by $z = \text{const}$ and are therefore planes parallel to the boundary. The surfaces of constant real phase are given by $x(1+\Delta)\sin\varphi'_s + zp\,(\cos\psi - \delta\sin\psi) = \text{const}$

and are planes whose normals make an angle φ_T'' with the z-axis, where

$$(26)\qquad \begin{cases} \cos\varphi_T'' = \dfrac{p(\cos\psi - \delta\sin\psi)}{[(1+\Delta)^2\sin^2\varphi_s' + p^2(\cos\psi - \delta\sin\psi)^2]^{\frac{1}{2}}} \simeq (1-\Delta\,\mathrm{tg}^2\varphi_s')\cos\varphi_s'\,, \\[2ex] \sin\varphi_T'' = \dfrac{(1+\Delta)\sin\varphi_s'}{[(1+\Delta)^2\sin^2\varphi_s' + p^2(\cos\psi - \delta\sin\psi)^2]^{\frac{1}{2}}} \simeq (1+\Delta)\sin\varphi_s'\,. \end{cases}$$

The simplified expressions appearing in eq. (26) correspond to assuming $\Delta \ll 1$, $\delta \ll 1$ and $\Delta\,\mathrm{tg}^2\varphi_s' \ll 1$. With these approximations, the planes of equal amplitude are still given by $z = \text{const}$ and the direction of the normal to the constant-phase planes is determined by Δ and not by δ. However δ is still responsible for the inhomogeneous character of the wave, owing to eq. (24). The following simplified expression for $\boldsymbol{k}_T\cdot\boldsymbol{r}$ holds:

$$(27)\qquad \boldsymbol{k}_T\cdot\boldsymbol{r} \simeq k_T'[x(1+\Delta)\sin\varphi_s' + z\cos\varphi_s'(1-\Delta\,\mathrm{tg}^2\varphi_s') + iz\delta/\cos\varphi_s']\,.$$

The intensities of the $\boldsymbol{k}_R$ and $\boldsymbol{k}_T$ fields are found by writing the continuity equations for the electric and magnetic fields at the boundaries. Furthermore, the continuity condition for the transverse components of the momenta, expressed by the generalized Snell equation given above, leads to the condition of real propagation of the fields. For the reflected field, that condition is $\sin\varphi_R' \ll 1$, leading to the following one: $\sin\varphi_s' \leqslant [n_T'(\omega)(1+\Delta)]^{-1}$. The amplitude of that wave is found to be

$$(28)\qquad E_R = \frac{[d_E' + d_Q'(\omega_p^2/D)](n_s\cos\varphi_s' - n_T\cos\varphi_T')}{(\varepsilon_s - \varepsilon_T)(n\cos\varphi_R' + n_T\cos\varphi_T')}\,|E_1E_2^*|\,.$$

In the vicinity of the reststrahl band the condition of real propagation for the $\boldsymbol{k}_R$-field leads us to consider only very small angles φ_s', φ_R', φ_T'. For normal reflection and for Δ, δ and n/n_T' much smaller than 1, the amplitude of the reflected field may be written in a simple and significant way:

$$(29)\qquad E_R \simeq \frac{d_E'|E_1E_2^*|}{2\varepsilon_\infty}\,\frac{D + C\omega_0^2}{D + \beta\omega_0^2}\,,$$

where $\beta = \omega_p^2/(\omega_0^2\varepsilon_\infty) = S/\varepsilon_\infty$, and $C = Sd_Q'/d_E'$ is the nonlinear parameter we have defined above. We verify that, with the given approximations, the reflected field is proportional to the ratio of the nonlinear contribution to the dielectric constant at frequency ω_3 and of the same (linear) dielectric constant. A measurement of the intensity and of the phase of the nonlinearly reflected wave as a function of ω_3 would lead to a direct measurement of the parameter $C(\omega_3)$ that characterizes the nonlinear response of the crystal. By introducing the field $E_s = [d_E'(1 + C\omega_0^2/D)/(\varepsilon_s - \varepsilon_T)]|E_1E_2|$, a nonlinear reflectivity

$R = |E_R|^2/|E_s|^2$ can be defined that is formally identical to the usual expression of the linear reflectivity of linear optics [16], provided we reinterpret, in an obvious way, the quantities n_s and φ_s' appearing in eq. (28). The simple expression (29) obviously does not hold for ω_3 lying in the reststrahl band in which $\Delta \gg 1$. It must be replaced by a more involved and less transparent expression. Of course, as far as nonlinear reflectivity is concerned, the reststrahl band keeps most of its well-known linear properties in the nonlinear regime. We shall consider later in the paper the polariton modes that propagate in the restrahl band (surface polaritons). The wave travelling into the nonlinear medium comes from the interference of the solution $\boldsymbol{E}_s$ given by eq. (23) with the transmitted (inhomogeneous) wave E_T' which is the solution of the homogeneous propagation eq. (21). In the case in which the nonlinear polarization is orthogonal to the incidence plane defined by $\boldsymbol{k}_1$ and $\boldsymbol{k}_2$ and for $\Delta \ll 1$, we are led to the complete solution of eq. (21) through the continuity equations for the electric and magnetic fields at the boundary. If we write the transmitted wave in the form of a plane wave, the complete solution may be given in the following form:

$$\boldsymbol{E}_q(\omega) = \frac{\boldsymbol{p}d_E'(1 + C\omega_0^2/D)|E_1 E_2^*|}{2n_T'(\Delta + i\delta)} \exp[i\boldsymbol{k}_s \cdot \boldsymbol{r}] \cdot \{1 - A_T \exp[-k_T'(\delta + i\Delta)(z/\cos\varphi_s')]\}, \tag{30}$$

where

$$A_T = \frac{n\cos\varphi_R' + n_s\cos\varphi_s'}{n\cos\varphi_R' + n_T\cos\varphi_T'} \simeq \frac{n_s + n - [n_s^2/(2n)]\sin^2\varphi_s'}{n_T + n - [n_s^2/(2n)]\sin^2\varphi_s'}. \tag{31}$$

Equations (30) and (31) correspond to the condition $|\varphi_s'| \ll \pi/2$. We can verify in (30) that the surfaces of constant amplitude of the wave $\boldsymbol{E}_q$ are planes parallel to the boundary. We can also verify that the phase-matching parameter $\Delta \ll 1$, which affects in a quasi-resonant way the undamped transmitted wave, appears in the exponential argument of the free solution and in the expression of A_T. The approximate expression of A_T given in (31), valid for φ_s' small, may be further simplified by setting $\varphi_s' = 0$, $n = 1$, $\delta \ll 1$, $\Delta \simeq 0$. In that case it reduces to the simple equation $A_T \simeq 1 - i\delta$, which corresponds to the simplified expression of $\boldsymbol{E}_R$ given in (28). As we remarked above, it is interesting to take into consideration the particular wave solution $\boldsymbol{E}_s$ of the propagation equation, because, in the approximation that neglects parametric gain effects and for an interaction length L much larger than $(k_T'')^{-1}$, it gives the amplitude of the e.m. part of the polariton wave generated in the crystal. In eq. (30) we can verify that, in the range of polariton frequencies in which phase matching of the interaction can be achieved and for $\delta \ll 1$, the driven-wave intensity E_s^2 is a Lorentzian function of Δ if k_T'' is independent of k_T'. This behaviour has recently suggested to us the use of the coherent excitation of the polaritons as a means of studying the optical linear and nonlinear responses of the crystal near the reststrahl band; a zone in which the large optical absorption of the medium does not in practice allow accurate measurements of the optical parameters by the usual methods of the linear spectroscopy of solids. Obviously, the above considerations refer to a behaviour of the parametric mixing processes which is quite general in nonlinear optics. They apply to experiments involving all kinds of e.m. active elementary excitations, such as optical phonons, acoustic phonons, magnons, plasmons, excitons, etc. In the case of perfect phase matching $\Delta = 0$, the amplitude of the driven solution appearing in eq. (30) may also be written in the following simple form:

$$E_s = i\,\frac{d'_E|E_1 E_2|}{\varepsilon_\infty \beta \Gamma \omega_3 \omega_0^2}\,D^*(D + C\omega_0^2)\,, \tag{32}$$

where $\beta = S/\varepsilon_\infty$. Thus far we have discussed the propagation of the general solution of the Maxwell equation for the e.m. field created in the medium, disregarding the final process of the emission of that field out of crystal. The study of this process does not present particular complications. As far as the free solution is concerned, the usual Fresnel refraction theory applies without modifications [16]. The same theory is nevertheless not valid in general for the driven solution and it must be substituted by the more involved analysis we have previously adopted in connection with the nonlinear refraction at the entrance boundary. Thus, we are led to consider, in correspondence with the driven solution $\boldsymbol{E}_s$ at the exit boundary, a wave $\boldsymbol{E}'_T$ that is transmitted out of the crystal and a damped inhomogeneous wave reflected back in the medium. In conclusion, the e.m. field, which is created in the crystal slab in a single-pass interaction, comes from the interference of an undamped driven wave with wave vector $\boldsymbol{k}_s$ and two (propagating or evanescent) inhomogeneous waves originating at the boundaries and travelling in the crystal with wave vectors having moduli equal to $n'_T(\omega_3/c)$.

The direction of propagation of the transmitted e.m. field $\boldsymbol{E}_T(\omega)$ is still given by a set of generalized Snell equations analogous to the ones written for the entrance boundary. In addition, analogous considerations on the continuity at the boundary of the electric and magnetic fields lead to the amplitude of the refracted fields. We limit ourselves to giving here, for the sake of completeness, the expression of the field $|\boldsymbol{E}'_T(\omega)|$ that is radiated in a linear medium of refractive index n in a direction making an angle φ with the normal $\boldsymbol{z}'$ to the exit boundary:

$$\boldsymbol{E}'_T(\omega) = \boldsymbol{E}_s(\omega)\,\frac{n_s \cos\varphi_s + n_T \cos\varphi_T}{n\cos\varphi + n_T\cos\varphi_T}\,. \tag{33}$$

By analogy with our previous discussion, φ_s and φ_T are now the angles (lying in the range $-\pi/2 \rightarrow \pi/2$) made by the wave vectors of the corresponding beams with $\boldsymbol{z}'$.

We have assumed so far that the phase matching of the three-wave interaction can be achieved throughout the polariton region. Actually that process is generally a critical one and can be achieved only in special conditions, if the crystal does not show a suitable optical anisotropy to allow the corresponding well-known method of phase matching [1, 17]. If that condition is not present (*e.g.* in cubic crystals), momentum conservation of the interaction may still be provided, even for collinear kinematics, by the so-called « dispersion » phase matching which is based on the combined effects of the optical dispersion affecting the pump frequencies ω_1, ω_2 and the near-resonance frequency ω_3. We refer the reader to [2] for a detailed discussion of that effect in DFG. We only note here that this process ensures phase matching for ω_3 lying in the lower branch of the polariton dispersion curve for DFG and in the upper branch for SFG, if ω_1 and ω_2 lie in a zone of normal optical dispersion. Dispersion phase matching in the upper branch for DFG and in the lower branch for SFG is possible, but implies that one or both pump frequencies lie in a zone of anomalous dispersion in the crystal, *e.g.* near a strong exciton or impurity resonance.

3'1. *Nonlinear gain.* – Thus far we have neglected the effect of the nonlinear gain on the dynamics of the process of polariton generation. Nevertheless we have seen that this effect becomes important when high fields and long interaction lengths are present. An expression for the gain may be derived from eqs. (7*b*) and (9*b*) by applying the classical arguments we have adopted in the present section. We assume here that $\tilde{\Gamma}'(\omega)=\tilde{\Gamma}(\omega)$, $\tilde{\Gamma}(\omega_0)\equiv\Gamma$ and we write $\tilde{\Gamma}(\omega)=\Gamma\beta\omega_0^2\omega^2\times[D'^2+\beta\omega_0^4]^{-1}$, owing to (6) and the Huang dispersion relation for $v_g(\omega)$ and $\alpha(\omega)$ [6]. We find that the gain g affecting the polariton wave may be expressed in the following form:

$$g(\omega, k_s)=g_{\max}\frac{\omega_p^4\omega^6\Gamma^4}{4c^4(D')^4}\frac{1}{(\Delta\omega)^2+[\frac{1}{2}\tilde{\Gamma}(\omega)]^2}\frac{1}{|k_s^2-(\omega/c)^2(\varepsilon_\infty+\omega_p^2/D)|^2}, \tag{34}$$

where

$$g_{\max}=\frac{h\omega_1\omega_2\omega_p^2}{n_1^2n_2^2\omega\Gamma}\left|\frac{D'}{\omega_p^2}d'_E+d'_Q\right|^2 \tag{35}$$

is the expression for the maximum gain, which is reached when $\Delta\omega=\omega_1\pm\omega_2-\omega$ and $|\boldsymbol{k}_s|\equiv|\boldsymbol{k}_1\pm\boldsymbol{k}_2|=|\boldsymbol{k}'_T|$, *viz.* when $\{(\omega_1\pm\omega_2), k_s\}$ lie on the undamped polariton dispersion curve. A similar conclusion has been reached in ref. [5] for the gain g_2 affecting the $\boldsymbol{k}_2$-wave in stimulated scattering. In that work, an expression for g_2 similar to our $g_{\max}$ is given. We note that, for $\omega\to\omega_0$, the effect of the electronic nonlinearity d'_E disappears and we are left with an expression corresponding to the resonant 2-photon absorption coefficient for SFG and to the Raman gain for DFG. Our present theory is valid only if spatial dispersion affecting the excitation associated with the excited polariton wave is considered to be a small effect [18]. If it is not, substantial complications arise in the theory of wave propagation due to the occurrence of « anomalous » waves [19]. We shall develop the theory of gain in the presence of spatial dispersion in an *ad hoc* paper.

4. – 4-photon resonant coherent scattering.

The presence of the coherent polariton wave created by SFG or DFG in the medium may be probed either by direct detection of the generated e.m. field at frequency $\omega_3 = \omega_1 \pm \omega_2$ and transmitted out of the medium, or by means of a resonant 4-photon coherent inelastic-scattering process. In spite of the higher-order nonlinear process involved, this new kind of experiment generally proves to be easier, for the purpose of the polariton wave detection, than the corresponding direct SF or DF 3-wave mixing experiment, because surface effects are generally negligible and, more important, because we can by this process up or down convert the polariton wave in a zone of the optical spectrum where absorption of the crystal is negligible and where fast and sensitive detectors are operating. Furthermore, we should stress the intrinsic physical interest of this effect, which allows the study of some linear and nonlinear properties of the crystal not generally accessible by other methods of spectroscopy. This process also suggests entirely new methods of nonlinear spectroscopy such as the ones devised for the study of the near-resonance dispersion of the nonlinear optical susceptibilities

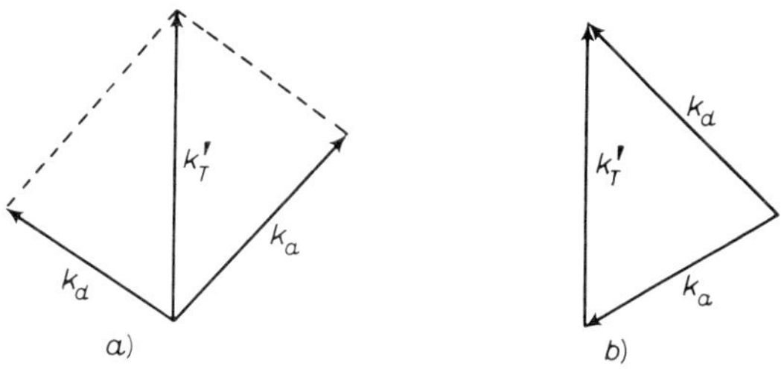

Fig. 3. – Kinematics of the polariton wave $\boldsymbol{k}'_T$ coherent-scattering schemes. The $\boldsymbol{k}'_T$-wave may be excited in turn by either the sum or difference frequency interaction for either scheme. The four possible overall configurations are considered in sect. **5**. *a*) $\boldsymbol{k}'_T = \boldsymbol{k}_d + \boldsymbol{k}_a$, *b*) $\boldsymbol{k}'_T = \boldsymbol{k}_d - \boldsymbol{k}_a$.

of crystals [2, 9, 20] and the solid-state spectroscopy *in momentum space* that has been applied for the first time in optics by the Quantum Optics Group in Orsay [20]. The diagrams of the possible kinematical configurations for this process are shown in fig. 3 for SF and DF excitations.

The analysis of the present effect is not particularly involved if we make use of the simplifying approximations we have already introduced in the paper, *viz.* transverse plane-wave approximation, monochromaticity for the fields and the scalar representation of the tensor components. Obviously, this last simplifying condition broadens the scope of the present discussion, as it covers the general behaviour of a nonlinear medium, but care is advised in the application of the theory to the case of a particular crystal, even of cubic symmetry.

If we assume that the probing field $\boldsymbol{E}_d(\omega_d, \boldsymbol{r}) = \boldsymbol{E}_d \exp[-i\boldsymbol{k}_d \cdot \boldsymbol{r}]$ is not depleted or amplified along the interaction path, the simple expression for the free-energy density U^{NL} must be rexritten in the following form:

$$U^{\text{NL}} = -\{\underline{d}_E :(\boldsymbol{E}_1 \boldsymbol{E}_2^* \boldsymbol{E}_q^* + \boldsymbol{E}_d \boldsymbol{E}_a^* \boldsymbol{E}_q^*) + N\underline{d}_Q :(\boldsymbol{E}_1 \boldsymbol{E}_2^* \boldsymbol{Q}^* + E_d \boldsymbol{E}_a^* Q^*)\}, \tag{36}$$

while the polariton e.m. field may be written after (23) in the form

$$\boldsymbol{E}_q(\omega_3) = \frac{d'_E + d'_Q \omega_p^2/D}{\Delta\varepsilon} \left(|E_1 E_2^*| + |E_d E_a^*|\right)$$

with $\boldsymbol{E}_d(\omega_a \cdot \boldsymbol{r}) = \boldsymbol{E}_a \exp[-i\boldsymbol{k}_a \cdot \boldsymbol{r}]$ the (weak) optical scattered field generated in the medium and $\Delta\varepsilon \equiv n_s^2 - (\varepsilon_\infty + \omega_p^2/D)$. Equation (36) is written for down-conversion scattering of the field $\boldsymbol{E}_d$ and for DFG. Generalization to other kinematical scattering schemes and for SFG is immediate and does not affect the final results of the theory. We can now apply (21*b*) in order to evaluate the nonlinear polarization $P^{\text{NL}}(\omega_a, \boldsymbol{r}) = P_a^{\text{NL}} \exp[-i\boldsymbol{k}_a \cdot \boldsymbol{r}]$. We finally find the simple expression

$$P_a^{\text{NL}} = \frac{d_E'^2}{4\pi} \left\{ \frac{(1 + C\omega_0^2/D^*)^2}{(\Delta\varepsilon)^*} + \frac{C^2\omega_0^2}{SD^*} \right\} E_a |E_d|^2 + \\ + \frac{d_E'^2}{4\pi} \left\{ \frac{(1 + C\omega_0^2/D^*)^2}{(\Delta\varepsilon)^*} + \left(\delta + \frac{C^2\omega_0^2}{SD^*}\right) \right\} E_1^* E_2 E_d \,. \tag{37}$$

In eq. (37) $\delta = \chi_{\text{NR}}^{(3)}/(4\pi[d'_E]^2)$ brings into the theory the contribution of the third-order susceptibility that accounts for the overall coupling of the 4 photons $\boldsymbol{k}_1, \boldsymbol{k}_2, \boldsymbol{k}_d, \boldsymbol{k}_a$. It is not affected by the polariton resonance and can be assumed to be a real quantity, as for d'_E, if the interacting optical frequencies are far from other resonances of the medium. The experimental value of δ has been found to be ~ 6 for GaAs [21] and ~ 180 for $LiNbO_3$ [22]. The large difference among the two values is not unexpected as δ can be shown, by quantum-mechanical arguments, to be approximately proportional to the inverse of the absorption coefficient of the crystal in the transparency region [23]. Inspection of (37) and of the Maxwell equation for $\boldsymbol{E}_a(\omega_a, \boldsymbol{r})$ leads us to consider a nonlinear correction to the refractive index $\bar{n}'_a = \sqrt{\bar{\varepsilon}_a}$ of the crystal at the frequency ω_a and a nonlinear gain for the scattered field. We find

$$n'_a = \sqrt{\bar{\varepsilon}_a + d_E'^2 \frac{C^2\omega_0^2}{SD'} |E_d|^2}\,,$$

$$\text{gain} = \frac{\omega_a}{c} \frac{d_E'^2 (\bar{D}')^2}{2S\omega_0^3 \omega_3 \gamma_0 \bar{n}'_a}\,,$$

where $\bar{D}' \equiv \omega_0^2(1+C) - \omega^2$ and $\gamma_0 = \Gamma/\omega_0$. For $\omega_3 \to \omega_0$ the nonlinear correction to $\bar{n}'_a$ and the gain correspond to the well-known behaviour of the anti-Stokes generation in coherent Raman scattering [1, 24, 25]. The intensity of the $\boldsymbol{k}_a$-beam detected outside the medium may be expressed in the following form:

$$I_a = \frac{4(2\pi\omega_3)^4\omega_a^2}{c^6 n_a'^2} I_1 I_2 I_d |\tilde{d}|^2 \left[\sin^2\frac{|\Delta\boldsymbol{k}_a|l}{2}\Big/\left(\frac{|\Delta\boldsymbol{k}_a|l}{2}\right)^2\right] l^2 \tag{38}$$

with

$$|\tilde{d}|^2 = \left| d'_E \left\{ \frac{(1 + C\omega_0^2/D^*)^2}{[k_s^2 - (\omega/c)^2(\varepsilon_\infty + \omega_p^2/D^*)]} + \left[\frac{C^2\omega_0^2}{SD^*} + \delta\right]\right\}\right|^2 \tag{39}$$

being the nonlinear coefficient that accounts for the coupling. We can verify that, for $k_s \approx k'_T$, $|k_s^2 - (\omega/c)^2(\varepsilon_\infty + \omega_p^2/D^*)|^{-2}$ approaches a Lorentzian function of $k_s - k'_T$ having its maximum for $k_s = k'_T$. The dependence of I_a on the phase matching of the 3-wave polariton generation process is at the basis of the mentioned spectroscopy in momentum-space [20, 26]. We should nevertheless notice that the second term appearing at the r.h.s. of (39) is generally not negligible compared with the phase-matching resonant one in the same expression. In particular, the two terms are of the same size and $\pi/2$ out of phase for $\omega = \omega_0$ if $\delta = 0$. The effect of this additional term could explain the asymmetry of some of the momentum-space spectral lines we have found in the experiments [20]. The moduli and phases of the two contributions to $|\tilde{d}|^2$ may be easily evaluated on the basis of the nonlinear coefficients of the crystal for each polariton frequency excited and the experimental spectra may therefore be easily reduced to give the correct spectra leading to the study of the polariton dispersion and damping. We note that I_a is proportional to the product of the intensities of the interacting beams and to the function $[\sin\frac{1}{2}|\Delta\boldsymbol{k}_a|l/(\frac{1}{2}|\Delta\boldsymbol{k}_a|l)]^2$ that accounts for the overall phase mismatch $\Delta\boldsymbol{k}_a$ of the four optical waves $\boldsymbol{k}_1$, $\boldsymbol{k}_2$, $\boldsymbol{k}_d$, $\boldsymbol{k}_a$. The value of the function depending on $\Delta\boldsymbol{k}_a$ is always near unity under our experimental conditions [20].

5. – Nonlinear spectroscopy of bulk polaritons.

The present work is not a review paper and therefore we shall not give an exhaustive account of the experimental situation of the branch of nonlinear spectroscopy we are dealing with here. However, before focusing briefly on the author's own work, it seems worth-while to mention some key works performed on the subject.

5'1. *Sum frequency generation (resonant second-harmonic generation).* – The first experiment of near-resonance SFG was reported by MAHR and co-workers at Cornell University [27]. In that experiment, a tunable dye laser was used to excite resonantly by SFG the first exciton line in CuCl at 3.215 eV. The result of that experiment shows a resonant behaviour in the nonlinear susceptibility of the crystal in correspondence with the exciton resonance. This work parallels the investigation of the same polariton resonance by 2-photon absorption [28, 29]. An extensive presentation of earlier works on 2-photon absorption spectroscopy may be found in [30].

The 4-photon mixing associated with a SF excitation of the above polariton resonance in CuCl has been reported by KRAMER, PARSONS and BLOEMBERGEN, which investigated experimentally the effect of interference of the third-order light mixing and second-harmonic exciton-polariton generation [31]. We have presented the theory of this effect in sect. **4**. For further details on this work see the lecture of BLOEMBERGEN in this volume.

5·2. *Difference frequency generation.* – The first DF mixing experiment of relevance in nonlinear spectroscopy has been reported by FAUST and HENRY [9]. In that work the dispersion of the nonlinear optical susceptibility near the 0-phonon resonances in GaP has been thorougly studied. Further works on this process have been reported: they correspond to mixing experiments [2, 32] and to stimulated scattering experiments from 0-phonon polaritons [33].

In the author's work [2], two coherent beams generated by tunable lasers were injected in a GaP crystal and the coherent IR radiation corresponding to the 0-phonon polariton excitation was detected outside the crystal. Various measurements were performed in the experiment. The angular distribution

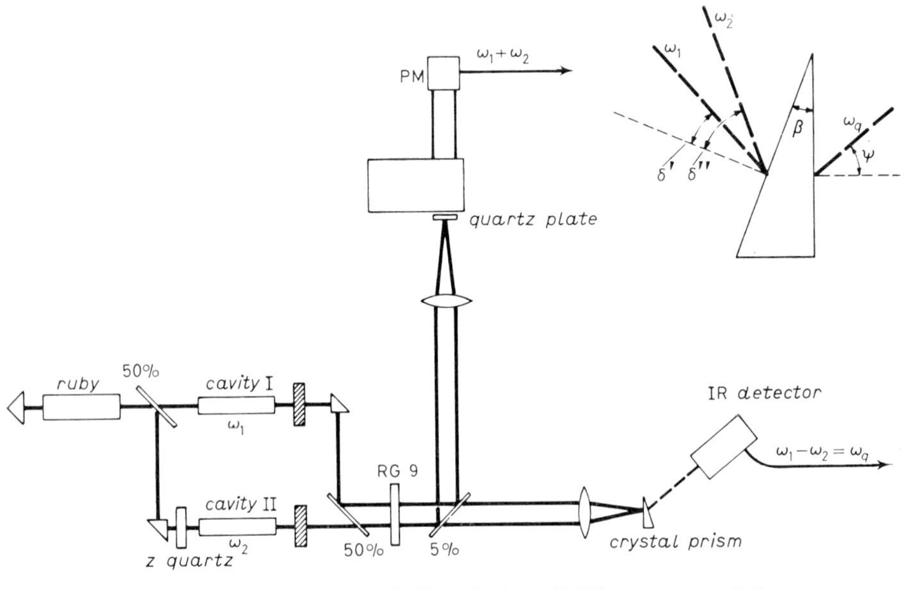

Fig. 4. – Schematic diagram of the bulk polariton DFG apparatus [2].

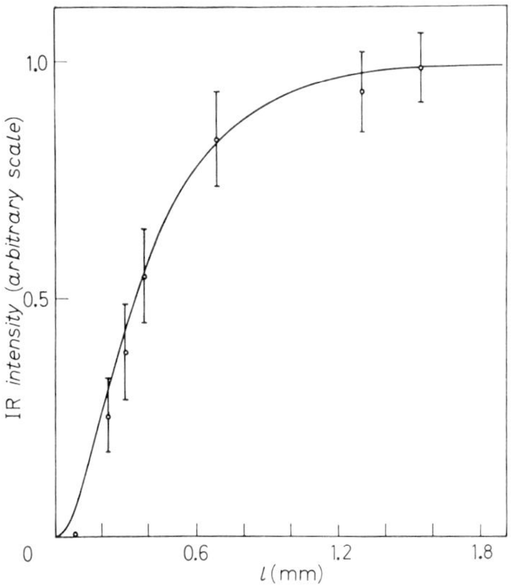

Fig. 5. – Infra-red intensity emitted at $\lambda = 34.6\ \mu m$ in the phase-matching condition as a function of the interaction length l in GaP. The curve directly shows the spatial distribution of the e.m. field in the crystal. $\alpha = 72\ cm^{-1}$, 300 K.

of the emitted intensity I_{IR} and the study of the evolution of the e.m. field in the crystal (fig. 4, 5) allowed a set of measurements of the absorption coefficient of GaP in a zone of large absorption ($\lambda_{IR} \approx 35\ \mu m$). Furthermore, the frequency dependence of I_{IR} has led us to the study of the dispersion of the nonlinear susceptibility of GaP in the lower branch of the polariton dispersion curve.

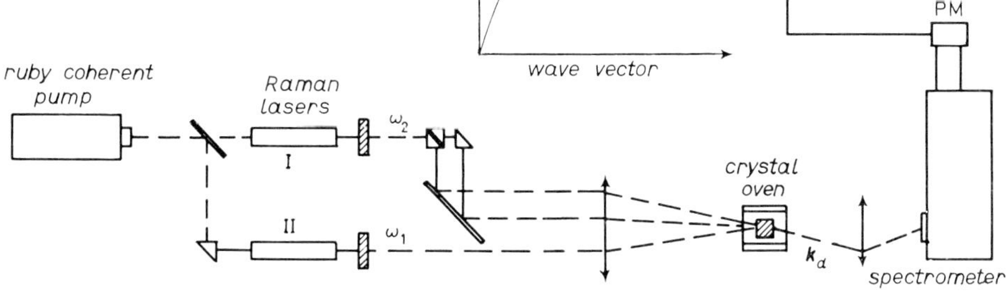

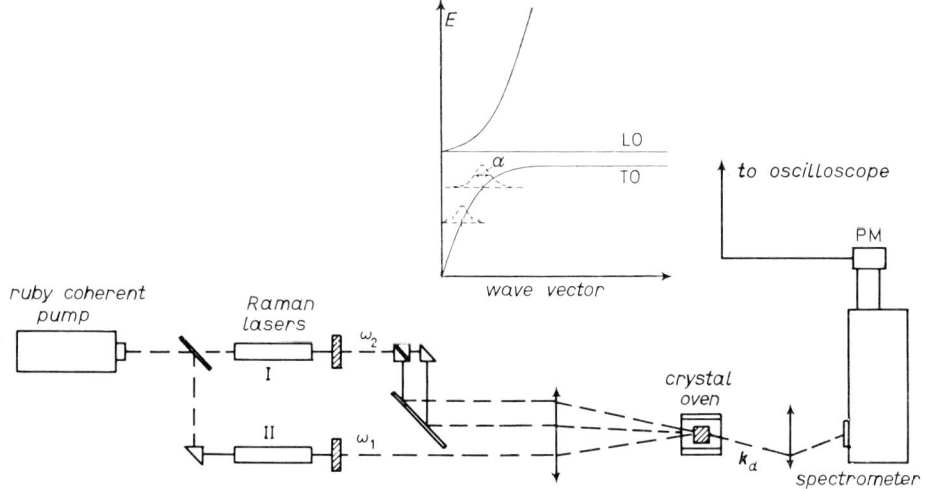

Fig. 6. – Schematic diagram of the $\boldsymbol{k}$-space spectroscopy apparatus [20, 34].

Momentum-space spectroscopy has been performed in GaP at various temperatures (fig. 6) (ref. [20, 26, 34]). At 300 K the lower-branch polariton dispersion curve has been experimentally plotted with the high accuracy that is intrinsic in the method [26]. Furthermore, the absorption coefficient of the crystal has been measured in the near-reststrahl high-reflectivity region, which is hardly accessible to linear methods as far as measurement of this quantity is concerned (fig. 7). The temperature dependence of $|\boldsymbol{k}'_T|$ and α near the TO

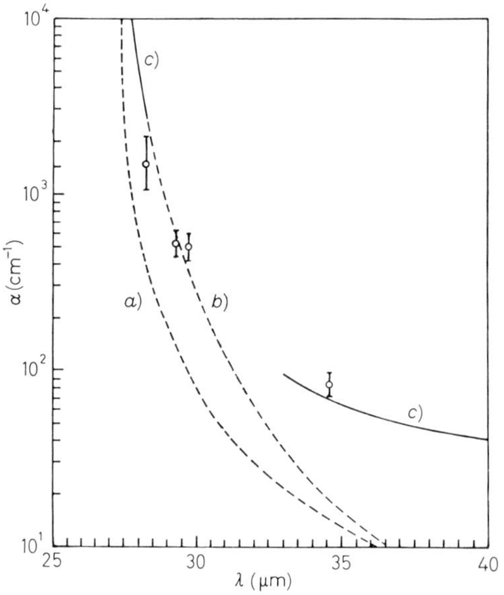

Fig. 7. – Absorption coefficient of GaP (300 K) *vs.* wavelength. Curve *a*) has been drawn on the basis of the IR data reported in ref. [36]. Curve *b*) has been drawn on the basis of the multiple-oscillator model for the damping of the 0-phonon in GaP [37]. Curve *c*) shows the results of a previous experimental investigation on $\alpha(\lambda)$ using conventional linear methods. Upper branch: Kramers-Krönig analysis of reflection experiments. Lower branch: transmission data [38]. Curve *c*) is reported without structure.

resonance has been considered for Gap in ref. [34]. It has been found that near $T = 450$ K these two quantities, as well as ω_{TO} and Γ, show a rapid change. That temperature coincides with the Debye temperature of GaP calculated on the basis of the elastic constants [35].

The 4-photon coherent-scattering process has been applied to the study of the dispersion of the nonlinear susceptibility in $LiNbO_3$ [22].

The reader can find further discussions on the SFG, DFG, 4-photon scattering experiments and a comprehensive bibliography in the works of BLOEMBERGEN, SHEN and LOUDON published in this volume.

6. – Nonlinear excitation of surface polaritons.

There are situations in which the dielectric constant of the medium, at the frequency ω at which the nonlinear polarization is excited, is negative [36]. The most common situation of this sort is presented by a metal for $\omega < \omega_p$ (plasma frequency) and by a semiconductor crystal for ω lying in the (phonon or exciton) reststrahl band [2, 18]. When this is the case, the driven e.m. fields (*i.e.* the particular solution of the inhomogeneous Maxwell equation which is associated with the driving NL polarization wave) can still propagate in the bulk of the crystal, with a wave vector $\boldsymbol{k}_s = \boldsymbol{k}_1 \pm \boldsymbol{k}_2$ and without attenuation, if the attenuation of the fundamental fields is absent. However, the free solution, which arises at the boundaries, cannot propagate in the bulk because $k''_T(\omega) > k'_T(\omega)$. The only propagating mode in the medium corresponds to an evanescent field which is bound to the surface, propagates with a wave vector $\boldsymbol{k}_x$ parallel to the surface itself and decreases exponentially in the direction z normal to the surface being proportional to [16]

$$\exp[-\bar{\alpha}z]\,, \qquad \bar{\alpha} = \sqrt{k_x^2 - k_0^2\bar{\varepsilon}\bar{\mu}}\,, \qquad k_0 = \omega/c\,.$$

We can verify that for $\bar{\varepsilon} < 0$ (and $\bar{\mu} > 0$) $\bar{\alpha}$ is real for every k_x. Suppose now that the boundary we are considering is an interface between two media: a nonlinear one characterized by $\bar{\varepsilon} < 0$ and a linear one with $\varepsilon > 0$. The free wave can propagate in the bulk of the linear medium if $k_x^2 < k_0^2\varepsilon$. If this equation is not satisfied, a second evanescent wave with $\alpha = \sqrt{k_x^2 - k_0^2}$ travels at the surface in the linear medium with the same wave vector k_x of the wave in medium $\bar{\varepsilon}$. Let us study this situation on the basis of the Maxwell equations written in the neighbourhood of the boundary. Let us consider the excitation of free TM waves on either side of the interface. That one is defined by the (x, y) reference plane, where x is assumed to be the direction of wave propagation. The z-axis is assumed directed toward the $\bar{\varepsilon}$-medium. Assuming harmonic ω-dependence of the fields and $D^{\text{NL}} = 4\pi P^{\text{NL}} = 4\pi\tilde{P}^{\text{NL}}\exp[i\boldsymbol{k}^s\cdot\boldsymbol{r}]$, the relevant Maxwell equations are written in the following form:

$$\text{(40)}\qquad \begin{cases} \dfrac{\partial H_y}{\partial z} - ik_0\bar{\varepsilon}E_x = ik_0 D_x^{\text{NL}}\,, \\[2ex] \dfrac{\partial H_y}{\partial x} + ik_0\bar{\varepsilon}E_z = -ik_0 D_z^{\text{NL}}\,, \\[2ex] \dfrac{\partial E_x}{\partial z} - \dfrac{\partial E_z}{\partial x} - i\bar{\mu}k_0 H_y = 0\,, \end{cases}$$

for the nonlinear medium $\bar{\varepsilon}$. The equations for medium ε are identical to (40) with $D^{\text{NL}} = 0$, $\bar{\mu} \to 1$ and interchange $\bar{\varepsilon} \leftrightarrows \varepsilon$. The solution of the boundary condition problem (continuity across the boundary of E_x and H_y) leads to the following expressions for the fields in the nonlinear medium:

$$(41)\quad \begin{cases} \bar{E}_x(z,x) = \dfrac{-1}{\bar{\varepsilon}[(k_z^s)^2+\bar{\alpha}^2]}\cdot \\ \qquad \cdot\left\{[k_z^s k_x^s D_z^{NL} + \bar{\alpha}^2 D_x^{NL}] - \dfrac{\bar{\alpha}}{\bar{\alpha}+\eta\alpha} A \exp[-\bar{\alpha}z]\exp[ik_x^s x]\right\}, \\ \bar{E}_z(z,x) = \dfrac{-1}{\bar{\varepsilon}[(k_z^s)^2+\bar{\alpha}^2]}\cdot \\ \quad \cdot\left\{[(k_z^s)^2-(k_x^s)^2+\bar{\alpha}^2]D_z^{NL} + [k_x^s k_z^s]D_x^{NL} - \dfrac{ik_x^s}{\bar{\alpha}+\eta\alpha} A \exp[-\bar{\alpha}z]\exp[ik_x^s x]\right\}, \\ \bar{H}_y(z,x) = \dfrac{k_0}{(k_z^s)^2+\bar{\alpha}^2}\cdot \\ \qquad \cdot\left\{[(k_z^s) D_x^{NL} - (k_x^s) D_z^{NL}] - \dfrac{i}{\bar{\alpha}+\eta\alpha} A \exp[-\bar{\alpha}z]\exp[ik_x^s x]\right\}, \end{cases}$$

where

$$\eta \equiv \bar{\varepsilon}/\varepsilon$$

and

$$A \equiv [k_x^s(k_z^s + i\alpha\eta) D_z^{NL} + (\bar{\alpha}^2 - i\alpha\eta k_z^s) D_x^{NL}]\exp[-i\boldsymbol{k}^s\cdot\boldsymbol{r}].$$

Expressions (41) show that in the medium $\bar{\varepsilon}$ the fields are the sum of a driven solution and of a « free » solution ($\propto A$). In the linear medium ε only the « free » solution exists:

$$(42)\quad \begin{cases} E_x(x,z) = \dfrac{-\alpha B}{\varepsilon[(k_z^s)^2+\bar{\alpha}^2][\bar{\alpha}+\eta\alpha]}\exp[\alpha z]\exp[ik_x^s x], \\ E_z(x,z) = \dfrac{ik_x^s B}{\varepsilon[(k_z^s)^2+\bar{\alpha}^2][\bar{\alpha}+\eta\alpha]}\exp[\alpha z]\exp[ik_x^s x], \\ H_y(x,z) = \dfrac{-ik_0 B}{[(k_z^s)^2+\bar{\alpha}^2][\bar{\alpha}+\eta\alpha]}\exp[\alpha z]\exp[ik_x^s x] \end{cases}$$

with B obtained from A witht he change (not interchange) $\alpha\eta \to \bar{\alpha}$. We verify that the « free » solutions on either side of the interface diverge for $\bar{\alpha}+\eta\alpha = 0$. It is easy to recognize that this equation reduces to the following expression:

$$(43)\qquad k_x^2 = k_0^2 \frac{\bar{\varepsilon}\varepsilon}{\varepsilon+\bar{\varepsilon}},$$

which is the dispersion relation for surface polaritons (SP) at the boundary [39-43]. In actual cases $\bar{\varepsilon}$ and/or ε are complex and the expression

$[\bar{\alpha} + \eta\alpha]^{-1}$ can be given in a Lorentzian form having a finite width and a finite maximum at $(k_x^s)^2 = k_0^2 \operatorname{Re}(\bar{\varepsilon}\varepsilon/(\bar{\varepsilon}+\varepsilon))$, reproducing the behaviour of the process of bulk polariton nonlinear excitation, eq. (23). In this connection an important point must be clarified. The « free » waves we have just considered and that coexist at the interface with the driven solution *do not* form a surface polariton because they are not « free » of choosing their own wave vector, which is determined by the optical properties of the medium, eq. (43). Their wave vector is in fact determined by the driving NL polarization wave being equal to k_x^s. This situation is entirely different from the one presented by the process of NL excitation of bulk polaritons considered in sects. **2** and **3** of this paper: in that case the k_T-waves are real free waves, *i.e.* polaritons. In order to consider NL generation of real surface polaritons we can think of the realistic situation of a NL polarization wave with finite cross-section at the boundary. Assuming sharp transverse boundaries at the interface, we can consider a new set of e.m. boundary conditions at the surface of the generalized ideal cone which intersects the physical interface between the two media and that represents the separation surface between the regions in which $P^{\text{NL}} = 0$ and $P^{\text{NL}} \neq 0$. This boundary condition problem is quite complex, because it must be solved for a tridimensional boundary. However, with the transformations {space→time, k-vectors→frequency}, the situation is similar to the case of the excitation of a harmonic oscillator by an oscillatory force represented by a square pulse in time. Suppose that $\omega_0 = \omega_0' + i\omega_0''$ is the complex eigenfrequency of the oscillator and $\omega \sim \omega_0'$ is the real exciting carrier frequency. Just after the time boundary defined by the sharp trailing edge of the force pulse, the frequency of the oscillator jumps from ω to ω_0 and the oscillation intensity, which is resonantly dependent on the frequency matching $\omega^2 - \omega_0'^2$, starts decreasing exponentially with a time constant $\tau \sim 1/\omega_0''$. Physically the picture is quite adequate to represent qualitatively the behaviour of the surface polariton at the boundary. The SP intensity is determined at the boundary by the phase matching $(k_x^s)^2 = k_0^2 \operatorname{Re}(\bar{\varepsilon}\varepsilon/(\bar{\varepsilon}+\varepsilon))$, and the SP travelling away from the boundary is damped with an attenuation constant k_x'', which is determined by the following set of equations:

$$\text{(44)}\qquad \begin{cases} k_x \equiv k_x' + ik_x'' \quad \text{(SP wave vector)} \\ \bar{\varepsilon} = \bar{\varepsilon}' + i\bar{\varepsilon}'' , \\ (k_x')^2 - (k_x'')^2 = k_0^2 \operatorname{Re} \dfrac{\varepsilon\bar{\varepsilon}}{\varepsilon + \bar{\varepsilon}} , \\ 2k_x' k_z'' = k_0^2 \operatorname{Im} \dfrac{\varepsilon\bar{\varepsilon}}{\bar{\varepsilon} + \varepsilon} . \end{cases}$$

We can see that k_x'' is determined by the damping of the crystal excitation, *i.e.* by $\bar{\varepsilon}''$. In actual cases k_x'', evaluated by eq. (44), represents the lower limit of the damping constant of the SP wave. That one is also affected by some physical properties of the surface, like the surface roughness, etc. The damping length $l = 1/k''_x$ is of the order of several cm for SP on a metal surface, in the far infra-red, and of a fraction of a mm for SP associated with reststrahlung phonon or exciton bands of a semiconductor.

So far we have considered the process of NL generation of coherent surface polaritons. The driving wave is the NL polarization which can be created, for two-wave interaction, by SFG or DFG. The detection of the coherent polariton can be obtained by a straightforward application of the NL four-photon mixing scheme considered in sect. **4** or by direct coupling of the radiation associated with the polariton at the surface. In this connection let us consider the crystal-vacuum interface, $\varepsilon = 1$. We recall that, owing to eq. (43), $k'_x > k_0$. The SP radiation in the vacuum is associated with an evanescent wave and, for a perfectly plane interface, it can be coupled to a detector away from the surface only by a reverse frustrated total internal reflection (FTR) using a transparent prism placed on the surface at a distance ξ. The FTR method, which has been first applied by OTTO [44, 45], is the only linear method for exciting and detecting surface polaritons on a plane surface. It transforms a bulk wave into an evanescent wave with $k_x > k_0$, as required for the excitation of SP, and the evanescent wave associated with a SP into a bulk wave to be detected. We may note that the method is inefficient, because the field energy transferred to (or from) the surface from (or to) the prism is proportional to

$$\exp\left[-2\sqrt{k_x^2 - k_0^2}\,\xi\right] \sim \exp\left[-\frac{4\pi}{\lambda_0}\xi\right]$$

for $k_x \gg k_0$. For $\xi \sim \lambda_0$ the energy transfer is of the order of 10^{-5}. We could prove that the same or a higher efficiency can be obtained with our NL method in which, in the parametric approximation [1], almost the entire energy transferred by the fundamental beams to the NL polarization is in turn transferred to the surface polariton. The theory of the nonlinear excitation of surface polaritons has been worked out by DE MARTINI and SHEN [46].

7. – Nonlinear spectroscopy of surface polaritons.

The first experiment on coherent excitation and detection of surface phonon-polaritons (SP) has been reported recently by DE MARTINI, MATALONI, PALANGE and SHEN [47]. In that work the dispersion of the SP wave vector $K_x(\omega) = K'_x(\omega) + iK''_x(\omega)$ in the III-V semiconductor GaP has been investigated by the methods of 4-wave mixing and K-space spectroscopy [20, 26].

The experimental set-up, similar to the one in ref. [20], is shown in fig. 8. A Q-switched ruby laser with a 30 ns pulse width provided a beam at $\omega_1 =$

$= 14\,403\ \text{cm}^{-1}$. It was used to pump simultaneously two dye lasers emitting two tunable beams at ω_2 and ω_3 with a line width of $1.5\ \text{cm}^{-1}$. In the experiment, ω_3 (the probe frequency) was held fixed at $13\,333\ \text{cm}^{-1}$, while ω_2 was tuned in the range between $14\,006$ and $14\,035\ \text{cm}^{-1}$, so that optical mixing of ω_1 and

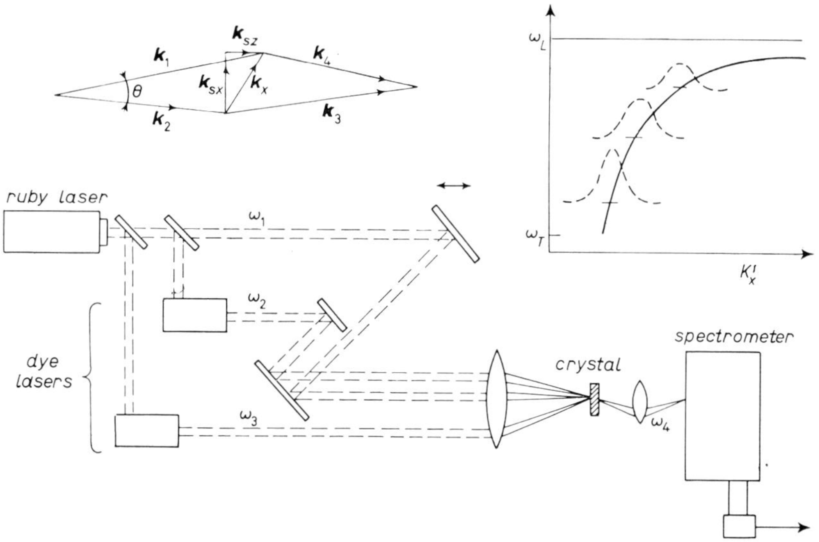

Fig. 8. – Experimental set-up for surface polariton nonlinear excitation.

ω_2 beams could excite surface polaritons in the reststrahl band of the crystal in the range between 368 and 397 cm^{-1}. Part of the ruby laser beam and the two dye laser beams were focused by a common achromatic lens with 20 cm focal length on the surface of a 2 mm GaP slab. Small diaphrams were used in front of the lens to reduce the convergence angle of each beam at the crystal to $4 \cdot 10^{-3}$ rad. The focal spot was about 0.5 mm. The output at $\omega_4 = \omega_3 - \omega_1 + \omega_2$ around the phase-matched direction $\boldsymbol{k}_4 = \boldsymbol{k}_3 - \boldsymbol{k}_1 + \boldsymbol{k}_2$ was collected by a large-aperture lens and analysed by a double-stage Hilger D330/1 monochromator and detected by a RCA C31025C Quantacon photomultiplier. With 50 kW incident beams we detected a resonant output peak power at ω_4 of 0.1 μW. For each given excited SP frequency $\omega_1 - \omega_2$ we measured $I(\omega_4)$ as a function of k_x^s, which is the projection of $\boldsymbol{k}^s = \boldsymbol{k}_1 - \boldsymbol{k}_2$ on the surface of the slab. We found that the experimental curves $I(k_x^s)$ at different ω_1 fitted well a set of Lorentzian curves, as expected from the theory [46]. In fact when $k_x^s \approx K'(\omega)$, the surface wave in the crystal may be expressed in the form [46]

$$(45)\qquad E(\omega) = A\,\frac{1}{\Delta k_x - iK_x''}\,(\hat{x}k_{b_z} - \hat{z}k_x^s)\,P_x^{\text{NL}}(\omega)\exp\left[i(k_x^s x - \omega t) + ik_{b_z} z\right],$$

where

$$
\begin{aligned}
A &= -\,[2\pi(k_{bz}+\bar{\varepsilon}k_{az})/K'\,\bar{\varepsilon}(1-\bar{\varepsilon}^2)(k^{s^2}-k_b^2)(\bar{\varepsilon}k_{az}k_z^s-k_{bz}^2)]\,,\\
k_b^2 &\equiv \bar{\varepsilon}(\omega/c)^2, \qquad \Delta k_x \equiv k_x^s - K_x'\,,\\
k_{az}^2 &\equiv k_0^2 - k_x^{s^2},\\
k_{bz}^2 &\equiv k_b^2 - k_x^{s^2},\\
k_0 &\equiv \omega/c\,.
\end{aligned}
$$

$\boldsymbol{P}_x^{\text{NL}}(\omega)=[\chi^{(2)}(\omega):\boldsymbol{E}_1(\omega_1)\,\boldsymbol{E}_2(\omega_2)]_x$ is the x-component of the nonlinear polarization at the difference frequency $\omega=\omega_1-\omega_2$. One can use a third laser beam $E(\boldsymbol{k}_3,\omega_3)$ to probe the excited surface wave [46]. Optical mixing of this probing field with the surface wave induced a nonlinear polarization $P^{\text{NL}}(\omega_4=\omega_3-\omega)$, which in turn generates a new wave $E(\omega_4)$ with an intensity given by

$$
(46)\qquad I(\omega_4,\Delta k_x)\propto|\boldsymbol{P}^{\text{NL}}(\omega_4)|^2=\frac{|A|^2}{\Delta k_x^2+K_x''^2}\,|\chi^2(\omega_4):\boldsymbol{E}(\omega_3)\,\chi^{(2)*}(\omega):\boldsymbol{E}^*(\omega_1)\,\boldsymbol{E}(\omega_2)|^2;
$$

$I(\omega_4)$ is a Lonretzian function of k_x having its maximum for $k_x^s=K_x'$ and half-width $2K_x''$. Our experiment gives therefore full information on the linear properties of the surface polariton. In fig. 9 and 10 the experimental data in dispersion of K_x' and K_x'' are shown. The solid theoretical curves are obtained

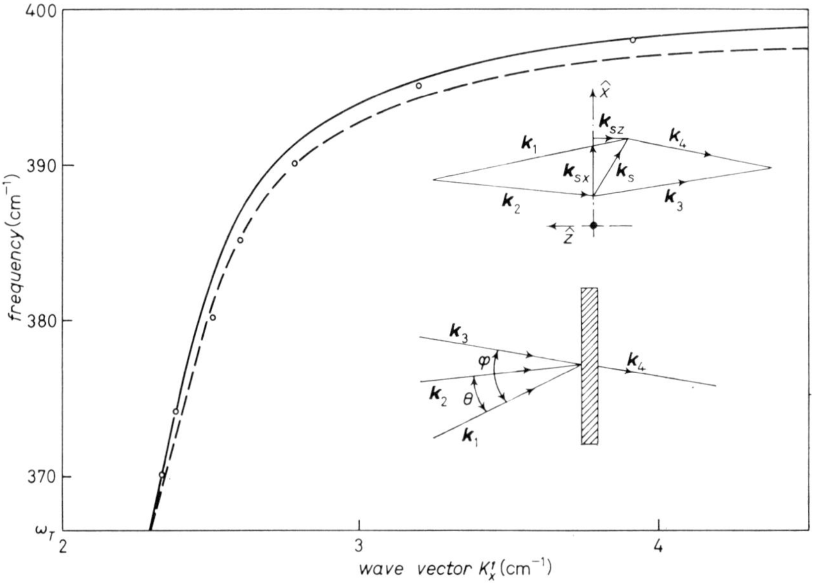

Fig. 9. – Frequency dependence of the surface polariton wave vector $K'(\omega)$.

from a single-oscillator model with the following values of the parameters: $\omega_{TO} = 367.3\ cm^{-1}$, $\omega_{LO} = 403\ cm^{-1}$, $\varepsilon_\infty = 9.091$, $\Gamma = 0.0035\,\omega_{TO}\ cm^{-1}$. The dashed curves correspond to a three-oscillator model, which accounts for a selective damping process involving the two-phonon band $TA(X)+LA(X)$ peaked at 357 cm^{-1} [48]. The resolution (in $\boldsymbol{k}$-space) of our present apparatus

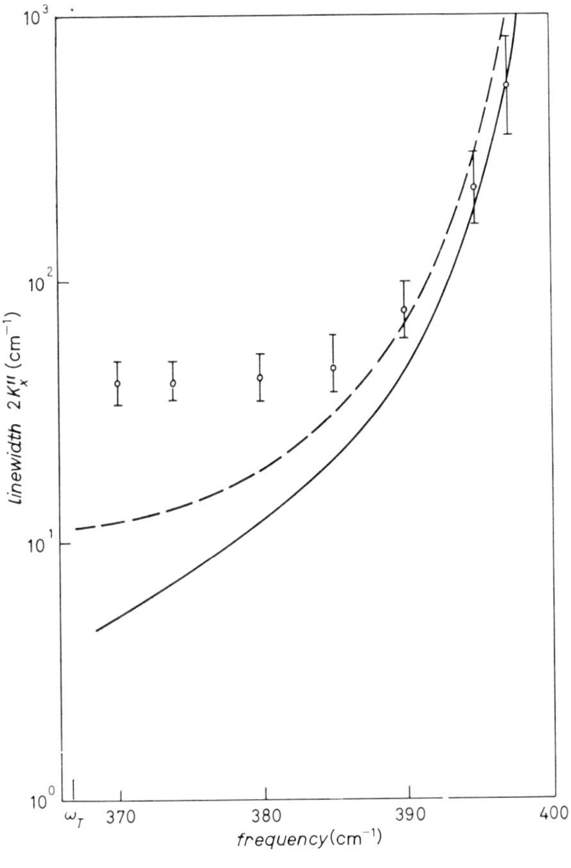

Fig. 10. – Frequency dependence of the $\boldsymbol{K}$-space line width $2K''(\omega)$.

is not yet large enough for us to deduce the true values of $K''(\omega)$ below $\omega = 385\ cm^{-1}$. This is mainly due to the sizable angular spread of the focused beams ω_1 and ω_2 (4 mrad). The same lack of accuracy in the same frequency region affects the measurement of the nonlinear optical susceptibility of the crystal $\chi^{(2)}(\omega)$. More accurate measurements of $\chi^{(2)}(\omega)$ in the reststrahl band and of $K''_x(\omega)$ with an apparatus having a 5 times larger resolution are presently in progress in our laboratory, for GaP an other semiconductors.

A technique similar to the one we have just described can be applied to the investigation of the exciton surface polaritons by the method of 4-wave mixing or by sum frequency generation (sect. **3**). Some numerical examples relative to the Z_3 exciton resonance in CuCl are given in [46].

Note added in proofs.

The nonlinear excitation of SP in the *C*-exciton reststrahl band in ZnO ($\hbar\omega_T = 3.421$ eV) has been recently achieved [49]. The field has been excited by second-harmonic generation and the (SP) UV radiation has been coupled out by the prism method (FTR).

REFERENCES

[1] N. Bloembergen: *Nonlinear Optics* (New York, N. Y., 1965). Throughout the present paper we shall make frequently use of some results reported in N. Bloembergen and P. S. Pershan: *Phys. Rev.*, **128**, 606 (1962).

[2] F. De Martini: *Phys. Lett.*, **30** A, 319, 547 (1969); *Phys. Rev. B*, **4**, 4556 (1971). The comment reported in ref. [43] of this paper is incorrect.

[3] Y. R. Shen: *Phys. Rev.*, **138**, A 1741 (1965).

[4] P. N. Butcher and T. P. McLean: *Proc. Phys. Soc.*, **81**, 219 (1963).

[5] C. H. Henry and C. G. B. Garrett: *Phys. Rev.*, **171**, 1058 (1968).

[6] R. Loudon: *Proc. Roy. Soc.*, A **275**, 218 (1963); *Rendiconti S.I.F.*, Course XLII, edited by R. Glauber (New York, N. Y., 1969).

[7] K. Huang: *Proc. Roy. Soc.*, A **208**, 352 (1951).

[8] C. Kittel: *Quantum Theory of Solids* (New York, N. Y., 1964), p. 42.

[9] W. L. Faust and C. H. Henry: *Phys. Rev. Lett.*, **17**, 1265 (1956); W. L. Faust, C. H. Henry and R. H. Eick: *Phys. Rev.*, **173**, 781 (1968).

[10] W. Heitler: *Quantum Theory of Radiation*, sect. **7** (Oxford, 1964).

[11] P. N. Butcher: *Nonlinear Optical Phenomena* (Columbus, O., 1965), p. 43.

[12] M. Abragam: *The Principles of Nuclear Magnetism* (Oxford, 1961).

[13] M. Goeppert-Mayer: *Ann. der Phys.*, **9**, 273 (1931).

[14] J. J. Hopfield, J. M. Worlock and K. Park: *Phys. Rev. Lett.*, **11**, 414 (1963).

[15] J. J. Hopfield: *Phys. Rev.*, **182**, 945 (1968).

[16] M. Born and E. Wolf: *Principles of Optics* (New York, N. Y., 1964), p. 14.

[17] J. A. Giordmaine: *Phys. Rev. Lett.*, **8**, 19 (1962).

[18] R. S. Knox: *Theory of Excitons* (New York, N. Y., 1963), p. 163.

[19] S. I. Pekar: *Žurn. Ėksp. Teor. Fiz.*, **33**, 1022 (1957); *Phys. Chem. Sol.*, **5**, 11 (1958).

[20] J. P. Coffinet and F. De Martini: *Phys. Rev. Lett.*, **22**, 60 (1969); *Phys. Rev. Lett.*, **22**, 752 (1969) (Erratum).

[21] E. Yablonovitch, C. Flytzanis and N. Bloembergen: *Phys. Rev. Lett.*, **29**, 864 (1972).

[22] J. J. Wynne: *Phys. Rev. Lett.*, **29**, 650 (1972).

[23] P. N. Butcher, R. Loudon and T. P. McLean: *Proc. Phys. Soc.*, **85**, 565 (1965).

[24] J. Ducuing: *Rendiconti S.I.F.*, Course XLII, edited by R. Glauber (New York, N. Y., 1969).

[25] F. DE MARTINI: *Nuovo Cimento*, **51** B, 16 (1967).
[26] F. DE MARTINI: *Rendiconti S.I.F.*, Course LII, edited by E. BURSTEIN (New York, N. Y., 1972).
[27] D. C. HAUSEISEN and H. MAHR: *Phys. Rev. Lett.*, **26**, 838 (1971).
[28] D. FRÖLICH, B. STAGINNUS and E. SCHÖNHERR: *Phys. Rev. Lett.*, **19**, 1032 (1967).
[29] A. BIVAS, C. MARANGE, J. B. GRUN and C. SCHWAB: *Opt. Comm.*, **6**, 142 (1972).
[30] A. GOLD: *Rendiconti S.I.F.*, Course XLII, edited by R. GLAUBER (New York, N.Y., 1969).
[31] S. D. KRAMER, F. G. PARSONS and N. BLOEMBERGEN: *Phys. Rev. B*, **9**, 1853 (1974); see also S. D. KRAMER and N. BLOEMBERGEN: *Optical Properties of Highly Transparent Solids*, edited by S. S. MITRA and B. BENDOW (New York, N.Y., 1975); S. KRAMER: *Ph.D. Thesis* (Harvard University, 1976), unpublished; H. LOTEM, R. LYNCH, S. KRAMER and N. BLOEMBERGEN: *Opt. Comm.* (1976), to be published.
[32] J. M. AREF'EV: *JETP Lett.*, **8**, 84 (1968); F. ZERNICKE and P. R. BERMAN: *Phys. Rev. Lett.*, **15**, 999 (1965); M. D. MARTIN and E. L. THOMAS: *IEEE Journ. Quant. Elec.*, QE-**2**, 196 (1966); T. YAJIMA and K. INOUE: *IEEE Journ. Quant. Elec.*, QE-**5**, 140 (1969); D. W. FARIES, K. A. GEHRING, P. L. RICHARDS and Y. R. SHEN: *Phys. Rev.*, **180**, 363 (1969); F. ZERNICKE: *Phys. Rev. Lett.*, **22**, 931 (1969); S. BIRAUD and G. CHARTIER: *Phys. Lett.*, **30** A, 177 (1969).
[33] S. K. KURTZ and J. GIORDMAINE: *Phys. Rev. Lett.*, **22**, 192 (1969); J. M. YARBOROUGH, S. S. SUSSMAN, H. E. PUTHOFF, R. H. PANTELL and C. JOHNSON: *Appl. Phys. Lett.*, **15**, 102 (1969).
[34] F. DE MARTINI and J. LEROY: *Sol. State Comm.*, **9**, 1779 (1971).
[35] See work of V. PIESBERGEN: in *Semiconductors and Semimetals*, edited by WILLARDSON and BEER, Vol. **2** (New York, N. Y., 1966).
[36] D. A. KLEINMAN and W. G. SPITZER: *Phys. Rev.*, **118**, 110 (1960).
[37] A. S. BARKER: *Phys. Rev.*, **118**, 118 (1960).
[38] A. S. BARKER: *Phys. Rev.*, **165**, 917 (1968).
[39] A. SOMMERFELD: *Ann. der Phys.*, **28**, 665 (1909).
[40] U. FANO: *Journ. Opt. Soc. Amer.*, **31**, 213 (1941).
[41] E. LIFSHITZ and L. N. ROSENZWEIG: *Žurn. Ėksp. Teor. Fiz.*, **18**, 1012 (1948).
[42] *Proceedings of the Taormina Research Conference « Polaritons »*, edited by E. BURSTEIN and F. DE MARTINI (New York, N. Y., 1974).
[43] R. W. GOULD and A. W. TRIVELPIECE: *Proc. Int. Electr. Eng. (London)*, B **105**, *Suppl.* **10**, 516 (1958).
[44] A. OTTO: *Zeits. Phys.*, **216**, 398 (1968).
[45] N. MARSCHALL and B. FISHER: *Phys. Rev. Lett.*, **28**, 811 (1972).
[46] F. DE MARTINI and Y. R. SHEN: *Phys. Rev. Lett.*, **36**, 216 (1976).
[47] F. DE MARTINI, P. MATALONI, E. PALANGE and R. Y. SHEN: to be published in *Phys. Rev. Lett.* and in *Proceedings of XIII International Conference on the Physics of Semiconductors* (Roma, 1976), to be published.
[48] A. S. BARKER: *Phys. Rev.*, **165**, 917 (1968).
[49] M. COLOCCI, F. DE MARTINI, S. KOHN and Y. R. SHEN: to be published.

Nonlinear Interaction between Excitons and Coherent Light.

H. HAKEN

Institut für Theoretische Physik, Universität Stuttgart
Pfaffenwaldring 57, 7000 *Stuttgart*-80 (*Vaihingen*)

1. – What are excitons?

a) *Wannier excitons.* Let us consider an electron moving in the periodic potential of a lattice. According to quantum theory the wave function of the electron is a periodically modulated plane wave. The characteristic energy levels are grouped into the so-called bands, which are separated by energy gaps. We focus our attention on insulators or semiconductors where the valence band is filled by electrons according to the Pauli principle, whereas the next higher band, the conduction band, is empty. When we shine light on such a crystal, absorption occurs if the photon energy is bigger than the gap. Thus the absorption spectrum of fig. 1 is expected, which is indeed found in many crystals. However, in many cases, further discrete lines appear which are not caused by impurities but belong to intrinsic electronic states of the ideal crystal (see fig. 2).

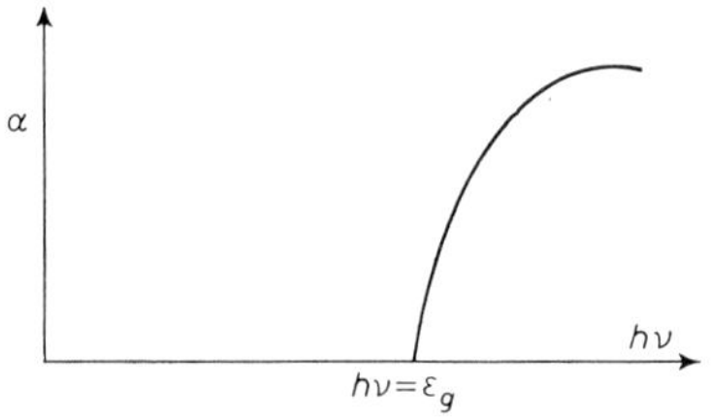

Fig. 1.

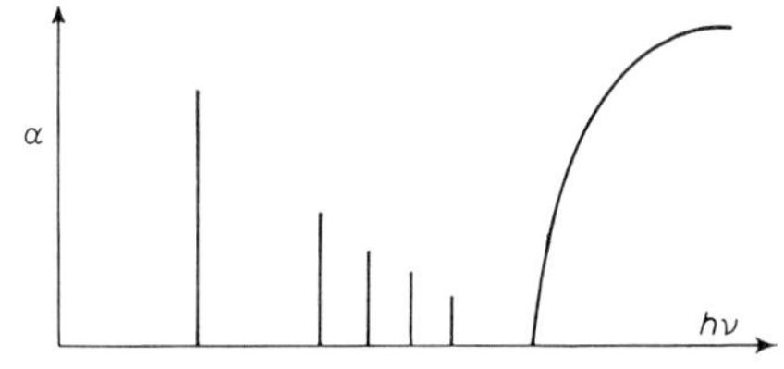

Fig. 2.

Fig. 1. – Absorption coefficient *vs.* light quantum energy $h\nu$.

Fig. 2. – Exciton lines (schematic).

To identify their physical nature let us consider the optical transition more closely. When an electron is brought from the valence band to the conduction band it leaves a « hole » in the valence band. Or, if we use a local description, it leaves a positive charge in the crystal (fig. 3). It can be shown that the hole behaves like a particle with the effective mass m_h and a positive charge. On the other hand, the electron in the conduction band behaves like a particle with the effective mass m_{el} and a negative charge. Thus, these particles attract each other by a Coulomb force with the potential e^2/r, which is modified by a dielec-

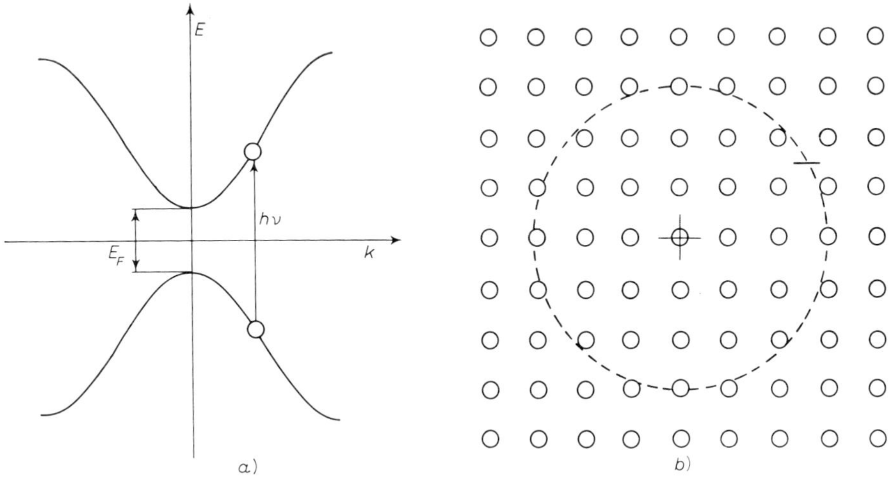

Fig. 3. – Creation of an electron-hole pair *a*) in the energy level diagram, *b*) in a local description. The circles indicate atoms of the crystal lattice, the cross indicates the position of the positive charge.

tric constant ε. Due to dynamic polarization effects the interaction potential may have a still more complicated form, which we shall not discuss here, however. The bound electron-hole pair is called an exciton. It can form bound states in analogy to the hydrogen atom. The exciton can move through the crystal. Since m_{el}, m_h and ε can be varied in large regions by the choice of different kinds of crystals, the orbit of an exciton can vary from several to many hundred Å. Thus it is possible to create, so to speak, an artificial kind of matter. When high-intensity light is shone on a crystal, excitons may be created at high concentrations; this leads to new kinds of bound states: exciton molecules, Bose condensed excitons, electron-hole droplets, etc. Though these effects are very interesting, we shall not be concerned with them here but rather with effects which arise from the coherent interaction between excitons and light.

Besides the so-called Wannier excitons, where the orbit is several lattice constants, there exists a second type of excitons, the so-called Frenkel excitons, in which the excited electron remains at the same lattice site as the hole.

b) *Frenkel excitons.* There are essentially two types of causes for the propagation of Frenkel excitons, which may best be described in an atomic

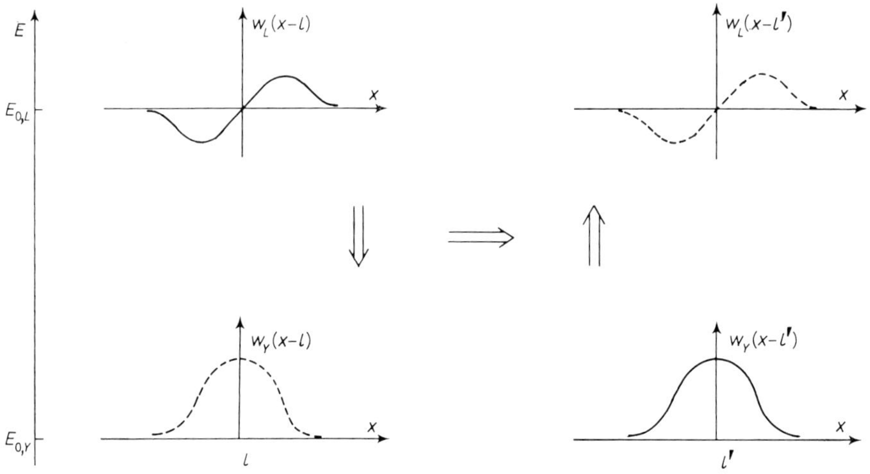

Fig. 4. – Motion of Frenkel exciton by Coulomb-exchange interaction. One electron recombines at site l, another electron at site l' is simultaneously excited or, in other words, the exciton at site l is transferred to site l'.

or molecular picture: 1) the excited electron and the hole may jump from one site to the next one simultaneously on account of overlapping wave functions; 2) on account of the Coulomb interaction between electrons at different sites an exciton may recombine creating simultaneously an exciton at a different site (see fig. 4). As we will show below Frenkel excitons have features strongly reminescent of spin systems.

Interaction between excitons and light. The interaction between excitons and light causes different kinds of transitions, which may be classified as follows:

	Initial state	Final state	Process
1)	crystal ground state (no exciton)	one exciton	single-photon absorption
2)	crystal ground state (no exciton)	one exciton	multiple-photon absorption
3)	one exciton	crystal ground state (no exciton)	single-photon emission
4)	one exciton	crystal ground state (no exciton)	multiple-photon emission
5)	exciton in state (n)	exciton in state (n')	single- or multiple-photon emission or absorption

Further real transitions may occur with the participation of other elementary excitations. In the following we shall discuss resonant interaction effects based on the transitions 1) and 3). There are several selection rules:

1) *Momentum conservation*, *i.e.* selection rule for centre-of-gravity motion $\boldsymbol{K}_i + \boldsymbol{k} = \boldsymbol{K}_f$. Here $\boldsymbol{K}_i$, $\boldsymbol{K}_f$ are the total wave vectors of the crystal and $\boldsymbol{k}$ the wave vector of the photon. Since in the initial state i, $\boldsymbol{K}_i = 0$, and the $\boldsymbol{k}$ of the photon is very small, we find $\boldsymbol{K}_f \approx 0$.

2) *Internal selection rules.* The exciton wave function is composed of atomic wave functions belonging to the excited and ground state of the atom, and an envelope function (fig. 5). In the case of a Frenkel exciton the envelope function is simply a δ-function.

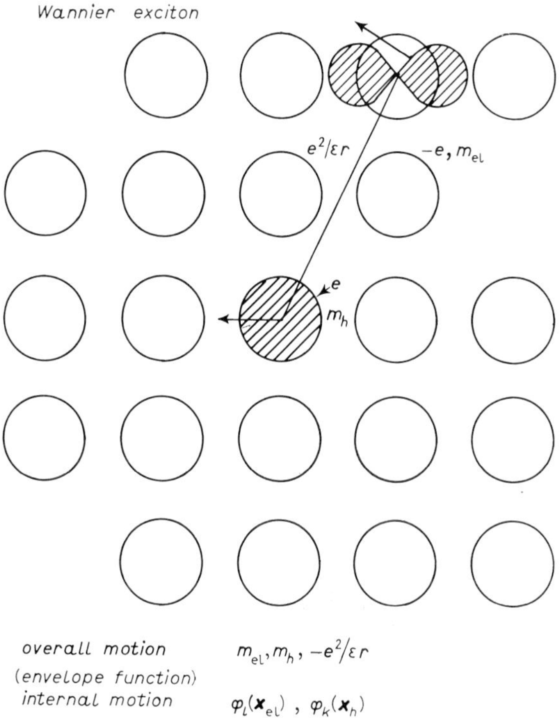

Fig. 5. – The structure of the exciton wave function in a local description. The internal motion of the electron and the hole is indicated by dashed charge distributions; external motion: the electron and hole circle around each other.

2. – Some basic features of exciton theory.

In order to be able to show where nonlinearities come in and why they are so important we must deal a little bit with exciton theory. The most elegant way to do this is the use of second quantization. We formulate this formalism here in a way that should be understandable also for readers otherwise not familiar with this method. Let us start with the Frenkel exciton.

a) Frenkel exciton. We denote the total ground state of the crystal in which all electrons are in their ground states by Ψ_0. By application of certain operators on Ψ_0 we may describe states in which, *e.g.*, an electron is lifted from its atomic ground state at site $\boldsymbol{l}$ to its excited state. Considered from a more formal point of view this process consists of two steps: 1) an electron in its ground state is annihilated, 2) it is created in the upper atomic level. This is achieved by applying the operators $d_{\boldsymbol{l}}^\dagger a_{\boldsymbol{l}}^\dagger$ to the ground state Ψ_0. For the experts we remark that $a_{\boldsymbol{l}}$ or $d_{\boldsymbol{l}}$ and their Hermitian conjugates obey Fermi commutation relations. We will further use the creation operator of a localized exciton $B_{\boldsymbol{l}}^\dagger = a_{\boldsymbol{l}}^\dagger d_{\boldsymbol{l}}^\dagger$. Whether a state is occupied or not can be tested with the occupation number operator for electrons, $a_{\boldsymbol{l}}^\dagger a_{\boldsymbol{l}}$, or for holes, $d_{\boldsymbol{l}}^\dagger d_{\boldsymbol{l}}$, thus, *e.g.*, $a_{\boldsymbol{l}}^\dagger a_{\boldsymbol{l}} \Psi = \Psi$, if the state Ψ is occupied, or $= 0$, if the state is not occupied by an electron at site $\boldsymbol{l}$. With the aid of the creation and annihilation operators we may write down a Hamiltonian for Frenkel excitons. If we denote the energy of the electron ground state by $-E^h$ and that of the excited state by E^e, the energy operator of the noninteracting lattice sites can be written in the form

$$\sum_{\boldsymbol{l}} E^e a_{\boldsymbol{l}}^\dagger a_{\boldsymbol{l}} + \sum_{\boldsymbol{l}} E^h d_{\boldsymbol{l}}^\dagger d_{\boldsymbol{l}} \,.$$

Furthermore, the transport of excitons due to the interactions described above can be taken into account by a term which describes the annihilation of an exciton at site $\boldsymbol{l}_2$ and its subsequent creation at site $\boldsymbol{l}_1$. Since this interaction between excitons at sites $\boldsymbol{l}_1$ and $\boldsymbol{l}_2$ depends on the distance of the lattice sites, we have an additional term for the Hamiltonian in the form

$$\hbar \sum_{\boldsymbol{l}_1 \boldsymbol{l}_2} a_{\boldsymbol{l}_1}^\dagger d_{\boldsymbol{l}_1}^\dagger d_{\boldsymbol{l}_2} a_{\boldsymbol{l}_2} W(\boldsymbol{l}_1 - \boldsymbol{l}_2) \,.$$

As an example for W we quote the Coulomb-exchange interaction matrix element defined by

$$\hbar W(\boldsymbol{l}_1 - \boldsymbol{l}_2) = \iint w_c^*(\boldsymbol{x} - \boldsymbol{l}_1) w_v^*(\boldsymbol{x}' - \boldsymbol{l}_2) \frac{e^2}{|\boldsymbol{x} - \boldsymbol{x}'|} w_c(\boldsymbol{x}' - \boldsymbol{l}_2) w_v(\boldsymbol{x} - \boldsymbol{l}_1) \, \mathrm{d}^3x \, \mathrm{d}^3x' \,,$$

where w_c, w_v are localized atomic or molecular wave functions for the electrons in the ground state (v) or in the excited state (c) and at sites $\boldsymbol{l}_1$, $\boldsymbol{l}_2$.

Finally we take into account the interaction with the light field. This term describes the creation or annihilation of an exciton at sites $\boldsymbol{l}$, so that it contains the operators $a_{\boldsymbol{l}}^\dagger d_{\boldsymbol{l}}^\dagger$ or $d_{\boldsymbol{l}} a_{\boldsymbol{l}}$. Furthermore, it depends on the light field, which is described by its vector potential $\boldsymbol{A}_{\boldsymbol{l}}^\dagger$ and its complex conjugate. In what follows we will treat the light field as a classical quantity.

Putting all terms together we find the Hamiltonian in the form

$$(2.1)\qquad H=\sum_{\boldsymbol{l}} E^{e} a_{\boldsymbol{l}}^{\dagger} a_{\boldsymbol{l}}+\sum_{\boldsymbol{l}} E^{h} d_{\boldsymbol{l}}^{\dagger} d_{\boldsymbol{l}}+\hbar \sum_{\boldsymbol{l}_1 \boldsymbol{l}_2} a_{\boldsymbol{l}_1}^{\dagger} d_{\boldsymbol{l}_1}^{\dagger} d_{\boldsymbol{l}_2} a_{\boldsymbol{l}_2} W(\boldsymbol{l}_1-\boldsymbol{l}_2)+ \\ +\frac{ie\hbar}{mc}\left\{\sum_{\boldsymbol{l},i} a_{\boldsymbol{l}}^{\dagger} d_{\boldsymbol{l}}^{\dagger}\int \mathrm{d}^3x\, w_c^*(\boldsymbol{x}-\boldsymbol{l}) A_i^{+} \nabla_i w_v(\boldsymbol{x}-\boldsymbol{l})-\text{h.c.}\right\}+H_F\,.$$

H_F is the Hamiltonian of the light field. Equation (2.1) can be derived from first principles by a rather lengthy calculation, so that we must refer the reader to a detailed representation (cf., *e.g.*, H. HAKEN: *Quantenfeldtheorie des Festkörpers* (Stuttgart, 1973)).

b) *Wannier excitons.* In the case of Wannier excitons the first three sums occurring in (2.1) must be replaced by the more general Hamiltonian

$$(2.2)\qquad H_{El}=\sum_{\boldsymbol{l}\boldsymbol{l}'} H^{c}_{\boldsymbol{l},\boldsymbol{l}'} a_{\boldsymbol{l}}^{\dagger} a_{\boldsymbol{l}'}+\sum_{\boldsymbol{l}\boldsymbol{l}'} H^{v}_{\boldsymbol{l},\boldsymbol{l}'} d_{\boldsymbol{l}}^{\dagger} d_{\boldsymbol{l}'}-\sum_{\boldsymbol{l}\boldsymbol{l}'} W\begin{pmatrix}\boldsymbol{l} & \boldsymbol{l}' & \boldsymbol{l}' & \boldsymbol{l}\\ c & v & v & c\end{pmatrix} a_{\boldsymbol{l}}^{\dagger} d_{\boldsymbol{l}'}^{\dagger} d_{\boldsymbol{l}'} a_{\boldsymbol{l}}+ \\ +\sum_{\boldsymbol{l}\boldsymbol{l}'} W\begin{pmatrix}\boldsymbol{l} & \boldsymbol{l}' & \boldsymbol{l}' & \boldsymbol{l}\\ v & c & v & c\end{pmatrix} a_{\boldsymbol{l}'}^{\dagger} d_{\boldsymbol{l}}^{\dagger} d_{\boldsymbol{l}} a_{\boldsymbol{l}}+\sum_{\boldsymbol{l}\boldsymbol{l}'}\frac{1}{2} W\begin{pmatrix}\boldsymbol{l} & \boldsymbol{l}' & \boldsymbol{l}' & \boldsymbol{l}\\ c & c & c & c\end{pmatrix} a_{\boldsymbol{l}}^{\dagger} a_{\boldsymbol{l}'}^{\dagger} a_{\boldsymbol{l}'} a_{\boldsymbol{l}}+ \\ +\sum_{\boldsymbol{l}\boldsymbol{l}'}\frac{1}{2} W\begin{pmatrix}\boldsymbol{l} & \boldsymbol{l}' & \boldsymbol{l}' & \boldsymbol{l}\\ v & v & v & v\end{pmatrix} d_{\boldsymbol{l}}^{\dagger} d_{\boldsymbol{l}'}^{\dagger} d_{\boldsymbol{l}'} d_{\boldsymbol{l}}$$

with W defined by

$$W\begin{pmatrix}\boldsymbol{l} & \boldsymbol{l}' & \boldsymbol{l}' & \boldsymbol{l}\\ i & j & k & l\end{pmatrix}=\iint w_i^*(\boldsymbol{x}-\boldsymbol{l})\, w_j^*(\boldsymbol{x}'-\boldsymbol{l}')\,\frac{e^2}{|\boldsymbol{x}-\boldsymbol{x}'|}\, w_k(\boldsymbol{x}'-\boldsymbol{l}')\, w_l(\boldsymbol{x}-\boldsymbol{l})\,\mathrm{d}^3x\,\mathrm{d}^3x'\;;$$

$H^{c}_{\boldsymbol{l},\boldsymbol{l}'}$, $H^{v}_{\boldsymbol{l},\boldsymbol{l}'}$ are the transition matrix elements (in the Hartree-Fock approximation) from site $\boldsymbol{l}'$ to site $\boldsymbol{l}$ for the electron in the conduction band and the hole in the valence band, respectively.

The Schrödinger equation belonging to (2.1) can be easily solved if only one exciton is present and if we drop all terms of the Hamiltonian containing the light field. The solution reads

$$(2.3)\qquad \Phi_{\boldsymbol{k}}=\frac{1}{\sqrt{\mathcal{N}}}\sum_{\boldsymbol{l}} \exp[i\boldsymbol{k}\boldsymbol{l}]\, a_{\boldsymbol{l}}^{\dagger} d_{\boldsymbol{l}}^{\dagger}\,\psi_0\,,$$

where $\mathcal{N}$ is the total number of lattice sites. The corresponding energy is given by the Fourier transform and reads thus (besides the constant $E^{e}+E^{h}$)

$$(2.4)\qquad E(\boldsymbol{k})=\hbar\sum_{\boldsymbol{l}} \exp[i\boldsymbol{k}\boldsymbol{l}]\, W(\boldsymbol{l})\,.$$

Writing the right-hand side of (2.3) in the form

$$\Phi_{\boldsymbol{k}} = B_{\boldsymbol{k}}^{\dagger} \Psi_0 , \tag{2.5}$$

we may define the creation operator of an exciton in the propagation state $\boldsymbol{k}$ by $B_{\boldsymbol{k}}^{\dagger}$.

Because $B_{\boldsymbol{k}}^{\dagger}$ is composed of a sum of the products of two Fermi operators, it is tempting to assume that $B_{\boldsymbol{k}}^{\dagger}$ is now a Bose operator. The commutation relation between $B_{\boldsymbol{k}}^{\dagger}$ and its Hermitian conjugate for different $\boldsymbol{k}'$ is, however, given by

$$B_{\boldsymbol{k}} B_{\boldsymbol{k}'}^{\dagger} - B_{\boldsymbol{k}'}^{\dagger} B_{\boldsymbol{k}} = \delta_{\boldsymbol{k}\boldsymbol{k}'} - \frac{1}{\mathcal{N}} \sum_{\boldsymbol{l}} \exp\left[i(\boldsymbol{k}' - \boldsymbol{k})\,\boldsymbol{l}\right](a_{\boldsymbol{l}}^{\dagger} a_{\boldsymbol{l}} + d_{\boldsymbol{l}}^{\dagger} d_{\boldsymbol{l}}) . \tag{2.6}$$

Because $a_{\boldsymbol{l}}^{\dagger} a_{\boldsymbol{l}}$ and $d_{\boldsymbol{l}}^{\dagger} d_{\boldsymbol{l}}$ represent the density of electrons and holes, respectively, the boson character is more radically violated the higher the density of excitons.

Because it must be expected that, under strong light excitation, a macroscopic number of excitons is created at least locally, it is necessary to take into account the nonboson character of excitons.

3. – Nonlinear resonant interaction between excitons and coherent light.

A great difference between excitons and usual atoms consists in the fact that excitons can be created and annihilated. Furthermore, there is a strong interaction W among excitons. Thus we shall discuss in detail the impact of the interaction term containing W in (2.1) on resonant phenomena. The simplest way to deal with the interaction between excitons and light consists in deriving Heisenberg equations of motion which can be obtained from the Hamiltonian by

$$\frac{\mathrm{d}}{\mathrm{d}t} \Omega = \frac{i}{\hbar} [H, \Omega] , \tag{3.1}$$

where Ω is an arbitrary operator. In this section we neglect damping terms, *i.e.* we imply that all coherent processes occur so quickly that damping can be neglected. It has turned out that the most useful way to exploit (3.1) is to derive equations for the following operators:

$$B_{\boldsymbol{l}}^{\dagger} = a_{\boldsymbol{l}}^{\dagger} d_{\boldsymbol{l}}^{\dagger} , \tag{3.2}$$

$$B_{\boldsymbol{l}} = d_{\boldsymbol{l}} a_{\boldsymbol{l}} , \tag{3.3}$$

$$N_{\boldsymbol{l}} = \tfrac{1}{2} (a_{\boldsymbol{l}}^{\dagger} a_{\boldsymbol{l}} + d_{\boldsymbol{l}}^{\dagger} d_{\boldsymbol{l}}) . \tag{3.4}$$

We thus obtain our fundamental equations of motion for Frenkel excitons interacting with the light field:

$$\frac{\mathrm{d}}{\mathrm{d}t}B_{\boldsymbol{l}}^{\dagger}=\frac{i}{\hbar}E_0B_{\boldsymbol{l}}^{\dagger}-i\sum_{\boldsymbol{l}'}W(\boldsymbol{l}'-\boldsymbol{l})B_{\boldsymbol{l}'}^{\dagger}(2N_{\boldsymbol{l}}-1)-(2N_{\boldsymbol{l}}-1)A^{-}(\boldsymbol{l})\,g\,, \tag{3.5}$$

$$\frac{\mathrm{d}}{\mathrm{d}t}N_{\boldsymbol{l}}=i\sum_{\boldsymbol{l}'}W(\boldsymbol{l}'-\boldsymbol{l})(B_{\boldsymbol{l}'}^{\dagger}B_{\boldsymbol{l}}-B_{\boldsymbol{l}}^{\dagger}B_{\boldsymbol{l}'})+g(B_{\boldsymbol{l}}^{\dagger}A_{\boldsymbol{l}}^{\dagger}+\text{h.c.})\,, \tag{3.6}$$

$$\left\{\Delta-\frac{1}{c^2}\frac{\mathrm{d}^2}{\mathrm{d}t^2}\right\}A^{+}(\boldsymbol{x},t)=\text{const}\,B_{\boldsymbol{l}}\,. \tag{3.7}$$

The quantities appearing in these equations have the following meaning:

$$E_0=E^{e}+E^{h}_{\mathbf{\Delta}}\,, \tag{3.8}$$

$$g=\frac{e}{mc}\int w_c^*(\boldsymbol{x})\nabla_i\,w_v(\boldsymbol{x})\,\mathrm{d}^3\boldsymbol{x}\,, \tag{3.9}$$

$$\text{const}=-i\frac{4\pi\hbar}{\Delta V}g\,, \tag{3.10}$$

where ΔV is the volume of the unit cell.

It should be noted that so far all equations are quantum-mechanical equations. They are nonlinear in B and N due to some terms on the right-hand side of (3.5), (3.6). In the following we fully keep this nonlinearity, but we treat eqs. (3.5) and (3.6) as classical ones. This can be achieved by taking on both sides of (3.5), (3.6) expectation values and approximating the expectation values of a product by the product of expectation values, *i.e.* $\langle BN\rangle=\langle B\rangle\langle N\rangle$. This procedure can be justified by the method of quantum classical correspondence.

We now discuss some special cases.

a) The « old » polariton. To this end we take $N_{\boldsymbol{l}}\equiv 0$. Equation (3.5) then reduces to

$$\frac{\mathrm{d}}{\mathrm{d}t}B_{\boldsymbol{l}}^{\dagger}=\frac{i}{\hbar}E_0B_{\boldsymbol{l}}^{\dagger}+i\sum_{\boldsymbol{l}'}W(\boldsymbol{l}'-\boldsymbol{l})B_{\boldsymbol{l}'}^{\dagger}+gA_{\boldsymbol{l}}^{-}\,. \tag{3.11}$$

If we assume for $A_{\boldsymbol{l}}^{-}$ the form of a running wave, (3.11) is readily diagonalized by

$$B_{\boldsymbol{k}}^{\dagger}=\frac{1}{\sqrt{\mathcal{N}}}\sum_{\boldsymbol{l}}\exp[i\boldsymbol{k}\boldsymbol{l}]B_{\boldsymbol{l}}^{\dagger}\,, \tag{3.12}$$

so that (3.11) transforms into

$$\frac{\mathrm{d}}{\mathrm{d}t}B_{\boldsymbol{k}}^{\dagger}=\frac{i}{\hbar}(E_0+E(\boldsymbol{k}))B_{\boldsymbol{k}}^{\dagger}+gA_{\boldsymbol{k}}^{-}\,, \tag{3.13}$$

where $E(\boldsymbol{k})$ has been defined in (2.4). If we use for $A_{\boldsymbol{k}}^-$ the slowly varying amplitude approximation, (3.7) transforms into

(3.14) $$\frac{\mathrm{d}}{\mathrm{d}t} A_{\boldsymbol{k}}^- = i\omega_0 A_{\boldsymbol{k}}^- + \text{const}\, B_{\boldsymbol{k}}^\dagger\,.$$

Equations (3.13) and (3.14) are evidently the usual polariton equations, if nonresonant terms are neglected. $B_{\boldsymbol{l}}^\dagger$ depends on $\boldsymbol{l}$ in the form

(3.15) $$B_{\boldsymbol{l}}^\dagger(t) = \exp[-i\boldsymbol{kl}]\, B^\dagger(t)\,.$$

b) Analogy with spin systems. Equations (3.5) and (3.6) contain another interesting special case. If we drop the sum containing the exciton interaction, *i.e.* terms containing W, eqs. (3.5) and (3.6) are decoupled and reduce to equations for the dipole-moment operator and the occupation number of the upper level $N_{\boldsymbol{l}}$ of ordinary two-level atoms. These equations are equivalent to the equations for operators for a spin $\frac{1}{2}$ if we identify $B^\dagger$, B and N with the spin-flip operators σ^+, σ^- and the operator for the z-component of the spin $s_z + \frac{1}{2}$ (see, *e.g.*, H. Haken: *Laser Theory*). This means that all phenomena which can be observed in spin resonance must be present, at least in principle, also for two-level atoms.

We leave the details of this correspondence to the reader as an exercise. All one has to do is to convince oneself that the B's and N's obey the same commutation relations as the spin operators and that eqs. (3.5) and (3.6) with $W = 0$ have the same form as the corresponding spin equations.

c) Exciton induction. What is new in our treatment is the occurrence of the sum over $\boldsymbol{l}'$ in (3.5) and (3.6). If we recollect the above treatment, we realize that these terms give rise to exciton motion or, in other words, to energy transfer. It is known, *e.g.* from photon echo experiments in ruby, that such interaction terms spoil the echo. Thus small concentrations of chromium atoms in ruby are required. Therefore, we want to focus our attention on the effect of these terms and we want to show that, under certain experimental conditions, we can compensate for the effect of the W's. We consider optical induction, *i.e.* we apply a pulsed coherent light field and look for the effect it has on the excitons. We assume that the field is spatially homogeneous so that we may anticipate that $N_{\boldsymbol{l}}$ is independent of the lattice site $\boldsymbol{l}$. We then multiply eq. (3.5) by $\exp[i\boldsymbol{kl}]$, sum over $\boldsymbol{l}$ and introduce operators according to (3.12). Using further the formulae

(3.16) $$\frac{1}{\sqrt{\mathcal{N}}} \sum_{\boldsymbol{k}} \exp[-i\boldsymbol{kl}] B_{\boldsymbol{k}}^\dagger = B_{\boldsymbol{l}}^\dagger$$

and

(3.17) $$\sum_{\boldsymbol{l}} W(\boldsymbol{l}) \exp[-i\boldsymbol{kl}] = \frac{1}{\hbar} E_{\boldsymbol{k}} = \nu_{\boldsymbol{k}}\,,$$

we obtain the equation

$$\dot{B}_{\boldsymbol{k}}^{\dagger} = i\nu B_{\boldsymbol{k}}^{\dagger} - i\nu_{\boldsymbol{k}}(2N-1)\,B_{\boldsymbol{k}}^{\dagger} - (2N-1)\,A_0^{-}\,g\delta_{\boldsymbol{k}_1\boldsymbol{k}_0}\,, \tag{3.18}$$

where it was assumed that the vector potential has the form

$$A_{\boldsymbol{l}}^{-} = \exp\left[-\,i\boldsymbol{k}_0\,\boldsymbol{l}\right]A_0^{-}\,\frac{1}{\sqrt{\mathcal{N}}}\,. \tag{3.19}$$

In a similar way we obtain, instead of eq. (3.6),

$$\dot{N}_{\boldsymbol{l}} = g\,\frac{1}{\sqrt{\mathcal{N}}}\sum_{\boldsymbol{k}}\exp\left[-\,i\boldsymbol{k}\boldsymbol{l}\right]B_{\boldsymbol{k}}^{\dagger}\cdot\exp\left[i\boldsymbol{k}_0\,\boldsymbol{l}\right]A_0^{+}\,\frac{1}{\sqrt{\mathcal{N}}} + \text{h.c.} \tag{3.20}$$

According to (3.18) only $B_{\boldsymbol{k}}^{\dagger} \neq 0$ for $\boldsymbol{k} = \boldsymbol{k}_0$, unless we incoherently pump the crystal to produce further excitons. In the following we drop the index $\boldsymbol{k}_0$. We further put

$$\nu_{\boldsymbol{k}_0} = \bar{\nu} \tag{3.21}$$

and

$$A_0^{-}\,g = \bar{A}^{*}\,. \tag{3.22}$$

Equations (3.18)-(3.20) then acquire the form

$$\dot{B}^{*} = i\nu B^{*} - i\bar{\nu}(2N-1)\,B^{*} - (2N-1)\,\bar{A}^{*} \tag{3.23}$$

and

$$\dot{N} = \frac{1}{\mathcal{N}}\,(\bar{A}^{*}B + \bar{A}B^{*})\,. \tag{3.24}$$

From (3.23) and (3.24) we obtain the following conservation law:

$$\frac{\mathrm{d}}{\mathrm{d}t}\left\{\frac{1}{\mathcal{N}}\,B^{*}B + \frac{1}{4}\,(2N-1)^{2}\right\} = 0\,. \tag{3.25}$$

Equation (3.25) can be immediately integrated. The integration constant can be determined by the requirement that, at the initial time, all atoms are in their ground state, *i.e.* no excitons are present, $N = 0$, and that there is no local dipole moment, $B = 0$. According to this conservation law

$$\frac{1}{\mathcal{N}}\,B^{*}B + \frac{1}{4}\,(2N-1)^{2} = \frac{1}{4}\,, \tag{3.26}$$

we can express N by B. We now put

(3.27) $$2N - 1 = \cos\theta$$

and

(3.28) $$B = \sqrt{\mathcal{N}}\,\tfrac{1}{2}\sin\theta\exp[-i\varphi]\,,$$

which transforms the complex equation (3.23) into the equations

(3.29) $$\dot{\theta} = -a\cos(\varphi - \chi)$$

and

(3.30) $$\dot{\varphi} = \nu - \bar{\nu}\cos\theta + a\,\mathrm{ctg}\,\theta\sin(\varphi - \chi)\,,$$

where we have used the abbreviations

(3.31) $$\bar{A}^* = \exp[i\chi]\,\hat{A}$$

and

(3.32) $$2\frac{1}{\sqrt{\mathcal{N}}}\hat{A} = a\,.$$

The angles θ and φ have a simple meaning. $\cos\theta$ is a measure for the occupation, while φ measures the phase of the dipole moment. In (3.31) χ is the phase, whereas $\hat{A}$ is the modulus. We now assume the phase χ in the form

(3.33) $$\chi = \omega t$$

and make the replacements

(3.34) $$\varphi \to \varphi + \omega t$$

and

(3.35) $$\hat{\nu} = \nu - \omega$$

in (3.29) and (3.30). This leaves us with the basic equations

(3.36) $$\dot{\theta} = -a\cos\varphi$$

and

(3.37) $$\dot{\varphi} = \hat{\nu} - \bar{\nu}\cos\theta + a\,\mathrm{ctg}\,\theta\sin\varphi\,.$$

We first discuss ordinary spin resonance. If we have complete resonance, we may put $\hat{\nu} = 0$. Furthermore, due to the lack of spin-spin interaction, $\bar{\nu} = 0$, and eq. (3.37) is satisfied by $\varphi = 0$. Equation (3.36) then has the solution

$\theta = \pi - at$. In the language of spins, the spin is erected; in the language of atoms, the atom is driven coherently from the lower state into the upper state.

If $\bar{\nu}$ or $\hat{\nu}$ is unequal to zero, φ changes with time according to (3.37), so that $\cos\varphi$ in (3.36) decreases. Or, in other words, the effect of the electric field $\sim a$ is decreased and the erection of the spin, or the inversion of the atoms, may become impossible.

It is therefore necessary to discuss the conditions under which this may happen in more detail. We base our considerations on eqs. (3.36) and (3.37). From these equations we eliminate the time derivatives using as an intermediate step the equations

$$\text{(3.38)} \qquad d\theta = -a\cos\varphi\, dt$$

and

$$\text{(3.39)} \qquad d\varphi = [\hat{\nu} - \bar{\nu}\cos\theta + a\,\text{ctg}\,\theta\sin\varphi]\, dt\,.$$

Using the abbreviation

$$\text{(3.40)} \qquad \sin\varphi = z\,,$$

we obtain

$$\text{(3.41)} \qquad \frac{dz}{d\theta} = -\frac{\hat{\nu}}{a} + \frac{\bar{\nu}}{a}\cos\theta - z\,\text{ctg}\,\theta\,,$$

which is a linear first-order differential equation for the unknown function $z(\theta)$. The general solution reads

$$\text{(3.42)} \qquad \sin\varphi = z = \frac{C}{\sin\theta} + \frac{\hat{\nu}}{a}\,\text{ctg}\,\theta - \frac{\bar{\nu}}{4a}\,\frac{\cos 2\theta}{\sin\theta}\,.$$

If we take as initial conditions, for $t = 0$, $\varphi = 0$, $\theta = \pi$, the constant turns out to be

$$\text{(3.43)} \qquad C = \frac{\hat{\nu}}{a} + \frac{\bar{\nu}}{4a}\,,$$

and the solution (3.42) now reads

$$\text{(3.44)} \qquad \sin\varphi = \frac{\hat{\nu}}{a}\,\text{ctg}\,\frac{\theta}{2} + \frac{\bar{\nu}}{2a}\sin\theta\,.$$

For what follows it is convenient to introduce, instead of θ, a new variable (*)

$$\text{(3.45)} \qquad \phi = \pi - \theta\,.$$

(*) The following considerations are based on a paper by GOLL and HAKEN (to be published).

Then (3.44) reads

$$\sin\varphi = \frac{\hat{\nu}}{a}\,\mathrm{tg}\,\frac{\phi}{2} + \frac{\bar{\nu}}{2a}\sin\phi\,, \tag{3.46}$$

whereas eq. (3.36) acquires the form

$$\dot{\phi} = a\cos\varphi\,. \tag{3.47}$$

With the help of (3.47) we can now distinguish between two cases. If $\cos\varphi$ has no change of sign for $0\leqslant\phi\leqslant\pi$, the total inversion can be fully reached. We refer to this case as the inversion case. On the other hand, the full inversion is not reached and the spin returns before being completely flipped, if for $\phi<\pi$ the necessary condition

$$|\sin\varphi| = 1 \tag{3.48a}$$

and the sufficient condition

$$\text{change of sign of } \cos\varphi, \tag{3.48b}$$

are fulfilled.

To discuss condition (3.48) we consider (3.46) and ask that the value of the right-hand side be smaller than unity. Because for $\phi\leqslant\pi$ the sine function remains restricted, but tg $(\phi/2)$ tends to infinity, the right-hand side of (3.46) becomes bigger than unity for $\hat{\nu}\neq 0$. This means that ϕ cannot reach the value π, or, in other words, the oscillatory case is realized. Though the full inversion cannot be established, there may be cases, even for $\bar{\nu}\neq 0$, in which the inversion can be reached partially, *i.e.* $\phi\geqslant\pi/2$. To investigate which conditions are favourable to that case we use the fact that for the considered ϕ, $0\leqslant\phi\leqslant\pi$, the following inequality holds:

$$\left|\frac{|\hat{\nu}|}{a}\,\mathrm{tg}\,\frac{\phi}{2} + \frac{|\bar{\nu}|}{2a}\sin\phi\right| > \left|\frac{|\hat{\nu}|}{a}\,\mathrm{tg}\,\frac{\phi}{2} - \frac{|\bar{\nu}|}{2a}\sin\phi\right| \qquad \text{for } \hat{\nu},\,\bar{\nu}\neq 0\,. \tag{3.49}$$

Here we have made use of the fact that $\hat{\nu}$ and $\bar{\nu}$ can take both positive and negative signs depending on the system under consideration and on the experimental conditions with respect to tuning. Since the right-hand side is smaller than the left-hand side, the limit (3.46) $=1$ is reached for a bigger maximal inversion, if the signs of $\hat{\nu}$ and $\bar{\nu}$ are opposite.

A closer inspection which we will not re-present here shows that, for $|\sin\varphi|=1$, $\cos\varphi$ changes its sign, so that the turning point occurs there.

We now seek the time-dependent solution of eqs. (3.36) and (3.37). To this end we write $\cos\varphi=\sqrt{1-\sin^2\varphi}$ and we substitute eq. (3.44) for $\sin\varphi$ in eq. (3.47). We obtain a first-order differential equation in ϕ. Separating the variables and integrating we find

$$t = \frac{1}{a}\int\limits_0^{\Phi} \frac{\mathrm{d}\phi'}{\sqrt{1-\{(\hat{\nu}/a)\,\mathrm{tg}\,(\phi'/2)+(\bar{\nu}/2a)\sin\phi'\}^2}} \tag{3.50}$$

for

$$0 \leqslant \phi \leqslant \pi .$$

Let us discuss some special cases.

1) *Two-level atom (or spin* $\frac{1}{2}$*) with exact resonance* ($\bar{\nu}=0$, $\hat{\nu}=0$). From (3.37) we see that $\varphi = 0$ for all times provided $\varphi = 0$ for $t = 0$. From (3.36) it follows that $\theta = \pi - at$. The inversion can be fully reached.

2) *Two-level atom (or spin* $\frac{1}{2}$*).* No resonance ($\bar{\nu} = 0$, $\hat{\nu} \neq 0$). The turning point $\phi = \Theta$ is given by

$$\mathrm{tg}\,\frac{\Theta}{2} = \left|\frac{a}{\hat{\nu}}\right| . \tag{3.51}$$

3) *Excitons* ($\bar{\nu} \neq 0$, $\hat{\nu} = 0$). Here eq. (3.46) reduces to

$$\sin\varphi = \frac{\bar{\nu}}{2a}\sin\phi . \tag{3.52}$$

Since the right-hand side of eq. (3.52) is $\leqslant |\bar{\nu}/2a|$, the condition

$$|\sin\varphi| \leqslant 1 \tag{3.53}$$

is fulfilled provided

$$\left|\frac{\bar{\nu}}{2a}\right| \leqslant 1 \tag{3.54}$$

holds.

To realize a full inversion the condition (3.54) means that the light-pulse intensity must be big enough, where a measure for the size is the interaction energy (expressed in frequency of the excitons). The integral (3.50) reduces to

$$t = \frac{1}{a}\int\limits_0^{\phi} \frac{\mathrm{d}\phi'}{\sqrt{1-(\bar{\nu}/2a)^2\sin^2\phi'}} = \frac{1}{a}F\left(\left|\frac{\bar{\nu}}{2a}\right|, \phi\right), \tag{3.55}$$

which is nothing but the incomplete elliptic integral of the first kind. We now consider the oscillatory case for which

$$\left|\frac{\bar{\nu}}{2a}\right| > 1 \tag{3.56}$$

must hold. In this case we may deduce for the differential equation for ϕ

$$\frac{\mathrm{d}\phi}{\mathrm{d}t} = a\sqrt{1-\left(\frac{\bar{\nu}}{2a}\right)^2 \sin^2\phi} \tag{3.57}$$

up to the turning point $\phi = \Theta$, *i.e.* that point where the square root vanishes. The corresponding turning time can be found from (3.50) as the integral

$$\tau = \frac{1}{a}\int_0^\Theta \frac{\mathrm{d}\phi}{\sqrt{1-(\bar{\nu}/2a)^2 \sin^2\phi}}, \tag{3.58}$$

or

$$\tau = \frac{2}{|\bar{\nu}|} K\left(\left|\frac{2a}{\bar{\nu}}\right|\right), \tag{3.59}$$

which is the complete elliptic integral of the first kind. A more detailed discussion which we will not perform here, but which we have indicated already above, shows that a maximal ϕ can be realized even if (3.54) is not fulfilled but if $\hat{\nu}$ is appropriately chosen.

I hope I have shown by the above considerations that there are, at least, certain ranges for the light field intensity and the detuning $\hat{\nu}$ for which full or partial inversion of the excitonic system can be reached, even in spite of the mutual interaction between excitons.

The inverted system can now give rise to superradiant emission, where the treatment can now be performed in close analogy to superradiance or to the photon echo effect which is discussed in other lectures.

We now deal with

d) Self-induced transparency. We closely follow the papers of HAKEN and SCHENZLE with a recent modification suggested by GOLL (see below). We treat a slightly more general case than eq. (3.5), which applies not only to Frenkel excitons but also to Wannier excitons:

$$\dot{B}_{\boldsymbol{l}}^\dagger = \frac{i}{\hbar}\sum_{\boldsymbol{l}'} W(\boldsymbol{l}-\boldsymbol{l}')B_{\boldsymbol{l}'}^\dagger - \frac{2i}{\hbar}\sum_{\boldsymbol{l}'}(\tilde{W}_{\mathrm{c}}^{\mathrm{e\text{-}h}}(\boldsymbol{l}-\boldsymbol{l}') - \tilde{W}^{\mathrm{e\text{-}e}}(\boldsymbol{l}-\boldsymbol{l}'))B_{\boldsymbol{l}}^\dagger N_{\boldsymbol{l}'} - \\ - \frac{2i}{\hbar}\sum_{\boldsymbol{l}'}\tilde{W}_{\mathrm{ex}}^{\mathrm{e\text{-}h}}(\boldsymbol{l}-\boldsymbol{l}')B_{\boldsymbol{l}'}^\dagger N_{\boldsymbol{l}} - gA_{\boldsymbol{l}}^-(2N_{\boldsymbol{l}}-\psi(0)), \tag{3.60}$$

$$\dot{N}_{\boldsymbol{l}} = \frac{i}{\hbar}\sum_{\boldsymbol{l}'}\overline{W}_{\mathrm{ex}}^{\mathrm{e\text{-}h}}(\boldsymbol{l}-\boldsymbol{l}')(B_{\boldsymbol{l}'}^\dagger B_{\boldsymbol{l}} - B_{\boldsymbol{l}}^\dagger B_{\boldsymbol{l}'}) + g(A_{\boldsymbol{l}}^+ B_{\boldsymbol{l}}^\dagger + A_{\boldsymbol{l}}^- B_{\boldsymbol{l}}), \tag{3.61}$$

$$\left(\frac{\partial}{\partial t} + c\frac{\partial}{\partial \boldsymbol{l}}\right)A_{\boldsymbol{l}}^- = -\Lambda^{-2}\psi(0)gB_{\boldsymbol{l}}^\dagger \tag{3.62}$$

with

$$\Lambda^{-2} = \frac{2\pi\hbar c^2}{ck}. \tag{3.63}$$

The W's are essentially electron-electron and electron-hole Coulomb and Coulomb-exchange matrix elements. $\psi(0)$ is the exciton envelope wave function at the same position of electron and hole. c is the light velocity in the noninverted medium.

If we drop the second sum on the right-hand side of eq. (3.60), we are back to eq. (3.5). For the further treatment we split the exciton dipole moment and the vector potential into a rapidly varying space- and time-dependent factor and a slowly space- and time-dependent amplitude:

$$B_l^\dagger = \tilde{B}_l^* \exp[-ikl + i\Omega t], \qquad A_l^- = \tilde{A}_l^- \exp[-ikl + i\Omega t]. \tag{3.64}$$

While in the original paper by HAKEN and SCHENZLE Ω was taken as the frequency of a light wave with wave vector k, HANAMURA uses for Ω polariton frequencies. In the present treatment we follow GOLL in chosing Ω as an adjustable frequency, which secures that the slowly varying amplitudes on the right-hand sides of eqs. (3.64) do not depend on a phase factor proportional to the time t. We confine our treatment to a first approximation, which is obtained when we expand the right-hand sides of eqs. (3.60) and (3.61) with respect to $l' - l$. We anticipate pulselike solutions for the amplitudes $\tilde{B}$, $\tilde{A}$, so that we require that these quantities depend only on the co-ordinate $\tau = t - l/v$, where v is the pulse velocity. The original partial differential equations are thus transformed into ordinary differential equations:

$$\dot{\tilde{B}}_l^* = -g\tilde{A}_l^-(2N_l - \psi(0)) + \frac{i}{\hbar}(E_1 - \hbar\Omega + E_2 N_l)\tilde{B}_l^*, \tag{3.65}$$

$$\dot{N}_l = g(\tilde{A}_l^* \tilde{B}_l^* + \tilde{A}_l^- \tilde{B}_l), \tag{3.66}$$

$$\left(1 - \frac{c}{v}\right)\frac{\partial}{\partial t}\tilde{A}_l^- = -\Lambda^{-2}\psi(0) g \tilde{B}_l^* + ick\tilde{A}_l^- - i\Omega\tilde{A}_l^-, \tag{3.67}$$

with the abbreviations

$$\left\{\begin{aligned} E_1 &= E_g - \varepsilon_b + \frac{\hbar^2 k^2}{2M} = E_0 + \frac{\hbar^2 k^2}{2M}, \\ E_2 &= -2\sum_{l'}(\tilde{W}_c^{\text{e-}h}(l-l') - \tilde{W}^{\text{e-e}}(l-l')) - 2\sum_{l'}\tilde{W}_{\text{ex}}^{\text{e-}h}(l-l')\exp[ik](l-l'), \end{aligned}\right. \tag{3.68}$$

M is the total effective mass of the exciton, ε_b the exciton binding energy, E_g the band gap energy.

We now split the vector potential into a modulus and a phase factor. It should be noted that the treatment can be also performed if there are derivatives of (3.65) and (3.66) with respect to space co-ordinates.

The following conservation laws follow from eqs. (3.65)-(3.67):

$$\tilde{B}^* \tilde{B} + N^2 - \psi(0) N = C_1 \tag{3.69}$$

and

$$N = \frac{1}{\psi(0)} \lambda^2 |\tilde{A}|^2 + C_2 , \tag{3.70}$$

where $\lambda^2 = (c/v - 1)\Lambda^2$.

The initial conditions lead to $C_1 = C_2 = 0$. Using (3.69), (3.70) and

$$\tilde{A}^{\pm} = A \exp [\pm i\phi] , \tag{3.71}$$

we obtain the following differential equation for the phase:

$$\dot{\phi}(\tau) = -\frac{1}{2\hbar}\left[(E_1 - \hbar\Omega) + \frac{1}{2} E_2 \frac{1}{\psi(0)} \lambda^2 A^2 - (\hbar\Omega - \hbar ck)\left(1 - \frac{c}{v}\right)^{-1}\right]. \tag{3.72}$$

Ω is now determined by the condition that $\dot{\phi}$ does not contain a constant τ-independent contribution. This leads to the dispersion law for the central frequency of the pulses

$$\hbar\Omega = \frac{-\alpha(E_g - \varepsilon_b + k^2\hbar^2/2M) + \hbar ck}{1 - \alpha} \tag{3.73}$$

with

$$\alpha = \frac{c}{v} - 1 . \tag{3.74}$$

Equation (3.73) fixes a relation between Ω and the wave number k or, if Ω is prescribed, we obtain the relation

$$\hbar k = \frac{Mc}{\alpha}\left[1 \pm \sqrt{1 - \frac{2\alpha}{Mc^2}\{\hbar\Omega(1-\alpha) + \alpha E_0\}}\right], \tag{3.75}$$

where the lower sign of the root must be chosen, as can be seen by the discussion of certain limiting cases. We obtain for the field amplitude the equation

$$\dot{A}^2 = \left[\frac{1}{\alpha}\frac{g^2}{\Lambda^2}\psi^2(0) - \frac{(\Omega - ck)^2}{\alpha^2}\right] A^2 - g^2 A^4 - \dot{\phi}^2 A^2 - \frac{2\dot{\phi} A^2(\Omega - ck)}{\alpha} , \tag{3.76}$$

and, on account of the consistency requirement (3.73) for $\dot{\phi}$,

$$\dot{\phi}(\tau) = -\frac{1}{4\hbar\psi(0)} E_2 \lambda^2 A^2 . \tag{3.77}$$

Inserting $\dot{\phi}$ into eq. (3.76) we obtain a differential equation of the form

$$\dot{A}^2 = T^{-2}A^2 + C_1' A^4 + C_2' A^6 \tag{3.78}$$

with the solution

$$A^2 = \frac{\text{const}}{a \cosh (2/T)(\tau - \tau_0) + b}\,. \tag{3.79}$$

According to (3.66) the pulse width is given by

$$T^{-2} = \frac{1}{\alpha}\frac{2\pi\hbar c g^2 \psi^2(0)}{k} - \frac{1}{\hbar^2}\left[\hbar\Omega - \left(E_0 + \frac{k^2\hbar^2}{2M}\right)\right]^2, \tag{3.80}$$

where k is given by (3.75).

From (3.80) we see that the pulse width is a positive quantity provided we remain in the neighbourhood of $\Omega = E_0/\hbar$. The pulse velocity v can be determined by means of the given pulse width and central frequency.

Experimental evidence of self-induced transparency of excitons has been given by BRUCKNER, DNESTROVSKII and KOSHCHUG.

They used semiconducting single crystals $\text{Cd}\,\text{S}_{0.75}\,\text{Se}_{0.25}$ at 90 K. The sample was exposed to a train of second-harmonic ultra-short pulses from a mode-locked neodymium laser ($\tau = 5\cdot 10^{-12}$ s, $\lambda = 0.53$ μm, $\Delta\lambda = 5$ Å). The chosen composition of the mixed crystal and the chosen temperature allowed for a resonant excitation of the excitons.

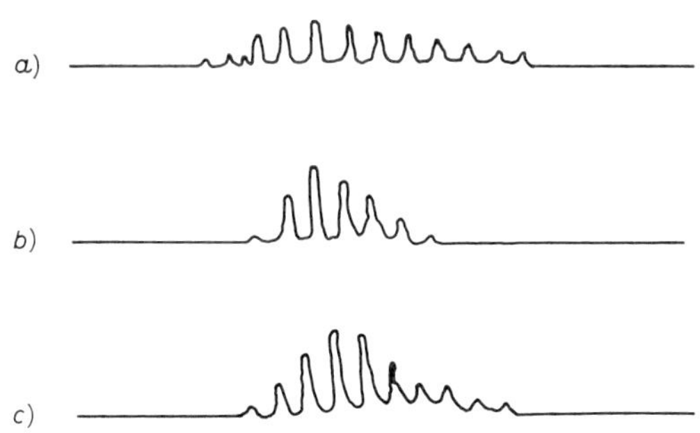

Fig. 6. – Oscillograms of USP train entering the sample (a)) and leaving the sample at $W_0 = 10^{-3}$ J (b)) and $W_0 = 1.5\cdot 10^{-3}$ J (c)). (After F. BRUCKNER *et al.*)

Typical oscillograms of a pulse train entering and leaving the sample are shown in fig. 6. Note that the most powerful and shortest pulses are separated; additional pulses past the sample appear when the pump level is increased. Figure 7 shows the dependence of W/W_0 (W_0, W are the energies of incident or transmitted radiation) on W_0. Evidently, the transmission increases sharply, after a critical level W_0 is reached in accordance with the theoretical expectation.

Further experiments permitted the measurement of T_1 and T_2. For a final discussion of whether the above indicated results are a conclusive proof for self-induced transparency, note the following.

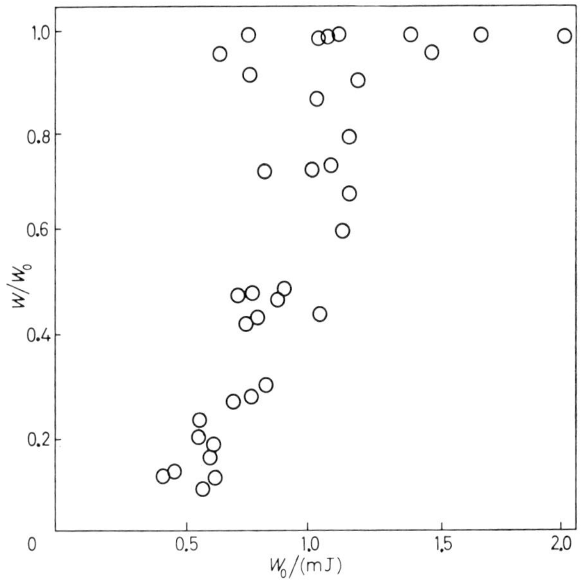

Fig. 7. – Dependence of sample transmission on the radiation energy at the entrance to the sample. (After F. BRUCKNER *et al.*)

At least fig. 7 can be explained by some kind of band-filling effect of excitons. According to it, the absorption line is shifted to higher and higher frequencies when more and more excitons are present. Thus, when the frequency of the impinging radiation is kept fixed, the absorption should rapidly decrease. This explanation for fig. 7 will be excluded, if the absorption spectrum is measured simultaneously and no absorption line shift occurs.

4. – Spatially homogeneous solution.

We now look for solutions which come closest to the old polariton concept, that is those which merely propagate like a plane wave through the crystal. To this end we set

$$B_l^\dagger = \exp[-ikl]\exp[i\omega_0 t]\hat{B}^\dagger\,, \tag{4.1}$$

$$A^+ = \exp[+ikl]\exp[-i\omega_0 t]\hat{A}^+\,, \tag{4.2}$$

$$\hat{A},\ \hat{B}^\dagger \sim \exp[i\Omega t]\,, \tag{4.3}$$

$$\hat{A}^+ = A\exp[i\phi]\exp[-i\Omega t]\,. \tag{4.4}$$

Because the « slowly varying amplitude » no longer depends on space, spatial derivatives must be set equal to 0; this is equivalent to applying the case treated in the preceding paragraph $v = \infty$, so that we must replace $\lambda^2 \to -\Lambda^2$. The elimination technique is essentially the same as before via the two intermediate steps

$$N = N_0 - \Lambda^2 A^2 \tag{4.5}$$

and

$$N(N-1) + B^\dagger B = 0\,,$$

and leads to the following two coupled equations:

$$\Omega_{1,2} = \frac{1}{2}(\Delta\omega + \omega_k 2N_0) - \omega_k \Lambda^2 A^2 \pm \frac{1}{2}\left\{(\Delta\omega + 2\omega_k N_0)^2 + \left(\frac{2g}{\Lambda}\right)^2 (1 - 2N_0) + \right. \\ \left. + 4\Lambda^2 A^2 \left(\Lambda^2 \omega_k^2 A^2 + 2\left(\frac{g}{\Lambda}\right)^2 - \omega_k(\Delta\omega + \omega_k 2N_0)\right)\right\}^{\frac{1}{2}}, \tag{4.6}$$

$$\Lambda^2 A^4 - A^2\left[2\left(N_0 - \frac{1}{2}\right) - \left(\frac{\Lambda}{g}\right)^2 \Omega^2\right] - \frac{1}{\Lambda^2} N_0 (1 - N_0) = 0\,. \tag{4.7}$$

For small excitation levels, *i.e.* $N_0 \approx 0$ and A^2 very small, (4.6) reduces to the well-known polariton dispersion relation (provided that nonresonant terms are neglected). Equations (4.6) and (4.7) can be looked at in different ways: *a*) The field amplitude A is prescribed. Then the inversion N_0 and the frequency Ω are fixed by (4.7) and (4.6). It is clear that the dispersion is now power dependent. *b*) A physically less realistic case is given if one keeps the inversion N_0 fixed and lets the field amplitude A adjust according to eqs. (4.6) and (4.7).

5. – A brief reminder of conventional laser theory.

To prepare the following paragraphs on laser action by excitons, let us first consider the conventional theory of laser action by atoms. Laser theory is a lucid example of the duality of light, which may manifest itself as particles (photons) or waves.

Accordingly laser equations may be formulated by means of photon numbers n or by means of wave amplitudes b.

a) *Laser equations for the photon number* n. Let us consider a set of two-level atoms which are pumped by an external source from the lower level 1 to the upper level 2. We denote the corresponding occupation numbers of the atoms by N_1 and N_2. The number of spontaneously emitted photons per second is given by GN_2, where G is proportional to the squared optical matrix element. According to EINSTEIN the stimulated emission rate is obtained from that for spontaneous emission by multiplying the latter one by the photon number n. On the other hand, the photon number is decreased by absorption with rate $-GN_1 n$. Since photons eventually leave the laser, we must take into account a loss rate which we write in the form $-2\varkappa n$. Taking the gain and loss terms together, we obtain the laser equation

$$\dot{n} = G(N_2 - N_1)n + GN_2 - 2\varkappa n . \tag{5.1}$$

b) Laser equations for the wave amplitude b. We decompose the electric-field strength in the form of plane waves

$$E(x, t) = b(t)\mathscr{N} \exp[ikx] + \text{c.c.} \tag{5.2}$$

The quantity of interest is the time-dependent wave amplitude b. According to laser theory the equation for the time dependence of b reads

$$\dot{b} = -i\omega b + i \sum_{\mu} p_\mu g_\mu - \varkappa b . \tag{5.3}$$

The first term on the right-hand side describes the free oscillation, which stems from the homogeneous equation. The second term stems from the oscillating dipoles with dipole moments p_μ, which drive the mode amplitude. g_μ are coupling coefficients, which are proportional to the optical matrix element and which depend on the wave amplitude at space point x_μ, where x_μ is the position of the μ-th atom. The last term in (3.3) describes the damping of the mode due to the emission through the mirrors at the laser end faces. Though eqs. (5.1), (5.3) look completely different, they are closely related to each other. It will become clear later on in our article that (5.1) follows from (5.3) when we put $n = b^* b$. Note that (5.3) contains more information than (5.1) because the complex amplitude b contains both amplitude and phase information, while (5.1) does not contain any phases.

Equations (5.1) or (5.3) do not completely describe the laser process, because in the course of the laser process the occupation numbers N_j or the dipole moments p_μ change. We present the additional equations for the two pictures (photon picture and wave picture) separately:

a) Photon picture. Equation (5.1) must be supplemented as follows. The occupation number of the upper level changes due to the pump from the lower level with rate $w_{21} N_1$, by nonlasing transitions from 2 to 1 with the rate $-w_{12} N_2$ and finally due to the interaction with the photons under consideration:

$-G(N_2-N_1)n$. Thus the occupation number changes according to

(5.4) $$\dot{N}_2=w_{21}N_1-w_{12}N_2-G(N_2-N_1)n\,.$$

If we take into account that $N_1+N_2=N$, it follows from (5.4) in the stationary state that

(5.5) $$N_2\approx N_{2,0}-\text{const}\cdot n\,,$$

where $N_{2,0}$ is the so-called unsaturated inversion, which is achieved by pumping and nonlasing processes. When laser action occurs, this unsaturated inversion is diminished by a quantity essentially proportional to the photon number. The resulting N_2 is the so-called saturated inversion. Using (5.5) in (5.1) one can show that the expression (5.5) is essential for the stabilization of the laser intensity. When it grows, *i.e.* when n grows, the emission rate occurring in (5.1) is lowered and the laser light intensity is decreased. An equivalent regulation mechanism holds in the case that the output intensity is too small. We now describe how to supplement eq. (5.3) in the case of the

b) *Wave picture.* The equation for the dipole moment has the form

(5.6) $$\dot{p}_\mu=-i\omega p_\mu-\gamma p_\mu-ig_\mu^*b(N_2-N_1)_\mu\,.$$

The first term on the right-hand side stems from the oscillatory motion, which can be described by a classical oscillator model or, in a more advanced theory, by the solution of the Schrödinger equation taking into account two atomic levels. The second term describes the damping which can be derived by means of a quantum statistical theory, which we shall not deal with here. The last term describes the driving force due to the field amplitude acting on the dipole moment. This force is proportional to the coupling constant g_μ, to the field amplitude b, and depends on the inversion in contrast to classical theory. If the inversion N_2-N_1 is positive the energy flux from the field to the atom is reversed; this means that the relative phase between p and b must be reversed. This is expressed by the change of sign of N_2-N_1. Actually there is a third equation for the occupation numbers which we neglect here. From (5.6) one finds

(5.7) $$p_\mu\propto b(N_2-N_1)_\mu\,,$$

which, when reinserted into (5.3), leads to a driving term for the field amplitude

(5.8) $$\dot{b}\propto|g|^2b(N_2-N_1)\,.$$

This is the other way in which stimulated emission can be described. The temporal change of the field amplitude is proportional to the field amplitude itself. Besides the photon number and the wave amplitude description of laser action there exists a fully quantum-mechanical one in which b, p, etc. are quantum-mechanical operators. We will come back to this description soon.

6. – Exciton laser processes.

a) *Theorem*: *laser emission by excitons is impossible.* This theorem holds as long as we treat excitons, as usual, as noninteracting bosons. The proof of this theorem is quite simple. Consider the production rate of photons when excitons (bosons) are annihilated. This rate is $GN(n+1)$, the probability for the reversed process is $G(N+1)n$. This follows from fundamental quantum-mechanical laws for the creation and annihilation of bosons. If we take into account the loss of photons as usual, the photon rate equation reads

$$\dot{n} = GN(n+1) - G(N+1)n - 2\varkappa n\,, \tag{6.1}$$

which can be simplified to

$$\dot{n} = GN - Gn - 2\varkappa n\,. \tag{6.2}$$

Evidently there is no production rate proportional to n, as is characteristic for stimulated emission. Thus laser action of excitons is impossible.

b) *Laser action by excitons is possible if other processes are involved.* Inspection of eq. (6.1) or (6.2) shows that laser action is prevented by the strong reabsorption. Thus, to achieve laser action, we must completely or partially forbid reabsorption processes. This can be achieved if we make G of the second term on the r.h.s. of (6.1) smaller than G of the first term on the r.h.s. of (6.1). Experimentally this can be achieved if the emission and absorption lines are shifted with respect to each other. Other mechanisms are provided by «wasting» exciton energy due to its interaction with other elementary excitations. These processes are, among others, simultaneous emission of a phonon, collision of 2 excitons, where one exciton is excited or ionized, laser emission from exciton molecules, where the binding energy is lost (provided the exciton concentration is not so high that the reverse process is probable), scattering of an electron (hole) by an exciton, whereby the electron (hole) picks up energy:

exciton ↗ photon, ↘ phonon $\qquad h\nu = E_{\text{ex}} - h\nu_{\text{ph}}\,,$

exciton + exciton ↗ photon, ↘ excited exciton $\qquad h\nu = E_{\text{ex}} - (\Delta E_{\text{ex}})\,,$

$$\text{exciton} - \text{molecule} \begin{array}{l} \nearrow \text{photon} \\ \searrow \text{exciton} + \text{binding energy} \end{array} \qquad h\nu = E_{\text{ex}} - E_{\text{binding}}\,,$$

$$\text{exciton} + \text{electron} \begin{array}{l} \nearrow \text{photon} \\ \searrow \text{electron} \end{array} \qquad h\nu = E_{\text{ex}} - \Delta E_{\text{electron}}\,,$$

$$\text{exciton} + \text{hole} \begin{array}{l} \nearrow \text{photon} \\ \searrow \text{hole} \end{array} \qquad h\nu = E_{\text{ex}} - \Delta E_{\text{hole}}\,.$$

7. – The quantum theory of exciton laser action.

We consider a single kind of light wave which has the classical amplitude b. We now adopt the usual quantum-mechanical approach in which b and its Hermitian conjugate $b^\dagger$ are the annihilation and creation operators of light quanta, respectively. We consider only one internal state of excitons, but take different total momenta $\boldsymbol{k}$ into account. In the following we direct our attention to optical transitions in which excitons are created and annihilated. Another possibility would be laser action due to transitions between internal states of excitons. Though these latter transitions represent an interesting experimental possibility, we will not deal with them here, because the corresponding theory is completely analogous to that of gas lasers. We denote the exciton creation and annihilation operators by $B_{\boldsymbol{k}}^\dagger$, $B_{\boldsymbol{k}}$, respectively, and treat them as Bose operators. We introduce elementary excitations such as phonons, electrons, holes, plasmons or dissociated electron-hole pairs. The corresponding creation and annihilation operators are denoted by $c_\mu^\dagger$, c_μ. The total Hamiltonian then acquires the form

$$H = \underset{\text{photons}}{\hbar\omega b^\dagger b} + \underset{\text{excitons}}{\sum_k \hbar\nu_k B_k^\dagger B_k} + \underset{\text{elementary excitation (s)}}{\sum_k \hbar\bar{\nu}_\mu c_\mu^\dagger c_\mu} + \underset{\text{exciton-photon}}{\hbar g(b^\dagger B_0 + \text{h.c.})} + \underset{\text{excitation-exciton}}{\sum_{Q,q_1 q_2} \hbar(V_{Q,q_1} C_{Q,q_1} B_{q_2-Q}^\dagger B_{q_2} + \text{h.c.})} \tag{7.1}$$

The first three sums are the Hamilton operators for the energies of the photons, of the excitons and of the elementary excitations. The fourth term describes the exciton-photon interaction, *i.e.* the annihilation of the exciton and the creation of a photon or the reverse process. The last term describes the scattering of an exciton from state q_2 to state $q_2 - Q$ caused by the interaction with elementary excitations. C contains operators describing the elementary excitation. V are matrix elements connected with this interaction. The different identifications of $c_\mu^\dagger$, c_μ, C_{Q,q_1} are given in table I.

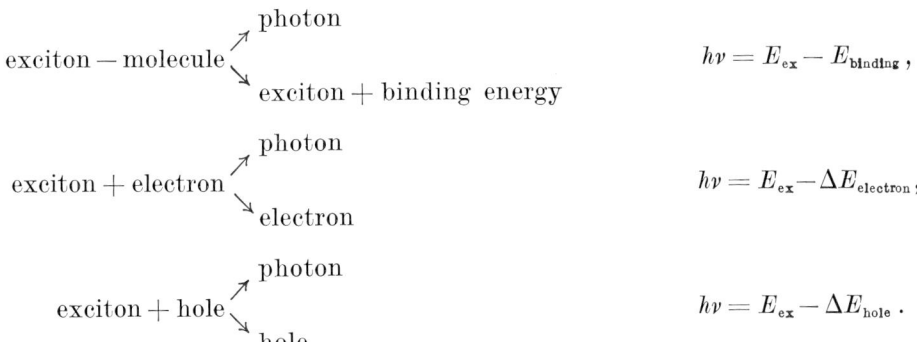

TABLE I.

excited or dissociated electron-hole pair	electrons
$c^\dagger_\mu = \hat{B}^\dagger_{\boldsymbol{k},j}$	$c^\dagger_\mu = a^\dagger_\mu$
$c_\mu = \hat{B}_{\boldsymbol{k},j}$	$c_\mu = a_\mu$
$C_{Q,q_1} = \hat{B}^\dagger_{q_1+Q} B_{q_1}$	$C_{Q,q_1} = a^\dagger_{q_1+Q} a_{q_1}$

holes	phonons
$c^\dagger_\mu = d^\dagger_\mu$	$c^\dagger_\mu = c^\dagger_{\boldsymbol{q}}$
$c_\mu = d_\mu$	$c_\mu = c_q$
$C_{Q,q_1} = d^\dagger_{q_1+Q} d_{q_1}$	

emission or absorption of phonons	phonon scattering
$C_{Q,q_1} = c^\dagger_Q$	$C_{Q,q_1} = c^\dagger_{q_1+Q} c_{q_1}$

It is well known that the Hamiltonian (7.1) is not sufficient to describe laser processes, since we must take into account damping and fluctuations. To incorporate them into quantum theory we use quantum-mechanical *Langevin equations*. They are obtained from the Heisenberg equations of motion by adding appropriate damping and fluctuation terms in a well-defined manner:

$$\dot{\Omega} = \frac{1}{i\hbar}[H, \Omega] + \text{damping and fluctuations}\,. \tag{7.2}$$

In (7.2) Ω is an arbitrary operator. The appropriate operators for our theory are $b^\dagger$, $B^\dagger_q$ and C_{q,q_1}. We then obtain the following equations:

photons:

$$\dot{b}^\dagger = (i\omega - \varkappa)\, b^\dagger + ig\, B^\dagger_0 + F_b\,, \tag{7.3}$$

excitons:

$\boldsymbol{q} = 0$

$$\dot{B}^\dagger_0 = (i\nu_0 - \gamma)\, B^\dagger_0 + ig\, b^\dagger + \sum \dots C^\dagger_{\boldsymbol{Q},\boldsymbol{q}} B^\dagger_{-Q} + F_{B,0}\,, \tag{7.4}$$

excitons:

$\boldsymbol{q} \neq 0$

$$\dot{B}^\dagger_{\boldsymbol{q}} = (i\nu_{\boldsymbol{q}} - \gamma_{\boldsymbol{q}})\, B^\dagger_{\boldsymbol{q}} + \sum \dots C^\dagger_{\boldsymbol{q},\boldsymbol{q}_1} B_0 + \sum \dots C^\dagger_{\boldsymbol{Q},\boldsymbol{q}_1} B^\dagger_{\boldsymbol{q}_2} \dots F_{B,\boldsymbol{q}}\,, \tag{7.5}$$

excitations:

$$\dot{C}^\dagger_{\boldsymbol{Q},\boldsymbol{q}_1} = (\Delta\nu_{\boldsymbol{Q},\boldsymbol{q}_1} - \Gamma_{\boldsymbol{Q},\boldsymbol{q}_1})\, C^\dagger_{\boldsymbol{Q},\boldsymbol{q}_1} + \sum \hat{C}_{\boldsymbol{Q},\boldsymbol{Q}',\boldsymbol{q}_1\boldsymbol{q}_2} B^\dagger_{\boldsymbol{q}_1'-\boldsymbol{Q}'} B_{\boldsymbol{q}_1'} + F_c\,, \tag{7.6}$$

where $\hat{C}$ is defined by the commutator

$$\hat{C}^{\dagger}_{Q,Q',q_1q_2} = [C^{\dagger}_{Q',q_2} + C_{-Q,q_2}, C^{\dagger}_{Q,q_1}] . \tag{7.6a}$$

The dots indicate lacking constants. $\varkappa$, γ, Γ are damping constants. The F's are fluctuating forces, which count, *e.g.*, for spontaneous emission. If we leave the fluctuating forces aside, eq. (7.3) is strongly reminiscent of eq. (5.3), $B^{\dagger}_0$ plays the role of the total effective atomic dipole moment $\sum_{\mu} p_{\mu} g_{\mu}$. We consider the process of fig. 8.

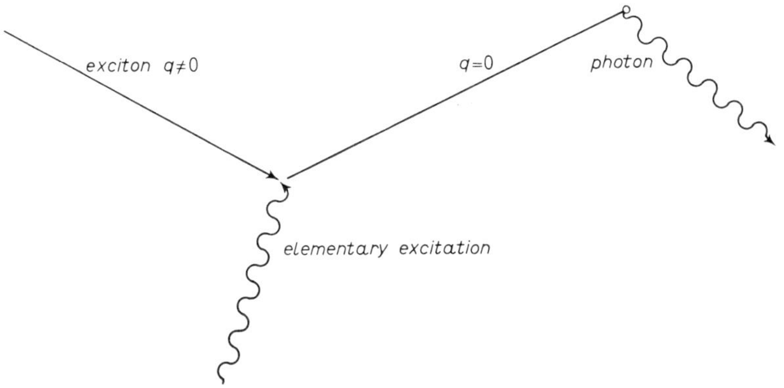

Fig. 8.

An exciton with $\boldsymbol{q} = 0$ interacts with an elementary excitation, so that an exciton with $\boldsymbol{q} = 0$ is created, which eventually decays into a photon. To treat such a process we assume that $B^{\dagger}_0$, $b^{\dagger}$ and all the other terms in (7.3), (7.4) oscillate with a frequency $\bar{\omega}$. This enables us to solve eq. (7.4) for $B^{\dagger}_0$, which is expressed by $b^{\dagger}$ and $C^{\dagger}_{Q,q} B^{\dagger}_{-Q}$. In the same way we express $C^{\dagger}$ by means of the other terms of eq. (7.6). Inserting B_0, $B^{\dagger}_0$ and $C^{\dagger}$ into the other equations, (7.3) and (7.4), and keeping only resonant terms, we find after a somewhat lengthy analysis equations of the following structure:

$$\dot{b}^{\dagger} = (i\omega' - \varkappa')\, b^{\dagger} + b^{\dagger} \sum \dots \hat{C}^{\dagger} \dots B^{\dagger}_{q_1} B_{q_2} , \tag{7.7}$$

$$\dot{B}^{\dagger}_{q} = (i\nu_{q} - \gamma_{q})\, B^{\dagger}_{q} + b^{\dagger} \sum \dots \hat{C}^{\dagger} \dots B^{\dagger}_{q_1} B^{\dagger}_{q_2} B_{q_3} + \text{pump} , \tag{7.8}$$

where fluctuating forces have been dropped for the time being. ω', $\varkappa'$ are « renormalized » frequency and damping constant, respectively. Since these equations are still too complicated to allow for an explicit treatment, we derive rate equations. We multiply (7.7) by b, add on both sides the complex conjugate and average over the fluctuations. In an analogous way we proceed with eq. (7.8) by multiplying it by B_q etc. Finally we make a random-phase approximation on the r.h.s. of the corresponding equations, keeping only terms $B_q^\dagger$, B_q. We thus obtain equations for the occupation numbers

$$\langle B_q^\dagger B_q \rangle = N_q\,, \tag{7.9}$$

$$\langle b^\dagger b \rangle = n\,. \tag{7.10}$$

They read

$$\dot{n} = -2\varkappa' n + \sum_{Q,q} \text{const}_1\, M_Q N_q (n+1) \tag{7.11}$$

$$\dot{N}_q = P_q - 2\gamma_q N_q - \sum \text{const}_2\, M_Q N_q (n+1)\,. \tag{7.12}$$

M_Q stems from the elementary excitation according to the following table II; see also table I.

TABLE II.

excited or dissociated electron-hole pairs: $M_Q \sim N_Q$	(number of excitons)
electrons: $M_Q \sim \langle a_Q^\dagger a_Q \rangle$	occupation number
holes: $M_Q \sim \langle d_Q^\dagger d_Q \rangle$	occupation number
phonon emission: $M_Q \sim \langle (c_q^\dagger c_q + 1) \rangle$	phonon number plus one

P_q is the pump strength, which has been introduced in (7.4) via appropriate fluctuating forces. The constant in (7.11) has the structure (example, electron scattering)

$$\text{const}_1 \propto |\text{Re}V_Q|^2 = \frac{g^2}{\gamma^2 + (\omega - \nu_0)^2} \left(\frac{1}{\tilde{\Gamma}_{Q+q,q}} - \frac{1}{\tilde{\Gamma}_{Q,q}} \right). \tag{7.13}$$

Assuming that the excitons equilibrate quickly we may adopt a relation of the form

$$N_q = \bar{N} f_N(T)\,, \tag{7.14}$$

where $\bar{N}$ is the total number of excitons, and f gives their distribution at an effective temperature T. By means of (7.14) and a suitable average for M_Q, (7.11) and (7.12) can be cast into the form

$$\dot{n} = -2\varkappa n + G_1 M N (n+1)\,, \tag{7.15}$$

$$\dot{N} = P - 2\gamma N - G_1 M N (n+1)\,. \tag{7.16}$$

(7.15) and (7.16) are now effective laser equations describing the change of the photon number and of exciton number under the impact of the interaction with an elementary excitation. In (7.15) G_1 is an effective gain constant, which is made up of terms of the form (7.13). P in (7.16) is an average pump rate of excitons. When several laser modes are present, a set of equations of the type (7.15) and (7.16) can be derived (see, *e.g.*, HAKEN and NIKITINE). It is a simple matter to solve (7.15) and (7.16) in the steady state. One obtains, *e.g.*, a critical number $N = N_{\text{threshold}}$

$$N_{\text{thr}} = \frac{2\varkappa}{G_1 M}, \tag{7.17}$$

above which (7.15) allows for nonvanishing solutions n. The photon number is then given by

$$n = \frac{P}{2\varkappa}, \tag{7.18}$$

i.e. the laser light intensity increases proportional to the pump. Time-dependent solutions, which are most probably necessary in the case of exciton lasers, can be obtained under several simplifying assumptions, see, *e.g.*, HAKEN and NIKITINE. A computer solution is presently performed. An interesting feature of sets of equations (7.15) and (7.16) is mode competition. If one has, *e.g.*, two different laser modes supported by exciton processes with different elementary excitations, one mode may survive and the other one dies out under stationary conditions, whereas in transient conditions both modes may coexist with different intensities.

* * *

I wish to thank Mr. J. GOLL for valuable discussions and for his assistance in preparing this manuscript.

BIBLIOGRAPHY

General background

H. HAKEN: *Quantenfeldtheorie des Festkörpers* (Stuttgart, 1973) (English translation to appear winter 1975 (Amsterdam)).

H. HAKEN: *Forts. Phys.*, **6**, 271 (1958).

R. S. KNOX: *Theory of excitons*, in *Solid State Physics*, Suppl. 5 (New York, N.Y., and London, 1963).

H. HAKEN and S. NIKITINE, Editors: *Excitons at high density*, in *Springer Tracts in Modern Physics*, Vol. **73** (1975).

For resonant effects in two-level atoms see the lectures at this summer school and H. HAKEN: *Encyclopedia of Physics*, Vol. **25**/2c, *Laser Theory* (Berlin, 1970); M. SARGENT III, M. O. SCULLY and W. E. LAMB: *Laser Physics* (Reading, Mass., 1974); L. ALLEN and J. H. EBERLY: *Optical Resonance and Two-Level Atoms* (New York, N. Y., 1975).

Exciton self-induced transparency

A. SCHENZLE and H. HAKEN: *Opt. Comm.*, **6**, 96 (1972).

H. HAKEN and A. SCHENZLE: *Zeits. Phys.*, **258**, 231 (1973).

H. HAKEN: *Zeits. Phys.*, **262**, 119 (1973).

V. V. SAMARTSEV, A. I. SIRAZIEV and YU. E. SHEIBUT: *Bull. Acad. Sci. USSR, Phys. Ser. (USA)*, **37**, 140 (1973).

H. HAKEN, J. GOLL and A. SCHENZLE: *Polaritons at high light intensities and in Bose-condensed exciton systems*, in *Springer Tracts in Modern Physics*, edited by H. HAKEN and S. NIKITINE, Vol. **73**, *Excitons at High Density* (1975).

E. HANAMURA: *Journ. Phys. Soc. Japan*, **37**, 1553 (1974).

F. BRUCKNER, V. S. DNESTROVSKII and D. G. KOSHCHUG: *JETP Lett.*, **20**, 4 (1974).

J. GOLL and H. HAKEN: to be published.

Exciton free induction and echo, exciton superradiance

O. N. GADOMSKII and V. V. SAMARTSEV: *Sov. Phys. Solid State*, **13**, 2354 (1972).

V. V. SAMARTSEV: *Bull Acad. Sci. USSR, Phys. Ser. (USA)*, **36**, 937 (1972).

V. V. SAMARTSEV: *Phys. Lett.*, **38** A, 363 (1972).

V. V. SAMARTSEV: *Bull. Acad. Sci. USSR, Phys. Ser. (USA)*, **37**, 123 (1973).

J. GOLL and H. HAKEN: unpublished, some of the results are reported in sect. **3**.

Laser action by excitons

See chapters by H. HAKEN and S. NIKITINE, K. L. SHAKLEE, and R. LEVY, J. B. GRUN and S. NIKITINE: *Excitons at high density*, in *Springer Tracts in Modern Physics*, edited by H. HAKEN and S. NIKITINE, Vol. **73** (1975).

H. HAKEN and A. KUCHELMEISTER: unpublished.

Structure of Biexcitons and Two-Photon Processes.

F. BASSANI

Istituto di Fisica dell'Università - Roma

J. J. FORNEY and A. QUATTROPANI

Laboratoire de Physique Théorique
Ecole Politechnique Federale - Lausanne, Suisse

1. – Introduction.

In 1958 LAMPERT [1] and MOSKALENKO [2] predicted the existence of the excitonic molecule or biexciton as the bound state of two Wannier excitons. In recent years new lines were observed in the luminescence spectra of crystals excited by high-power lasers: these lines were attributed to the dissociation of a biexciton into a free exciton and a photon. The experimental results in CuCl, CuBr [3], CdS and CdSe [4] have stimulated theoretical investigations on the binding energy of the biexciton.

On the basis of a variational calculation, SHARMA [5] predicted that the biexciton should be bound for all values of electron-hole mass ratio $\sigma = m_e/m_h$, except for $0.2 < \sigma < 0.4$. WEHNER [6] and ADAMOWSKI *et al.* [7] later proved that an excitonic molecule bound for $\sigma = 1$ must be bound for all values of σ. Improved calculations by AKIMOTO and HANAMURA [8], BRINKMAN *et al.* [9], HUANG [10] and HANDEL [11] have established that biexcitons are always bound, with binding energy decreasing monotonically in the range $0 \leqslant \sigma \leqslant 1$. The above authors use a model Hamiltonian describing two electrons and two holes interacting through their mutual Coulomb potential. They perform a variational calculation for the ground state using ground-state trial functions similar to those of the hydrogen molecule [12] and of the positronium molecule [13].

This description of the biexciton is incomplete in some respects:

1) The effects of the spin of the particles are partially neglected, though the ground state is assumed to be a singlet spin state.

2) Only the ground state of the molecule is considered, although other states of the molecule may be bound.

The fine structure observed in the luminescence spectra of CuCl [14] and attributed to the biexciton could not be described.

3) The effective Hamiltonian has to contain an electron-hole interaction (sometimes called electron-hole exchange) in order to include the effect of the Pauli principle on the excited electrons and on the $N-2$ valence electrons. In the case of the exciton this interaction is responsible for the singlet-triplet splitting.

As was shown by the controversies about the existence of the biexciton in Si, Ge [15] and Cu_2O [16], a clear identification of the luminescence lines involving transitions from biexciton states, as well as an accurate comparison with the calculated energies, is a difficult task. New criteria for the existence of the biexciton are necessary. As an example, the knowledge of the transformation rules of the biexciton states under the symmetry operations of the lattice would permit us to establish the optical selection rules of the various luminescence processes. This information cannot be extracted from the model used up to now.

In the first part of this seminar, we present the classification of the biexciton within a two-band model. The treatment includes the spin of the electrons and holes. We first establish an equivalent effective Hamiltonian which operates on a four-particle space (two electrons and two holes). The effective Hamiltonian reproduces all matrix elements of the total Hamiltonian between N-electron states with two electrons excited in the conduction band. The effective Hamiltonian contains, as in the case of the exciton [17], electron-hole exchange interactions which were neglected in all previous calculations [5-11]. In the case of our model, the total spin of the four particles is a good quantum number, and the eigenstates of the biexciton are products of a spin eigenstate and a spatial function. The spin eigenstates have definite symmetries with respect to the permutations of the electrons and of the holes; this imposes definite symmetries for the spatial part. We identify the biexciton ground state as a spin singlet which is associated with a totally symmetric spatial function. In the limit of small electron-hole mass ratio σ, the biexciton states can be related to the states of the hydrogen molecule. This comparison clearly shows the existence of other biexciton bound states in the limit $\sigma \ll 1$. This may be the explanation of the fine structure in the biexciton spectra [14]. A qualitative discussion of the states for all values of σ is given. We also discuss the correction to the ground-state binding energy due to the electron-hole exchange terms and we show its relevance for the evaluation of the binding energy.

In the second part of this seminar, we use a group-theoretical approach to classify the states of the biexciton, taking into account the details of the band structure and the effects of the spin-orbit coupling. This will allow us to assign definite crystalline symmetries to all possible biexciton states and to establish optical selection rules. This analysis will be applied to CuCl and CdS, for which biexcitons have been observed [3, 4]. The optical selection rules for two-photon excitation are shown to depend on the frequency and polarization of the photons. Nonlinear optical experiments are suggested for the cases of CuCl and CdS in order to establish the nature of the biexciton bound states.

2. – Many-electron problem and definition of excited states.

We adopt the Born-Oppenheimer approximation and make use of the following N-electron Hamiltonian:

$$H(\boldsymbol{r}_1, \dots, \boldsymbol{r}_N) = \sum_{i=1}^{N}\left(\frac{p_i^2}{2m} - \sum_{R}\frac{Ze^2}{|r_i - R|} + \sum_{j\neq i}\frac{e^2}{r_{ij}} + H_{\text{s.o.}}(\boldsymbol{r}_i, S_i) + H_{\text{rel}}(\boldsymbol{r}_i)\right). \tag{1a}$$

The first term gives the kinetic energy of the electrons; the second and third terms give the Coulomb interaction of the electrons with the nuclei at rest and among themselves; the fourth term

$$H_{\text{s.o.}}(\boldsymbol{r}_i, S_i) = \frac{-i\hbar^2}{4m^2c^2}(\boldsymbol{\nabla} V_i \wedge \boldsymbol{\nabla}_i)\cdot\boldsymbol{\sigma}_i\,, \tag{1b}$$

where V_i indicates the crystal potential and $\boldsymbol{\sigma}_i$ are the Pauli spin matrices, gives the spin-orbit interaction; the fifth term gives other relativistic corrections which do not depend on the spin [18]. The last two terms in eq. (1a) are separable and are to be considered in the band structure. We recall that in the Hartree-Fock approximation we treat the Hamiltonian (1a) as if it were separable, and we can write

$$H(\boldsymbol{r}_1, \dots, \boldsymbol{r}_N) = \sum_{i=1}^{N} H_{\text{H.F.}}(\boldsymbol{r}_i)\,. \tag{2a}$$

The band structure is given by the one-electron states of the independent Hartree-Fock Hamiltonians.

In the case of insulators and semiconductors we can separate the occupied energy levels $E_v(\boldsymbol{k})$ from the empty conduction levels $E_c(\boldsymbol{k})$, and the Bloch functions are the best one-electron wave functions which satisfy the equations

$$H_{\text{H.F.}}\,\psi_{\boldsymbol{k}} = E_v(\boldsymbol{k})\,\psi_{\boldsymbol{k}}\,, \tag{2b}$$

$$H_{\text{H.F.}}\,\varphi_{\boldsymbol{l}} = E_c(\boldsymbol{l})\,\varphi_{\boldsymbol{l}}\,, \tag{2c}$$

where, for convenience, we have denoted by $\boldsymbol{k}$ the quantum numbers for the valence bands and by $\boldsymbol{l}$ the quantum numbers for the conduction bands. In this approximation the N-electron wave function is given by the Slater determinant of the one-electron occupied states:

$$\Psi(\boldsymbol{r}_1, \dots, \boldsymbol{r}_N) \simeq (N!)^{-\frac{1}{2}} \det\{\psi_{\boldsymbol{k}_1}(\boldsymbol{r}_1)\,\psi_{\boldsymbol{k}_2}(\boldsymbol{r}_2) \dots \psi_{\boldsymbol{k}_N}(\boldsymbol{r}_N)\}\,. \tag{3}$$

The ground state $|\Phi_0\rangle$ is defined as the Slater determinant of the valence states. Any excited state is constructed from the ground state by substituting some valence states $\psi_k(i)$ by conduction states $\varphi_l(i)$.

To go behind the independent-particle approximation, we must consider the true states of Hamiltonian (1):

$$H\Psi(\boldsymbol{r}_1, \ldots, \boldsymbol{r}_N) = E\Psi(\boldsymbol{r}_1, \ldots, \boldsymbol{r}_N)\,, \tag{4}$$

where the wave functions are antisymmetric under permutation of any two electrons, $(\boldsymbol{r}_i, \boldsymbol{r}_j) \to (\boldsymbol{r}_j, \boldsymbol{r}_i)$. The ground state is separated from the excited states by a finite energy. The excited states can be classified by the number of electrons which are missing from the valence band. Excitons are those with one electron removed from the valence bands. Biexcitons are those with two electrons removed from the valence bands, and so on. We can expand the excited states in the excited states of the independent-particle model.

The exciton wave functions are defined as a linear combination of Slater determinants containing just one excited electron [19]:

$$|\Psi_{\text{ex}}\rangle = \sum_{\boldsymbol{l},\boldsymbol{k}} \tilde{A}(\boldsymbol{l}, \boldsymbol{k})|\Phi_{\boldsymbol{k}}^{\boldsymbol{l}}\rangle\,, \tag{5}$$

where $|\Phi_{\boldsymbol{k}}^{\boldsymbol{l}}\rangle$ denotes a Slater determinant in which the $\psi_{\boldsymbol{k}}$ valence electron has been substituted by the $\varphi_{\boldsymbol{l}}$ conduction electron. The stationary Schrödinger equation (4) determines the coefficients $\tilde{A}(\boldsymbol{l}, \boldsymbol{k})$ and the exciton energies.

The generalization of the above approach to the case of biexcitons is immediate. We have now to consider Slater determinants $|\Phi_{\boldsymbol{k}\boldsymbol{k}'}^{\boldsymbol{l}\boldsymbol{l}'}\rangle$ containing two conduction electron wave functions substituted for two valence wave functions. The biexciton wave function is then

$$|\Psi_{\text{biex}}\rangle = \sum_{\boldsymbol{l}\boldsymbol{l}'\boldsymbol{k}\boldsymbol{k}'} \tilde{A}(\boldsymbol{l}, \boldsymbol{k}, \boldsymbol{l}', \boldsymbol{k}')|\Phi_{\boldsymbol{k}\boldsymbol{k}'}^{\boldsymbol{l}\boldsymbol{l}'}\rangle\,, \tag{6}$$

and, by substitution into eq. (4), the expansion coefficients and the biexciton energies can in principle be determined.

3. – Hole states and reduction to a two-particle and a four-particle problem.

The above-described procedure can be better understood by reducing the problem to that of a small number of particles. This is possible through the introduction of the concept of « hole » as the particle associated with the missing electron in the valence band. By simple arguments one shows that the properties of the « hole » are related to those of the missing electron as follows:

$$\boldsymbol{k}_{\text{h}} = -\boldsymbol{k}\,, \quad e_{\text{h}} = -e\,, \quad \boldsymbol{v}_{\text{h}} = \boldsymbol{v}\,, \quad E_{\text{h}} = -E\,, \quad \boldsymbol{S}_{\text{h}} = -\boldsymbol{S}\,.$$

It has also been verified that there exists an antiunitary operator K which transforms the electron dynamical variables into the hole variables. We refer to a recent paper [20] for a detailed proof that this operator is the time-reversal operator [19]:

$$K = -i\sigma_y K_0 , \tag{7}$$

where K_0 is the complex conjugation operator and σ_i are the Pauli spin matrices. The hole wave function is defined as

$$(\psi|S\rangle)_\text{h} = K(\psi_{\boldsymbol{k}}(\boldsymbol{r})|S\rangle) , \tag{8}$$

where $|S\rangle$ are the eigenfunctions of S_z and are denoted by $|\alpha\rangle$ and $|\beta\rangle$ in accordance with the eigenvalues $\hbar/2$ and $-\hbar/2$. In particular we have

$$K\psi_{\boldsymbol{k}}(\boldsymbol{r})|\alpha\rangle = \psi^*_{\boldsymbol{k}}(\boldsymbol{r})|\beta\rangle , \tag{9a}$$

$$K\psi_{\boldsymbol{k}}(\boldsymbol{r})|\beta\rangle = -\psi^*_{\boldsymbol{k}}(\boldsymbol{r})|\alpha\rangle . \tag{9b}$$

To a Slater determinant $|\Phi^{\boldsymbol{l}}_{\boldsymbol{k}}\rangle$ describing a one-electron excited state (with spin $S = 1, 0$) we associate a triplet or singlet electron-hole state formed from the products of electron and hole functions. The electron-hole singlet state is, for instance,

$$\varphi_{\boldsymbol{l}}(\boldsymbol{r}_1)\psi_{\boldsymbol{k}}(\boldsymbol{r}_2)|0, 0\rangle = \varphi_{\boldsymbol{l}}(\boldsymbol{r}_1)\psi^*_{\boldsymbol{k}}(\boldsymbol{r}_2)\frac{1}{\sqrt{2}}\big(|\alpha(1)\rangle\,|\beta(2)\rangle - |\alpha(2)\rangle\,|\beta(1)\rangle\big) , \tag{10}$$

where states of even argument are hole states.

The N-electron problem for the exciton case can then be reduced to a two-particle problem, one electron and one hole. The Hamiltonian H defined by (1) must be replaced by an effective Hamiltonian H^ex which satisfies the condition

$$\langle T, t|\langle\varphi_{\boldsymbol{l}}\psi^*_{\boldsymbol{k}}|H^\text{ex}|\varphi_{\boldsymbol{l}'}\psi^*_{\boldsymbol{k}'}\rangle|T', t'\rangle = \langle\Phi^{\boldsymbol{l}}_{\boldsymbol{k}}|H|\Phi^{\boldsymbol{l}'}_{\boldsymbol{k}'}\rangle , \tag{11}$$

to be verified for all quantum numbers. It is shown in ref. [20] that eqs. (11) are satisfied by the following effective Hamiltonian:

$$H^\text{ex} = E_0 + H_\text{H.F.}(1) - KH_\text{H.F.}(2)\,K^\dagger - \frac{e^2}{r_{12}} + W(1, 2) , \tag{12a}$$

where E_0 is the ground-state energy of the N-electron system; $H_\text{H.F.}(1)$ and $E_0 - KH_\text{H.F.}(2)\,K^\dagger$ are the one-particle electron and hole Hamiltonians, respectively; the last two terms represent the residual interaction between electron and hole, and can be interpreted as a Coulomb attraction and an exchange repulsion between electron and hole. The electron-hole exchange term depends on the spin functions and is defined as follows:

$$W(1,2) = V(\boldsymbol{r}_1,\boldsymbol{r}_2)\,V_{\text{spin}}(1,2)\,, \tag{12b}$$

where

$$\begin{aligned}\langle\varphi_{\boldsymbol{l}}(\boldsymbol{r}_1)K_0\psi_{\boldsymbol{k}}(\boldsymbol{r}_2)|V(\boldsymbol{r}_1,\boldsymbol{r}_2)|\varphi_{\boldsymbol{l}'}(\boldsymbol{r}_1)K_0\psi_{\boldsymbol{k}'}(\boldsymbol{r}_2)\rangle = \\ = \iint \mathrm{d}^3r_1\,\mathrm{d}^3r_2(\varphi_{\boldsymbol{l}}(\boldsymbol{r}_1)K_0\psi_{\boldsymbol{k}}(\boldsymbol{r}_2))^*\,\delta(\boldsymbol{r}_1-\boldsymbol{r}_2)\cdot \\ \cdot\iint \mathrm{d}^3r'_1\,\mathrm{d}^3r'_2\,\frac{e^2}{|\boldsymbol{r}'_1-\boldsymbol{r}_1|}\,\delta(\boldsymbol{r}'_1-\boldsymbol{r}'_2)\varphi_{\boldsymbol{l}'}(\boldsymbol{r}'_1)K_0\psi_{\boldsymbol{k}'}(\boldsymbol{r}'_2)\end{aligned} \tag{12c}$$

and

$$\begin{aligned}V_{\text{spin}}(1,2) = |\alpha(1)\beta(2)\rangle\langle\alpha(1)\beta(2)| + |\beta(1)\alpha(2)\rangle\langle\beta(1)\alpha(2)| - \\ - |\alpha(1)\beta(2)\rangle\langle\beta(1)\alpha(2)| - |\beta(1)\alpha(2)\rangle\langle\alpha(1)\beta(2)|\,.\end{aligned} \tag{12d}$$

The only nonvanishing matrix element of $W(1,2)$ between the states defined by (11) is the matrix element between the singlet spin states:

$$\begin{aligned}\langle 0,0|\langle\varphi_{\boldsymbol{l}}(\boldsymbol{r}_1)\psi^*_{\boldsymbol{k}}(\boldsymbol{r}_2)|W(1,2)|\varphi_{\boldsymbol{l}'}(\boldsymbol{r}_1)\psi^*_{\boldsymbol{k}'}(\boldsymbol{r}_2)\rangle|0,0\rangle = \\ = \langle 0,0|V_{\text{spin}}(1,2)|0,0\rangle\langle\varphi_{\boldsymbol{l}}(\boldsymbol{r}_1)\psi^*_{\boldsymbol{k}}(\boldsymbol{r}_2)|V(\boldsymbol{r}_1,\boldsymbol{r}_2)|\varphi_{\boldsymbol{l}'}(\boldsymbol{r}_1)\psi_{\boldsymbol{k}'}(\boldsymbol{r}_2)\rangle = \\ = 2\langle\varphi_{\boldsymbol{l}}(\boldsymbol{r}_1)\psi_{\boldsymbol{k}'}(\boldsymbol{r}_2)|\frac{e^2}{r_{12}}|\psi_{\boldsymbol{k}}(\boldsymbol{r}_1)\varphi_{\boldsymbol{l}'}(\boldsymbol{r}_2)\rangle\,.\end{aligned} \tag{13}$$

In the previous treatment we have confined ourselves to one-electron excited states and to a two-band model. To relax this restriction we introduce polarization effects which produce a screening of the bare interaction between electron and hole [21]. The prescription is to replace the bare interaction $-e^2/r_{12}$ by $-e^2/\varepsilon r_{12}$, where ε is the static dielectric function. The exchange interaction is less affected by screening, since it represents a short-range interaction [21]. We here consider the exchange interaction to be totally unscreened.

The above-described procedure can be extended to the case of the biexciton by defining four-particle wave functions as combinations of products of two-electron states and two-hole states. These functions have to be antisymmetric under permutation of the two electrons and permutation of the two holes. It is shown [20] that the four-particle eigenfunctions can be classified as 2 singlets, 3 triplets and 1 quintuplet according to the eigenstates of the total spin operators T and T_z. They can also be chosen to have definite parities under permutation of the electrons and under permutation of the hole. The four-particle spin functions are enumerated by

$$|\mu,\nu;\,T,t\rangle\,, \tag{14}$$

where T can take the values 0, 1 and 2 and t is one of the corresponding eigenvalues of T_z. The symbols μ (ν) stand for the sign $+$ or $-$, to indicate the parity of the spin function under the permutation of the two electrons (holes). As an example, the function

$$|+-;1,-1\rangle = \frac{1}{\sqrt{2}}\big(|\beta(1)\beta(3)\alpha(2)\beta(4)\rangle - |\beta(1)\beta(3)\beta(2)\alpha(4)\rangle\big) \tag{15}$$

is symmetric under the permutation of the electrons and antisymmetric under the permutation of the holes: it belongs to a triplet $T=1$ with the eigenvalue of T_z $t=-1$. With this notation, the 16 spin functions are

$$\left\{\begin{array}{llll} \text{singlets} & |-,-;0,0\rangle, & |+,+;0,0\rangle; & \\ \text{triplets} & |+,-;1,t\rangle, & |-,+;1,t\rangle, & |+,+;1,t\rangle \\ & & & \text{with } t=-1,0,1; \\ \text{quintuplet} & |+,+;2,t\rangle & & \text{with } t=-2,-1,0,1,2\,. \end{array}\right. \tag{16}$$

We choose as basis states for the four-particle system wave functions which are eigenstates of $\boldsymbol{T}^2$ and T_z and are separately antisymmetric under the permutation of the two electrons and under the permutation of the two holes. Thus each spin function has to be multiplied by a spatial wave function with opposite parities. We adopt the following notation for the basis states:

$$\left|\bar{\mu},\bar{\nu};\begin{matrix}\boldsymbol{l}&\boldsymbol{l}'\\ \boldsymbol{k}&\boldsymbol{k}'\end{matrix}\right\rangle|\mu,\nu;T,t\rangle\,, \tag{17}$$

where the upper wave vectors $\boldsymbol{l}$ and $\boldsymbol{l}'$ refer to conduction electrons (arguments 1 and 3), the lower ones $\boldsymbol{k}$ and $\boldsymbol{k}'$ to holes (arguments 2 and 4). The ket $\left|\bar{\mu},\bar{\nu};\begin{matrix}\boldsymbol{l}\boldsymbol{l}'\\ \boldsymbol{k}\boldsymbol{k}'\end{matrix}\right\rangle$ corresponds to the wave function

$$\tfrac{1}{2}\big(\varphi_{\boldsymbol{l}}(\boldsymbol{r}_1)\varphi_{\boldsymbol{l}'}(\boldsymbol{r}_3)+\bar{\mu}\varphi_{\boldsymbol{l}'}(\boldsymbol{r}_1)\varphi_{\boldsymbol{l}}(\boldsymbol{r}_3)\big)\big(\psi^*_{\boldsymbol{k}}(\boldsymbol{r}_2)\psi_{\boldsymbol{k}'}(\boldsymbol{r}_4)+\bar{\nu}\psi^*_{\boldsymbol{k}'}(\boldsymbol{r}_2)\psi_{\boldsymbol{k}}(\boldsymbol{r}_4)\big) \tag{18}$$

with $\bar{\mu}=-\mu$ and $\bar{\nu}=-\nu$, where μ and ν are $+$ and $-$. Thus $\bar{\mu}$ and $\bar{\nu}$ indicate the symmetry or antisymmetry of the spatial part of the wave function. For example, the spin state (16) $|+,-;1,-1\rangle$ has to be multiplied by a spatial function which is antisymmetric for permutation of the electrons and symmetric for the permutation of the holes and which is denoted by $\left|-,+;\begin{matrix}\boldsymbol{l}\boldsymbol{l}'\\ \boldsymbol{k}\boldsymbol{k}'\end{matrix}\right\rangle$.

The N-electron Hamiltonian is replaced by the effective Hamiltonian

H^{biex} [20]:

$$H^{\text{biex}} = E_0 + H_{\text{H.F.}}(1) + H_{\text{H.F.}}(3) - K(2)H_{\text{H.F.}}(2)K^\dagger(2) - K(4)H_{\text{H.F.}}(4)K^\dagger(4) + \\ + \frac{e^2}{\varepsilon}\left\{\frac{1}{|\boldsymbol{r}_{13}|} + \frac{1}{|\boldsymbol{r}_{24}|} + \frac{-1}{|\boldsymbol{r}_{12}|} - \frac{1}{|\boldsymbol{r}_{14}|} - \frac{1}{|\boldsymbol{r}_{32}|} - \frac{1}{|\boldsymbol{r}_{34}|}\right\} + \\ + W(1,2) + W(1,4) + W(3,2) + W(3,4)\,, \tag{19}$$

where $W(i, j)$ is defined by eq. (12).

4. – Effective-mass equation for the biexciton.

With the procedure described in sect. **3** it is possible to rederive very simply the effective-mass equation for exciton states appropriate to semiconductors, where the electron-hole Coulomb interaction extends over many lattice cells, because the exciton function $A(l, k)$ is peaked in momentum space and very smooth in real space. In this way one reproduces the Onodera and Toyozawa [19] effective-mass equation with inclusion of electron-hole exchange for the exciton.

The same procedure can be used for the four-particle biexciton problem and makes it possible to obtain an effective-mass equation for this case, with inclusion of all electron-hole exchange contributions [20]. First of all we may observe that the translational invariance of the system implies that the total wave number Q is a good quantum number for the four-particle system. Then the space part of the biexciton state can be expanded in functions of type (17) as follows:

$$|\Phi'(\bar{\mu}, \bar{\nu}, \boldsymbol{Q})\rangle = \sum_{\boldsymbol{k},\boldsymbol{k}',\boldsymbol{q}} A(\boldsymbol{k}, \boldsymbol{k}', \boldsymbol{q}) \left| \bar{\mu}, \bar{\nu}; \begin{matrix} \boldsymbol{\lambda}\ \boldsymbol{\lambda}' \\ \boldsymbol{\varkappa}\ \boldsymbol{\varkappa}' \end{matrix} \right\rangle, \tag{20}$$

where

$$\boldsymbol{\lambda} = -\boldsymbol{k} + \frac{1}{2}\boldsymbol{q} + \frac{m_{\text{e}}}{M}\boldsymbol{Q}, \qquad \boldsymbol{\varkappa} = \boldsymbol{k}' + \frac{1}{2}\boldsymbol{q} - \frac{m_{\text{h}}}{M}\boldsymbol{Q},$$

$$\boldsymbol{\lambda}' = \boldsymbol{k} + \frac{1}{2}\boldsymbol{q} + \frac{m_{\text{e}}}{M}\boldsymbol{Q}, \qquad \boldsymbol{\varkappa}' = -\boldsymbol{k}' + \frac{1}{2}\boldsymbol{q} - \frac{m_{\text{h}}}{M}\boldsymbol{Q},$$

m_{e} = effective mass of the conduction band,

m_{h} = effective mass of the valence band and

$M = 2(m_{\text{e}} + m_{\text{h}})$.

One easily verifies that $\boldsymbol{\lambda} + \boldsymbol{\lambda}' - \boldsymbol{\varkappa} - \boldsymbol{\varkappa}' = \boldsymbol{Q}$.

We note that in expression (20) each function appears four times in the sum. Indeed,

$$(21)\qquad \left|\bar{\mu},\,\bar{\nu};\,\begin{matrix}\lambda\ \lambda'\\ \varkappa\ \varkappa'\end{matrix}\right\rangle = \bar{\mu}\left|\bar{\mu},\,\bar{\nu};\,\begin{matrix}\lambda'\ \lambda\\ \varkappa\ \varkappa'\end{matrix}\right\rangle = \bar{\nu}\left|\bar{\mu},\,\bar{\nu};\,\begin{matrix}\lambda\ \lambda'\\ \varkappa'\ \varkappa\end{matrix}\right\rangle = \bar{\mu}\bar{\nu}\left|\bar{\mu},\,\bar{\nu};\,\begin{matrix}\lambda'\ \lambda\\ \varkappa'\ \varkappa\end{matrix}\right\rangle .$$

At this point, we have either to limit the summation over $\boldsymbol{k}$, $\boldsymbol{q}$ and $\boldsymbol{k}'$ to linearly independent functions or to associate the same coefficient $A(\boldsymbol{k}, \boldsymbol{k}', \boldsymbol{q})$ to each of the four functions appearing in (21). Choosing the second alternative and denoting by $A^{\bar{\mu}\bar{\nu}}(\boldsymbol{k}, \boldsymbol{k}', \boldsymbol{q})$ the coefficients of $\left|\bar{\mu},\,\bar{\nu};\,\begin{matrix}\lambda\lambda'\\ \varkappa\varkappa'\end{matrix}\right\rangle$, we have to impose the following symmetry properties:

$$A^{+-}(\boldsymbol{k}, \boldsymbol{k}', \boldsymbol{q}) = A^{+-}(-\boldsymbol{k}, \boldsymbol{k}', \boldsymbol{q}) = -A^{+-}(\boldsymbol{k}, -\boldsymbol{k}', \boldsymbol{q}) = -A^{+-}(-\boldsymbol{k}, -\boldsymbol{k}', \boldsymbol{q}),$$

and similar ones for the other coefficients can be easily deduced from this example.

As in the case of the exciton, eq. (4) can be transformed into an equation for the Fourier transforms of the coefficients $A^{\bar{\mu}\bar{\nu}}(\boldsymbol{k}, \boldsymbol{k}', \boldsymbol{q})$:

$$(22)\qquad F^{\bar{\mu}\bar{\nu}}(\boldsymbol{r}, \boldsymbol{r}', \boldsymbol{R}) = \sum_{\boldsymbol{k}\boldsymbol{k}'\boldsymbol{q}} \exp\left[i(\boldsymbol{k}\cdot\boldsymbol{r} + \boldsymbol{k}'\cdot\boldsymbol{r}' + \boldsymbol{q}\cdot\boldsymbol{R})\right] A^{\bar{\mu}\bar{\nu}}(\boldsymbol{k}, \boldsymbol{k}', \boldsymbol{q}) .$$

Because of the particular choice (20) for $|\Phi\rangle$, the relative co-ordinates appearing in (22) are defined as follows:

$$(23)\qquad \boldsymbol{r} = \boldsymbol{r}_3 - \boldsymbol{r}_1 , \qquad \boldsymbol{r}' = \boldsymbol{r}_4 - \boldsymbol{r}_2 , \qquad \boldsymbol{R} = \tfrac{1}{2}(\boldsymbol{r}_1 + \boldsymbol{r}_3 - \boldsymbol{r}_2 - \boldsymbol{r}_4) ,$$

and can be visualized from fig. 1.

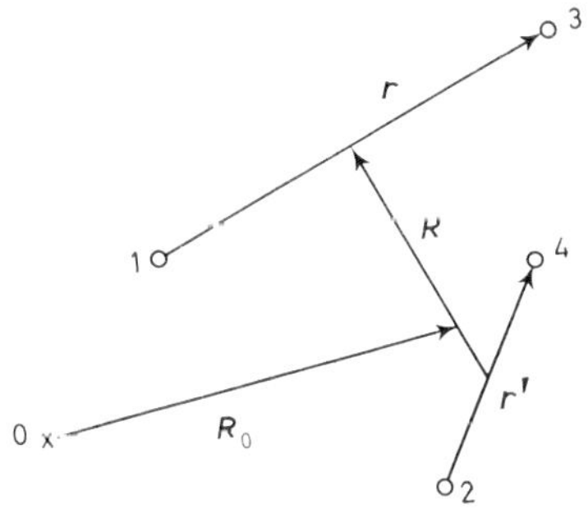

Fig. 1. – Co-ordinate system for the biexciton; 1 and 3 denote the electron positions, 2 and 4 denote the hole positions. $\boldsymbol{R}_0$ indicates the centre of mass and $\boldsymbol{R}$ is the average distance between electrons and holes.

Moreover, the total wave vector $\boldsymbol{Q}$ describes the motion of the centre of mass, whose co-ordinates are

$$\boldsymbol{R}_0 = \frac{m_e}{M}(\boldsymbol{r}_1 + \boldsymbol{r}_3) + \frac{m_h}{M}(\boldsymbol{r}_2 + \boldsymbol{r}_4)\,. \tag{24}$$

The effective-mass equation for the envelope function $F^{\bar{\mu}\bar{\nu}}(\boldsymbol{r}, \boldsymbol{r}', \boldsymbol{R})$ is the following:

$$\{E_0 + \delta E_{cv}(\nabla_{\boldsymbol{r}}, \nabla_{\boldsymbol{r}'}, \nabla_{\boldsymbol{R}}, \boldsymbol{Q}) + \mathscr{V}(\boldsymbol{r}, \boldsymbol{r}', \boldsymbol{R}) + \\ + Z(\mu, \nu; T)\mathscr{V}_{\text{exch}}(\boldsymbol{r}, \boldsymbol{r}', \boldsymbol{R}) - E_{\text{biex}}\} F^{\bar{\mu}\bar{\nu}}(\boldsymbol{r}, \boldsymbol{r}', \boldsymbol{R}) = 0\,, \tag{25a}$$

where

$$\delta E_{cv}(\nabla_{\boldsymbol{r}}, \nabla_{\boldsymbol{r}'}, \nabla_{\boldsymbol{R}}, \boldsymbol{Q}) F^{\bar{\mu}\bar{\nu}}(\boldsymbol{r}, \boldsymbol{r}', \boldsymbol{R}) \equiv \\ \equiv \left\{E_c\left(i\nabla_{\boldsymbol{r}} - \frac{i}{2}\nabla_{\boldsymbol{R}} + \frac{m_e}{M}\boldsymbol{Q}\right) + E_c\left(-i\nabla_{\boldsymbol{r}} - \frac{i}{2}\nabla_{\boldsymbol{R}} + \frac{m_e}{M}\boldsymbol{Q}\right) - \right. \\ \left. - E_v\left(-i\nabla_{\boldsymbol{r}'} - \frac{i}{2}\nabla_{\boldsymbol{R}} - \frac{m_h}{M}\boldsymbol{Q}\right) - E_v\left(i\nabla_{\boldsymbol{r}'} - \frac{i}{2}\nabla_{\boldsymbol{R}} - \frac{m_h}{M}\boldsymbol{Q}\right)\right\} F^{\bar{\mu}\bar{\nu}}(\boldsymbol{r}, \boldsymbol{r}', \boldsymbol{R})\,, \tag{25b}$$

$$\mathscr{V}(\boldsymbol{r}, \boldsymbol{r}', \boldsymbol{R}) = \frac{e^2}{\varepsilon}\left\{\frac{1}{r} + \frac{1}{r'} - \frac{1}{|\boldsymbol{R} + (\boldsymbol{r} + \boldsymbol{r}')/2|} - \right. \\ \left. - \frac{1}{|\boldsymbol{R} + (\boldsymbol{r} - \boldsymbol{r}')/2|} - \frac{1}{|\boldsymbol{R} - (\boldsymbol{r} + \boldsymbol{r}')/2|} - \frac{1}{|\boldsymbol{R} - (\boldsymbol{r} - \boldsymbol{r}')/2|}\right\}, \tag{25c}$$

$$Z(\mu, \nu; T) = \langle \mu, \nu; T, t|\mathscr{V}_{\text{spin}}|\mu, \bar{\nu}; T, t\rangle \tag{25d}$$

with

$$\mathscr{V}_{\text{spin}} = \tfrac{1}{4}\{V_{\text{spin}}(1, 2) + V_{\text{spin}}(1, 4) + V_{\text{spin}}(3, 2) + V_{\text{spin}}(3, 4)\}$$

and $V_{\text{spin}}(i, j)$ as defined in eq. (12).

Its numerical values are

$$\begin{cases} Z(+, +; 0) = \frac{3}{2}, \quad Z(+, +; 1) = 1\,, \quad Z(+, +; 2) = 0\,, \\ Z(-, -; 0) = Z(+, -; 1) = Z(-, +; 1) = \frac{1}{2}\,, \end{cases} \tag{25e}$$

$$\mathscr{V}_{\text{exch}}(\boldsymbol{r}, \boldsymbol{r}', \boldsymbol{R}) = \mathscr{J}\left\{\delta\left(\boldsymbol{R} + \frac{\boldsymbol{r} + \boldsymbol{r}'}{2}\right) + \right. \\ \left. + \delta\left(\boldsymbol{R} + \frac{\boldsymbol{r} - \boldsymbol{r}'}{2}\right) + \delta\left(\boldsymbol{R} - \frac{\boldsymbol{r} + \boldsymbol{r}'}{2}\right) + \delta\left(\boldsymbol{R} - \frac{\boldsymbol{r} - \boldsymbol{r}'}{2}\right)\right\}. \tag{25f}$$

An advantage of the present formulation is to attribute the permutational symmetry properties of $|\phi'(\bar{\mu}, \bar{\nu}, \boldsymbol{Q})\rangle$ to the envelope functions $F^{\bar{\mu}\bar{\nu}}(\boldsymbol{r}, \boldsymbol{r}', \boldsymbol{R})$.

We recall that this function has always to be associated with a spin function $|\mu, \nu; T, t\rangle$ of opposite symmetry. This proves that, from all possible spin states, one can construct biexciton states which satisfy the exclusion principle. This has been obtained by considering the fact that the functions $F^{\bar{\mu}\bar{\nu}}(\boldsymbol{r}, \boldsymbol{r}', \boldsymbol{R})$ result from the summation over an infinite set of k values.

Before discussing the solution of eq. (25) we recall the hypotheses on which it is based.

i) The semiconductor is represented by a two-band model with inclusion of spins but not of the spin-orbit interaction. For this reason biexciton states are eigenstates of the total spin.

ii) The biexciton states are expanded in Slater determinants with two electrons excited in the conduction band. This restriction reduces the general N-electron problem to a four-particle problem (two electrons and two holes).

iii) The eigenvalue equation (25) was derived in the effective-mass approximation and is valid in the limit of weak electron-hole coupling; the dielectric screening is essential in this respect.

5. – Discussion of the energy levels of the biexciton.

In the absence of the exchange interaction and for $Q = 0$ eq. (25) reads

$$\{E_0 + \delta E_{cv}(\nabla_r, \nabla_{r'}, \nabla_R, 0) + \mathscr{V}(\boldsymbol{r}, \boldsymbol{r}', \boldsymbol{R}) - E^{\bar{\mu}\bar{\nu}}\} F^{\bar{\mu}\bar{\nu}}(\boldsymbol{r}, \boldsymbol{r}', \boldsymbol{R}) = 0 . \tag{26}$$

This is identical to the equation previously discussed by many authors [5-11], except for a shift in energy $E_0 + 2E_g$, where E_g is the gap energy. These authors evaluate the ground-state energy of (26) using the variational method with a totally symmetric trial function. In our formalism their ground state corresponds to a wave function of the form $F^{++}(\boldsymbol{r}, \boldsymbol{r}', \boldsymbol{R})$. This state has to be associated with a spin state antisymmetric under the exchange of the electrons and antisymmetric under the exchange of the holes. Only the singlet spin state $|-, -; 0, 0\rangle$ has the required symmetries. To discuss the other possible states of the biexciton we use results known from the treatment of the H_2 molecule. Equation (26) differs from that of H_2 because of the different values of mass ratio $\sigma = m_e/m_h$; the results known for H_2 ($\sigma \ll 1$) and the invariance of (26) under the transformation $\sigma \to 1/\sigma$ imply

$$E^{\bar{\mu}\bar{\nu}}(\sigma) = E^{\bar{\mu}\bar{\nu}}(1/\sigma) , \tag{27}$$

where $E^{\bar{\mu}\bar{\nu}}(\sigma)$ is an eigenvalue of (26) associated with an eigenfunction $F^{\bar{\mu}\bar{\nu}}(\boldsymbol{r}, \boldsymbol{r}', \boldsymbol{R})$. In our discussion we restrict ourselves to solutions $F^{\bar{\mu}\bar{\nu}}(\boldsymbol{r}, \boldsymbol{r}', \boldsymbol{R})$ having the lowest eigenvalue for each symmetry $(\bar{\mu}, \bar{\nu})$.

We are interested in the bound states of the biexciton; they are solutions of (26) with

$$E^{\bar{\mu}\nu}(\sigma) < 2E_{ex}(\sigma) = E_0 + 2(E_g - G_{ex}(\sigma)) ,$$

where

$$G_{ex}(\sigma) = m_e e^4/2(1+\sigma)\hbar^2\varepsilon^2$$

is the binding energy of the free exciton, which is usually adopted as the energy unit. In the limit $\sigma \to 0$, the energies $E^{++}(0)$ and $E^{+-}(0)$ as well as $E^{-+}(0)$ and $E^{--}(0)$ are degenerate. This corresponds to the Born-Oppenheimer treatment of the hydrogen molecule, which separates the degrees of freedom of the nuclei from those of the electrons. In this case $E^{++}(0) = E^{+-}(0) < 2E_{ex}(0)$ are bound levels: they are singlet states only from the point of view of the electron spin. From the point of view of the total spin of the four particles, as we have shown, we have to distinguish a singlet $|-,-;0\,0\rangle$, associated with the totally symmetric spatial function $F^{++}(\boldsymbol{r}, \boldsymbol{r}', \boldsymbol{R})$, and a triplet $|-\,+;1,t\rangle$ associated with the function $F^{+-}(\boldsymbol{r}, \boldsymbol{r}', \boldsymbol{R})$. In the same limit, $E^{--}(0) = E^{-+}(0) > > 2E_{ex}(0)$ are unbound levels; the corresponding states are triplets only from the point of view of the electron spin (triplet state of H_2). From the point of view of the four particles, F^{-+} is associated with the triplet $|+,-;1,t\rangle$ and F^{--} is associated with the following spin states:

$$|+,+;0,0\rangle , \qquad |+,+;1,t\rangle , \qquad |+,+;2,t\rangle .$$

All these states are consequently degenerate.

By means of the relations (27) an analogous discussion can be repeated in the limit $\sigma \to \infty$.

For $\sigma \neq 0$ the levels E^{++} and E^{+-} as well as E^{-+} and E^{--} are no longer degenerate. ADAMOWSKI *et al.* [7] have shown that $E^{++}(\sigma)$ is a monotonic increasing function of σ in the interval $0 \leqslant \sigma \leqslant 1$. As the variational calculations indicate that $E^{++}(\sigma=1) < 2E_{ex}(\sigma=1)$, it follows that the ground states $F^{++}(\boldsymbol{r}, \boldsymbol{r}', \boldsymbol{R})|-,-;0,0\rangle$ represent a bound state for all values of σ in the approximation of eq. (26). This conclusion is not true for the level $E^{+-}(\sigma)$, which corresponds to a bound state for $\sigma \ll 1$, but which coincides with $E^{--}(\sigma) > > 2E_{ex}(\sigma)$ in the limit $\sigma \to \infty$. We conclude that the triplet state associated with F^{+-} is a bound state only in a limited range of σ. From the relation (27) it follows that $F^{-+}(\boldsymbol{r}, \boldsymbol{r}', \boldsymbol{R})|+,-;1,t\rangle$ are bound states for $\sigma \gg 1$.

The qualitative features of $E^{\bar{\mu}\bar{\nu}}(\sigma)$ are plotted as functions of σ in fig. 2. The curve $E^{++}(\sigma)$ is that calculated by BRINKMAN *et al.* [9]. The levels with $E > 2E_{ex}(\sigma)$ give a measure of the repulsion energy between two free excitons. The states represented by $F^{+-}(\boldsymbol{r}, \boldsymbol{r}', \boldsymbol{R})$ have a simple interpretation for $\sigma \ll 1$. As is known, the rotational states of the nuclei with odd quantum number of angular momentum are antisymmetric under the permutation of the nuclei.

The energy splitting between $E^{+-}(\sigma)$ and $E^{++}(\sigma)$ for $\sigma \ll 1$ can be estimated from the splitting of the lowest rotational levels of the H_2 molecule:

$$E^{+-}(\sigma) - E^{++}(\sigma) = 5.54 \text{ Ryd} .$$

This can be extended to wider range of σ values when the Born-Oppenheimer approximation holds ($\sigma \ll 1$), as is shown by NIKITINE [14].

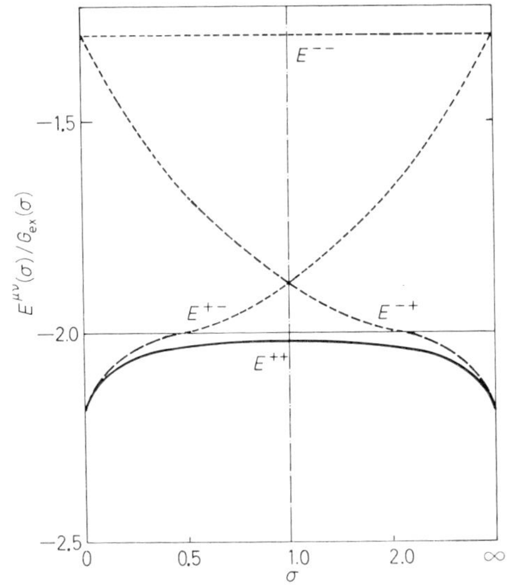

Fig. 2. – Qualitative scheme for $E^{\bar{\mu}\bar{\nu}}(\sigma)$ as a function of $\sigma = m_e/m_h$, without electron-hole exchange. The energy unit is the effective exciton rydberg $G^{ex}(\sigma) = m_e e^4/(2(1+\sigma)\hbar^2\varepsilon^2)$. ——— energy of the ground state of the biexciton $F^{++}(\boldsymbol{r}, \boldsymbol{r}', \boldsymbol{R})|-, -; 0, 0\rangle$ (after BRINKMAN *et al.* [9]), — — — energy of the states $F^{+-}(\boldsymbol{r}, \boldsymbol{r}', \boldsymbol{R})|-, +; 1, t\rangle$ and of the states $F^{-+}(\boldsymbol{r}, \boldsymbol{r}', \boldsymbol{R})|+, -; 1, t\rangle$, - - - repulsive energy of two free excitons.

6. – Electron-hole exchange corrections.

We now consider the effect of the electron-hole exchange terms on the energy levels of the biexciton. From eq. (30) we deduce the exchange contribution to the energy of the biexciton of spin $|\mu, \nu; T, t\rangle$ and envelope function $F^{\bar{\mu}\bar{\nu}}(\boldsymbol{r}, \boldsymbol{r}', \boldsymbol{R})$:

$$\Delta E^{\text{biex}}_{\text{exch}}(\mu, \nu; T) = Z(\mu, \nu; T)\tilde{J} I(\mu, \nu) , \tag{28a}$$

where

$$I(\bar{\mu}, \bar{\nu}) = \left\{\int \mathrm{d}^3r\, \mathrm{d}^3r'\, \mathrm{d}^3R |F^{\bar{\mu}\bar{\nu}}(\boldsymbol{r}, \boldsymbol{r}', \boldsymbol{R})|^2\right\}^{-1} \cdot \tag{28b}$$
$$\cdot \int \mathrm{d}^3r\, \mathrm{d}^3r'\, \mathrm{d}^3R |F^{\bar{\mu}\bar{\nu}}(\boldsymbol{r}, \boldsymbol{r}', \boldsymbol{R})|^2 \left[\delta\left(\boldsymbol{R} + \frac{\boldsymbol{r}+\boldsymbol{r}'}{2}\right) + \delta\left(\boldsymbol{R} + \frac{\boldsymbol{r}-\boldsymbol{r}'}{2}\right) + \right.$$
$$\left. + \delta\left(\boldsymbol{R} - \frac{\boldsymbol{r}+\boldsymbol{r}'}{2}\right) + \delta\left(\boldsymbol{R} - \frac{\boldsymbol{r}-\boldsymbol{r}'}{2}\right)\right] .$$

If we know $I(\bar{\mu}, \bar{\nu})$, $\Delta E^{\text{biex}}_{\text{exch}}(\mu, \nu; T)$ can be compared with the electron-hole exchange splitting of the exciton. In all cases in which the exchange integral $\tilde{J}$ is small compared with the binding energy of the exciton ($\tilde{J} \ll G_{\text{ex}}$), $\Delta E^{\text{ex}}_{\text{exch}}(T)$ and $\Delta E^{\text{biex}}_{\text{exch}}(\mu, \nu; T)$ can be evaluated by perturbation theory.

7. – Application to real crystals and symmetry of biexciton states.

In the previous discussion we have only considered a two-band model with spin degeneracy. And thus a single spatial wave function is associated with each band extremum. Furthermore, the effect of spin-orbit interaction has been neglected, so that a classification according to the eigenstates of the total spin was possible.

To apply this description to real substances, we have to modify the two-band model. In most semiconductors, the upper valence band is degenerate (in addition to the spin degeneracy). The actual band structure (including the spin variables) results from the splitting of this degenerate valence band under the effect of spin-orbit interaction. Since the spin-orbit term in the Hamiltonian does not commute with all components of the spin operators, the eigenstates of the Hamiltonian can no longer be chosen to be eigenstates of the total spin. A new classification scheme for the crystal eigenstates has to be adopted.

The natural way to classify the eigenstates of a crystal is to specify their transformation properties under the symmetry operations of the lattice. This requires the use of group-theoretical methods [22], which we will apply to the biexciton problem following the line of a previous paper of ours [23]. If we assume that the band extrema are located at $\boldsymbol{k} = 0$ and introduce the effective-mass approximation, the biexciton wave function reads

$$\Phi_{\text{biex}}(\boldsymbol{Q}) = \exp[i\boldsymbol{Q}\cdot\boldsymbol{R}_0]\, F(\boldsymbol{r}, \boldsymbol{r}', \boldsymbol{R})\, \psi(0, 0, 0, 0)\,, \tag{29}$$

where $F(\boldsymbol{r}, \boldsymbol{r}', \boldsymbol{R})$ is the triple Fourier transform of $A(\boldsymbol{k}, \boldsymbol{k}', \boldsymbol{q})$ and $\psi(0, 0, 0, 0)$ contains only Bloch functions at the band extrema. The vector $\boldsymbol{R}_0$ represents the co-ordinate of the centre of mass and $\boldsymbol{r}$, $\boldsymbol{r}'$ and $\boldsymbol{R}$ are the electron-electron, hole-hole and mean electron-hole distances, respectively. If we assume both valence and conduction bands at the extrema to be twofold degenerate in the double group, we can form 16 four-particle products of Bloch functions. These simple products can then be combined to form 16 functions belonging to one of the irreducible representations of the Γ-point. The possible representations are given by the decomposition

$$\Gamma_c \otimes \Gamma_c \otimes \Gamma_v^* \otimes \Gamma_v^* = \sum_l c_l \Gamma_l\,. \tag{30}$$

The symmetry operations of the crystal commute with the permutation operator of the two electrons P_{13} and with that of the two holes P_{24}. As shown in the case of the spin operators, all the matrix elements of the symmetry operations between four-particle functions of different parities vanish. One can thus impose that the basis functions of the Γ_i's appearing in the above decomposition have definite parities under permutation of the two electrons and under permutation of the two holes.

The Pauli principle requires that the total wave function be antisymmetric under both permutations. This is ensured by multiplying each of the sixteen functions $\psi(0, 0, 0, 0)$ by an envelope function of opposite parity with respect to the transformations

$$\boldsymbol{r} \to -\boldsymbol{r} \qquad \text{(permutation of the electrons)}$$

and

$$\boldsymbol{r}' \to -\boldsymbol{r}' \qquad \text{(permutation of the holes)}.$$

As in the case of the spin classification, we have to consider four different envelope functions: $F^{\mu\nu}(\boldsymbol{r}, \boldsymbol{r}', \boldsymbol{R})$ $(\mu, \nu; \pm)$, all being solutions of the same effective-mass equation. From the study of the hydrogen molecule, we know that the bound states correspond to $F^{++}(\boldsymbol{r}, \boldsymbol{r}', \boldsymbol{R})$ and to $F^{+-}(\boldsymbol{r}, \boldsymbol{r}', \boldsymbol{R})$. The totally symmetric function $F^{++}(\boldsymbol{r}, \boldsymbol{r}', \boldsymbol{R})$ describes the biexciton ground state; there exists only one totally antisymmetric combination of $k = 0$ Bloch functions associated with it.

$F^{+-}(\boldsymbol{r}, \boldsymbol{r}', \boldsymbol{R})$ describes the first excited states of the biexciton and can be interpreted, in the limit $m_e/m_h \ll 1$, as the first rotational states of the molecule; there exist three combinations of $k = 0$ Bloch functions associated with it. Thus four of the sixteen possible biexciton states will be bound in our model. The case of CuCl and CdS exactly corresponds to our band model. It is straightforward to generalize the procedure to more complicated band structures.

8. – Application to CuCl.

CuCl has the zincblende structure, which corresponds to the tetrahedral group T_d. We use the group theoretical notation of KOSTER *et al.* [24]. The spin-orbit splitting of the valence band has been discussed above (fig. 3). In CuCl this splitting is of the order to 69 meV. In our analysis we shall only consider holes belonging to the Γ_7 upper valence band, completely neglecting the lower Γ_8 band. This approximation is justified by the experimental results showing two well-separated exciton series.

The conduction Bloch functions Γ_6 can be written as

$$(31a) \qquad c_{\frac{1}{2}} = \varphi_c|\alpha\rangle\,, \qquad c_{-\frac{1}{2}} = \varphi_c|\beta\rangle\,,$$

where φ_c can be identified with the $4s$-functions of Cu. The valence Bloch functions Γ_7 are of the form

$$v_{-\frac{1}{2}} = \psi_{-1}|\alpha\rangle - \psi_0|\beta\rangle \quad \text{and} \quad v_{\frac{1}{2}} = \psi_1|\beta\rangle + \psi_0|\alpha\rangle \,, \tag{31b}$$

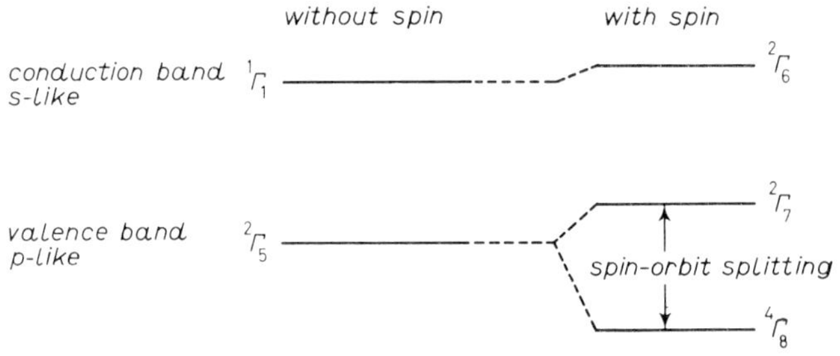

Fig. 3. – Spin-orbit splitting and symmetries of the upper valence band and lowest conduction band in CuCl: group T_d, group-theoretical notations according to ref. [22]. The degeneracies of the representations are indicated by a superscript.

where, to a good approximation, the top valence wave functions may be identified with the $3p$-functions of Cl:

$$\psi_{\pm 1} = (p_x \pm ip_y)/\sqrt{3} \,, \qquad \psi_0 = p_z/\sqrt{3} \,, \tag{31c}$$

and p_x, p_y, p_z are separately normalized. In this approximation the Γ_7-states coincide with those of total angular momentum $J = \frac{1}{2}$ of the atom ($J_z = \frac{1}{2}$ and $J_z = -\frac{1}{2}$). The hole states are obtained by application of the time-reversal operator to the two Γ_7-functions. The lowest exciton and biexciton states are constructed from properly symmetrized products of electron functions Γ_6 and hole functions Γ_7.

The symmetry of the $1s$-excitons ($\Gamma_{\text{env}}(1s) = \Gamma_1$) is given by $\Gamma_6 \otimes \Gamma_7 = \Gamma_2 + \Gamma_5$; Γ_2 corresponds to a total-angular-momentum state $J = 0$ and can be shown to be a superposition of pure triplet electron-hole spin states. Consequently, it is not affected by the electron-hole exchange interaction. Γ_5 corresponds to $J = 1$ and is a superposition of singlet and of triplet electron-hole spin states, so that its energy is increased by the electron-hole exchange by an amount $\Delta E^{\text{ex}}_{\text{exch}} = \frac{2}{3}\tilde{J}|F(0)|^2$, where $\tilde{J}$ is defined in (16).

The crystal symmetries of the biexciton states are given by the following products of irreducible representations:

$$\Gamma_6 \otimes \Gamma_7^* \otimes \Gamma_6 \otimes \Gamma_7^* = (\Gamma_2 + \Gamma_5)(\Gamma_2 + \Gamma_5) = 2\Gamma_1 + 3\Gamma_4 + \Gamma_3 + \Gamma_5 \,. \tag{32}$$

In the total-angular-momentum classification, Γ_1 corresponds to $J = 0$, Γ_4 to $J = 1$ and $\Gamma_3 + \Gamma_5$ results from the splitting of $J = 2$. The corresponding wave functions have been chosen to have definite parities under the permutation of electrons and the permutation of holes. As discussed above, they must be multiplied by an envelope function with opposite permutational parities. The ground state of the molecule will be the totally symmetric Γ_1-state with envelope function $F^{++}(\boldsymbol{r}, \boldsymbol{r}', \boldsymbol{R})$. The next higher state is also bound in CuCl, because of the very small electron-hole mass ratio; it has symmetry Γ_4 with envelope function $F^{+-}(\boldsymbol{r}, \boldsymbol{r}', \boldsymbol{R})$, all other states being unbound. The numerical factor Z affecting the exchange contribution must be recomputed, because each of the present states is a combination of different spin multiplets. Using the appropriate functions one obtains, for the ground state, $Z(\Gamma_1^{\text{biex}}) = \frac{1}{2}$ and, for $F^{+-}(\Gamma_4)$, $Z(\Gamma_4^{\text{biex}}) = \frac{1}{2}$. All the above results are displayed in fig. 4.

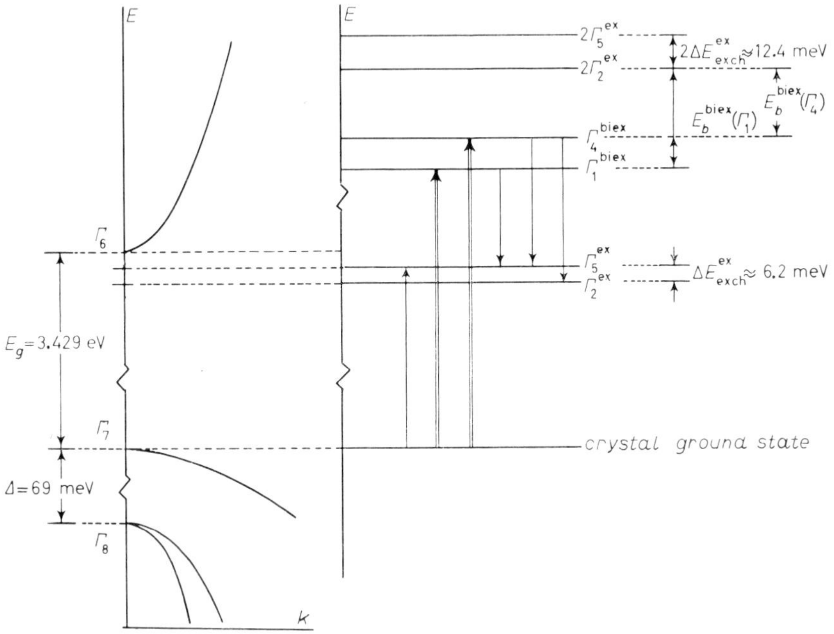

Fig. 4. – Schematic representation of the band structure of the exciton states and of the biexciton states in CuCl. On the right-hand side of the figure the lowest exciton levels and the bound biexciton states are indicated. E_b^{biex} is the binding energy of the molecule. $\Delta E_{\text{exch}}^{\text{ex}}$ is the electron-hole exchange splitting for the exciton. The single (double) arrows indicate one-photon (two-photon) allowed transitions.

9. – Application to CdS.

A similar analysis is performed for the case of the wurtzite structure, which has an uniaxial symmetry (group C_{6v}). We consider the case appropriate to CdS, where the bottom of the conduction band is the state Γ_7 at $k = 0$, and the top of the valence band is the state Γ_9 at $k = 0$. Two other states, both of symmetry Γ_7, are split off from the Γ_9-state by the anisotropy of the crystal field and the spin-orbit interaction (fig. 5). The conduction Bloch function Γ_7 can be written as

$$c_{\frac{1}{2}} = \varphi_c|\alpha\rangle\,, \qquad c_{-\frac{1}{2}} = \varphi_c|\beta\rangle\,,$$

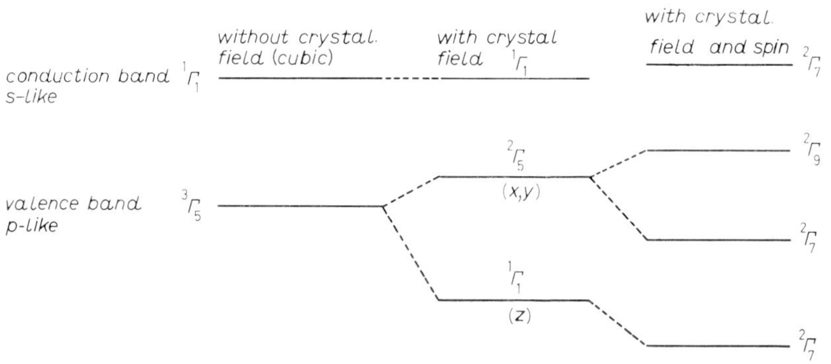

Fig. 5. – Spin-orbit splitting, crystal field splitting and symmetries of the upper valence band and lowest conduction band in CdS (group C_{6v}).

where φ_c can be identified with the $5s$-function of Cd. The top valence Bloch functions of Γ_9 are of the form

$$v_{\frac{3}{2}} = \psi_1|\alpha\rangle\,, \qquad v_{-\frac{3}{2}} = \psi_{-1}|\beta\rangle\,,$$

where

$$\psi_{\pm 1} = (p_x \pm ip_y)/\sqrt{2}\,,$$

and p_x, p_y are, to a good approximation, two of the $2p$-functions of S. We may notice that, because of the anisotropy of the crystal, these states are no longer eigenstates of total J but of J_z, the states with the same absolute value of the eigenvalue of J_z being degenerate, because of Kramer's theorem.

The lowest exciton and biexciton states are constructed from properly symmetrized electron functions Γ_7 and hole functions Γ_9. Two excitons are formed with symmetry Γ_5 and Γ_6. $\Gamma_6(J_z = \pm 2)$ corresponds to a superposition of pure triplet spin states, while $\Gamma_5(J_z = \pm 1)$ contains both triplet and singlet spin states. Γ_5 is shifted by an amount $\Delta E^{\text{ex}}_{\text{exch}} = \tilde{J}|F(0)|^2$ to higher energy by the electron-hole exchange interaction. The wave functions are given in ref. [22].

The crystal symmetries of the biexciton states are given by the following product of irreducible representations:

$$\Gamma_7 \otimes \Gamma_9^* \otimes \Gamma_7 \otimes \Gamma_9^* = (\Gamma_5 + \Gamma_6) \otimes (\Gamma_5 + \Gamma_6) = \tag{33}$$
$$= 2\Gamma_1 + 2\Gamma_2 + 2\Gamma_3 + 2\Gamma_4 + 2\Gamma_5 + 2\Gamma_6\,.$$

The corresponding wave functions are also given in ref. [22]. As before, the permutational symmetry has been explicitly imposed, and, consequently, the permutational parities of the envelope function are also determined. The ground state of the molecule is the totally symmetric Γ_1-state with the envelope function $F^{++}(\boldsymbol{r}, \boldsymbol{r}', \boldsymbol{R})$. It is also possible, though not certain in this case, to have two higher bound states with envelope function $F^{+-}(\boldsymbol{r}, \boldsymbol{r}', \boldsymbol{R})$, corresponding to one state of symmetry Γ_2 and a doubly degenerate state of symmetry $\Gamma_3 + \Gamma_4$. The separation of this state is due to the anisotropy of CdS. The states of the excitonic molecule for CdS are visualized in fig. 6.

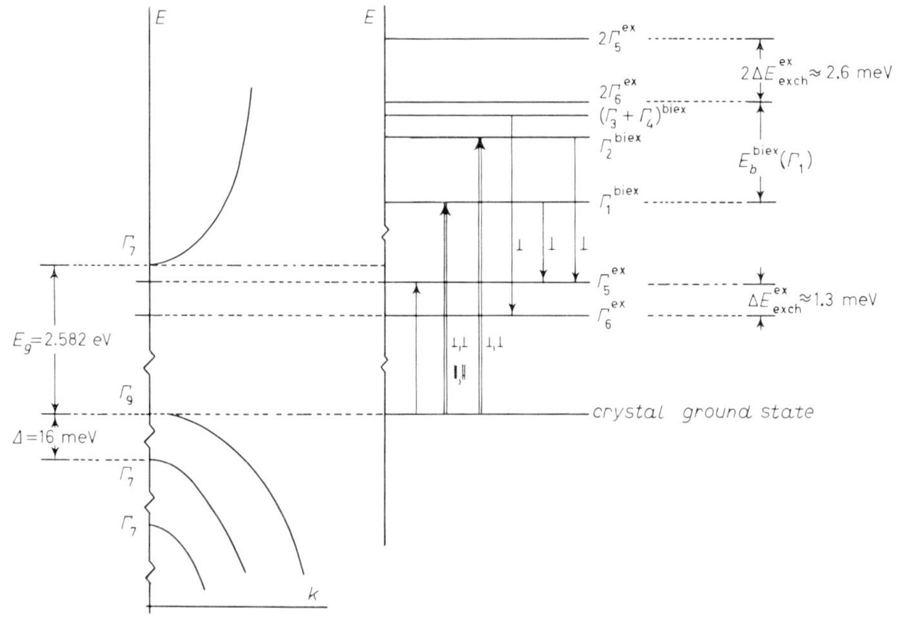

Fig. 6. – Schematic representation of the band structure of the exciton states and of the biexciton states in CdS. On the right-hand side of the figure the lowest exciton levels and the bound biexciton states are indicated. E_b^{biex} is the binding energy of the molecule. $\Delta E_{\text{exch}}^{\text{ex}}$ is the electron-hole exchange splitting for the exciton. The single (double) arrows indicate one-photon (two-photon) allowed transitions with indication of the polarization direction of the electric field with respect to the c-axis. The $(\Gamma_3 \oplus \Gamma_4)^{\text{biex}}$ and Γ_2^{biex} biexcitons are indicated as bound states, though no evidence for their existence is so far available.

10. – Optical transitions and selection rules.

Like all electronic excitations, the biexciton decays to the crystal ground state by spontaneous emission. The process occurs in two steps: first by emission of a photon and formation of an exciton, second by the decay of this exciton.

The first process gives rise to new luminescence lines occurring on the low-energy side of the free-exciton emission. According to our discussion of selection rules, this transition is allowed if the decomposition of $\Gamma_{\text{ex}} \otimes \Gamma_{\text{p-e}} \otimes \Gamma_{\text{biex}}$ contains Γ_1, or, equivalently, if the decomposition of $\Gamma_{\text{ex}} \otimes \Gamma_{\text{p-e}}$ contains Γ_{biex}.

In the zincblende symmetry the electron-photon interaction transforms like Γ_5. The selection rules for the decay of the biexciton ground state Γ_1 and of the excited state Γ_4 are obtained from

$$\left\{\begin{array}{l}\Gamma_2 \otimes \Gamma_5 \to \Gamma_4\,,\\ \Gamma_5 \otimes \Gamma_5 \to \Gamma_1 + \Gamma_3 + \Gamma_4 + \Gamma_5\,.\end{array}\right. \tag{34}$$

Consequently, the biexciton ground state Γ_1 decays into the Γ_5 exciton, and the excited biexciton state Γ_4 can decay into both Γ_2 and Γ_5 excitons. These selection rules are illustrated in fig. 4.

In the wurtzite symmetry the electron-photon interaction transforms like Γ_1 for photons polarized along the c-axis, and like Γ_5 for photons polarized perpendicular to the c-axis. Only the latter polarization direction allows a decay of bound molecular states into one of the two existing Γ_5 and Γ_6 1s-excitons. For the grounds state Γ_1 the only allowed transition is obtained from

$$\Gamma_5 \otimes \Gamma_5 \to \Gamma_1 + \Gamma_2 + \Gamma_6\,, \tag{35}$$

while for the possible excited molecular states (Γ_2, Γ_3, Γ_4) the allowed transitions are given by (35) for Γ_2 and by

$$\Gamma_6 \otimes \Gamma_5 \to \Gamma_3 + \Gamma_4 + \Gamma_5$$

for $\Gamma_3 + \Gamma_4$.

These selection rules are illustrated in fig. 6.

Since, to first order, the electromagnetic interaction cannot excite two electrons, it is necessary to consider two-photon processes to create biexcitons by optical absorption. The amplitude A of the second-order transition probability is given, in the N-electron space, by the expression [25]

$$A \propto \sum_n \frac{\langle \phi^{\text{biex}} | \sum_{i=1}^{N} \boldsymbol{A}(\boldsymbol{r}_i)\cdot \boldsymbol{p}_i | \phi_n^{\text{ex}}\rangle \langle \phi_n^{\text{ex}} | \sum_{i=1}^{N} \boldsymbol{A}(\boldsymbol{r}_i)\cdot \boldsymbol{p}_i | \phi^0\rangle}{E_n^{\text{ex}} - E_0 - \hbar\omega}\,, \tag{36}$$

where ϕ^{biex} and ϕ_0 are the biexciton state and the crystal ground state, respectively. According to HANAMURA [25], only resonant exciton intermediate states ϕ_n^{ex} of energy E_n have been considered. E_0 and $\hbar\omega$ are the crystal ground-state energy and the photon energy, respectively. The energy denominators become very small for resonant intermediate states and for small biexciton binding energy.

In the dipole approximation the symmetry of the operator connecting initial and final states in second order is that of the direct product of the irreducible representations of the two photons. Consequently, in the zincblende structure, selection rules for the two-photon transitions from the crystal ground state Γ_1 are given by the product of the two-photon symmetries:

$$\Gamma_5 \otimes \Gamma_5 \to \Gamma_1 + \Gamma_3 + \Gamma_4 + \Gamma_5 , \tag{37}$$

so that the ground (Γ_1) and the excited (Γ_4) states of the biexciton can be produced. In the wurtzite symmetry, one has to consider the two polarization directions of the photons. The relevant selection rules are given by

$$\begin{cases} \Gamma_5 \otimes \Gamma_5 \to \Gamma_1 + \Gamma_2 + \Gamma_6 , \\ \Gamma_1 \otimes \Gamma_1 \to \Gamma_1 , \\ \Gamma_1 \otimes \Gamma_5 \to \Gamma_5 ; \end{cases} \tag{38}$$

this means that the molecular ground state Γ_1 can be reached with two photons both parallel and perpendicular to the *c*-axis. The two-photon selection rules for zincblende and wurtzite symmetry are also illustrated in fig. 4 and 6.

Besides the above-mentioned selection rules, which are based on the general symmetry of the states and of the electron-photon interaction, more stringent selection rules for two-photon excitation processes can be obtained by considering polarized photons and by including time-reversal symmetry. This general problem has been considered first by INOUE and TOYOZAWA [26] and extended and applied to the case of simple excitons by a number of authors [27]. Recently DONI *et al.* [28] have been able to separate, in two-photon processes, a contribution which depends on the frequency of the two photons (dynamical) from a contribution which depends on the polarization of the two photons (geometrical). Both contributions give rise to selection rules and give information on the relative strength of transitions to different final states. The dynamical factor gives relevant contributions only to states which are contained in the symmetric product of the momentum representations. The states which are contained in the antisymmetric product are forbidden if the two photons have the same frequency, and are weighted by a much weaker factor in any case. The geometrical factor gives a contribution which depends only on the symmetry of the final state.

In the case of CuCl it turns out that the state Γ_1 is contained in the symmetric product of the two-photon symmetries, while Γ_4 is contained in the antisymmetric product. As a consequence, though both states Γ_1 and Γ_4 are allowed by two-photon processes as shown in fig. 4, the state Γ_4 is much weaker and is strictly forbidden if the photons have the same frequency. Furthermore, the probability of the ground state Γ_1 is proportional to the factor $\frac{1}{3}|\boldsymbol{e}_1 \cdot \boldsymbol{e}_2|^2$, which has a maximum for parallel polarization, while the probability of the state Γ_4 is proportional to $\frac{1}{2}|\boldsymbol{e}_1 \wedge \boldsymbol{e}_2|^2$, which has a maximum when the two photons are polarized perpendicularly with respect to one another [28].

Similar selection rules have been obtained for CdS crystals. They allow, in principle, an experimental determination of the symmetry of the biexciton states produced by nonlinear spectroscopy.

11. – Final remarks on the experimental results.

A detailed interpretation of the experimental results on biexcitons is not yet available. The results obtained by BIVAS, LÉVY, NIKITINE and GRUN on CuCl are shown in fig. 7. The biexciton luminescence line M shows a fine structure $M_1 - M_2$. We may interpret this fine structure as related to the allowed processes

$$\Gamma_1^{\text{biex}} \to \Gamma_5^{\text{ex}} + \text{photon } (M_1)$$

and

$$\Gamma_4^{\text{biex}} \to \Gamma_5^{\text{ex}} + \text{photon } (M_2)\,,$$

if both Γ_1 and Γ_4 molecular states are involved. The energy difference between the two photons would be in this case

$$\hbar\omega_{M_1} - \hbar\omega_{M_2} = E^{\text{biex}}(\Gamma_1) - E^{\text{biex}}(\Gamma_4)\,,$$

which would give a direct measure of the rotational energy of the molecule.

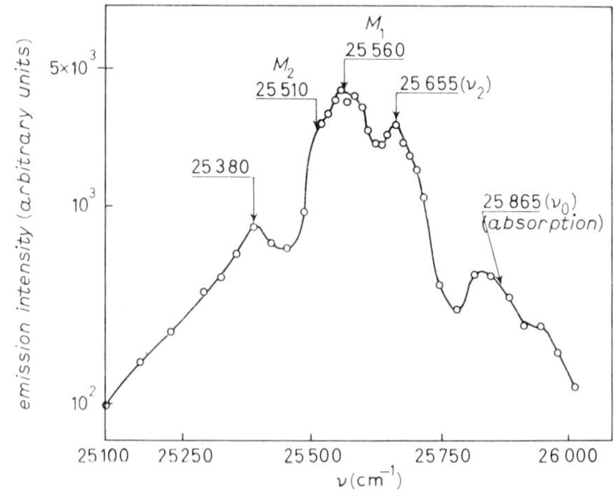

Fig. 7. – Luminescence of CuCl under high excitation at 4.2 K. ν_0 indicates the energy of the Γ_5-exciton absorption peak. The biexciton luminescence consists of two lines denoted M_1 and M_2. (Taken from A. BIVAS, R. LÉVY, S. NIKITINE and J. GRUN: *J. Physique*, **31**, 227 (1970).)

Figure 8 shows the biexciton luminescence of CdS as measured by VOIGT and MAUERSBERGER [4]. The corresponding allowed process is

$$\Gamma_1^{\text{biex}} \to \Gamma_5^{\text{ex}} + \text{photon } (\boldsymbol{E}\perp c\text{-axis})\,.$$

The ratio of the luminescence intensities with $\boldsymbol{E}\perp c$-axis and with $\boldsymbol{E}\parallel c$-axis is about 15. This can be considered as a good confirmation of the polarization selection rules we established in the preceding section.

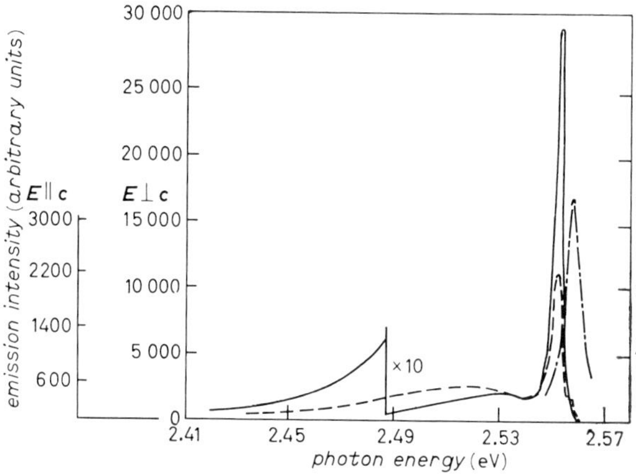

Fig. 8. – Luminescence of CdS under high excitation at 4.2 K. ——— luminescence with $\boldsymbol{E}\perp c$-axis, — — — luminescence with $\boldsymbol{E}\parallel c$-axis, —·—·— absorption with $\boldsymbol{E}\perp c$-axis. The main luminescence peak is due to recombination of biexcitons. The absorption peak is due to Γ_5 excitons. (Taken from J. VOIGT and G. MAUERSBERGER: *Phys. Stat. Sol.*, b **60**, 678 (1973).)

Finally, GALE and MYSYROWICZ [29] have measured two-photon absorption in CuCl. They found that this absorption occurs at the energy of the Γ_1 biexciton states, but the polarization dependence of the transition was not investigated.

Biexcitons have been searched for in other semiconductors, and it is certain that CuCl and CdS are not the only materials where they can be found. A good candidate, on the basis of the electron-hole effective-mass ratio σ, seemed Cu_2O, where a very-well-resolved exciton series had been observed. The existing experimental evidence is not very clear, with GROSS *et al.* [30] claiming evidence for biexcitons and PETROFF *et al.* [31] disclaiming this evidence. To clarify the situation we performed a detailed theoretical calculation of the biexciton ground-state energy in Cu_2O [32]. We found that, because of the relevance of the electron-hole exchange contribution, the binding energy, if of the right sign, is smaller than 0.1 meV, so that biexcitons cannot exist in Cu_2O.

In the case of indirect gap materials like Ge and Si, luminescence due to exciton aggregates has been observed [33], but is interpreted as due to liquid-like drops rather than to molecules [34]. In reality, excitonic molecules should exist also in indirect gap materials, with nonnegligible binding energies [35]. However, in this case, the exciton concentration may be larger and the condensed state may be favoured. Only two-photon excitation experiments could prove in this case that biexcitons exist.

REFERENCES

[1] M. A. LAMPERT: *Phys. Rev. Lett.*, **1**, 450 (1958).

[2] S. A. MOSKALENKO: *Sov. Phys. Opt. Spectr.*, **5**, 147 (1958).

[3] A. MYSYROWICZ, J. GRUN, R. LÉVY, A. BIVAS and S. NIKITINE: *Phys. Lett.*, **26** A, 615 (1968); S. NIKITINE, A. MYSYROWICZ and J. GRUN: *Helv. Phys. Acta*, **41**, 1058 (1968); A. MYSYROWICZ: Thesis, Strasbourg (1968) (unpublished); A. BIVAS, R. LÉVY, S. NIKITINE and J. GRUN: *J. Physique*, **31**, 227 (1970); A. BIVAS: Thesis (3ème cycle), Strasbourg (1969) (unpublished); R. KNOX, S. NIKITINE and A. MYSYROWICZ: *Opt. Comm.*, **1**, 19 (1969); J. GRUN, S. NIKITINE, A. BIVAS and R. LÉVY: *Journ. Lumin.*, **1**, **2**, 241 (1970); R. LÉVY: Thesis, Strasbourg (1973) (unpublished); S. NIKITINE: in *Proceedings of the International Symposium on Excitons at High Density and Polaritons*, Baiersbronn-Tonbach (Germany), to appear; H. SOUMA, H. KOIKE, K. SUZUKI and M. UETA: *Journ. Phys. Soc. Japan*, **31**, 1285 (1971); S. SUGA and T. KODA: *Phys. Stat. Sol.*, b **61**, 291 (1974).

[4] H. KURODA, S. SHIONOYA, H. SAITO and E. HANAMURA: *Sol. State Comm.*, **12**, 553 (1973); H. SAITO, S. SHIONOYA and E. HANAMURA: *Sol. State Comm.*, **12**, 227 (1973); S. SHIONOYA, H. SAITO, E. HANAMURA and O. AKIMOTO: *Sol. State Comm.*, **12**, 223 (1973); H. KURODA, S. SHIONOYA, H. SAITO and E. HANAMURA: *Journ. Phys. Soc. Japan*, **35**, 534 (1973); H. KURODA and S. SHIONOYA: *Sol. State Comm.*, **13**, 1195 (1973); J. F. FIGUEIRA and H. MAHR: *Phys. Rev. B*, **7**, 4520 (1973); J. VOIGT and G. MAUERSBERGER: *Phys. Stat. Sol.*, b **60**, 679 (1973).

[5] R. R. SHARMA: *Phys. Rev.*, **170**, 770 (1968).

[6] R. K. WEHNER: *Sol. State Comm.*, **7**, 457 (1969).

[7] J. ADAMOWSKI, S. BEDNAREK and M. SUFFCZYNSKI: *Sol. State Comm.*, **9**, 2037 (1971).

[8] O. AKIMOTO and E. HANAMURA: *Sol. State Comm.*, **10**, 253 (1972); *Journ. Phys. Soc. Japan*, **33**, 1537 (1972). For a lower bound to binding energy see also E. D. GOUTLIANSKI and V. E. KHARTSIEV: *Sol. State Comm.*, **12**, 1087 (1973).

[9] W.F. BRINKMAN, T. M. RICE and B. BELL: *Phys. Rev. B*, **8**, 1570 (1973).

[10] W.T. HUANG and V. SCHRÖDER: *Phys. Lett.*, **38** A, 507 (1972); W. T. HUANG: *Phys. Stat. Sol.*, b **60**, 309 (1973).

[11] P. H. HANDEL: *Phys. Rev. B*, **7**, 5183 (1973).

[12] T. INUI: *Proc. Phys. Math. Soc. Japan*, **20**, 770 (1938); **23**, 992 (1941).

[13] E. A. HYLLERAAS and A. ORE: *Phys. Rev.*, **71**, 493 (1947).

[14] S. NIKITINE: in *Proceeding of the International Symposium on Excitons at High Density and Polaritons*, Baiersbronn-Tonbach (Germany), to appear.

[15] J. R. HAYNES: *Phys. Rev. Lett.*, **17**, 860 (1966); YA. POKROVSKII: *Phys. Stat. Sol.*, a **11**, 385 (1972); C. BENOÎT À LA GUILLAUME, F. SALVAN and M. VOOS: *Journ. Lumin.*, **1**, **2**, 315 (1970); *Phys. Rev. B*, **5**, 3079 (1972).

[16] Y. Petroff, P. Y. Yu and V. R. Shen: *Phys. Rev. Lett.*, **29**, 1558 (1972).
[17] K. Dos, A. Haug and P. Rohner: *Phys Stat. Sol.*, **30**, 619 (1968); P. Rohner: *Phys. Rev. B*, **3**, 433 (1971).
[18] A. Messiah: *Quantum Mechanics*, Vol. **2**, Chap. XV, Sect. **18** (Amsterdam, 1969).
[19] See, for instance, F. Bassani and G. Pastori-Parravicini: *Electronic States and Optical Transitions in Solids*, Chap. 6 (Oxford, 1975).
[20] J. J. Forney, A. Quattropani and F. Bassani: *Nuovo Cimento*, **22** B, 153 (1974).
[21] W. Kohn: *Phys. Rev.*, **110**, 857 (1958); A. Morita, M. Azuma and H. Nara: *Journ. Phys. Soc. Japan*, **17**, 1570 (1962); Y. Abe, Y. Osaka and A. Morita: *Journ. Phys. Soc. Japan*, **17**, 1576 (1962).
[22] F. Bassani, J. J. Forney and A. Quattropani: *Phys. Stat. Sol.*, b **65**, 591 (1974).
[23] For more detail on group theory and its application to solid-state physics, see ref. [19].
[24] K. F. Koster, J. O. Dimmock, R. G. Wheeler and H. Statz: *Properties of the Thirty-Two Point Groups* (Cambridge, Mass., 1963).
[25] E. Hanamura: *Sol. State Comm.*, **12**, 951 (1973); F. Bassani: *Rendiconti S.I.F.*, Course LII (New York, N. Y., 1972).
[26] M. Inoue and Y. Toyozawa: *Journ. Phys. Soc. Japan*, **20**, 363 (1965).
[27] T. R. Bader and A. Gold: *Phys. Rev.*, **171**, 997 (1968); D. Fröhlich, B. Staginnus and S. Thurm: *Phys. Stat. Sol.*, **40**, 287 (1970); M. M. Denisov and V. P. Makarov: *J. Phys. C*, **5**, 273 (1972).
[28] E. Doni, R. Girlanda and G. Pastori-Parravicini: *Phys. Stat. Sol.*, b **65**, 203 (1974); E. Doni, R. Girlanda and G. Pastori-Parravicini: *Sol. Stat. Comm.*, **17**, 189 (1975).
[29] G. M. Gale and A. Mysyrowicz: *Proceedings of the XII International Conference on the Physics of Semiconductors* (Stuttgart, 1974), p. 133; G. M. Gale and A. Mysyrowicz: *Phys. Rev. Lett.*, **32**, 727 (1974).
[30] F. F. Gross and F. I. Kreingold: *JETP Lett.*, **12**, 68 (1970).
[31] Y. Petroff, P. Y. Yu and Y. R. Shen: *Phys. Rev. Lett.*, **29**, 1558 (1972).
[32] F. Bassani and M. Rovere: *Solid State Comm.*, to be published.
[33] Ya. Pokrovskii: *Phys. Stat. Sol.*, a **11**, 385 (1972); Ya. Prokovskii and K. I. Svistunova: *Sov. Phys. Semicon.*, **4**, 409 (1970); A. S. Kaminskii, Ya. E. Pokrovskii and N. V. Alkeev: *Sov. Phys. JETP*, **32**, 1048 (1971); C. Benoit à la Guillaume, M. Voos and F. Salvan: *Phys. Rev. B*, **5**, 3079 (1972).
[34] M. Combescot and P. Nozières: *J. Phys. C*, **5**, 2369 (1972); W. F. Brinkmann and T. M. Rice: *Phys. Rev. B*, **7**, 1508 (1973).
[35] J. J. Forney, A. Quattropani and F. Bassani: *Phys. Stat. Sol.*, to appear.

Ultra-Fast Dynamical Investigations of Vibrational Relaxation and Energy Transfer in Polyatomic Liquids.

W. KAISER and A. LAUBEREAU

Technische Universität München - München, West Germany

During the past several years we have been interested in elementary processes of polyatomic liquids. We have developed experimental methods to study the dephasing time, the population (or energy) relaxation time and energy transfer processes of well-defined vibrational normal modes of polyatomic molecules. Our investigations were made in the electronic ground state.

Our knowledge of relaxation processes of molecular vibrations in liquids is rather scarce. Ultra-sonic waves interact predominantly with the lowest vibrational modes [1]; they do not provide relaxation values of the higher normal modes of polyatomic molecules. The main source of information on vibrational relaxation times is deduced from line width measurements of infra-red and Raman bands [2-4]. These investigations are frequently difficult to interpret because different physical processes contribute to the observed line shape. For instance, phase relaxation, energy relaxation, rotational motion and isotope splitting are line-broadening factors.

In our investigations a well-defined normal mode of the molecule is first excited above its equilibrium value and the return to thermal equilibrium is studied as a function of time. Since vibrational relaxation processes proceed very rapidly, ultra-fast measuring techniques are required. Recent advances in the generation of intense and reproducible light pulses on a time scale of 10^{-12} s are essential for the investigations discussed here.

At the beginning of this paper several brief remarks should be made concerning the quality of picosecond light pulses. In addition, a recent technique to generate tunable infra-red pulses will be discussed.

In order to obtain reliable experimental data, it is most important to work with well-defined single picosecond pulses. One must know the pulse duration, the pulse shape, the peak intensity and the frequency band width of the pulses used. These pulse parameters have to be kept constant during the experimental run. The application of the whole mode-locked pulse train should be avoided, since in high-gain solid-state systems the important parameters, pulse duration, pulse shape, peak intensity and frequency band width, vary from the beginning to the end of the pulse train [5, 6].

In fig. 1 an experimental system for the generation of single picosecond pulses is schematically depicted. The laser oscillator contains a switching dye

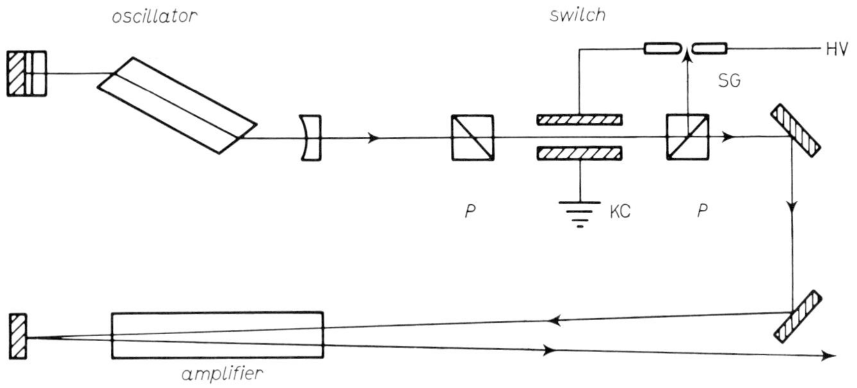

Fig. 1. – Schematics of the experimental set-up for the generation of single picosecond light pulses; Kerr cell KC, polarizer P, laser-triggered spark gap SG, high voltage HV.

in close contact with a mirror. A mode-locked Nd-glass laser operates at a frequency of $\nu_L = 9455\ \text{cm}^{-1}$. A single picosecond pulse is cut from the leading part of the pulse train by means of an electro-optic switch. This switch consists of a high-pressure spark gap in conjunction with an optical Kerr cell [7]. A subsequent optical amplifier increases the peak pulse intensity by a factor of approximately one hundred. In fig. 2 an oscilloscope trace of a pulse train and

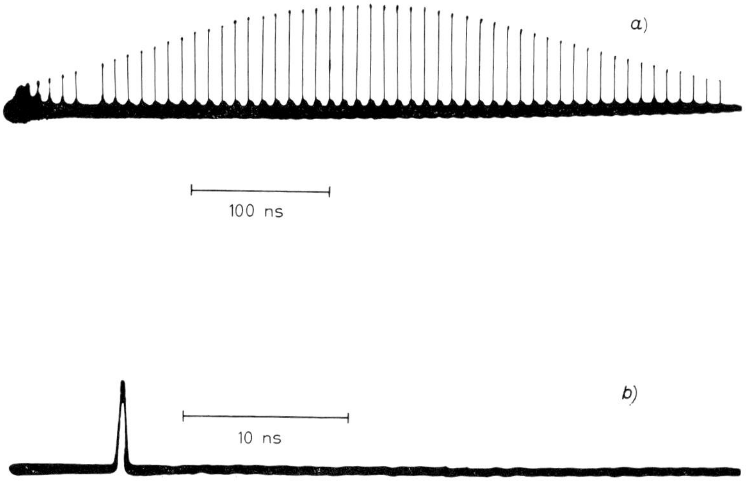

Fig. 2. – Oscilloscope display of the pulse train (a)) and of the selected pulse (b)); the rise time of the detection system is not sufficient to resolve the picosecond pulses.

a switched-out single pulse is shown. Pictures of this sort represent necessary but not sufficient conditions for useful picosecond pulses. We have studied our single picosecond pulses in detail. The various parameters are listed in fig. 3. Of special importance are i) a reproducible pulse duration (6 ps), ii) no

Pulse duration	$t_p = 6$ ps
Contrast ratio	3 ± 0.1
Frequency width	$\Delta\tilde{\nu} = 2.9\ \text{cm}^{-1}$
$t_p \times \Delta\nu$	0.6
Peak-to-background ratio	$\sim 10^4$
Pulse intensity	$5 \cdot 10^8\ \text{W/cm}^2$
Pulse energy after amplification	$4 \cdot 10^{-3}$ J

Fig. 3. – Table of the properties of a single picosecond pulse.

satellite structure, iii) a band width limited pulse and iv) a large peak-to-background ratio.

Solid-state mode-locked laser systems generate picosecond pulses at fixed frequencies. For spectroscopic investigations tunable light sources are fre-

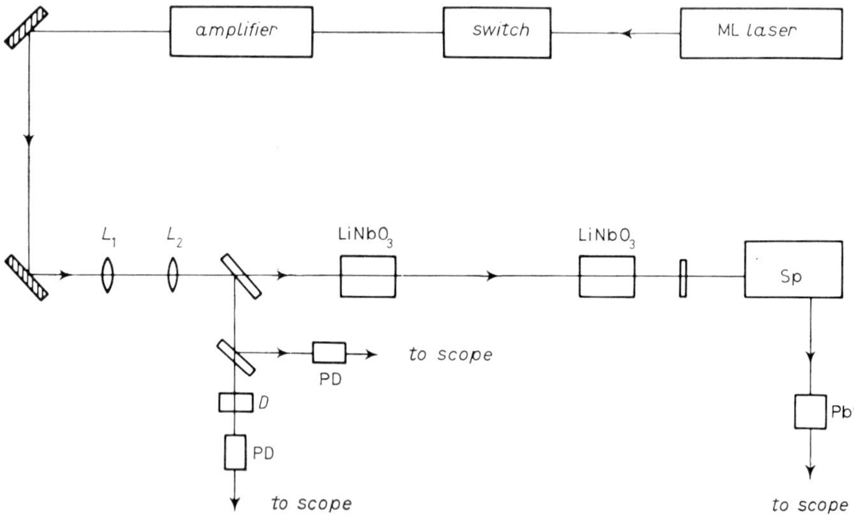

Fig. 4. – Schematics of the experimental system; generator crystals $LiNbO_3$, lens L, filter F, IR spectrometer Sp, IR detector PbS, nonlinear absorber D, fast photodiode PD.

quently desired. As a first step in this direction, we have investigated three-photon parametric amplification at high power levels as an infra-red light source [8]. This technique has subsequently been applied in the visible part of the spectrum.

In fig. 4 the schematics of the part of our experimental system which generates intense tunable pulses in the infra-red is depicted. A single picosecond pulse is switched out of the pulse train of a mode-locked Nd-glass laser and amplified by double passage through an optically pumped Nd-glass rod. Signal and idler pulses are generated when the intense laser pulse travels through a nonlinear $LiNbO_3$ crystal. If we known the dielectric properties of $LiNbO_3$, it is possible to predict the frequencies as a function of crystal orientation [9]. In fig. 5 experimentally

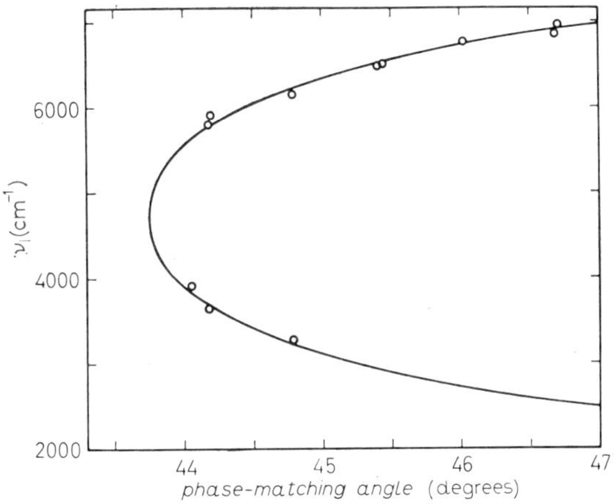

Fig. 5. – Angular tuning curve of the picosecond parametric generator; experimental points and calculated curve; $LiNbO_3$, $T = 293$ K.

determined frequencies are compared with the calculated phase-matching curve for $LiNbO_3$ at 293 K [8]. The agreement between experimentally observed frequencies and the calculated curve is excellent. The frequency range extends from 2500 cm^{-1} to 7000 cm^{-1}. The lower frequency limit is determined by the beginning of infra-red absorption in the $LiNbO_3$ crystal. In addition to the tuning range, the conversion efficiency of input power into signal power is important for practical applications. The signal power is plotted as a function of input peak intensity in fig. 6. It is readily seen from the figure that the signal power first rises exponentially over several orders of tens and then begins to saturate for very high input peak intensities. Power conversion efficiencies of several percent have been achieved [8]. This high-power conversion makes our travelling-wave parametric generator very attractive for the generation of picosecond pulses of new tunable frequencies.

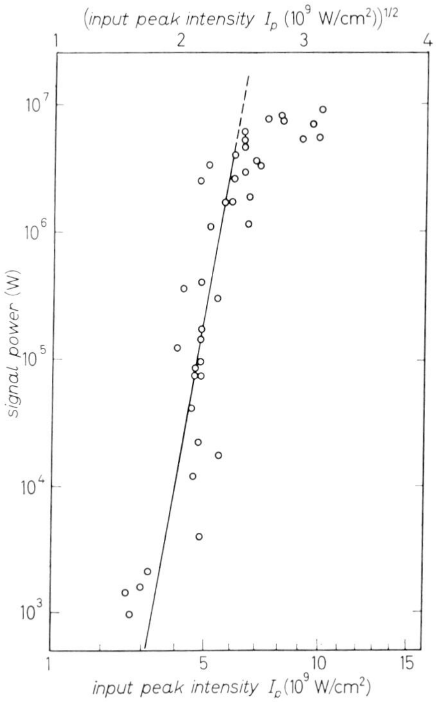

Fig. 6. – Peak power of signal pulses at 6500 cm^{-1} as a function of pump intensity I_p. Line calculated by using experimental parameters. $LiNbO_3$, $\tilde{\nu}_s = 6500$ cm^{-1} ($\tilde{\nu}_i = 2950$ cm^{-1}), $t_p = 3$ ps.

In fig. 7 various important parameters of our infra-red pulses are summarized. The pulse duration of the IR pulses is shorter than the duration of the pump pulse on account of the highly nonlinear generation process. The slopes of the infra-red pulses are expected to rise very steeply with time. This effect is of interest for a high time resolution (see below).

Tunability	(2500 ÷ 7000) cm^{-1}
Frequency width	40 cm^{-1}
Divergence	$1 \cdot 10^{-2}$ rad
Intensity	$\sim 10^9$ W/cm^2
Energy conversion	> 5%
Pulse duration	(2 ÷ 3) ps
IR photons	10^{15}

Fig. 7. – Pulse properties of tunable infra-red picosecond pulses.

A final experimental remark should be made on the slope of our picosecond light pulses. It will be apparent from our later discussion that the shape of the pulses is important for the quality of our relaxation measurements. In previous investigations we have shown that our single mode-locked laser pulse is well approximated by a Gaussian pulse shape [10]. Additional pulse shaping is possible when the pulse passes through a saturable absorber system (*e.g.* a dye solution) [11]. In fig. 8 calculated curves are presented. The normalized

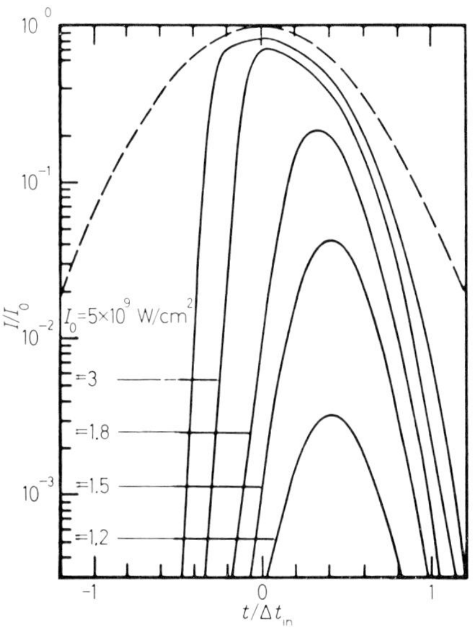

Fig. 8. – Changes of the pulse shape after a single pass through a saturable absorber. Broken curve: normalized intensity of the input pulse ($\Delta t_{in} = 8$ ps, Gaussian pulse shape). Solid curves: normalized intensity of the transmitted pulse for several values of the input peak intensity I_0. Dye parameters: $T_0 = 10^{-7}$, $\sigma = 1.84 \cdot 10^{-16}$ cm^2, $\tau = 9.1$ ps.

intensity of the input pulse (broken curve) and of various transmitted pulses is plotted *vs.* time. The dye parameters chosen for this example are the initial transmission $T_0 = 10^{-7}$, the lifetime of the excited state of the dye $\tau = 9.1$ ps and the absorption cross section of the dye molecule $\sigma = 1.8 \cdot 10^{-16}$ cm^2. It is readily seen from fig. 8 that the shape of the transmitted pulse is substantially altered compared to the input pulse for a peak input intensity of $I_0 = 5 \cdot 10^9$ W/cm^2. The leading part of the pulse is strongly absorbed by the bleachable dye, while the peak and trailing parts of the pulse are less affected. A pulse shortening by a factor of two is readily obtained by a single passage. It should be noted that the pulse at $I_0 = 5 \cdot 10^9$ W/cm^2 is asymmetric with a a steeply rising leading edge. Such pulses will be advantageous for an improved time resolution (see below).

We now turn to the discussion of different dynamical investigations of vibrational molecular processes in liquids. Our various techniques and results are summarized in fig. 9. Two excitation processes for well-defined normal modes were successfully used in our investigations: first, stimulated Raman scattering, which is discussed in more detail in the following part of this paper, and, second, direct infra-red excitation with ultra-short infra-red pulses. Three

Excitation	Probing	Information
	coherent scattering	dephasing time
stimulated Raman		collective beating
	incoherent scattering	energy transfer
IR absorption	fluorescence	energy relaxation

Fig. 9. – Physical processes for the excitation and the probing of molecular vibrations in liquids. The dynamical information obtained is listed on the right.

probing techniques lead to different physical information. Coherent light scattering requires a well-defined scattering geometry among the coherent material excitation, the probe pulse and the measured coherent scattering signal. The phase-matching angles are calculated from the colour dispersion of the medium to be investigated. Two investigations were made with coherently scattered probe pulses. The dephasing time of normal modes of vibrations was measured in a variety of polyatomic molecules and a beating phenomenon of collectively excited molecular vibrations was observed. Both investigations will be discussed below.

After the excitation of molecular vibrations, a certain number of molecules is promoted to the first excited vibrational state $v = 1$. Spontaneous anti-Stokes scattering, which represents an incoherent-scattering process, is directly proportional to the momentary degree of excitation of the system. Measuring the spontaneous anti-Stokes signal as a function of time after the initial excitation allows a direct determination of the population (or energy) relaxation of the vibrational mode. The same incoherent probing technique gives information on the decay routes of an excited mode and on energy transfer processes between different molecules and within one polyatomic molecule.

Finally, we have developed an experimental technique whereby the energy relaxation time of a well-defined molecular vibration can be determined by using a transition to an excited electronic state with observation of a fluorescence signal. The latter system is very sensitive and allows investigations of highly diluted systems. In the following part of this paper characteristic examples of the different investigations are presented.

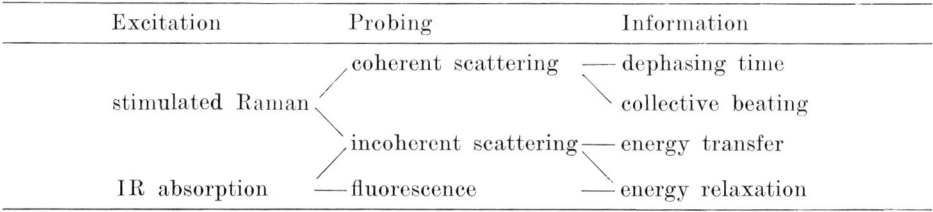

In fig. 10 the excitation of molecular vibrations via stimulated Raman scattering is indicated schematically. When an intense light pulse with peak intensity I_L traverses the medium, a positive gain at the Stokes frequency builds up an intense Stokes wave (I_S) at the end of the sample. In the past the generated Stokes pulse was the subject of extensive investigations [12-14].

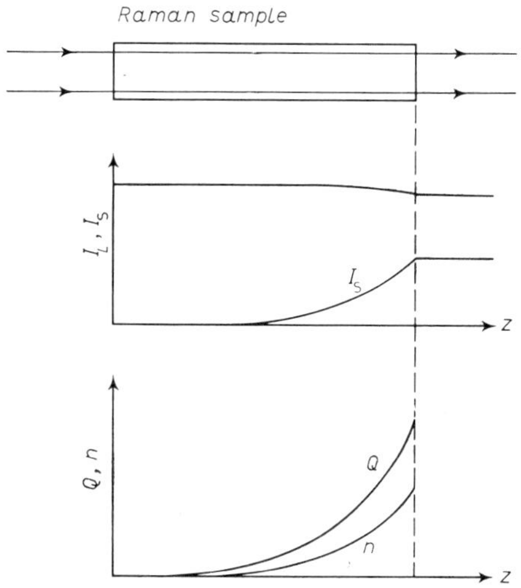

Fig. 10. – Light pulses and material excitation in stimulated Raman scattering as a function of the interaction length z.

In this paper we are interested in the material excitation. As indicated in the lower part of fig. 10, there are two types of material excitation. First, a coherent material excitation of amplitude Q is built up in the stimulated-scattering process. As a result, there exists a well-defined phase relationship between molecules in different parts of the excited volume. Second, an excess population n is generated in the first excited vibrational state $v = 1$, which is highest at the end of the interaction path. The differential equations which govern the stimulated Raman process are presented in fig. 11. From the top

$$\frac{\partial^2 E}{\partial z^2} - \frac{1}{c^2}\frac{\partial^2(\varepsilon E)}{\partial t^2} = \frac{4\pi}{c^2}\frac{\partial^2 P}{\partial t^2},$$

$$P = N\frac{\partial\alpha}{\partial Q}QE,$$

$$\frac{\partial^2 Q}{\partial t^2} + \frac{1}{\tau}\frac{\partial Q}{\partial t} + \omega_0^2 Q = \frac{1}{2m}\frac{\partial\alpha}{\partial Q}E^2(1-2n),$$

$$\frac{\partial n}{\partial t} + \frac{1}{\tau'}n = \frac{1}{2\hbar\omega_0}\frac{\partial\alpha}{\partial Q}E^2\frac{\partial Q}{\partial t}.$$

Fig. 11. – Differential equations of transient stimulated Raman scattering.

to the bottom of the picture we find i) the wave equation for the total electromagnetic field, ii) a polarization equation which contains the change in polarizibility with normal co-ordinate $\partial\alpha/\partial Q$, iii) a damped oscillator equation for the vibrational mode with coherent amplitude Q and iv) a relaxation equation for the excess population n. It is important to emphasize the two different time constants τ and τ', which are called dephasing time and population relaxation time, respectively. The experimental determination and the understanding of these time constants is the essential part of this work.

We have made a detailed theoretical study of the differential equations of fig. 11 [15]. Of special interest for our investigations is the transient excitation where the pulse duration t_p is of the order of the dephasing time τ. In this case the maximum of the material excitation (Q and n) occurs after the maximum of the incident pulse has passed a certain position within the

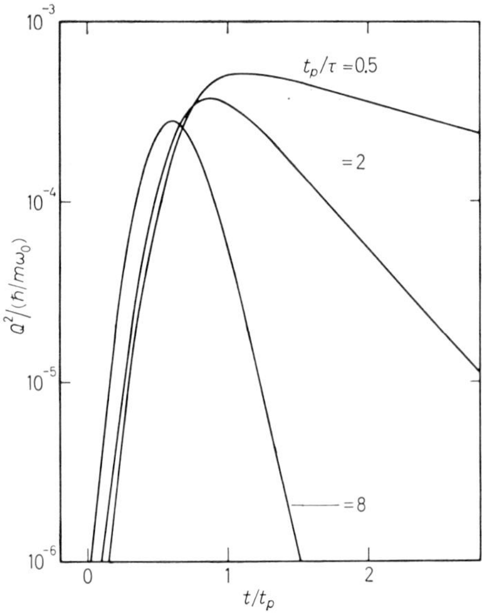

Fig. 12. – Coherent vibrational excitation Q^2 in stimulated Raman scattering as a function of time for three values of the ratio of pulse duration t_p to dephasing time τ.

sample. Some numerical results of Q^2 as a function of time are depicted in fig. 12 for three values of t_p/τ. It is readily seen from the figure that the coherent excitation rises rapidly during the pumping process and decays during the subsequent free relaxation with the characteristic time constant τ. Figure 13 shows—on a linear plot—the pump pulse of duration t_p (short-dashed line), the rise and decay of coherent excitation Q^2 (solid line) and the probing pulse (long-dashed curve) which interrogates the instantaneous degree of excitation

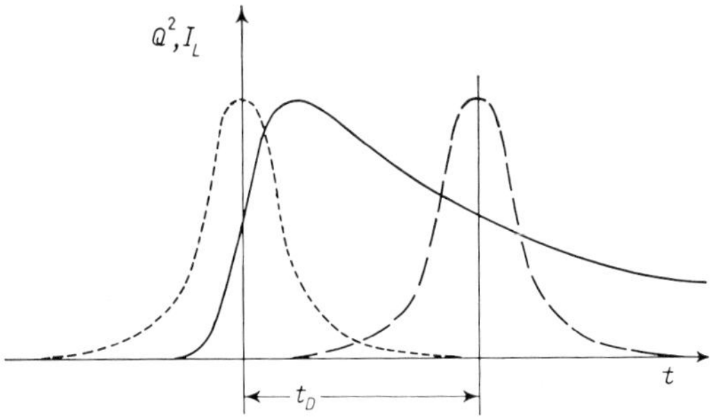

Fig. 13. – Schematics of the probe scattering process; exciting light pulse (short-dashed line), coherent excitation Q^2 (solid curve), probing light pulse (long-dashed line). $t_p/\tau \gg 1$.

after a delay time t_D. The measured probe scattering signal is a convolution of the probe pulse with the coherent excitation Q^2 at time t_D.

The experimental system for the determination of the dephasing time τ is schematically depicted in fig. 14 [10]. The incident powerful laser pulse ex-

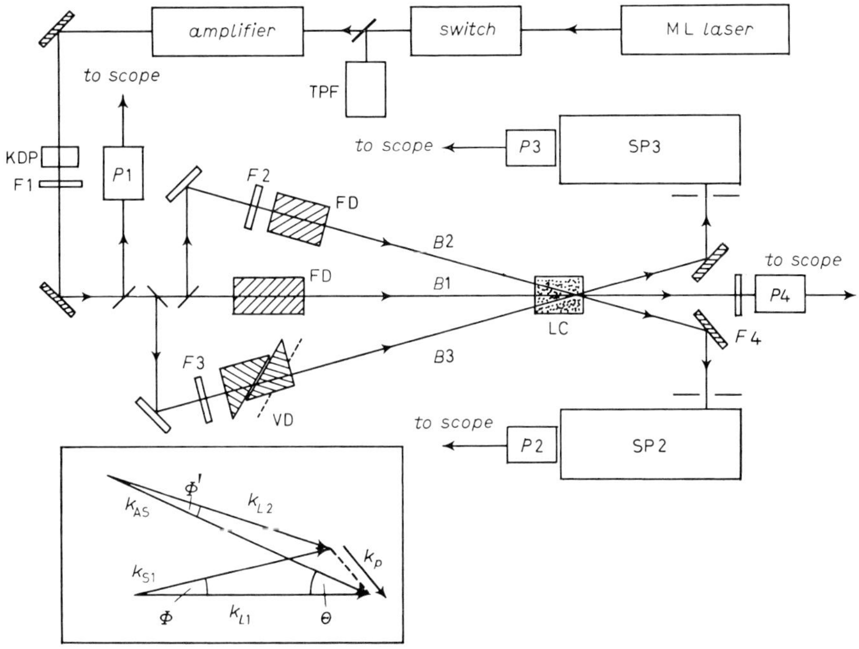

Fig. 14. – Schematics of the experimental system measuring the vibrational dephasing time. Three light beams $B1$, $B2$, $B3$ interact in the sample LC; fixed delays FD, variable delay VD, filters F, spectrometer SP, photodetector P, two-photon fluorescence system TPF. Inset: scheme of noncollinear phase-matched probe anti-Stokes scattering. k_{L1} and k_{L2} give the directions of the exciting and probing pulse, respectively.

cites one normal mode of the material in the liquid cell LC, while a weak probing pulse of variable delay VD monitors the excited volume. A second probe pulse with constant delay serves as a reference beam to improve the accuracy of the system. The insert in the lower left-hand corner of the figure indicates the phase-matching geometry. The pump laser and Stokes pulses generate a coherent material excitation of wave vector k_p. The probe pulse with wave vector k_{L2} scatters on the material excitation producing a coherent anti-Stokes pulse of wave vector k_{AS}. In practical cases the angles between the various wave vectors are small, *i.e.* the coherent-scattering signal is observed close to the forward direction.

We have measured the coherent-scattering signal of a variety of molecules [16-18]. One normal mode of vibration was excited for each molecular system. The mode with the largest scattering cross-section and, as a result, with the largest gain dominates in the stimulated-excitation process. Two typical experimental results are presented in fig. 15 and 16. The coherent anti-Stokes scattering signal is plotted as a function of the delay time between pump and probe pulses. Time zero marks the maximum of the pump pulse. The

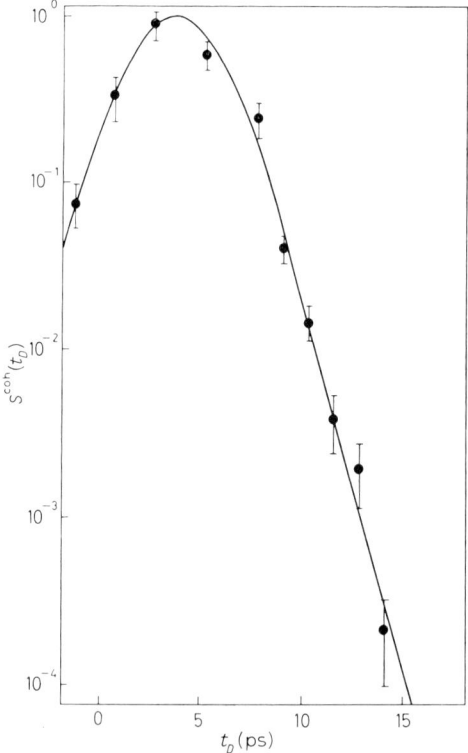

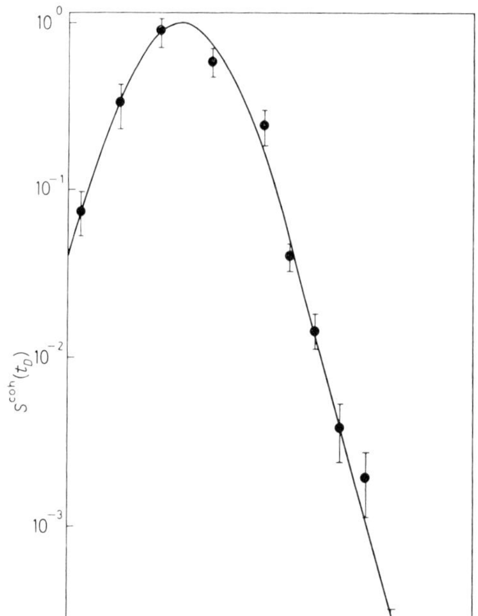

Fig. 15. – Coherent anti-Stokes probe scattering signal *vs.* delay time between pump and probe pulse. Symmetric CH_3 stretching vibration of CH_3CCl_3 at $\tilde{\nu} = 2939\ cm^{-1}$. $T = 295$ K, $\tau = (1.0 \pm 0.2)$ ps.

scattered probe signal rises to a delayed maximum and decays with a time constant characteristic of the specific molecular vibration. Relaxation times of 75 ps and 3.6 ps are deduced from the experiments for the fundamental vibration of N_2 at 2326 cm^{-1} and for the symmetric CH_3 stretching vibration of

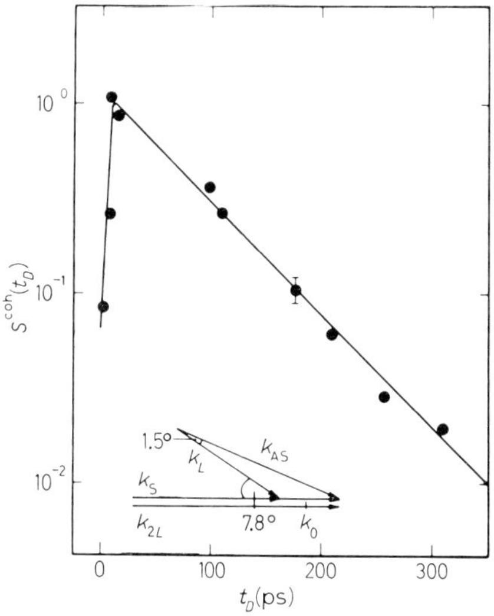

Fig. 16. – Anti-Stokes probe scattering signal $S^{coh}(t_D)$ *vs.* delay time for the fundamental vibration of N_2 (polarized scattering component); inset: wave vector diagram of the pump and probe scattering processes (schematic). $T = 77$ K, $\tau = (75 \pm 8)$ ps.

CH_3CCl_3 at 2939 cm^{-1}, respectively. In fig. 17 dephasing times τ obtained from coherent probe scattering are summarized for seven different molecules. The vibrational frequencies cover a range from 370 cm^{-1} to $\sim$3000 cm^{-1}. For polyatomic molecules we find time constants of several picoseconds at room temperature. It is interesting to compare our dephasing times τ with time constants calculated from the line width of spontaneous Raman scattering ac-

	$\tilde{\nu}_0$ (cm^{-1})	τ (10^{-12} s)	
		from line width	measured
N_2 (77 K)	2326	79 $\pm$ 8	75 $\pm$ 8
$SnCl_4$	368	2.5 $\pm$ 0.5	2.8 $\pm$ 0.3
$SiCl_4$	425	2.8 $\pm$ 0.5	3.0 $\pm$ 0.5
CCl_4	459	3.8 $\pm$ 0.5	3.6 $\pm$ 0.5
CH_3CCl_3	2939	1.1 $\pm$ 0.1	1.0 $\pm$ 0.2

Fig. 17. – Comparison of experimental data on the vibrational dephasing time with values estimated from spontaneous Raman line width data for a number of liquids.

cording to the equation $\delta\tilde{\nu} = (2\pi c\tau)^{-1}$. The good agreement between the two time values indicates that the Raman line width of the vibrations listed here is determined by dephasing processes. Very recently a mechanism for phase relaxation based on semi-classical collision theory was studied by FISCHER and LAUBEREAU [19]. Due to molecular interaction in elastic collisions the vibrating molecules are shown to experience a change of the vibrational frequency during the collision. The quasi-elastic collision model was treated quantum mechanically. Comparison between numerical phase relaxation times and experimental data for τ give good agreement within a factor of approximately two for a number of liquids.

It was pointed out above that, for relatively long pump pulses of $t_p > \tau$, the Raman excitation process is highly selective. Only the molecular vibration with the largest scattering cross-section is excited. The situation is different for the highly transient stimulated Raman process where $t_p \lesssim \tau$. In this case it is possible to excite simultaneously vibrational modes of small frequency differences $\delta\omega$ [18]. It can be shown that neighbouring vibrational modes are excited with equal amplitude and phase when the duration of the incident light pulse is sufficiently short, *i.e.* $t_p \ll 2\pi/\delta\omega$. After the short pump pulse has passed the medium, the collectively vibrating (isotope) components relax freely oscillating with their individual resonance frequencies. A beating between the vibrating systems is predicted, where maxima and minima occur after a time interval of $T = 2\pi/\delta\omega$.

Recently, we have observed this beating phenomenon of isotope components by measuring the coherent probe scattering as a function of delay time t_D. The probe pulse generates signals scattered in slightly different directions. The scattering directions result from the different vibrational frequencies of the isotope components. There are two possibilities to investigate the collectively

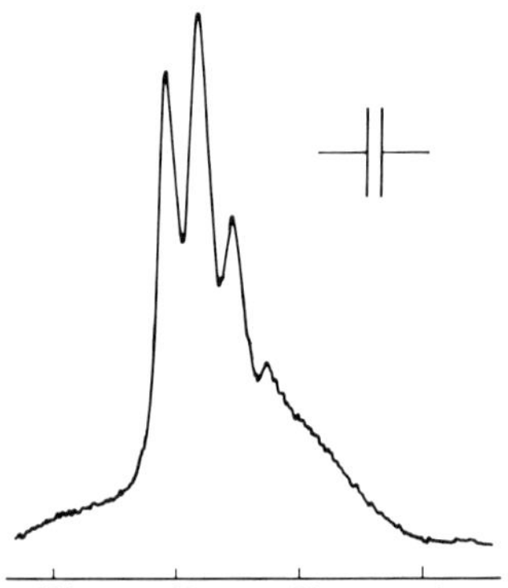

Fig. 18. – Isotope splitting of the totally symmetric tetrahedron vibration of CCl_4 in the spontaneous Raman spectrum. $C^{35}Cl_4$ 0.772, $C^{35}Cl_3{}^{37}Cl$ 1.0, $C^{35}Cl_2{}^{37}Cl_2$ 0.486, $C^{35}Cl^{37}Cl_3$ 0.105.

vibrating molecular systems: 1) a beam direction is selected where one vibrational component has a predominant scattering signal, 2) a direction is chosen where two (or more) beams have scattering intensities of comparable magnitude. In this case we observe, as a function of delay time t_D, strong beats of the coherent-scattering signal. This result was expected since each coherent probe beam reflects the phase information of the respective vibrational component.

In fig. 18 the spontaneous Stokes Raman spectrum of the totally symmetric tetrahedral vibration of common CCl_4 is depicted. The three major peaks correspond to the three most abundant molecular components, which are listed in the caption. A line splitting is observed of $\delta\omega/2\pi c \simeq 3\ cm^{-1}$. We have excited common CCl_4 by transient stimulated Raman scattering using a pump pulse of approximately 3.5 ps duration. In fig. 19 the coherent probe scattering signal

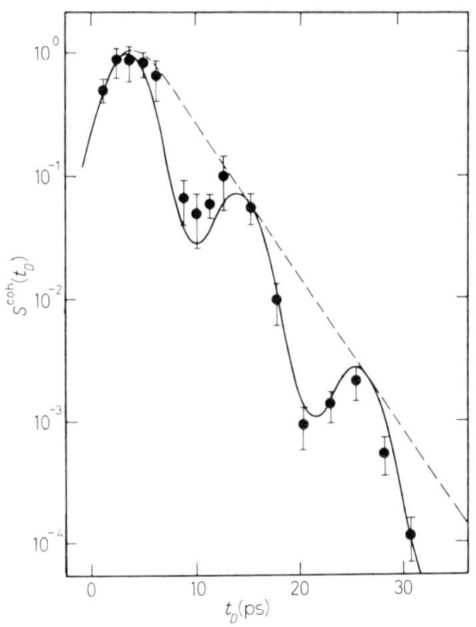

Fig. 19. – Coherent-scattering signal of common CCl_4 (295 K) representing the beating of the isotope components separated by 2.9 cm^{-1}. The broken line indicates the dephasing of one isotope component. $\tilde{\nu} \sim 459\ cm^{-1}$.

of CCl_4 is presented, which shows clearly the collective beating of the isotope species with $T = 11.5$ ps [18]. The beat frequency is deduced to be $\delta\omega/2\pi c = (2.9 \pm 0.15)\ cm^{-1}$ in good agreement with spontaneous data. The curve is calculated from the known parameters of CCl_4 and a dephasing time $\tau = (3.6 \pm 0.5)$ ps which was determined in an independent experiment where one isotope component was investigated separately. The good agreement between the calculated curve and our experimental points should be noted.

We now turn to the discussion of a series of experiments to study the population lifetime τ' of the $v = 1$ state of a normal mode of vibration in the electronic ground state. In a first group of investigations, the molecules were excited via stimulated Raman scattering and the degree of excitation was monitored by spontaneous anti-Stokes scattering of a delayed probe pulse. In a second experimental system ultra-short infra-red pulses excite directly the vibrational mode of interest and a second properly delayed probe pulse monitors the excitation by means of a transition to a fluorescent electronic state. The second technique is described below.

An experimental set-up for measuring the population relaxation time τ' is depicted schematically in fig. 20 [16]. A single laser pulse enters the system

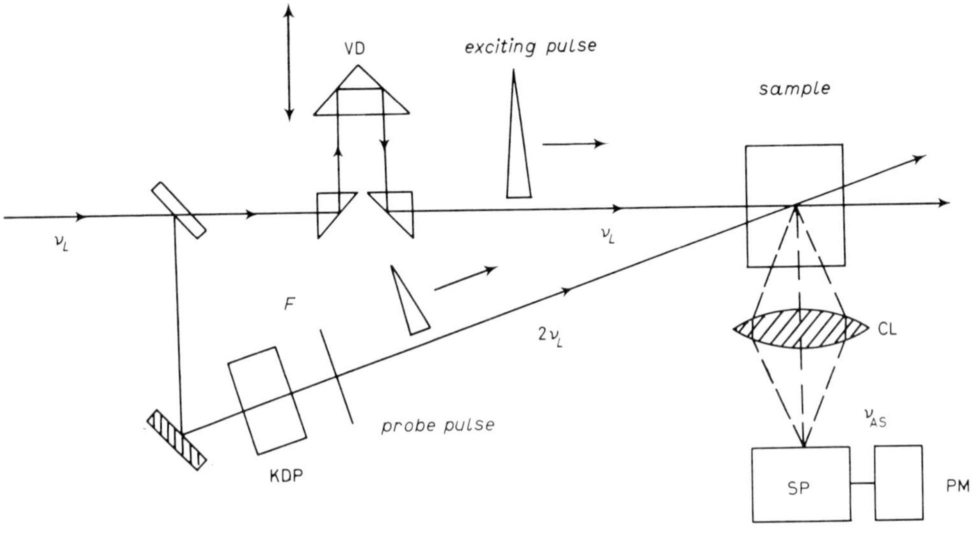

Fig. 20. – Experimental system for the measurement of the energy relaxation time τ' of a molecular vibration in liquids: variable delay VD, frequency-doubling crystal KDP, collecting lens CL for the incoherent probe scattering, spectrometer SP and photomultiplier PM.

from the left and produces vibrational excitation of the first excited $v = 1$ state. A second pulse generated by the beam splitter serves as the probe pulse to interrogate the excited volume after a delay time t_D. Spontaneous anti-Stokes scattering is measured at a scattering angle of 90° with the help of a spectrometer and of a photomultiplier. This signal is directly proportional to the momentary degree of excitation of the $v = 1$ state.

In fig. 21 the (incoherent) spontaneous anti-Stokes signal is plotted (full points) *vs.* t_D for ethanol. An energy relaxation time τ' of 22 ps is deduced from the data [16, 20]. It is readily seen from the figure that the coherent (phase matched) scattering signals (open circles) decrease much faster with time, *i.e.* the dephasing time is much shorter than the energy relaxation of the same CH_3 stretching vibration of ethanol.

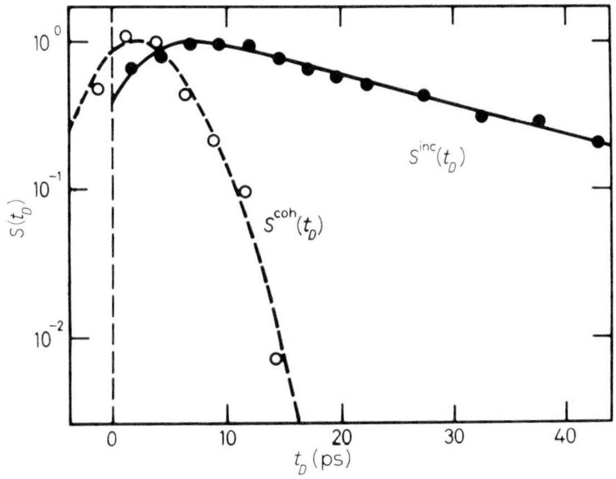

Fig. 21. – Measured incoherent scattering $S^{inc}(t_D)/S^{inc}_{max}$ (closed circles) and coherent scattering $S^{coh}(t_D)/S^{coh}_{max}$ (open circles) *vs.* delay time t_D for ethyl alcohol. The solid and dashed curves are calculated.

The concentration dependence of the energy (population) relaxation time τ' of the ν_H stretching mode was studied in several $CH_3Cl_3:CCl_4$ mixtures with mole fractions $x = 1$, 0.8, 0.6 and 0.4 of CH_3CCl_3 (see fig. 22). It was found that the energy relaxation time of the ν_H mode of CH_3CCl_3 at $\nu \sim 3000\ cm^{-1}$

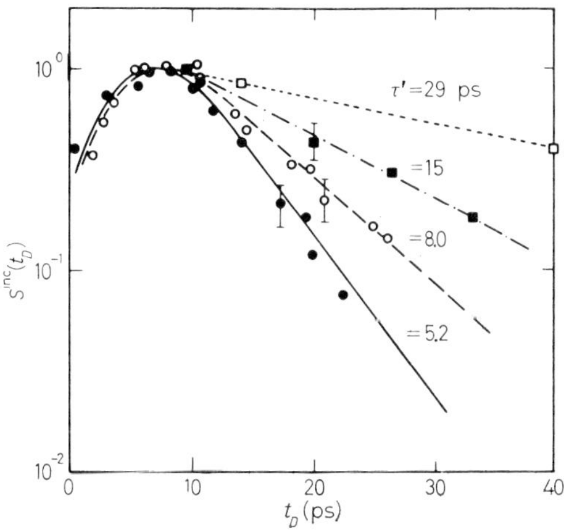

Fig. 22. – Incoherent probe signals of the ν_H vibration *vs.* delay time t_D in various mixtures $CH_3CCl_3:CCl_4$. Mole fractions of CH_3CCl_3, $x = 0.4$ (□), 0.6 (■), 0.8 (○) and 1.0 (●). $\tilde{\nu}_H = 2939\ cm^{-1}$.

increases strongly from 5.2 ps to 29 ps when x decreases from 1.0 to 0.4 (see fig. 22) [21]. Since the vibrational ($v=1$) levels of CCl_4 are small ($\nu \leqslant 800\ cm^{-1}$) compared with the excited ν_H vibration ($\nu \simeq 2900\ cm^{-1}$), the solvent CCl_4 is considered to be an energetically « inert » molecule in the mixture. It is interesting to note that the Raman line width of the ν_H mode of CH_3CCl_3 (a measure of the dephasing time in this molecule) does not change by more than 10% in the same mixtures studied, *i.e.* line width measurements do not allow the determination of the strong concentration dependence of τ' found in these investigations.

The question now arises as to what physical processes determine the energy relaxation τ'. Figure 23 serves to illustrate possible decay routes and energy transfer processes. The normal-mode frequencies of these molecules are shown.

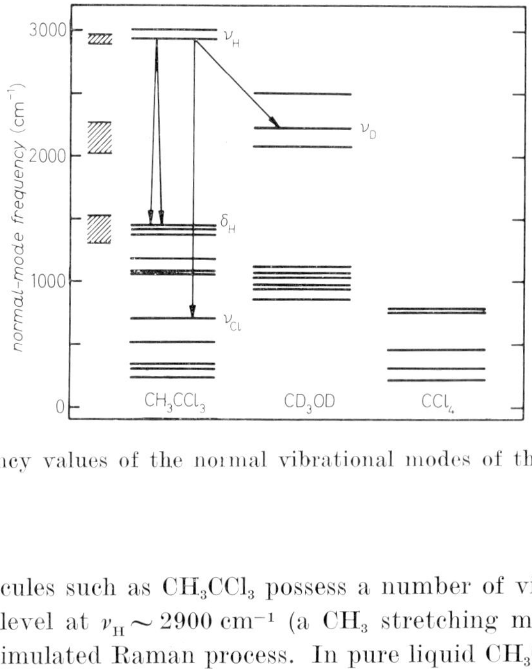

Fig. 23. – Frequency values of the normal vibrational modes of the liquids CH_3CCl_3, CD_3OD and CCl_4.

Polyatomic molecules such as CH_3CCl_3 possess a number of vibrational modes. The vibrational level at $\nu_H \sim 2900\ cm^{-1}$ (a CH_3 stretching mode) is primarily excited by the stimulated Raman process. In pure liquid CH_3CCl_3 this ν_H mode couples effectively to the CH_3 bending mode δ_H at approximately half the energy. The decay process $\nu_H \to 2\delta_H$ is indicated by the arrows in the figure. This energy decay route generates a subsequent excess population of the δ_H modes, which was experimentally observed [21]. The spectrometer was tuned to the corresponding anti-Stokes frequency position in our incoherent-scattering experiment.

Direct information on intermolecular energy transfer was obtained in the mixture CH_3CCl_3:CD_3OD. Figure 23 shows the energy levels of the solvent CD_3OD. The CD_3 stretching vibration of this molecule at $\nu_D = 2227\ cm^{-1}$ and the frequency $\nu_{Cl} = 713\ cm^{-1}$ of a CCl_3 vibration of CH_3CCl_3 add up with good accuracy to the ν_H mode of CH_3CCl_3, which is experimentally excited in the mixture. An interaction between CH_3CCl_3 and CD_3OD molecules of the form $\nu_H \to \nu_D + \nu_{Cl}$ is suggested by energy resonance arguments (see the arrows in fig. 23).

Experimental data on the intermolecular energy transfer from the laser-excited ν_H vibration of CH_3CCl_3 to the ν_D vibration of the solvent molecule are presented in fig. 24 [21]. The data of the ν_H vibration (open circles) of CH_3CCl_3

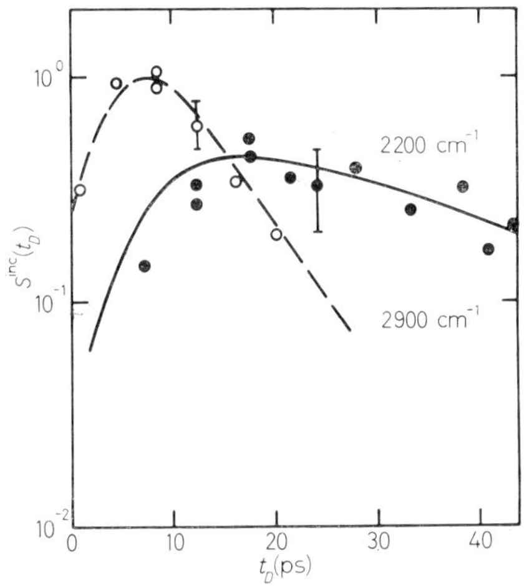

Fig. 24. – Measurement of the intermolecular energy transfer in the mixture CH_3CCl_3:CD_3OD; incoherent anti-Stokes scattering (measuring the population of the first excited vibrational state) is plotted *vs.* delay time t_D. Open circles denote probe scattering from the CH_3 valence bond vibration (2939 cm^{-1}) of CH_3CCl_3, which is excited by the pump pulse. Full points represent scattering observed from the CD_3 valence bond vibration (~ 2200 cm^{-1}) of CD_3OD, which directly indicates intermolecular transfer of vibrational energy.

have been discussed above. Of special interest is the incoherent-scattering signal observed at a frequency shift of ~ 2200 cm^{-1} (full points), which originates from the subsequent excess population of the ν_D vibration of CD_3OD. The data display a delayed maximum with a slow decay of (25 ± 10) ps. This experiment gives convincing evidence of energy transfer from the CH_3CCl_3 to the CD_3OD molecules. The results of fig. 24 represent the first observation of intermolecular transfer of vibrational energy in liquids. The curves in fig. 24 are calculated from the theory of transient stimulated Raman scattering exciting the ν_H mode (broken line) and from a rate equation model for the energy transfer to the ν_D vibration (solid curve). If we know the values of the spontaneous cross-sections of the two vibrations, a comparison of the scattering signals yields a quantum efficiency of approximately 60% for the intermolecular transfer process. Our study of the concentration dependence of the energy relaxation of

CH_3CCl_3 in mixtures with CCl_4 and CD_3OD show that the $\nu_H \rightarrow 2\delta_H$ interactions have an efficiency of 40 %.

Figure 25 shows some of the normal modes of ethanol. In the frequency range around 3000 cm^{-1} we have five CH stretching vibrations and around 1500 cm^{-1} there are four CH bending modes of the molecule. The $v = 2$ states of the CH bending modes are also indicated in the figure. Now we wish to answer

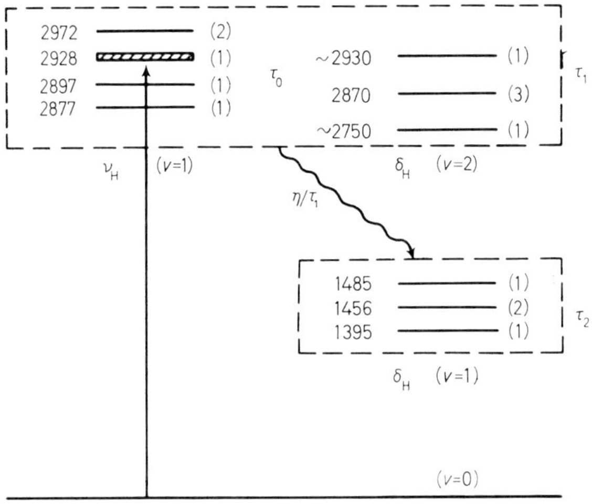

Fig. 25. – Schematics of relevant vibrational states of ethanol CH_3CH_2OH.

the following question. When one CH stretching mode is excited above its equilibrium value, is there energy transfer to neighbouring energy states? In particular, is there energy transfer to the $v = 2$ states of the CH bending modes? In order to answer this question we have performed three independent experiments which will be discussed next. In fig. 26 (top) the spontaneous Raman Stokes spectrum of ethanol in the range around 3000 cm^{-1} is presented. This spectrum is readily obtained with an argon laser in conjunction with a spectrometer. Three distinct energy states are clearly visible, a fourth one is hidden around the shift of 2900 cm^{-1}, a fifth vibrational state is a degenerate state. We have applied our picosecond probe technique to measure the degree of vibrational excitation of different modes in the frequency range of 3000 cm^{-1}. In fig. 26 (bottom) the spontaneous anti-Stokes spectrum of probe pulses delayed by $t_D = 11$ ps with respect to the maximum of the input pulse is depicted [20]. Considering the duration of the excitation pulse (here 8 ps) and the transient build-up of the population of $v = 1$, we estimate that the spectrum was taken several picoseconds after the maximum excitation of the pumped mode at 2928 cm^{-1}. It is readily seen from fig. 26 (bottom) that the spectrum extends over a frequency range of approximately 150 cm^{-1}, indicating the population of neighbouring vibrational energy states. The curve drawn through our experimental points is calculated under the assumption that the five vibrational modes are populated. The spectral resolution of the specific experiments was taken into account. The good agreement between experimental points and calculated curve should be noted. In fig. 27 we discuss the spectral range around

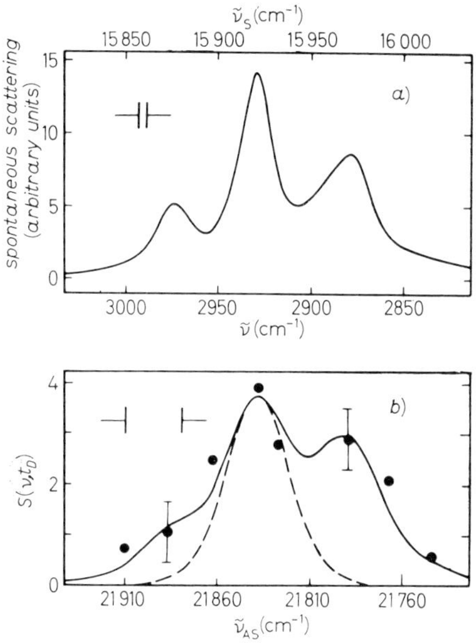

Fig. 26.

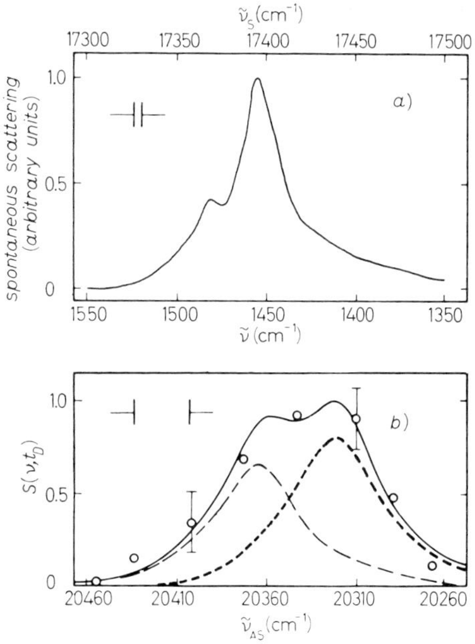

Fig. 27.

Fig. 26. – *a*) Spontaneous Stokes spectrum of the CH stretching vibrations. *b*) Anti-Stokes probe scattering of CH stretching vibrations at $t_D = 11$ ps. The broken line indicates the spectral profile of the pumped mode at $2928\ \text{cm}^{-1}$. The solid curve is calculated for quasi-equilibrium between levels around $2900\ \text{cm}^{-1}$.

Fig. 27. – *a*) Spontaneous Stokes spectrum of the CH bending modes δ_H ($v = 1$). *b*) Anti-Stokes probe scattering around $1400\ \text{cm}^{-1}$. $t_D = 19$ ps. Calculated curves for δ_H ($v = 1$) scattering (long-dashed line) and δ_H ($v = 2$) scattering (short-dashed line).

$1500\ \text{cm}^{-1}$. At the top the spontaneous Raman spectrum of the CH bending modes is presented with the mode $1456\ \text{cm}^{-1}$ having the largest scattering cross-section. The experimental anti-Stokes spectrum obtained by our probe pulses with a time delay of 19 ps after the excitation pulse exhibits an interesting spectral distribution [20]. The observed frequency profile is broad with a shift towards smaller frequencies. For an analysis we calculated the convolution of the spectrum of fig. 27 (top) with our instrumental function (broken curve). Obviously there is additional scattering around $1400\ \text{cm}^{-1}$, which is taken as evidence for anti-Stokes scattering from the δ_H ($v = 2$) states. The shift to smaller frequencies results from the anharmonicity of the δ_H bending mode. The scattering cross-section of $v = 2 \to v = 1$ scattering is known to be twice that of $v = 1 \to v = 0$ scattering. We conclude from this experiment that, approximately 20 ps after the excitation, the $v = 2$ and $v = 1$ states have a definite excess population due to energy transfer and relaxation processes. A third experiment concerning the energy transfer problem in ethanol is depicted in fig. 28 [22]. Here the anti-Stokes probe signal around 2900 cm^{-1} is plotted *vs.* delay time for a $CH_3CH_2OH:CCl_4$ solution with 0.04 M concentration of ethanol.

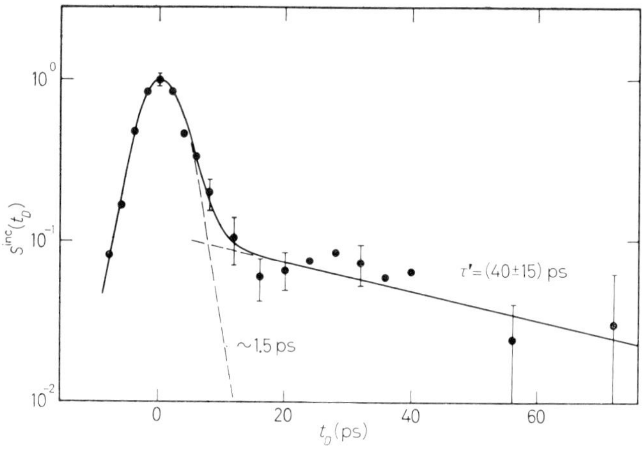

Fig. 28. – Incoherent probe scattering signal of ethanol (solvent CCl_4) around 2900 cm^{-1} *vs.* delay time (295 K). The CH_3 mode at 2928 cm^{-1} is primarily excited by an infra-red pulse of approximately the same frequency. $x = 0.04$ M.

The probe signal was measured over a total band width of 90 cm^{-1}. The CH_3 mode at 2928 cm^{-1} was primarily excited by a short (3 ps) infra-red pulse of the same centre frequency. The scattered signal shows a very rapid decay with a time constant of one to two ps and a slower decay with $\tau' = 40$ ps. We interpret this result as follows. The rapid decrease of the scattered signal is due to the fast energy transfer to neighbouring stretching ($v = 1$) and bending ($v = 2$) levels, which gives rise to a reduced light scattering around 3000 cm^{-1} of the magnitude observed. The longer time constant of 40 ps corresponds to energy decay to the lower bending modes around 1500 cm^{-1} [23], the time constant of which was observed to be 22 ps in pure ethanol [20]. We conclude about our observations on ethanol that fast energy transfer between neighbouring energy states appears to occur in this molecule with a slower energy decay to lower energy states.

Very recently we have devised an experimental system for the study of molecular vibrations in highly diluted systems. This technique is particularly suited for investigations of vibrational modes in the electronic ground state of fluorescent molecules. A scheme of our set-up is depicted

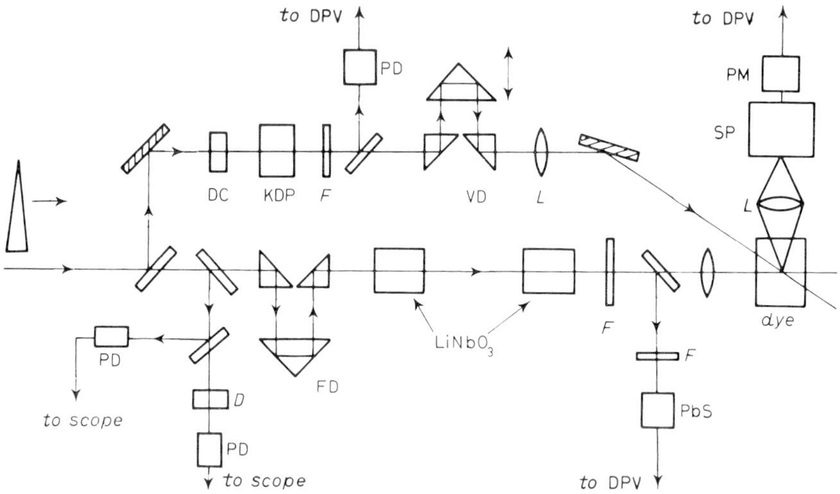

Fig. 29. – Schematics of the experimental system measuring the energy relaxation time of a molecular vibration after IR excitation with the help of the fluorescence probing technique.

in fig. 29 [24]. The single picosecond laser pulses enter the system from the left and generate the desired infra-red pulse traversing two properly oriented $LiNbO_3$ crystals. A small part of the input pulse produces a green probe pulse at $2\nu_L$ in the KDP crystal. The fluorescence of the sample is measured with spectrometer and photomultiplier. The relevant transitions involved in our experiment are schematically illustrated in fig. 30. A first short infra-red pulse

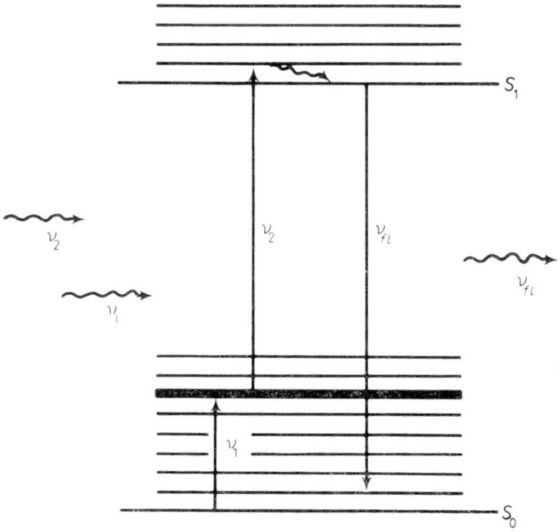

Fig. 30. – Fluorescence probing technique. Schematics of the molecular energy levels and transitions during the vibrational excitation and probing process.

excited the molecular system via infra-red absorption. The frequency ν_1 of the infra-red pulse is properly selected in order to interact only with one well-defined vibrational mode. Our pump pulse has a duration of approximately 3 ps and contains 10^{14} infra-red quanta at a frequency of approximately 3000 cm^{-1}. The molecular system is subsequently interrogated by a second pulse of frequency ν_2. The second probe pulse promotes molecules which are vibrationally excited to a level close to the vibrational ground state of the first excited singlet state S_1. The fluorescence originating from this singlet state is experimentally observed. It should be noted that energy levels smaller than $h\nu_1$ (in the electronic ground state) do not interact with the probing pulse of frequency ν_2. The time-integrated fluorescent radiation serves as a direct measure of the instantaneous vibrational excitation of the energy level $h\nu_1$. The fluorescence signal is measured as a function of delay time t_D between the infra-red excitation pulse and the probing light pulse. With this technique it is possible to study well-defined vibrational excitations even in large molecules with complicated vibrational spectra. Of special interest is the high sensitivity of this method for molecules with high fluorescent quantum efficiency.

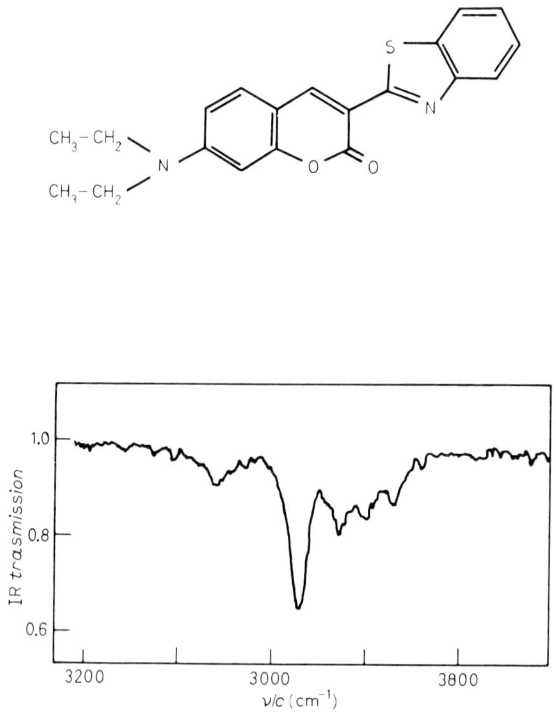

Fig. 31. – Top: the coumarin 6 molecule. Bottom: infra-red absorption spectrum of coumarin 6 in CCl_4 around 3000 cm^{-1} (4×10^{-4} M). The absorption band at 2970 cm^{-1} is excited in the picosecond experiment.

We have applied this technique to dynamic investigations of dye solutions in a large concentration range of 10^{-6} M to 10^{-3} M [24]. As an example, we discuss results on the molecule coumarin 6 in the solvent CCl_4. The coumarin 6 molecule is depicted in the upper part of fig. 31. Two C_2H_5 groups are bonded to the conjugate ring system. The infra-red transmission spectrum of coumarin 6 around 3000 cm^{-1} is presented in the lower part of fig. 31. Several absorption peaks are clearly resolved in the spectrum. The bands between 2865 cm^{-1} and 2970 cm^{-1} are readily interpreted as normal vibrational modes of the two ethyl groups of coumarin 6. The CH valence bond vibrations of the ring system produce the absorption peak at 3055 cm^{-1}. Most important for our investigations is the absorption maximum at 2970 cm^{-1}, which corresponds to the asymmetric CH_3 mode. It is this vibrational mode which is populated by the resonant short infra-red pulse. The degree of vibrational excitation is subsequently monitored as a function of time by our fluorescence technique. As frequency of the probing pulse we choose $\nu_2 = 2\nu_L = 18\,910$ cm^{-1} (0.53 μm), *i.e.* the second harmonic of the Nd-glass laser. In fig. 32 the absorption and fluorescence spectra of coumarin 6 are presented; the probing frequency ν_2 is marked at the abscissa.

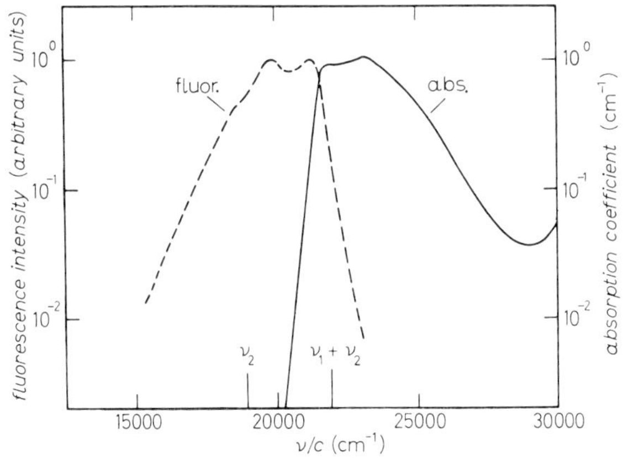

Fig. 32. – Absorption and fluorescence spectra of coumarin 6 (10^{-5} M in CCl_4). The frequency of the probe pulse ν_2 and the sum frequency $\nu_1 + \nu_2$ are indicated.

The figure shows that the frequency of the probing pulse is off the steeply rising absorption edge. As a result the probe pulse gives rise to a small fluorescent background signal.

Experimental results of our two-pulse techniques are presented in fig. 33 for coumarin 6 in CCl_4 for a concentration of 3×10^{4} M. The observed fluorescence signal $S(t_D)$ initiated by the probe pulse is plotted as a function of the delay time t_D for two temperatures, 253 K and 295 K. The time scale of a few ps should be noted. The fluorescence signal sharply increases within several ps to a maximum value at $t_D \simeq 2$ ps. The build-up of excess population in the vibration spectrum at $\nu_1 = 2970\ \text{cm}^{-1}$ is directly seen from the rise of the signal curve. Of special interest are the decaying parts of the signal curves which extend over two orders of ten. Fast relaxation of the vibrational excitation with slightly different slopes for the two temperatures is clearly indicated by the data. At room temperature a time constant of (1.3 ± 0.4) ps is directly obtained from the exponential slope (open circles). At $-20\ °\text{C}$ the energy relaxation time

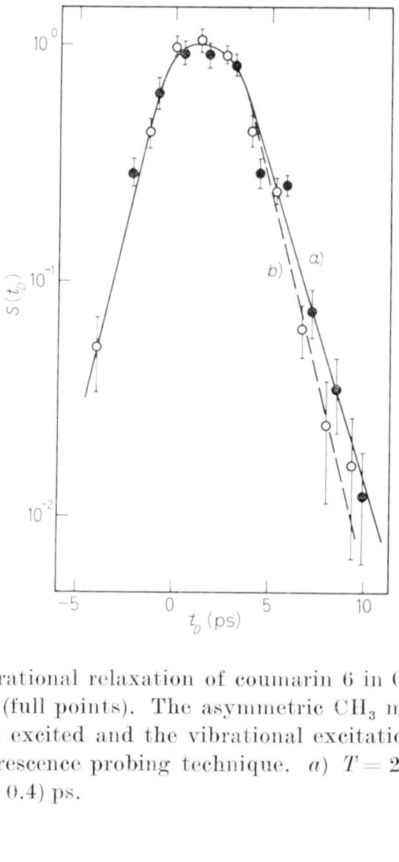

Fig. 33. – Ultra-fast vibrational relaxation of coumarin 6 in CCl_4 at 295 K (3×10^{-5} M) (open circles) and 253 K (full points). The asymmetric CH_3 mode at $2970\ \text{cm}^{-1}$ in the electronic ground state is excited and the vibrational excitation is observed as a function of time with a fluorescence probing technique. *a*) $T = 253$ K, $\tau' = (1.7 \pm 0.3)$ ps; *b*) $T = 295$ K, $\tau' = (1.3 \pm 0.4)$ ps.

is measured to be $\tau' = (1.7 \pm 0.3)$ ps. We have ascertained that these time constants represent true relaxation times. The liquid cell containing the coumarin molecules was replaced by a nonlinear crystal. The intensity of the sum frequency at $\nu_1 + \nu_2$ was measured as a function of the delay time between the two pulses of frequency ν_1 and ν_2. In this way the time resolution of our system was tested and found to be (0.4 ± 0.2) ps.

At the present time the energy relaxation process is not yet definitely determined. There are two possible physical mechanisms which lead to a decay of our fluorescence signal: 1) energy transfer to vibrational states of approximately equal energy with smaller absorption cross-sections to the excited electronic state, 2) energy decay to lower vibrational states which cannot interact with the probe pulse. Work is in progress to trace the relaxation process in more detail.

In summary we wish to emphasize that different experimental techniques are now available for the study of ultra-fast dynamical processes in liquids. In this paper, investigations are reviewed of dephasing times, of collective vibrational processes, of energy relaxation times and of intermolecular vibrational transfer of polyatomic molecules.

REFERENCES

[1] K. F. Herzfeld and T. A. Litovitz: *Absorption and Dispersion of Ultrasonic Waves* (New York, N. Y., 1959).
[2] R. G. Gordon: *Journ. Chem. Phys.*, **40**, 1973 (1964); **42**, 3658 (1965); **43**, 1302 (1965).
[3] S. Bratos, J. Rios and Y. Guissany: *Journ. Chem. Phys.*, **52**, 439 (1970).
[4] F. J. Bartoli and T. A. Litovitz: *Journ. Chem. Phys.*, **56**, 404 (1972); **56**, 413 (1972).
[5] D. von der Linde, O. Bernecker and W. Kaiser: *Opt. Comm.*, **2**, 149 (1970); D. von der Linde: *IEEE Journ. Quant. Electr.*, QE-**8**, 328 (1972).
[6] R. C. Eckardt, C. H. Lee and J. N. Bradford: *Appl. Phys. Lett.*, **19**, 420 (1971).
[7] D. von der Linde, O. Bernecker and A. Laubereau: *Opt. Comm.*, **2**, 215 (1970).
[8] A. Laubereau, L. Greiter and W. Kaiser: *Appl. Phys. Lett.*, **25**, 87 (1974).
[9] J. A. Giordmaine and R. C. Miller: *Phys. Rev. Lett.*, **14**, 973 (1965).
[10] D. von der Linde, A. Laubereau and W. Kaiser: *Phys. Rev. Lett.*, **26**, 954 (1971).
[11] A. Penzkofer, D. von der Linde, A. Laubereau and W. Kaiser: *Appl. Phys. Lett.*, **20**, 351 (1972); A. Penzkofer: *Opto-Electronics*, **6**, 87 (1974).
[12] R. L. Carman, F. Shimizu, C. S. Wang and N. Bloembergen: *Phys. Rev. A*, **2**, 60 (1970).
[13] S. A. Akhmanov: *Mater. Res. Bull.*, **4**, 455 (1969); S. A. Akhmanov, K. N. Drabovich, A. P. Sukhorukov and A. S. Chirkin: *Sov. Phys. JETP*, **32**, 266 (1971).
[14] For a review, see W. Kaiser and M. Maier: in *Laser Handbook*, edited by F. T. Arecchi and E. O. Schulz-Du Bois (Amsterdam, 1972).
[15] A. Laubereau and W. Kaiser: to be published.
[16] A. Laubereau, D. von der Linde and W. Kaiser: *Phys. Rev. Lett.*, **28**, 1162 (1972).
[17] A. Laubereau: *Chem. Phys. Lett.*, **27**, 600 (1974).
[18] G. Wochner, A. Laubereau and W. Kaiser: *Opt. Comm.*, **17**, 91 (1976).
[19] S. F. Fischer and A. Laubereau: *Chem. Phys. Lett.*, **35**, 6 (1975).
[20] A. Laubereau, G. Kehl and W. Kaiser: *Opt. Comm.*, **11**, 74 (1974).
[21] A. Laubereau, L. Kirschner and W. Kaiser: *Opt. Comm.*, **9**, 189 (1973).
[22] K. Spanner, A. Laubereau and W. Kaiser: to be published.
[23] R. R. Alfano and S. L. Shapiro: *Phys. Rev. Lett.*, **29**, 1655 (1972).
[24] A. Laubereau, A. Seilmeier and W. Kaiser: *Chem. Phys. Lett.*, **36**, 232 (1975).

Laser Spectroscopy of the F-Centres (*).

F. De Martini, D. Frigione, G. Giuliani, P. Mataloni and F. Simoni
Istituto di Fisica dell'Università - 00185 Roma, Italia

1. – Introduction.

The quantum structure and the physical properties of the colour centres in ionic alkali halides have been extensively investigated in the past four decades, mostly by conventional methods of linear optical spectroscopy in the presence of static or slowly varying electric or magnetic fields. (For a thorough presentation of previous work we refer the reader to [1-3].) In the present lectures we report the results of the first experimental investigation of the colour centres by a nonlinear optical method and the study of the stimulated (laser) emission by these centres. In the present section, we outline some aspects of the physics of F-centres that can be of interest for application in quantum optics, namely some aspects of the optical properties that make this system promising for laser applications. We shall devote our attention mainly to the « prototype » colour centre and the most simple one: the F-centre. In fact, we can say that most of the colour centres that are not related to the presence of impurities (*i.e.* foreign atoms) in the crystal can be considered as clusters of various sizes of F-centres or F-like centres with a more complex electronic structure (see fig. 1). The F-centre is an hydrogenlike system consisting of an electron bound to a Schottky defect of the crystal, *i.e.* a lattice vacancy or α-centre. Energetically the ground state of the centre lies ~ 3 eV below the conduction band of the crystal and it can be resonantly coupled by radiation to a discrete set of quantum states, which, for low orbital quantum number n, lie in the transparency band of the crystal. States with $n > 4$ generally overlap the conduction band and give rise to the so-called K and L bands of the centre [1]. The transition $1s$-$2p$ is the one which corresponds to the most intense absorption band of the spectrum, the F-absorption band. This band is the one that determines the typical hue of the crystal after colouration. Another bell-shaped Stokes-shifted band

(*) Work supported by Gruppo di Ricerca Elettronica Quantistica e Plasmi del C.N.R.

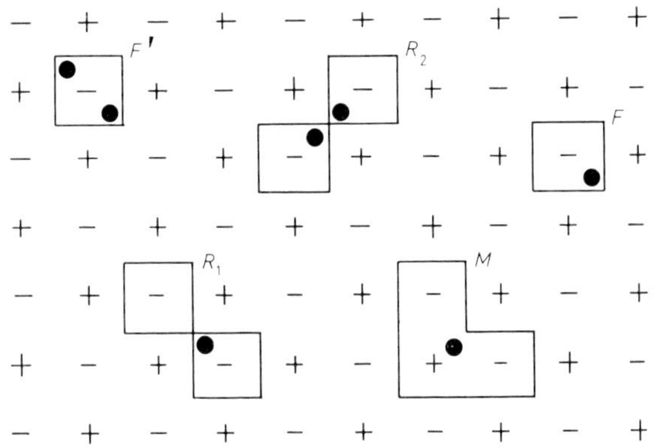

Fig. 1. – Structure of some colour centres in a ionic crystal with no impurities. A lattice vacancy is represented by a square surrounding a missing charge (ion).

characterizes the optical properties of the centre: it is the F-emission band which corresponds to the radiative de-excitation from the relaxed excited states (RES) to the unrelaxed electronic ground state following a « vertical » transition in a Franck-Condon-Seitz diagram of the centre. The absorption and emission spectra of the F and of the F_A centres in KCl are shown in fig. 2. The F_A-centre is also considered here, because it is the quantum system that

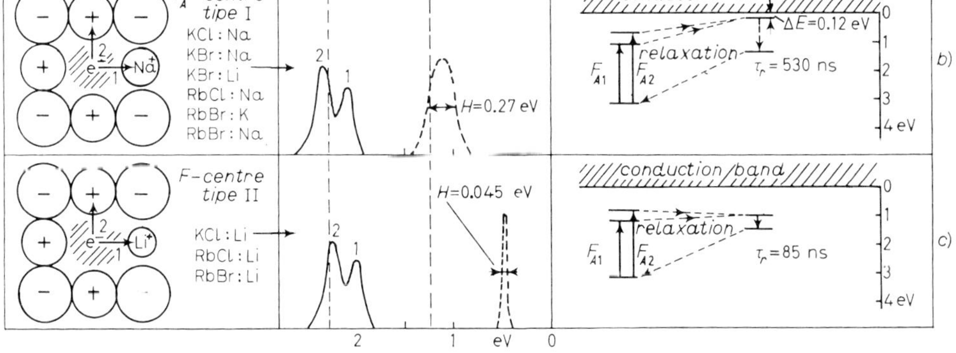

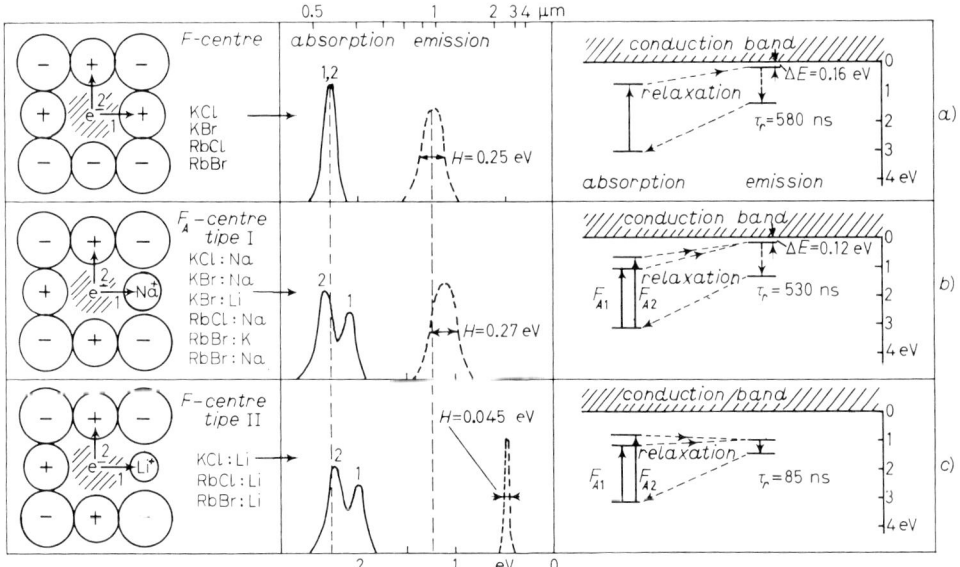

Fig. 2. – Structure and typical spectra of F, F_A (I), F_A (II) centres (from LÜTY in ref. [1]).

has given rise to the first laser action from colour centres and the one which so far is associated to the largest laser gain [4, 5]. Typically, the F-absorption band peaks in the visible spectrum and the emission band in the near infrared. In the table below we list the energies of the absorption and emission bands of the F-centre E_A, E_E together with the corresponding zero-point half-widths W_A, W_E and the fluorescence times τ_s from the RES in several crystals.

TABLE I. – *Optical parameters of the F-centre.*

	E_A	E_E	W_A	W_E	τ_s
NaF	3.72	1.665	0.37	0.39	0.1
NaCl	2.77	0.975	0.255	0.337	1.00
KF	2.85	1.66	0.23	0.39	0.21
KCl	2.30	1.2	0.163	0.26	0.55
KBr	2.06	0.915	0.160	0.215	1.11
KI	1.875	0.827	0.155	0.185	2.22
RbF	2.43	1.33	0.20	0.335	0.42
RbCl	2.05	1.090	0.145	0.237	0.60

The values of E_A, E_E, W_A, W_E are given in eV. The values of τ_s are given in microseconds. Further information and references are given in ref. [1], p. 627.

The dynamics of the F-centre can be described to a first approximation in terms of a Hamiltonian which is very similar to the one of a hydrogen atom. However, the similitude with an atomic system in the usual sense cannot be pushed too far in this case. The electron of an F-centre is in fact bound to a structural anomaly of the crystal itself with a classical orbit extending over several lattice constants. As a consequence, a very strong coupling exists between the motions of the electron and of the lattice with a large mutual dynamical interference. As a consequence, a perturbative approach to the dynamical problem is not adequate, as it is in general in the case of impurity atoms in a crystal, and only a complete vibronic theory can properly describe the dynamics of the system. The related mathematical problem is a formidable one and can be solved only when a significant approximation is introduced: the lattice motion is assumed to be slow in the time scale of the electron motion. This allows one to consider vertical transitions in radiative interactions, and the adiabatic theorem [6] is supposed to be valid at various degrees of approximation, when the simultaneous motions of the electron and of the lattice are considered. The assessment of a particular dynamical regime for the excited F-centre and, consequently, of the validity of a set of approximations allowing the solution of the Schrödinger equation for the vibronic system is of particular interest for the F-centre, because some anomalies affecting the process of spontaneous radiation emission from the RES are still partially unexplained (see the work of FOWLER in [1] and ref. [7]). In fact the structure of the relaxed excited states from which the fluorescent or stimulated decays originate is not a simple one, as will be shown in the following. It is strongly affected by the type of electron-phonon interactions and by the size of the vibronic couplings.

The above considerations outline a few aspects of an experimental and theoretical problem of solid-state physics that we believe to be important and quite fascinating.

2. – Nonlinear spectroscopy of the F-centre in KCl.

In the present section we report the first investigation on the relaxed excited states (RES) of the F-centre in an alkali-halide crystal by a 2-photon excitation technique [8]. The structure of the F^*-centre (F-centre in RES) in KCl is of particular interest, because of the rather complex processes that are found to take place during the crystal relaxation. The experiments of CHIAROTTI *et al.* [9], KUHNERT [10] and BOGAN *et al.* [11] have shown the evidence that in KCl the $2s$-state, which in absorption lies 0.1 eV above the $2p$-state, crosses the $2p$-state during the crystal relaxation and ends up lower in energy. This model is consistent with Bogan's assumption that the RES structure consists of nearly degenerate states $|2s'\rangle$ and $|2p'\rangle$, which result from an admixture of the $2s$ and $2p$ states, with $|2s'\rangle$ lower in energy. The exact nature and the strength of the mixing as well as the validity of the model itself could not have been directly ascertained by previous linear-spectroscopy experiments. We believe that a nonlinear method can supply new information on the system and contribute to clarifying its structure.

A 30 ns $\lambda = 1.06$ μm pulse generated by a Q-sw Nd-glass laser was focused on the surface of an additively coloured KCl crystal (dimensions $(1\times1\times1)$ cm^3) cooled to liquid-N_2 temperature (LNT) (fig. 3). The fluorescence arising from the F^* de-excitation to the ground ($1s$) state, following a Franck-Condon (F-C) scheme, was detected at 90° by a Bausch-Lomb 338625 spectrometer and a 56 CVP photomultiplier. The output signal was suitably gated before being processed by a photon-counting technique in order to discriminate against the large amount of laser photons scattered into the detection system. The laser intensity was kept during the experiment well under the dielectric break-down threshold for the crystal [12]. In this respect frequent checks were made using clear KCl specimens or with different F-centre concentrations N. It has been found that the fluorescence intensity is proportional to N, as expected. The crystal was frequently quenched at 700 K with rapid cooling at LNT in order to rule out any effect due to centers other than F. The number of fluorescence photons emitted over the entire band in a 4π solid state angle and following a 10 kW laser excitation (intensity 100 kW/cm^2)

has been evaluated to be $2 \cdot 10^8$ for $N = 1.3 \cdot 10^{16}$ cm^{-3}. This figure is in order-of-magnitude agreement with the theoretical results [8]. Details on the 2-photon excitation theory in an F-centre system will be reported elsewhere.

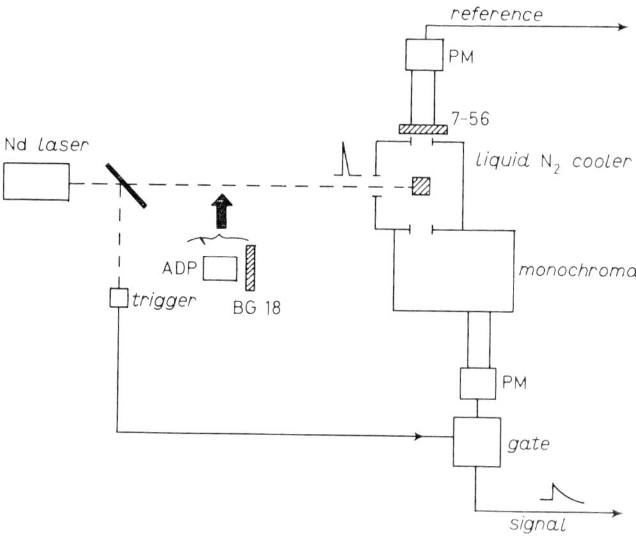

Fig. 3. – Experimental apparatus for two-photon spectroscopy of F-centres in KCl.

Two different experiments have been performed at successive times: *a*) the 2-photon experiment we have just described, in which the 1*s*-state is selectively coupled in centrosymmetric KCl to the 2*s*-state; *b*) the 1-photon absorption experiment, in which coupling occurs to the 2*p*-state by resonant excitation of the crystal in the F-band, which peaks at 2.3 eV at LNT [13]. A nonlinear ADP crystal inserted in the laser path generated a 2nd-harmonic pulse at $\lambda = 5300$ Å (2.34 eV). The insertion of the NL crystal and of a Shott BG 18 (green) filter in the laser path were the only changes we introduced between the two sets of measurements. The 1- and 2-photon resonance conditions were approximately satisfied because of the large line widths (~ 0.1 eV) of the nearly degenerate 2*s* and 2*p* states [9]. Experiment *b*) is similar to the one on the measurement of the F-centre laser gain in KCl we have recently reported [7].

In fig. 4 the fluorescence spectra obtained in experiments *a*) and *b*) are shown. In the 2-photon experiment, the peak power of the laser was 10 kW with an intensity of 10 kW/cm^2, while in the 1-photon case the peak power and intensity of the 2nd-harmonic pulse were approximately 100 times less. The 1-photon spectrum reproduces the well-known fluorescence spectrum of the F-centre [13], while the 2-photon spectrum is here reported for the first time.

Figure 4 shows a marked energy displacement between the two spectra, corresponding, in the case of vertical alignment in the F-C diagram of the minima of the potential wells from which the fluorescence occurs, to a large energy splitting $E_{sp} = 0.052$ eV between the corresponding relaxed states. The lack of any apparent interference between the two spectra (no bumps appear in the 1- and 2-photon bands, respectively at $\lambda \sim 0.96$ μm and $\lambda \sim 1$ μm) is

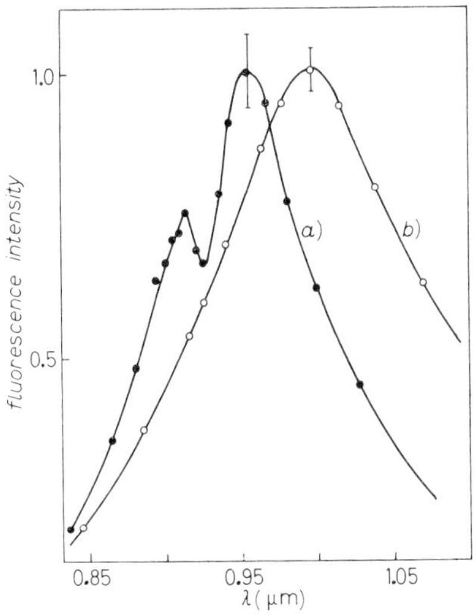

Fig. 4. – Fluorescence spectra obtained in a two-photon absorption experiment (curve a)) and in a one-photon absorption experiment (curve b)) in KCl ($T = 77°$ K). The two spectra are normalized to the same height, F-centre concentration $N = 1.3 \cdot 10^{16}$ cm^{-3}.

also physically significant, as it shows that: a) The probability of radiative decay from the 2-photon populated RES, referred to as $|2p'\rangle$-state, is much larger for transition to $|1s\rangle$ than to the nearby $|2s'\rangle$. Apart from the different values of the corresponding electric-dipole matrix elements, this effect may be simply understood on the basis of the dependence of the spontaneous emission probability on the transition frequency [1]. b) The amount of non-radiative transitions occurring between the states $|2s\rangle$ and $|2p\rangle$ during the crystal relaxation is surprisingly small in spite of the presence of an intense LO phonon field.

The 2-photon spectrum shows the evidence of an additional transition at $\lambda \sim 0.93$ μm, that may be attributed to a $3p$-$1s$ decay, if a new process is introduced which explains the 2-photon excitation of a relaxed $3p$-state. This process could be similar to the one proposed for $n = 2$ and consists of a mixing of $3s$ and $3p$ states [11]. This assignment is substantiated by the good agreement between the value of the $3p$-$2s'$ energy splitting recently measured [14] by IR spectroscopy, $E_{3p\text{-}2s'} = 0.102$ eV, and the one given by our experiment, $E_{2p\text{-}2s'} = 0.09$ eV.

Additional information on the RES structure is supplied by the direct measurement of the fluorescence decay times τ_p and τ_s, respectively, in *a*) and *b*) experiments (see fig. 5, inset). Our results are $\tau_s = 550$ ns and $\tau_p = (253 \pm 6)$ ns.

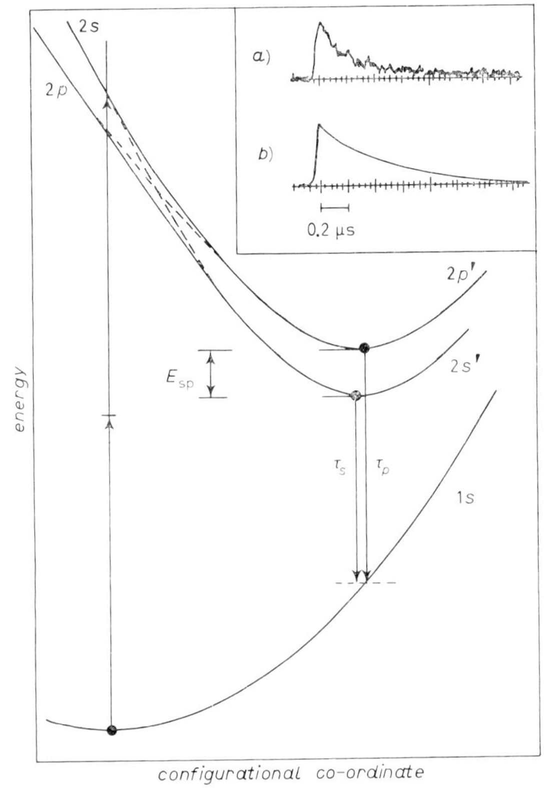

Fig. 5. – Franck-Condon-Seitz diagram of the *F*-centre in KCl. The two oscilloscope signals in the inset of the figure correspond to fluorescent decay from RES after two-photon exciton (trace *a*)) and one-photon excitation (trace *b*)).

Our value of τ_s coincides with the well-known result of linear spectroscopy [15], while τ_p is reported for the first time.

We discuss our results on the basis of the F-C diagram of fig. 5. After the selective excitation of the centre to the state $|2p\rangle$ (or to $|2s\rangle$), the crystal relaxes with emission of LO phonons. The interaction of the bound electron with the phonon field is recognized to be of crucial importance for the establishment of the RES structure [16-19]. Two kinds of electron-phonon interactions must be considered: the linear interaction of the triplet states $|2p_i\rangle$ with vibrations caused by the dynamic Jahn-Teller effect and the coupling with the p-mode phonon field (T_{1u}-symmetry), which mixes the $2s$ and $2p$ states. These effects may be formalized in a rather complex Hamiltonian, which requires a numerical solution of the dynamic problem [16-19]. When the $2s$-$2p$ coupling through the p-mode phonons is strong, the degeneracy of the $2s$ and $2p$ states during the relaxation is removed and a situation of state anticrossing occurs. This effect results in the relaxation of an atom, which is initially prepared in the pure $2p$ (or $2s$)-state, to a mixed $2s'$ (or $2p'$)-state. This process of « dynamic » mixing is obviously an irreversible one and it raises subtle questions about the definition of a state during a rapid crystal relaxation. In general, we can say that the anticrossing process will be effective on the population distribution in the RES, if the $2s$-$2p$ coupling energy during relaxation is larger than $h\bar{\tau}^{-1}$, with $\bar{\tau}$ the crystal relaxation time. In our case, the substantial lack of interference between the two spectra of fig. 4 shows that the anticrossing process is indeed effective and that the coupling should be strong. Assuming the value of the coupling energy E_g evaluated later in the paper, we obtain $\bar{\tau} > 7\cdot 10^{-14}$ s. It would be interesting to investigate this relaxation process by subpicosecond time resolution spectroscopy.

The experimental evidence for the anticrossing-mixing process we have just described is of extreme interest in solid-state physics, as we do not know of any other example showing such a clean effect for an electronic process in a solid. In our work we were able to determine for the first time a case in which the adiabatic limit holds *strictly*.

The strength of the vibronic coupling is confirmed by our measurements of τ_s and τ_p. If we assume the phenomenological Bogan's model for the RES for crystal distortions in the x, y, z directions, $|2s'\rangle = (1+\alpha^2)^{-\frac{1}{2}}\times(|2s\rangle + \alpha|2p\rangle)$, $|2p'\rangle = (1+\alpha^2)^{-\frac{1}{2}}\times(|2p\rangle - \alpha|2s\rangle)$, $|2p\rangle \equiv a_1|2p_x\rangle + a_2|2p_y\rangle + a_3|2p_z\rangle$, the radiative lifetimes are given by

$$\tau_s \propto [(1+\alpha^2)/\alpha^2]\times\left|\langle 1s|\boldsymbol{r}|2p\rangle\right|^{-2}, \qquad \tau_p \propto [1+\alpha^2]\times\left|\langle 1s|\boldsymbol{r}|2p\rangle\right|^{-2}.$$

The ratio $\tau_p/\tau_s = \alpha^2$ provides direct information on the amount of the (s-p)-mixing. We obtain the large value $\alpha = 0.67$.

Also the value of E_{sp} is relevant for the evaluation of the strength of the electron-phonon coupling. Write the coupling Hamiltonian

$$H_{el} = G\{Q_x\varrho_x + Q_y\varrho_y + Q_z\varrho_z\}$$

in terms of the phonon and electron operators Q_i and $\varrho_i = |2p_i\rangle\langle 2s| + |2s\rangle\langle 2p_i|$. In the case of strong coupling, the coupling energy $E_g = G^2/(2\mu\omega^2)$ (μ, ω = lattice effective mass and phonon frequency) may be obtained from the diamagnetic part of the magnetic circular polarization of the luminescence [17]: $\Delta_d(0) = -g_L\mu_B H/E_g$. For KCl it has been found [20] that $\Delta_d(0)/H = -(9\pm 1)\cdot 10^{-8}$.

Taking [21] $g_L = 0.95 \pm 0.1$, we obtain $E_g = 60$ meV. Therefore Ham's criterion for the strong-coupling limit, $E_g > \frac{1}{4}|E_{sp}|$, is verified in our case [17].

In conclusion, our work demonstrates in a very direct way that the effect of crystal relaxation on the F-centre structure in KCl cannot be accounted for in a satisfactory way by a perturbative approach [18] but rather by a complete vibronic theory in the strong-coupling limit [17, 19]. Furthermore, our work provides the numerical values of the relevant parameters, that were still not available for an exact solution of the dynamic problem.

The conclusions we have reached in the discussion of our results hold, both qualitatively and quantitatively, in the case of nonvertical alignment of the minima of the RES potential wells of fig. 5. The only exception is that the energy shift of the spectra of fig. 4 cannot be entirely attributed to the splitting E_{sp}. The different spectral widths shown in fig. 4 suggest indeed that the minimum of the $2p'$ vibronic well is slightly displaced toward the left in the diagram of fig. 5, in agreement with previous indirect measurements of E_{sp}. If we assume the value of E_{sp} reported in ref. [11], we are able, on the basis of our data of fig. 4, to draw the exact configuration of the RES structure including the shape and the location of the potential wells in the configurational space. We shall thoroughly consider these important details in a following, more extended paper.

3. – Laser effects.

Apart from details in the RES structure, the Franck-Condon-Seitz configuration shown in fig. 5 is typical for most of the colour centres as well as for all electronic bound systems interacting with slowly relaxing molecular or lattice systems. If we neglect again details of the RES, fig. 5 shows that the colour centres can be used as active quantum systems in a 4-level laser emitting in the infra-red under pumping in the optical or near infra-red part of the spectrum [22] (see, for F-centres, table I). In principle, this 4-level laser cycle appears to be an ideal one because of the high pumping efficiency of the absorption band and of the large ratio between the fluorescence decay times τ_s of the most common centres and the phonon-assisted crystal relaxation times $\bar{\tau} = (10^{-12} \div 10^{-13})$ s. As a consequence, any energy transfer to the excited electronic state must give rise to a net population inversion, *i.e.* gain. Some other considerations make this system very appealing for laser applications.

a) The emission and absorption bands do not overlap, as is shown in fig. 2 for F and F_A centres. Consequently we do not expect for this system the optical reabsorption effects which lower considerably the efficiency of the common dyes used in tunable lasers.

b) The absorption and emission bands are quite broad (typically $\Delta\lambda \sim 2200$ Å for F-centres). This ensures in general good efficiency under broad-band (flash) pumping and tunability over a wide frequency range.

c) The emission frequencies cover a region of the spectrum (λ ranging from about 0.9 μm to 3.5 μm) where the liquid dye laser action is very inefficient or totally absent (for $\lambda \gtrsim 1.1$ μm). Note that this spectral region is of particular interest for molecular spectroscopy of light molecules.

In spite of these favourable properties of the colour centres, it is easy to show that the laser gain for most centres is small and its effect can be easily overwhelmed by parasitic absorption processes in the crystal, like resonant absorption by F' or N centres [2], photoionization of the active centres, etc. If we assume amplification along the x-direction, the relative amplification (gain) of the e.m. energy flux is given by [22]

$$g = \frac{dS}{dx}\frac{1}{S} = \frac{h\nu}{v} S(\nu)(n_1 B_{12} - n_2 B_{21}) , \tag{1}$$

where $S(\nu)$ is the normalized shape factor, that, for Gaussian line shape, is given by

$$S(\nu) = \sqrt{\frac{\ln 2}{\pi}}\frac{1}{\Delta\nu_D}\exp\left[-\left[\frac{\nu - \nu_0}{\Delta\nu_D}\right]^2 \ln 2\right] ,$$

$\Delta\nu_D$ being the line width of the emission band and v the speed of light in the crystal.

If we assume a simple RES structure, the Einstein B-factors may be obtained in terms of $A = 1/\tau_s$ [1]:

$$B = \frac{1}{\tau_s}\frac{\lambda_E^3}{16\pi^2 \hbar n^3} , \tag{2}$$

n being the refractive index of the « host » crystal at the amplification wavelength λ_E. Inserting (2) in (1), making use of the characteristic parameters for the F-centre in KCl, assuming a Gaussian emission line shape [2], we find $g = 1.65 \cdot 10^{-8}\, N^*$ (cm^{-1}), N^* being the population inversion density. Assuming total inversion with F-centre density $N = 10^{16}$ cm^{-3}, we find $g = 1.65 \cdot 10^{-9}$ (cm^{-1}), which is a small value [23].

There are colour centres for which the value of the gain can be two orders of magnitude larger than the one we have just evaluated. These are the type II of the F_A-centres: a tetragonal structure which can be considered to be obtained from a (cubic) F-structure when a positive ion adjacent to a vacancy is replaced by a smaller impurity cation (*e.g.* Na^+ or Li^+). We note in fig. 2 that the absorption band of F_A(I) and F_A(II) centres exhibits two peaks arising from the Stark splitting of the $2p$ triplet due to the strong local electric field generated by the impurity ion. The structure of these centres, which accounts for their optical properties, is somewhat complex, requiring a careful discussion of the theoretical and phenomenological aspects of the problem. For that we refer the reader to the excellent review work of LÜTY [23]. We can nevertheless understand on simple arguments the origin of the large gain.

The lifetime τ_s of the F_A(II)-centre in KCl:Li has been measured to be $\tau_s = 8 \cdot 10^{-8}$ s, a value which is about seven times smaller than τ_s for F-centres in KCl. Furthermore, two other properties contribute to raise the value of the gain: the much narrower emission line width (5 times smaller) and the displacement of the F_A(II)-centre emission peak toward a large wavelength λ_E (fig. 2). We recall that the cube of this quantity enters in the expression for B (eq. (2)).

The first laser effect from a colour centre in alkali halides was in fact observed in KCl containing F_A(Li)-centres by FRITZ and MENCKE (ref. [4]). This was achieved by using a crystal rod with flat end mirrors, pumped at 77 K with light from a xenon flash tube, thus exciting both peaks of the absorption spectrum of the centres. The spectral range in which the laser emission was observed was of the order of 40 cm^{-1}. This is the range over which the laser frequency can be tuned.

The work of FRITZ and MENCKE has been forgotten for about 10 years. Only in 1974 scientists of Bell Telephone Laboratories revisited that idea and developed various types of lasers, using again F_A(II)-centres. These lasers are tunable, very efficient, and working in c.w. or pulsed operation under incoherent (flash lamps) or coherent pumping (Kr laser operating at 6471 Å) For the latest advances in that field we refer the reader to [25]. The gain of an F-centre laser amplifier has been measured in our laboratory in Rome [7].

We are not aware of investigations on laser oscillators or amplifiers using other types of colour centres.

REFERENCES

[1] *Physics of Color Centers*, edited by W. B. FOWLER (New York, N. Y., 1968). See in particular the work of W. B. FOWLER.

[2] J. J. MARKHAM: *F-Centers in Alkali Halides* (New York, N. Y., 1966).

[3] *Conference on Color Centers in Ionic Crystals, Extended Abstracts, Sendai, Japan, 1974* (Tohoku University, Sendai, Japan, 1974).

[4] B. FRITZ and E. MENCKE: *Solid State Comm.*, **3**, 61 (1965).

[5] L. F. MOLLENAUER and D. H. OLSON: *Appl. Phys. Lett.*, **24**, 386 (1974), and in ref. [3].

[6] M. BORN and J. R. OPPENHEIMER: *Ann. der Phys.*, **84**, 457 (1927).

[7] F. DE MARTINI, U. M. GRASSANO and F. SIMONI: *Opt. Comm.*, **11**, 8 (1974); *VIII International Quantum Electronics Conference, Digest of technical papers* (San Francisco, Cal., 1974), and in ref. [3].

[8] F. De Martini, G. Giuliani and P. Mataloni: *Phys. Rev. Lett.*, **35**, 1466 (1975).
[9] G. Chiarotti and U. M. Grassano: *Nuovo Cimento*, **46** B, 78 (1966).
[10] H. Kühnert: *Phys. Stat. Sol.*, **21**, K171 (1967).
[11] L. Bogan and D. Fichten: *Phys. Rev. B*, **1**, 4122 (1970); L. F. Stiles, M. P. Fontana and D. Fichten: *Phys. Rev. B*, **2**, 2077 (1970).
[12] W. L. Smith, J. H. Bechel and N. Bloembergen: to be published.
[13] F. Lüty: *Halbleiterprobleme*, **6**, 238 (1961).
[14] Y. Kondo and H. Kanzaki: *Phys. Rev. Lett.*, **34**, 664 (1975).
[15] L. Bosi, C. Bussolati and G. Spinolo: *Phys. Rev. B*, **1**, 890 (1970).
[16] T. Iida, K. Kurata and S. Muramatsu: *Journ. Chem. Phys. Sol.*, **33**, 1255 (1972).
[17] F. S. Ham: *Phys. Rev. B*, **8**, 2926 (1973).
[18] F. S. Ham and U. Grevsmühl: *Phys. Rev. B*, **8**, 2945 (1973).
[19] Y. Kayanuma and Y. Toyozawa: *Conference on Color Center, Proceedings* (Sendai, Japan, 1974). In that work the effect of the s-mode and d-mode phonons is also considered. The d-mode phonon is found to be of secondary importance, while the s-mode is found to be effective in causing the red Stark shift and the process of reduction of the orbital g-value.
[20] M. P. Fontana: *Phys. Rev. B*, **5**, 759 (1972).
[21] J. Margerie: *J. Physique, Suppl.*, **28**, 103 (1967).
[22] G. Birnbaum: *Optical Masers* (New York, N. Y., 1964).
[23] In work [7] we have reported an experimental value for the laser gain of F-centres in KCl which is more than one order of magnitude larger than the theoretical value. We are presently investigating the origin of this puzzling discrepancy.
[24] F. Lüty: *Physics of Color Centers*, edited by W. B. Fowler (New York, N. Y., 1968).
[25] L. F. Mollenauer: *Conference on Color Centers in Ionic Crystals. Extended Abstracts* (Sendai, Japan, 1974); *Laser Spectroscopy*, edited by Haroche, A. Pebay-Peyroula and T. Hänsh (Berlin, 1975).

Nonlinear Spectroscopy in the Rayleigh-Brillouin Region of the Spectrum of Light Scattered by Fluids.

A. BAMBINI, R. VALLAURI and M. ZOPPI

Laboratorio di Elettronica Quantistica del Consiglio Nazionale delle Ricerche
Via Panciatichi 56/30 - 50127 Firenze, Italia

This seminar can be divided into two parts: in the first we will review the results obtained both experimentally and theoretically in recent years by several research groups, and in the second we will examine a recent theoretical approach which allows one to describe the amplification of a signal wave interacting with a strong laser field through the density and temperature fluctuations which are present in a fluid.

Before proceeding to the stimulated effects, let us recall briefly the characteristic features of the Rayleigh-Brillouin region of the spectrum, in order to point out the physical processes which are responsible for the light scattered by fluids in this range of frequencies. In fig. 1 we have sketched the spectrum of the scattered intensity obtained when we illuminate a sample of fluid composed by isotropic molecules with a monochromatic beam at frequency ω_L, and collect the radiation scattered at a certain angle θ with respect to the direction of propagation of the incoming beam.

The sharp central peak of the spectrum, at a frequency equal to the incident one, is the Rayleigh component, while the two side peaks are determined by the Brillouin-scattering process. As is well known, a perfect homogeneous and isotropic medium would provide no scattering, due to the interference processes which would occur among the different waves scattered by each molecule. Fluctuations are easily recognized to be responsible for the appearance of the scattering; namely, nonpropagating fluctuations cause the Rayleigh scattering, which contributes to the spectrum with an unshifted narrow line. The line width Γ_R is of the order of $(10 \div 50)$ MHz and is connected with the decay constant of these fluctuations. As was early recognized, this decay constant is proportional to the heat conduction coefficient of the medium.

Propagating density fluctuations originate the Brillouin doublet shifted by a quantity

$$\omega_B = \pm 2\omega_L \frac{v}{c} n \sin\frac{\theta}{2},$$

where v is the velocity of sound waves, n the refractive index, c the velocity of light in vacuum, and θ is defined in fig. 1. For liquids the maximum value of ω_B (which is achieved when $\theta = \pi$) is approximately $(5 \div 30)$ GHz, while $\Gamma_B \simeq (100 \div 500)$ MHz. The change in frequency due to propagating fluctuations can easily be understood if we bear in mind that such scattering processes occur through a transfer of momentum from the medium to the electromagnetic field

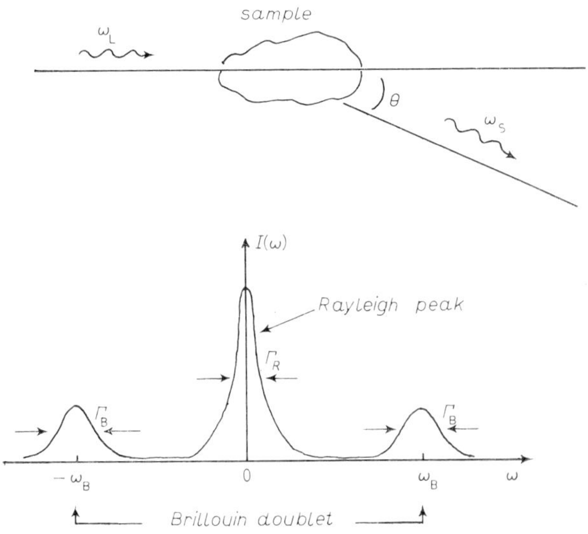

Fig. 1. – The spectrum of the intensity *vs.* frequency difference $\omega = \omega_L - \omega_S$ is sketched along with the experimental geometry. $\omega_B = \pm 2\omega_L(v/c)n \sin(\theta/2)$.

or *vice versa*. In a nondispersive medium the momentum transfer is linearly related with the frequency change, and taking into account the energy and momentum conservation laws one gets immediately the equation above. For a more complete discussion see ref. [1-3].

So far we have considered only those fluctuations which occur spontaneously in any medium at a finite temperature T. But if we increase the electromagnetic intensity, the fluctuations can be driven by the field itself, and in turn originate the scattering process. Correspondingly these processes are known as stimulated Rayleigh-Brillouin scattering.

Before going on to analyse the theoretical treatments that have been used in the past few years in order to describe the stimulated scattering of the electric field, it is better to review briefly what has been done from an experimental point of view. The typical experimental apparatus used for investigations of the amplification behaviour near the Brillouin peak is shown in fig. 2. The light of a giant pulse laser is focused in an oscillator cell filled with liquid and generates intense Brillouin radiation. The backward-travelling radiation is strongly attenuated by the use of a polarizer and a $\lambda/4$ plate and acts as a small signal which is amplified by the interaction with the incoming laser pulse in the amplifier cell. Here, the coupling of the two incoming fields through the medium causes the growth of the phonon waves and consequently the coupling of these waves with the electric field acts as a source for the electromagnetic

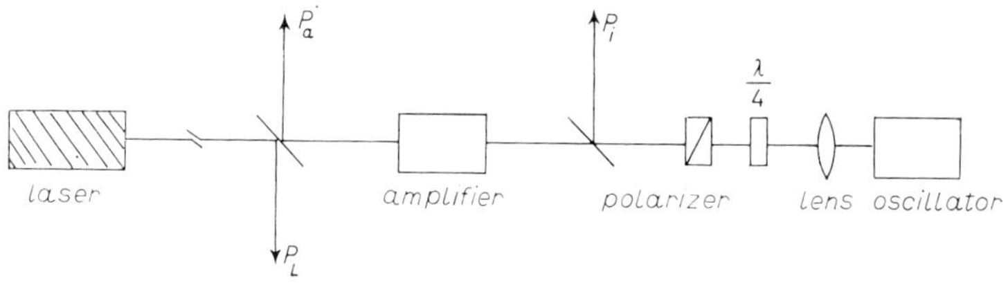

Fig. 2. – Typical experimental apparatus for gain measurements. The amplifier and oscillator cells are filled with liquids. The polarizer and $\lambda/4$ plate attenuate the strong Stokes radiation produced when the laser pulse is focused into the oscillator by the lens.

field itself. For each laser pulse three signals, the laser power P_L, the incoming power P_i and the amplified Brillouin power P_a, were measured with the same phototube (using appropriate time delays). The overall time constant of this photodetection system was approximately 0.3 ns. This kind of nonlinear spectroscopy was proposed and has been realized by KAISER and co-workers since 1968 [4-7]. With this apparatus two types of measurements are possible. Firstly one can observe the time behaviour of the amplification factor $g(\omega, t)$

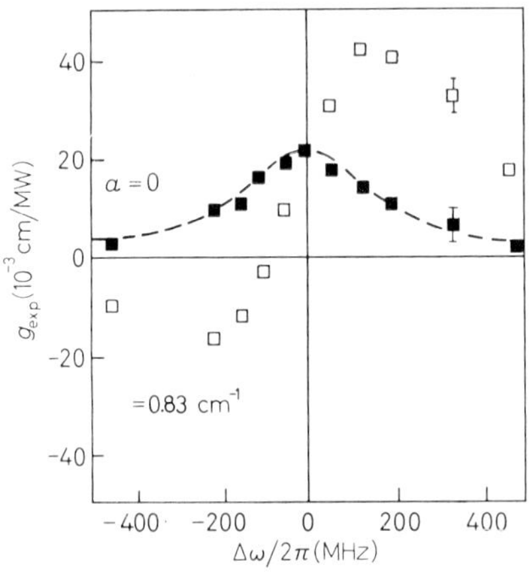

Fig. 3. – Gain function $g(\omega, t)$ *vs.* frequency around the Brillouin peak, in the steady-state regime. Full squares and open ones refer to the cases in which the optical absorption coefficient α is equal or different from zero respectively (ref. [6]).

defined as

$$g(\omega, t) = \big(P_a \exp[\alpha L] - P_i\big)/P_i$$

(where L represents the interaction length) during all the transient period till a stationary regime is reached. The characteristic time is connected with the lifetime of the acoustic phonons involved in the process.

Secondly one can describe the gain function in the steady-state regime as a function of frequency around the Brillouin peak. The tuning was experimentally achieved by varying the frequency of the signal wave. This change was obtained by filling up the oscillator cell with liquid mixtures at different concentrations, as it had been demonstrated that ω_B changes linearly with the concentration [8, 9]. The influence of the heat absorption was stressed in the experiments. Small quantities of dye were added in the amplifier in order to enhance the stimulated Rayleigh component. Experimental results are reported in fig. 3, where full squares refer to an absorption coefficient $\alpha = 0$, and the open ones to $\alpha \neq 0$.

A similar apparatus was also used to investigate the stimulated thermal

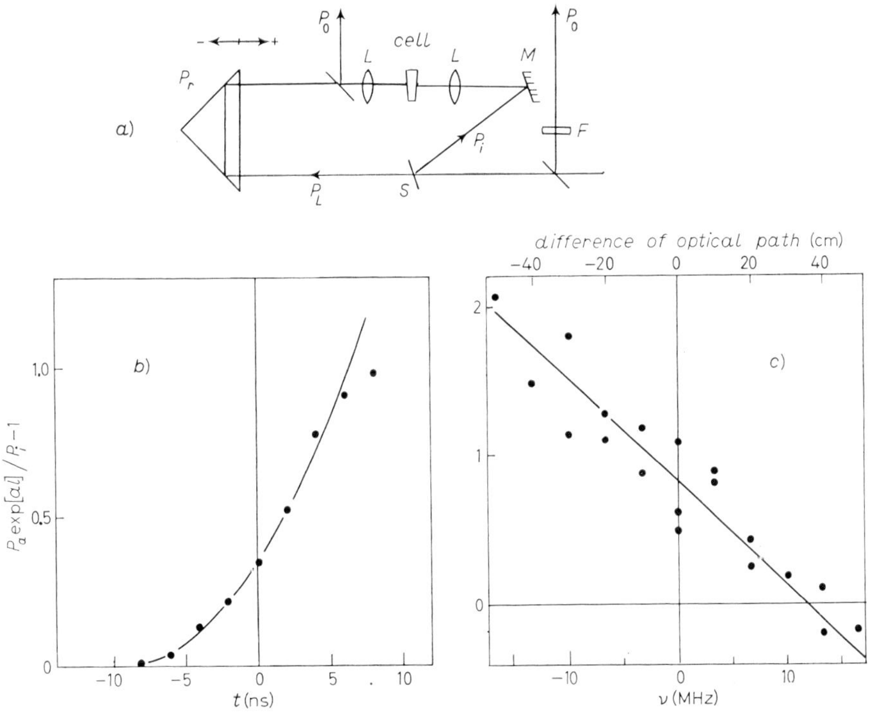

Fig. 4. – *a*) Typical experimental apparatus for the study of the amplification factor near the Rayleigh peak. *b*) Gain function *vs.* time. $t = 0$ refers to the maximum of the laser pulse. *c*) Gain function *vs.* frequency around the Rayleigh peak after ROTHER *et al.*, ref. [11, 12].

Rayleigh scattering, firstly predicted by HERMAN and GRAY [10]. The experimental set-up is reported in fig. 4*a*) [11, 12]. In this case the temperature fluctuations play an important role and the coupling with the electromagnetic field is enhanced by adding small quantities of dye to the liquid in the amplifying cell. The frequency tuning around the Rayleigh peak was achieved by delaying the pumping field with respect to the input signal in order to use the frequency drift experienced by the laser pulse during the emission. In fig. 4*b*) the full circles represent the experimental data related to the time behaviour of the gain function in the stimulated Rayleigh process when the frequency of the signal wave is the same as that of the laser: $t = 0$ refers to the time in which the pulse has its maximum value. As is apparent, a steady-state regime is never reached during the laser pulse duration ($\simeq 20$ ns). In this case the overall transient analysis is necessary. The amplification factor *vs.* frequency around $\omega = \omega_L - \omega_S = 0$ is reported in fig. 4*c*).

As far as the theoretical approach is concerned we refer mainly to the paper by POHL and KAISER [6] where some simplifying assumptions are discussed. According to the experimental geometry the electromagnetic field can be depicted as follows:

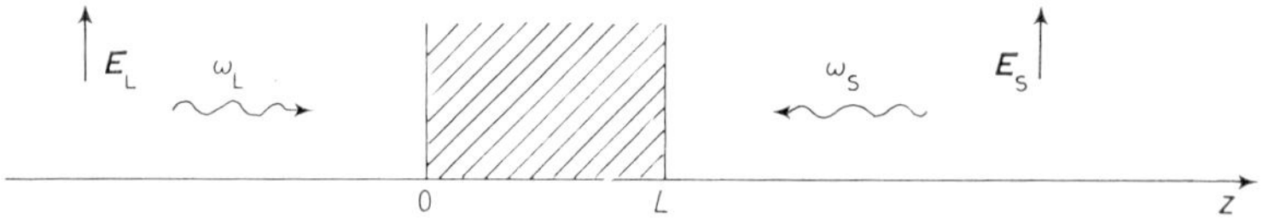

a small signal wave travelling along the $-z$-direction with amplitude E_S, frequency ω_S and wave vector k_S interacts with a strong laser wave travelling in the opposite direction with frequency ω_L, wave vector k_L and amplitude E_L, inside the medium confined between 0 and L.

The dynamics of the system, in the limit in which the medium can be considered as a continuum, is described by the hydrodynamical equations linearized in the small deviations of ϱ and T with respect to the mean value ϱ_0 and T_0. They read [3]

$$\frac{\partial \bar{\varrho}}{\partial t} + \varrho_0 \operatorname{div} \boldsymbol{V} = 0\,, \tag{1}$$

$$\varrho_0 \frac{\partial \boldsymbol{V}}{\partial t} + \frac{v^2}{\gamma} \nabla \bar{\varrho} + \frac{v^2 \beta_T \varrho_0}{\gamma} \nabla \bar{T} - \eta \nabla^2 \boldsymbol{V} = \frac{\gamma^e}{8\pi} \nabla E^2\,, \tag{2}$$

$$\varrho_0 c_V \frac{\partial \bar{T}}{\partial t} - \lambda_T \nabla^2 \bar{T} - \frac{c_V}{\beta_T} (\gamma - 1) \frac{\partial \bar{\varrho}}{\partial t} = \frac{nc\alpha}{4\pi} E^2\,, \tag{3}$$

where the variables are $\varrho = \varrho_0 + \bar{\varrho}$, $T = T_0 + \bar{T}$ and the velocity of the volume element $\boldsymbol{V}$. The quantities $\bar{\varrho}$, $\bar{T}$ and $\boldsymbol{V}$ are small, so that the products among them have been neglected.

The first equation is the linearized version of the continuity equation and defines the conservation of mass. The Navier-Stokes equation (2) represents the dynamical version of Newton's second law, where a frictional force has been introduced by a viscosity coefficient η and the force resulting by the hydrostatic pressure p has been represented by the temperature fluctuation $\bar{T}$ by the use of an equation of state. The third equation is the usual energy transport equation. If we set $E = 0$ in eqs. (2) and (3), we get the equations of motion for the evolution of the spontaneous fluctuations of the medium. So the field-dependent terms in eq. (2) and (3) act like source terms which drive the deviations of ϱ and T from the average values. The right-hand side term in eq. (2) represents the electrostrictive force due to the presence of a strong electromagnetic field, while the right-hand side term of eq. (3) accounts for an energy transfer from the electromagnetic field to the medium through a source term proportional to the optical absorption coefficient α. The other symbols represent

v = velocity of sound waves,

$\gamma = c_P/c_V$ is the ratio between the specific heats at constant pressure and volume,

β_T = thermal-expansion coefficient,

λ_T = heat conduction coefficient,

η = viscosity coefficient,

$\gamma^e = \varrho_0(\partial\varepsilon/\partial\varrho)_T$ (where ε represents the dielectric constant of the medium) is the electrostrictive coefficient which for liquids can be written as $\frac{1}{3}(n^2-1)(n^2+2)$.

These three differential equations are coupled with one describing the total electric field

$$\frac{\partial^2 \boldsymbol{E}}{\partial t^2} + \frac{\alpha c}{n}\frac{\partial \boldsymbol{E}}{\partial t} - \left(\frac{c}{n}\right)^2 \nabla^2 \boldsymbol{E} = -\frac{4\pi}{n^2}\frac{\partial^2 \boldsymbol{P}^{\mathrm{NL}}}{\partial t^2} = \frac{\gamma^e}{n^2\varrho_0}\frac{\partial^2}{\partial t^2}[\boldsymbol{E}\cdot\varrho]\,, \tag{4}$$

where the source term involving the nonlinear polarizability $\boldsymbol{P}^{\mathrm{NL}}$ has been written as

$$\boldsymbol{P}^{\mathrm{NL}} = (1/4\pi)(\partial\varepsilon/\partial\varrho)_T \boldsymbol{E}\cdot\varrho\,.$$

The nonlinearity comes out from the fact that in this case ϱ depends on $\boldsymbol{E}$. We note that terms proportional to $(\partial\varepsilon/\partial T)_\varrho$ have been neglected because

$$\bar{\varrho}(\partial\varepsilon/\partial\varrho)_T \gg \bar{T}(\partial\varepsilon/\partial T)_\varrho\,.$$

In the plane-wave approximation the total electric field can be written as the sum of E_S and E_L:

$$\boldsymbol{E} = \tfrac{1}{2}\boldsymbol{i}\left(E_S(z,t)\exp[i\omega_S + ik_S z]\exp[\tfrac{1}{2}\alpha z] + E_L(z,t)\exp[i\omega_L t - ik_L z]\exp[-\tfrac{1}{2}\alpha z] + \text{c.c.}\right).$$

The two fields are supposed to be linearly polarized along the x-axis whose unit vector is denoted by $\boldsymbol{i}$. The absorption is written out in the appropriate way for the two waves, so that E_L and E_S can be supposed independent of α. Further, the amplitudes E_L and E_S are supposed to be slowly varying functions both in the optical period and over one wavelength, *i.e.*

$$\left|\frac{1}{\omega_{S,L}}\frac{\partial E_{S,L}}{\partial t}\right| \quad \text{and} \quad \left|\frac{1}{k_{S,L}}\frac{\partial E_{S,L}}{\partial z}\right| \ll E_{S,L}.$$

The stimulated density and temperature fluctuations are described as waves travelling in the $+z$-direction with frequency ω and wave vector k:

$$\bar{\varrho}(z,t) = \tfrac{1}{2}\left(\varrho(z,t)\exp[i\omega t - ikz] + \text{c.c.}\right),$$
$$\bar{T}(z,t) = \tfrac{1}{2}\left(T(z,t)\exp[i\omega t - ikz] + \text{c.c.}\right).$$

Energy and momentum conservation requires

$$\omega = \omega_L - \omega_S,$$
$$k = k_L + k_S \simeq 2k_L.$$

Note that ϱ and T must not be confused with the total density and temperature fluctuations $\varrho_0 + \bar{\varrho}$ and $T_0 + \bar{T}$ respectively. They are now the amplitudes of the deviations $\bar{\varrho}$ and $\bar{T}$ from the average value. Furthermore, they are slowly varying functions of z in a length $1/k$ because almost all z-dependence is contained in the exponential factor $\exp[\pm ikz]$, due to the fact that the wavelength of the periodic waves in the medium involved in the process is comparable with the optical one. But they are not slowly varying functions of time in a period $2\pi/\omega$, since the acoustic period $2\pi/\omega_B$ is long compared with the optical one. By this approximation the system (1)-(4) reads

$$\left[\frac{\omega_B}{\gamma} - \omega^2 + i\omega\Gamma_B + (2i\omega + \Gamma_B)\frac{\partial}{\partial t} + \frac{\partial^2}{\partial t^2}\right]\varrho(z,t) + \frac{\omega_B\beta_T\varrho_0}{\gamma}T(z,t) = \frac{\gamma^e k^2}{8\pi}E_L E_S^*, \tag{5}$$

$$-(\gamma-1)\left(i\omega + \frac{\partial}{\partial t}\right)\varrho(z,t) + \beta_T\varrho_0\left(i\omega + \frac{1}{2}\gamma\Gamma_R + \frac{\partial}{\partial t}\right)T(z,t) = \frac{\gamma\gamma^a}{8\pi v}E_L E_S^*, \tag{6}$$

$$\frac{\partial E_S^*}{\partial z} = -i\frac{\gamma^e\omega_S}{4nc\varrho_0}E_L\varrho(z_1 t)\exp[-\alpha z], \tag{7}$$

where the parameters

$$\Gamma_{\mathrm{B}} = \frac{\eta k^2}{\varrho_0},$$

$$\Gamma_{\mathrm{R}} = \frac{2\lambda_T k^2}{\varrho_0 c_P}$$

were introduced. They represent the line width of the classical Brillouin and Rayleigh lines respectively.

γ^a is a thermo-optic coupling constant proportional to the coefficient α. The effect of propagation of the electric field E_{S} has been neglected, as it was experimentally confirmed that the spatial amplification is overwhelming, therefore we have retained only the spatial derivative in eq. (7). While the steady-state solution of eqs. (5)-(7) is very easy to derive, the behaviour over short time intervals is not so obvious; on the other hand the knowledge of the characteristic times involved in the transient period is important in order to justify the assumption that a steady-state solution is reached in the time duration of a giant laser pulse.

In the small-amplification approximation POHL and KAISER expanded E_{S}, ϱ and T in power series of the form

$$\varrho\;(z, t) = \varrho^{(0)}(t) + \varrho^{(1)}(t)\frac{z}{L} + \ldots,$$

$$E_{\mathrm{S}}(z, t) = E_{\mathrm{S}}^{(0)}(t) + E_{\mathrm{S}}^{(1)}(t)\frac{z}{L} + \ldots,$$

assuming E_{L} independent of z, thus neglecting any depletion of the pump field. L represents the interaction length. They solved the system to the first order and obtained good agreement with the experimental results on the amplification of a small signal with frequency ranging in the Stokes side of the Brillouin doublet. But when $\omega = \omega_{\mathrm{L}} - \omega_{\mathrm{S}} = 0$, the first-order approximation fails. The analysis was extended to the second order by ENNS and co-workers [13-16]. Figure 5*a*) shows the time behaviour of the amplification factor: different curves refer to different values of ω near $\omega = 0$. The characteristic time in which the stationary regime is reached is $1/\Gamma_{\mathrm{R}}$. Figure 5*b*) shows the amplification factor as a function of ω at different times. As is apparent in the steady-state regime the maximum is anti-Stokes shifted by a quantity $\omega_M = \frac{1}{2}\Gamma_{\mathrm{R}}$. A different approach used by ROTHER [17] is worth-while to be mentioned. He demonstrated that by separating the contributions at $\omega = 0$ and $\omega = \pm\omega_{\mathrm{B}}$ one can get an analytical solution for the Stokes field E_{S}, once the laser amplitude is supposed to be independent of time. His theoretical results were compared with experiments [12] confirming a t^2-dependence of the amplification at small times (near $\omega = 0$) in the transient regime and an α^2-dependence as well. If we take into account the time dependence of the laser intensity, the solution cannot be given in an analytical form. The relevant point is that this derivation is valid only when it is possible to separate the contri-

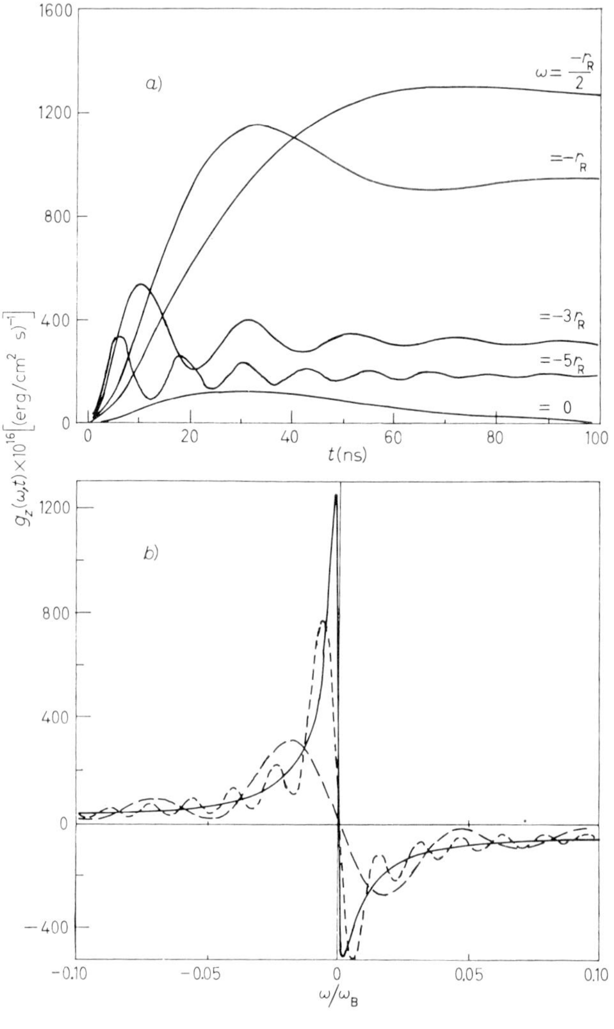

Fig. 5. – *a*) Gain factor *vs.* time at various frequency shifts near $\omega = 0$. *b*) Gain factor *vs.* ω/ω_B around $\omega = 0$ for various times after RANGNEKAR and ENNS (ref. [16]). — — — $t = 5$ ns, - - - $t = 15$ ns, ——— $t = 120$ ns.

butions of different components on the total spectrum, in other words when the density and temperature fluctuations do not influence each other.

So far for the first part of the seminar; in the second part we want to discuss a different theoretical approach proposed recently [18].

We start from the system of differential equations (5)-(7) and integrate it with the following conditions:

i) $\varrho(0) = 0$, $\dot{\varrho}(0) = 0$, $T(0) = 0$;

ii) we assume a step function for the Stokes field

$$E_S(L, t) = E_S^{(0)} = \text{constant};$$

iii) we assume a laser input field constant both in space and time.

We take the Laplace transform of the system (5)-(7) and solve the first two equations with respect to the density Laplace transform, then we substitute it into the third equation and integrate with respect to z between L and z $(0 \leqslant z \leqslant L)$. The slowly varying amplitude of the Stokes field turns out to be

$$(8)\qquad E_S^*(z, t) = \frac{1}{2\pi i}\int \frac{E_S^{(0)}}{p}\cdot$$
$$\cdot\exp\left[A(z)\left(p + \frac{1}{2}\gamma\Gamma_R - \frac{\gamma^a}{\gamma^e}\omega_B + i\omega\right)(p-p_1)^{-1}(p-p_2)^{-1}(p-p_3)^{-1}\right]\exp[pt]\,dp\,,$$

where

$$(9)\qquad A(z) = i(\gamma^e k E_L)^2\,\omega_S\big(\exp[-\alpha z] - \exp[-\alpha L]\big)/(32\pi n c \varrho_0 \alpha)\,.$$

The p_j are the roots of the characteristic equation derived from the Laplace transform of eqs. (5) and (6), and under the assumption $(\Gamma_{R,B}/\omega_B)^2 \ll 1$ are given by

$$p_1 = -i\omega - \tfrac{1}{2}\Gamma_R\,,$$
$$p_2 = -i(\omega + \omega_B) - \tfrac{1}{2}\Gamma'_B\,,$$
$$p_3 = -i(\omega - \omega_B) - \tfrac{1}{2}\Gamma'_B\,,$$

where

$$\Gamma'_B = \Gamma_B + \tfrac{1}{2}(\gamma - 1)\Gamma_R\,.$$

In order to get an analytical solution for the Stokes field E_S we must evaluate the residues of the four poles of the integrand in eq. (8). The residue in the pole $p = 0$ is easily evaluated and yields the steady-state contribution to the electromagnetic field. For the other three singularities we proceed as follows: first of all we expand the exponent in the integrand:

$$A(z)\left(p + \frac{1}{2}\gamma\Gamma_R - \frac{\gamma^a}{\gamma^e}\omega_B + i\omega\right)\Big/(p-p_1)(p-p_2)(p-p_3) =$$
$$= A(z)\left[\frac{B_1}{p-p_1} + \frac{B_2}{p-p_2} + \frac{B_3}{p-p_3}\right].$$

The coefficients can be easily calculated and we do not report their specific expression here. Then for each singularity we divide the exponential into two parts, one of which is singular while the other is regular. For the p_1-singularity we get

$$\text{(10)}\qquad \exp\left[A(z)\sum_{j=1}^{3}\frac{B_j}{p-p_j}\right]\exp[pt]=$$
$$=\exp\left[A(z)\frac{B_1}{p-p_1}+(p-p_1)t\right]\exp\left[A(z)\sum_{j=2}^{3}\frac{B_j}{p-p_j}+p_1t\right].$$

The singular part is expanded in a series of $p-p_1$ by the use of the generating function for Bessel function, *i.e.*

$$\exp\left[\tfrac{1}{2}x(y-1/y)\right]=\sum_{s=-\infty}^{+\infty}y^s J_s(x)\,.$$

The regular part is expanded in a Taylor series. Once we have made the product of these two expansions we can extract the term $(p-p_1)^{-1}$ whose coefficient is just the residue in the singularity considered. Such a procedure is applied for each of the three singularities and the amplitude E_S is evaluated. The Stokes field $\mathscr{E}_S$ (which contains the optical frequencies) turns out to be

$$\text{(11)}\qquad \mathscr{E}_S=\tfrac{1}{2}E_S^{(0)}\exp\left[\tfrac{1}{2}\alpha z\right]\{C_0(z)\exp[-i(\omega_S t+k_S z)]+$$
$$+C_1(z,t)\exp[-\tfrac{1}{2}\Gamma_R t]\exp[-i(\omega_L t+k_L z)]+$$
$$+C_2(z,t)\exp[-\tfrac{1}{2}\Gamma'_B t]\exp[-i[(\omega_L+\omega_B)t+k_S z]]+$$
$$+C_3(z,t)\exp[-\tfrac{1}{2}\Gamma'_B t]\exp[-i[(\omega_L-\omega_B)t+k_S z]]+\text{c.c.}\},$$

where

$$C_0(z)=\exp\left[A(z)\left((\gamma^a/\gamma^e)\,\omega_B-\tfrac{1}{2}\gamma\Gamma_R+i\omega\right)/p_1p_2p_3\right],$$
$$C_j(z,t)=\sum_{m=1}^{\infty}b_m(p_j)\left[-\frac{A(z)B_j}{t}\right]^{m/2}J_m\left[2(-A(z)B_jt)^{\frac{1}{2}}\right],$$

where $b_m(p_j)$ are t-independent coefficients.

The physical meaning of the result derived in eq. (11) is easily understandable. The first term represents the input wave spatially amplified in the steady-state regime. The other three components describe the transient behaviour as the sum of a Rayleigh process (term labelled by 1) and a Brillouin process (the anti-Stokes and Stokes components labelled by 2 and 3 respectively). The frequencies of these three components are imposed by the thermal fluctuations stimulated in the medium by the coupling of the two electromagnetic waves. As a check we note that when the temperature fluctuations are excluded (this can be achieved by setting α and λ_T equal to zero so that $\gamma^a=0$ and

$\Gamma_R = 0$), it turns out that the Stokes component increases more and more as the time t increases, as it is a sum of Bessel functions evaluated on the imaginary axis. On the contrary the anti-Stokes component goes to zero under the same conditions as the Bessel functions turn out to be evaluated on the real axis. This confirms that the anti-Stokes component cannot be excited by a stimulated process.

Out of the limit $\alpha = 0$ (and $\lambda_T = 0$) we study the behaviour of the amplified signal at $\omega = 0$. In the range of frequencies near the Rayleigh line the process is recognized to be wholly transient during the time duration of ex-

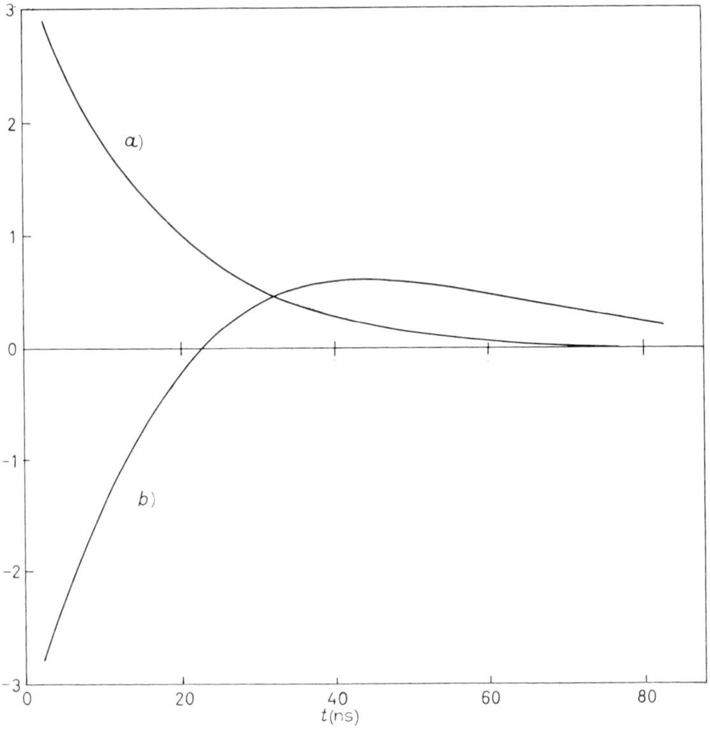

Fig. 6. – Various contributions to the amplified intensity of the signal field *vs.* time. Liquid-CCl_4 parameters were used for calculations (ref. [7]). Curves *a*) and *b*) show the behaviour of the $|C_1|^2 \exp[-\Gamma_R t]$ and $2\,\mathrm{Re}\,(C_0 C_1^* \exp[-\frac{1}{2}\Gamma_R t])$ terms, respectively (ref. [18]).

perimental pulses [11, 12]. We note that the present treatment makes it possible to distinguish the influence of different components on the total amplified intensity. The contribution is stressed in fig. 6 and 7. In fig. 6 the behaviour of $|C_1|^2 \exp[-\Gamma_R t]$ (curve *a*)) and of $2\,\mathrm{Re}(C_0 C_1^* \exp[-\frac{1}{2}\Gamma_R t])$ (curve *b*)) is reported when liquid-CCl_4 parameters are used for calculations. Other contributions to the gain factor are negligible. The reason is that Γ_B is much greater than Γ_R for liquid CCl_4 so that acoustic phonons are heavily damped and the transient behaviour is only due to transient temperature fluctuations. Figure 7 shows the results when liquid-CS_2 parameters are used. In this case the decay constant Γ_B is approximately equal to Γ_R. The beating between Brillouin Stokes and Rayleigh and stationary components is now important. Curve *c*) shows that this contribution is not negligible throughout the transient period.

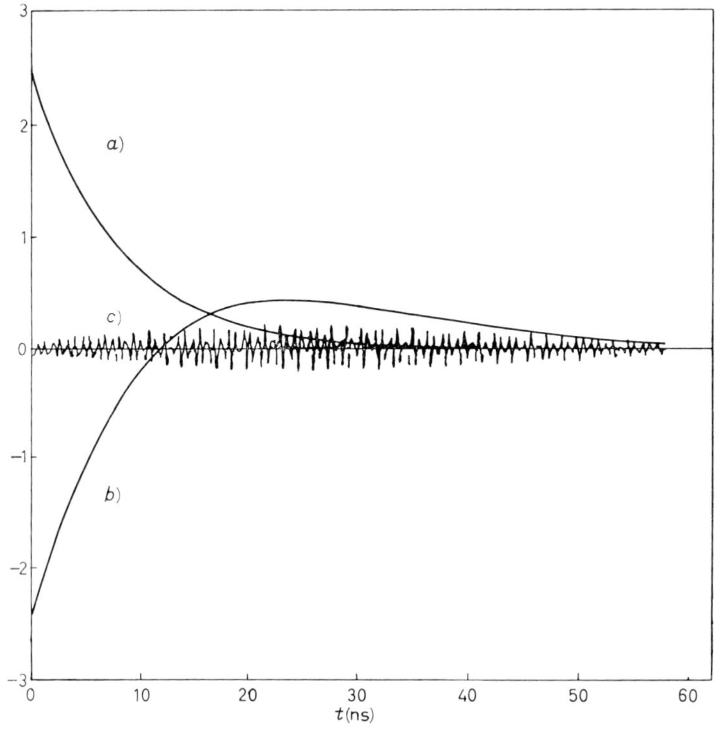

Fig. 7. – Contributions to the amplified intensity using liquid-CS_2 parameters (ref. [1]). Curve *c*) shows the beating between the Brillouin and the stationary and Rayleigh components (ref. [18]).

In conclusion we want to stress that the present treatment has some advantage with respect to the previous ones. Firstly the series expansion in powers of z/L is avoided, secondly the contribution of all the components of the total electric fields is considered in the intensity spectrum at any frequency ω.

As a final remark it can be said that this kind of nonlinear spectroscopy is a powerful tool for the study of the dynamics of thermal fluctuations. We think that both the transient and the stationary behaviour can supply a great deal of information useful for the study of nonequilibrium thermodynamics.

REFERENCES

[1] L. D. LANDAU and E. M. LIFSHITZ: *Electrodynamics of Continuum Media*, Chap. 14 (Reading, Mass., 1960).
[2] I. L. FABELINSKII: *Molecular Scattering of Light* (New York, N. Y., 1968).
[3] L. D. LANDAU and E. M. LIFSHITZ: *Fluid Mechanics* (Reading, Mass., 1960).
[4] D. POHL, M. MAIER and W. KAISER: *Phys. Rev. Lett.*, **20**, 366 (1968).
[5] D. POHL, I. REINHOLD and W. KAISER: *Phys. Rev. Lett.*, **20**, 1141 (1968).
[6] D. POHL and W. KAISER: *Phys. Rev. B*, **1**, 31 (1970).
[7] M. MAIER and G. RENNER: *Opt. Comm.*, **3**, 301 (1971).
[8] F. BAROCCHI, M. MANCINI and R. VALLAURI: *Nuovo Cimento*, **49** B, 233 (1967).
[9] F. BAROCCHI, M. MANCINI and R. VALLAURI: *Journ. Chem. Phys.*, **49**, 1935 (1968).
[10] R. M. HERMAN and M. A. GRAY: *Phys. Rev. Lett.*, **19**, 824 (1967).
[11] W. ROTHER, D. POHL and W. KAISER: *Phys. Rev. Lett.*, **22**, 915 (1969).
[12] W. ROTHER, H. MEYER and W. KAISER: *Zeits. Natur.*, **25** a, 1136 (1970).
[13] R. H. ENNS and I. P. BATRA: *Phys. Lett.*, **28** A, 591 (1969).
[14] I. P. BATRA and R. H. ENNS: *Can. Journ. Phys.*, **47**, 1283 (1969).
[15] R. H. ENNS and I. P. BATRA: *Can. Journ. Phys.*, **47**, 2265 (1969).
[16] S. S. RANGNEKAR and R. H. ENNS: *Can. Journ. Phys.*, **49**, 2307 (1971).
[17] W. ROTHER: *Zeits. Natur.*, **25** a, 1120 (1970).
[18] A. BAMBINI, R. VALLAURI and M. ZOPPI: *Phys. Rev. A*, **12**, 1713 (1975).

PROCEEDINGS OF THE INTERNATIONAL SCHOOL OF PHYSICS
« ENRICO FERMI »

Course I
Questioni relative alla rivelazione delle particelle elementari, con particolare riguardo alla radiazione cosmica
edited by G. PUPPI

Course II
Questioni relative alla rivelazione delle particelle elementari, e alle loro interazioni con particolare riguardo alle particelle artificialmente prodotte ed accelerate
edited by G. PUPPI

Course III
Questioni di struttura nucleare e dei processi nucleari alle basse energie
edited by G. SALVETTI

Course IV
Proprietà magnetiche della materia
edited by L. GIULOTTO

Course V
Fisica dello stato solido
edited by F. FUMI

Course VI
Fisica del plasma e applicazioni astrofisiche
edited by G. RIGHINI

Course VII
Teoria della informazione
edited by E. R. CAIANIELLO

Course VIII
Problemi matematici della teoria quantistica delle particelle e dei campi
edited by A. BORSELLINO

Course IX
Fisica dei pioni
edited by B. TOUSCHEK

Course X
Thermodynamics of irreversible processes
edited by S. R. DE GROOT

Course XI
Weak Interactions
edited by L. A. RADICATI

Course XII
Solar Radioastronomy
edited by G. RIGHINI

Course XIII
Physics of Plasma: Experiments and Tecniques
edited by H. ALFVÉN

Course XIV
Ergodic Theories
edited by P. CALDIROLA

Course XV
Nuclear Spectroscopy
edited by G. RACAH

Course XVI
Physicomathematical Aspects of Biology
edited by N. RASHEVSKY

Course XVII
Topics of Radiofrequency Spectroscopy
edited by A. GOZZINI

Course XVIII
Physics of Solids (Radiation Damage in Solids)
edited by D. S. BILLINGTON

Course XIX
Cosmic Rays, Solar Particles and Space Research
edited by B. PETERS

Course XX
Evidence for Gravitational Theories
edited by C. MØLLER

Course XXI
Liquid Helium
edited by G. CARERI

Course XXII
Semiconductors
edited by R. A. SMITH

Course XXIII
Nuclear Physics
edited by V. F. WEISSKOPF

Course XXIV
Space Exploration and the Solar System
edited by B. Rossi

Course XXV
Advanced Plasma Theory
edited by M. N. Rosenbluth

Course XXVI
Selected Topics on Elementary Particle Physics
edited by M. Conversi

Course XXVII
Dispersion and Absorption of Sound by Molecular Processes
edited by D. Sette

Course XXVIII
Star Evolution
edited by L. Gratton

Course XXIX
Dispersion Relations and Their Connection with Causality
edited by E. P. Wigner

Course XXX
Radiation Dosimetry
edited by F. W. Spiers and G. W. Reed

Course XXXI
Quantum Electronics and Coherent Light
edited by C. H. Townes and P. A. Miles

Course XXXII
Weak Interations and High-Energy Neutrino Physics
edited by T. D. Lee

Course XXXIII
Strong Interactions
edited by L. W. Alvarez

Course XXXIV
The Optical Properties of Solids
edited by J. Tauc

Course XXXV
High-Energy Astrophysics
edited by L. Gratton

Course XXXVI
Many-Body Description of Nuclear Structure and Reactions
edited by C. Bloch

Course XXXVII
Theory of Magnetism in Transition Metals
edited by W. Marshall

Course XXXVIII
Interaction of High-Energy Particles with Nuclei
edited by T. E. O. Ericson

Course XXXIX
Plasma Astrophysics
edited by P. A. Sturrock

Course XL
Nuclear Structure and Nuclear Reactions
edited by M. Jean

Course XLI
Selected Topics in Particle Physics
edited by J. Steinberger

Course XLII
Quantum Optics
edited by R. J. Glauber

Course XLIII
Processing of Optical Data by Organisms and by Machines
edited by W. Reichardt

Course XLIV
Molecular Beams and Reaction Kinetics
edited by Ch. Schlier

Course XLV
Local Quantum Theory
edited by R. Jost

Course XLVI
Physics with Storage Rings
edited by B. Touschek

Course XLVII
General Relativity and Cosmology
edited by R. K. Sachs

Course XLVIII
Physics of High Energy Density
edited by P. Caldirola and H. Knoepfel

Course IL
Foundations of Quantum Mechanics
edited by B. d'Espagnat

Course L
Mantle and Core in Planetary Physics
edited by J. Coulomb and M. Caputo

Course LI
Critical Phenomena
edited by M. S. Green

Course LII
Atomic Structure and Properties of Solids
edited by E. Burstein

Course LIII
Developments and Borderlines of Nuclear Physics
edited by H. Morinaga

Course LIV
Developments in High-Energy Physics
edited by R. R. Gatto

Course LV
Lattice Dynamics and Intermolecular Forces
edited by S. Califano

Course LVI
Experimental Gravitation
edited by B. Bertotti

Course LVII
Topics in the History of 20th Century Physics
edited by C. Weiner

Course LVIII
Dynamic Aspects of Surface Physics
edited by F. O. Goodman

Course LIX
Local Properties at Phase Transitions
edited by K. A. Müller

Course LX
C*-Algebras and their Applications to Statistical Mechanics and Quantum Field Theory
edited by D. Kastler

Course LXI
Atomic Structure and Mechanical Properties of Metals
edited by G. Caglioti

Course LXII
Nuclear Spectroscopy and Nuclear Reactions with Heavy Ions
edited by H. Faraggi and R. A. Ricci

Course LXIII
New Directions in Physical Acoustics
edited by D. Sette

Tipografia Compositori Bologna - Italy